Marine Actinomycetes

Karthik Loganathan • Zhiyong Li
Editors

Marine Actinomycetes

Diversity and Biotechnological Applications

Springer

Editors
Karthik Loganathan
Research and Development Department
Arqgene
Vellore, Tamil Nadu, India

Zhiyong Li
Shanghai Jiao Tong University
Minhang District, Shanghai, China

ISBN 978-981-95-9848-9 ISBN 978-981-95-9849-6 (eBook)
https://doi.org/10.1007/978-981-95-9849-6

This Springer imprint is published by the registered company Springer Nature Singapore Pte Ltd.
The registered company address is: 152 Beach Road, #21-01/04 Gateway East, Singapore 189721, Singapore

Preface

The marine environment represents one of the Earth's most dynamic and biologically rich ecosystems, yet its microbial diversity remains only partially understood. Among the vast assemblage of oceanic microorganisms, *marine actinomycetes* have emerged as one of the most promising and transformative resources for modern biotechnology. Once considered a niche group, these fascinating microbes are now recognized as some of the most prolific producers of structurally novel and functionally potent natural compounds.

The recent surge of discoveries from marine actinomycetes has reshaped expectations in bioprospecting and therapeutic innovation. New antibiotics active against multidrug-resistant pathogens, anticancer and antiviral lead molecules, immunomodulatory agents, neuroprotective compounds, biocatalysts, biosurfactants, and biomaterials are redefining the biomedical, pharmaceutical, and industrial landscapes. Their evolutionary adaptation to extreme and fluctuating marine conditions has given rise to unprecedented metabolic architectures—signifying a reservoir of molecular creativity unparalleled in nature.

Marine Actinomycetes: Rising Star of Biotechnology brings together emerging insights, foundational principles, and translational perspectives to build a comprehensive narrative around these extraordinary microorganisms. The chapters in this volume examine taxonomy, ecology, omics-driven discovery, biosynthetic gene clusters, fermentation and cultivation strategies, metabolic engineering, and downstream commercialization pathways. Together, they present a holistic framework that spans fundamental research to industrial deployment.

This book is shaped by a conviction that marine actinomycetes are not simply another group of microorganisms—they are catalysts for a new era of sustainable biotechnology. As global industries search for eco-friendly therapeutics, alternative biomaterials, climate-resilient bioprocesses, and innovative solutions for environmental remediation, marine actinomycetes stand at the center of opportunity. Their potential aligns perfectly with the future of the blue bioeconomy, where oceans become a source of innovation, not exploitation.

It is our hope that this work serves not only as a reference but also as an inspiration. We envision that students, researchers, industry leaders, and entrepreneurs will find in these pages the motivation to further explore, innovate, and translate the extraordinary capabilities of marine actinomycetes into tangible benefits for society. The time has come to unlock the oceans' microbial wealth—and marine actinomycetes are poised to be its brightest rising star.

Vellore, Tamil Nadu, India Karthik Loganathan
Minhang District, Shanghai, China Zhiyong Li

Contents

Editors and Contributors

About the Editors

Karthik Loganathan has an extensive expertise in actinobacteria, synthetic biology, and drug discovery strategies. He has a total of 17 years of research, 7 years of teaching, and 6 years industrial experience. He did his first postdoctoral training at the Shanghai Jiao Tong University, China. Further he has completed his second postdoctoral fellowship in synthetic biology, ECUST, Shanghai, China. He received his PhD degree in microbiology from the VIT University in 2013. He has published over 85 research papers in international and national peer-reviewed journals. He has also published 5 books and filed 6 patents. He is currently founder of Arqgene and Sasi Nutraceuticals, India.

Zhiyong Li obtained his PhD degree from South China University of Technology, and conducted his postdoctoral research at the State Key Laboratory of Bioreactor Engineering, East China University of Science and Technology. He is currently working as a Professor at the Shanghai Jiao Tong University (SJTU). From December 2007 to December 2008, he visited Scripps Institution of Oceanography at University of California, San Diego, USA, as a visiting professor. He is the founder and head of Marine Biotechnology Laboratory at SJTU and a PI at the State Key Laboratory of Microbial Metabolism, China. He has published 107+ SCI Journals papers and currently holds an h-index 44.

Contributors

Ravali Arugonda Department of Biological Sciences, School of Natural Science and Mathematics, University of Texas at Dallas, Richardson, TX, USA

Sohan Athmaraman Department of Biotechnology, School of Bio Sciences and Technology, Vellore Institute of Technology, Vellore, Tamil Nadu, India

Geetha Bharathi Independent Researcher (Medical Applications), Coimbatore, Tamil Nadu, India

K. V. Bhaskara Rao Marine Biotechnology Laboratory, Department of Bio-Medical Sciences, School of Bio Sciences and Technology, Vellore Institute of Technology, Vellore, Tamil Nadu, India

Hitarth B. Bhatt Department of Biological Sciences, P. D. Patel Institute of Applied Sciences, Charotar University of Science and Technology (CHARUSAT), Changa, Gujarat, India

Shruti P. Buddhdev Department of Life Sciences, Atmiya University, Rajkot, Gujarat, India

Mohit Chauhan Department of Microbiology and Biotechnology, Gujarat University, Ahmedabad, India

Kushi Poovaiah Chendira Department of Biotechnology, School of Bio Sciences and Technology, Vellore Institute of Technology, Vellore, Tamil Nadu, India

Diksha Dubey School of Bio Sciences and Technology, Vellore Institute of Technology, Vellore, Tamil Nadu, India

Lekshmi K. Edison Department of Comparative, Diagnostic, and Population Medicine, College of Veterinary Medicine, University of Florida, Gainesville, FL, USA

Carmen Pire Galiana Department of Biochemistry and Molecular Biology, Edaphology, and Agricultural Chemistry, Faculty of Sciences, University of Alicante, Alicante, Spain

Sangeeta D. Gohel Department of Biosciences, Saurashtra University, Rajkot, Gujarat, India

Aashika Jain Department of Microbiology, The Oxford College of Science (Affiliated to Bangalore University), Bangalore, Karnataka, India

Rishita Jain School of Bio Sciences and Technology, Vellore Institute of Technology, Vellore, Tamil Nadu, India

Riya S. Jasani Department of Life Sciences, Atmiya University, Rajkot, Gujarat, India

M. A. Jayasri Marine Biotechnology and Bioproducts Laboratory, Department of Biomedical Sciences, School of Bio Sciences and Technology, Vellore Institute of Technology, Vellore, Tamil Nadu, India

M. Kavin Raja Department of Biotechnology, School of Bio Sciences and Technology, Vellore Institute of Technology, Vellore, Tamil Nadu, India

Ramesh Kothari Department of Biosciences, Saurashtra University, Rajkot, Gujarat, India

A. Kowsalya Anna University, Chennai, Tamil Nadu, India

Saran Krishnamoorthy Center of Molecular Medicine and Diagnostics (COMManD), Department of Biochemistry, Saveetha Dental College and Hospitals,

Saveetha Institute of Medical and Technical Sciences, Saveetha University, Chennai, Tamil Nadu, India

Prerna Kumari School of Bio Sciences and Technology, Vellore Institute of Technology, Vellore, Tamil Nadu, India

B. Madhumitha Department of Biotechnology, School of Bio Sciences and Technology, Vellore Institute of Technology, Vellore, Tamil Nadu, India

Vaishali R. Majithiya Department of Biosciences, Saurashtra University, Rajkot, Gujarat, India

Faculty of Science, Atmiya University, Rajkot, Gujarat, India

Sankalp Mandal School of Bio Sciences and Technology, Vellore Institute of Technology, Vellore, Tamil Nadu, India

Rosa Maria Martinez-Espinosa Department of Biochemistry and Molecular Biology, Edaphology, and Agricultural Chemistry, Faculty of Sciences, University of Alicante, Alicante, Spain

Akshat Rajan Mehta School of Bio Sciences and Technology, Vellore Institute of Technology, Vellore, Tamil Nadu, India

Mithil Mohanraj Department of Biotechnology, School of Bio Sciences and Technology, Vellore Institute of Technology, Vellore, Tamil Nadu, India

Wasu Pathom-aree Department of Biology, Faculty of Science, Chiang Mai University, Chiang Mai, Thailand

B. G. Pavana Department of Biotechnology, Prathyush Engineering College, Tiruvallur, Tamil Nadu, India

Harsh Prajapati Department of Microbiology and Biotechnology, Gujarat University, Ahmedabad, India

Harishkumar Rajendran Department of Biotechnology, Sri Sankara College of Arts and Science, Kanchipuram, Tamil Nadu, India

Manikandan Rajendran Department of Biotechnology, Rathinam College of Arts and Science, Coimbatore, Tamil Nadu, India

Hasti Ramavat Department of Biosciences, Saurashtra University, Rajkot, Gujarat, India

Pharada Rangseekaew Department of Biology, Faculty of Science, Chiang Mai University, Chiang Mai, Thailand

Jenisha K. Ranpariya Department of Life Sciences, Atmiya University, Rajkot, Gujarat, India

Shivani V. Raval Department of Biotechnology, M.G. Science Institute (Autonomous), Ahmedabad, India

Vikram H. Raval Department of Microbiology and Biotechnology, Gujarat University, Ahmedabad, India

Swarnkumar Reddy Department of Biotechnology, Sri Sankara College of Arts and Science, Kanchipuram, Tamil Nadu, India

Kaviyarasi Renu Center of Molecular Medicine and Diagnostics (COMManD), Department of Biochemistry, Saveetha Dental College and Hospitals, Saveetha Institute of Medical and Technical Sciences, Saveetha University, Chennai, Tamil Nadu, India

Dibyani Rout School of Bio Sciences and Technology, Vellore Institute of Technology, Vellore, Tamil Nadu, India

A. Sai Ramesh Department of BioEngineering, Vels Institute of Science, Technology and Advanced Studies, Chennai, Tamil Nadu, India

G. Sainath Marine Biotechnology and Bioproducts Laboratory, Department of Biomedical Sciences, School of Bio Sciences and Technology, Vellore Institute of Technology, Vellore, Tamil Nadu, India

Vinodhini Shanmugam Department of Biotechnology, Sri Sankara College of Arts and Science, Kanchipuram, Tamil Nadu, India

Rushit Shukla Department of Microbiology, Christ College, Rajkot, Gujarat, India

Satya P. Singh Department of Biosciences, Saurashtra University, Rajkot, Gujarat, India

G. Sowmiya Department of Biotechnology, Prathyush Engineering College, Tiruvallur, Tamil Nadu, India

C. Subathra Devi Department of Biotechnology, School of Bio Sciences and Technology, Vellore Institute of Technology, Vellore, Tamil Nadu, India

Kothari Neeti Suresh School of Bio Sciences and Technology, Vellore Institute of Technology, Vellore, Tamil Nadu, India

K. Suthindhiran Marine Biotechnology and Bioproducts Laboratory, Department of Biomedical Sciences, School of Bio Sciences and Technology, Vellore Institute of Technology, Vellore, Tamil Nadu, India

S. Taejashwini Department of Biotechnology, School of Bio Sciences and Technology, Vellore Institute of Technology, Vellore, Tamil Nadu, India

Pratheep Thangaraj Department of Biotechnology, Rathinam College of Arts and Science, Coimbatore, Tamil Nadu, India

P. V. Tharani Marine Biotechnology Laboratory, Department of Bio-Medical Sciences, School of Bio Sciences and Technology, Vellore Institute of Technology, Vellore, Tamil Nadu, India

Anne Sahithi Somavarapu Thomas Centre for Global Health Research, Saveetha Medical College and Hospital, Chennai, Tamil Nadu, India

Nishtha R. Vaghela Department of Biosciences, Saurashtra University, Rajkot, Gujarat, India

Muneer Oladipupo Yaqub Department of Biological Sciences, School of Natural Science and Mathematics, University of Texas at Dallas, Richardson, TX, USA

Peng Zeng Colleges of Ecology and Environment, Central South University of Forestry and Technology, Changsha, China

Introduction to Marine Actinobacteria

A. Sai Ramesh, A. Kowsalya, B. G. Pavana, and G. Sowmiya

Abstract

Marine actinobacteria are a highly diverse group of microorganisms inhabiting oceanic niches ranging from shallow coastal waters to deep-sea sediments. Their ecological versatility allows them to play vital roles in organic matter degradation, nutrient recycling, and symbiotic interactions with marine flora and fauna. Over the past decades, these bacteria have gained recognition as prolific producers of secondary metabolites, many of which possess unique structural features and significant bioactivities, including antimicrobial, anticancer, and anti-inflammatory properties. With the growing need for novel therapeutics and sustainable biotechnological solutions, marine actinobacteria have emerged as a critical resource for natural product discovery and industrial innovation. This review provides a comprehensive introduction to their taxonomy, ecological functions, and metabolic capabilities, while also examining advances in cultivation techniques, genome mining, and bioprospecting strategies. Insights into current challenges and opportunities highlight their untapped potential for addressing global health and environmental demands.

A. Sai Ramesh (✉)
Department of BioEngineering, Vels Institute of Science, Technology and Advanced Studies, Chennai, Tamil Nadu, India

A. Kowsalya
Anna University, Chennai, Tamil Nadu, India

B. G. Pavana · G. Sowmiya
Department of Biotechnology, Prathyush Engineering College, Tiruvallur, Chennai, Tamil Nadu, India

K. Loganathan, Z. Li (eds.), *Marine Actinomycetes*,
https://doi.org/10.1007/978-981-95-9849-6_1

Graphical Abstract

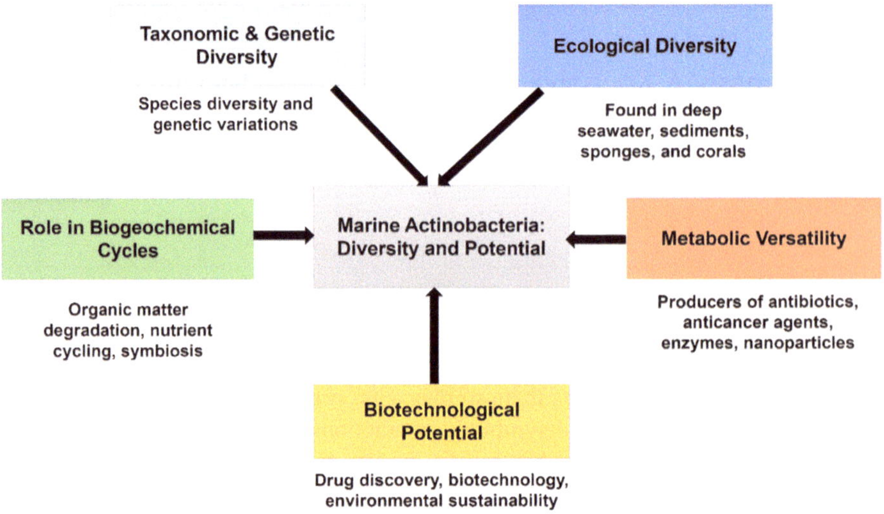

Keywords

Marine Actinobacteria · Marine biology · Secondary metabolism · Biosynthesis · Biotechnological applications

1 Introduction

Actinobacteria (Actinomycetota) constitute one of the oldest and most coherent groups of Gram-positive bacteria, distinguished by their characteristically high guanine–cytosine (G + C) content, often greater than 60–70 mol% (Pongen et al., 2023). Scientific interest in these microorganisms dates back to the late nineteenth and early twentieth centuries, when Cohn, Harz (1877), and later Waksman and Curtis (1916) described their filamentous form, noting its resemblance to fungi and their ability to generate bioactive compounds, especially antibiotics. Although long recognized as abundant soil and freshwater inhabitants, Actinobacteria are now increasingly documented in marine systems, where they adapt to diverse ecological niches, including sediments, mangroves, sponges, and algae (Subramani & Aalbersberg, 2012).

While less dominant in oceans compared to terrestrial environments, marine Actinobacteria are remarkably diverse. Independent surveys indicate their presence across global sediments, ranging from 0.1% to 3% in the Arctic, 0.7–2.4% in the Atlantic, and up to 10% in the Pacific, with deep-sea samples showing even higher relative abundance (Chen et al., 2016; López-García et al., 2003; Frontiers, 2023). Genera such as *Streptomyces, Micromonospora, Dietzia,* and particularly *Salinispora* have been consistently reported (Abdelmohsen et al., 2014). Salinispora species are of special interest because they are obligately marine and synthesize novel secondary metabolites, including salinosporamide A, a proteasome inhibitor first recovered from marine sediments (Maldonado et al., 2005; Fenical et al., 2009).

In recent decades, marine isolates have emerged as a promising source of structurally distinct compounds with diverse pharmacological activities, including antimicrobial, anticancer, antiviral, and anti-inflammatory, among them Fig. 4 (Manivasagan et al., 2014). Their significance is not limited to drug discovery. Many strains produce robust enzymes such as proteases, amylases, cellulases, and lipases, and these enzymes retain activity under extremes of salinity, temperature, and pressure. Such traits make them highly valuable for industries ranging from food and textiles to renewable energy. Ecologically, these organisms also contribute to nutrient cycling by degrading polymers such as chitin and cellulose in marine sediments (Barka et al., 2016).

This review emphasizes three key dimensions: the diversity and ecological distribution of marine Actinobacteria, the breadth of their chemical and pharmacological potential, and their emerging role in industrial biotechnology. It also discusses the challenges of cultivation, the opportunities presented by genome-mining and metagenomic strategies, and the importance of exploring under-investigated marine habitats for future natural product discovery.

2 Taxonomy and Diversity of Marine Actinobacteria

Actinobacteria's taxonomic positioning in the microbial tree of life underlines their ecological superpower status as members of a distinctly recognized lineage of Gram-positive bacteria. The data available from molecular phylogenetics studies and genome-based approaches have greatly enhanced our taxonomic precision of this group and have enabled us to gain a greater understanding of taxonomic richness, ecological function, and evolutionary associations of the Actinobacteria. The key content of the taxonomic hierarchy and the biological attributes of Actinobacteria are summarized in Fig. 1, which also highlights the marine-adapted genera.

Marine Actinobacteria have unique evolutionary and physiological adaptations that allow them to occupy a wide range of diverse and extreme marine environments—for example, deep-sea sediments, coral reefs, mangroves, and symbiotic interactions with marine invertebrates. These adaptations may include (not limited to) osmotic stress tolerance, specific enzymatic specializations, enhanced secondary metabolite biosynthetic gene clusters (BGCs), and co-evolutionary interactions with marine hosts. Table 1 summarizes the forms of these adaptations, their functional significance, and relevant references (Table 1).

Further, in many parts of the world, scientists have recently reported many marine-derived Actinobacteria, including new genera, species, and strains. These discoveries matter because marine Actinobacteria occur in special environments

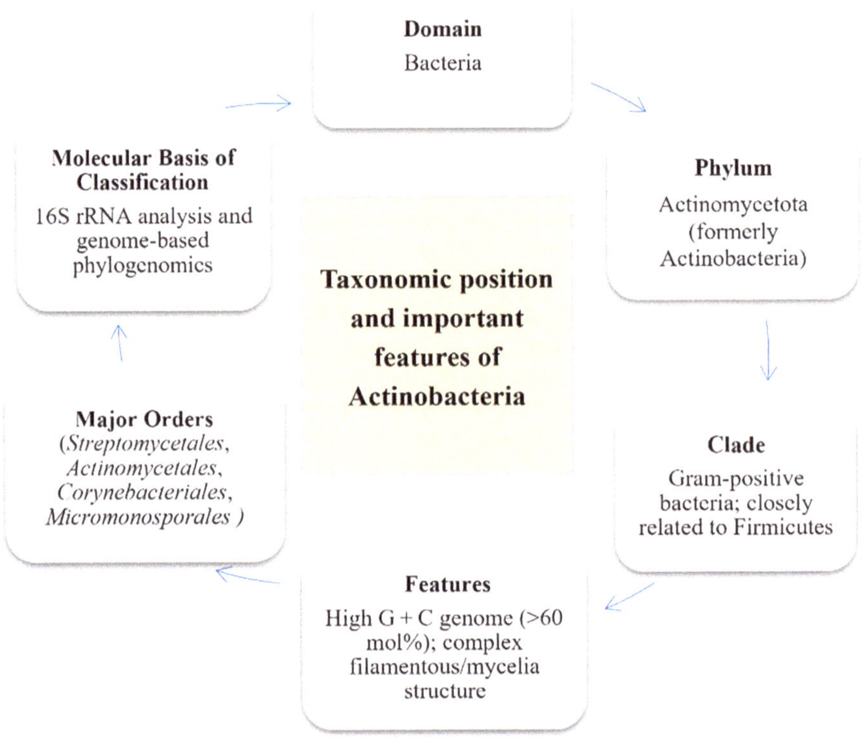

Fig. 1 Taxonomic position and important features of Actinobacteria in the microbial tree of life

and produce unique bioactive metabolites. The Table 2 below shows the novel marine Actinobacteria genera, species, and newly reported strains (2000–2024), which provides convenience and a broad overview.

3 Ecological Roles and Adaptations of Marine Actinobacteria

Marine Actinobacteria are regarded as one of the most adaptable microbial groups in oceanic ecosystems. They inhabit a wide spectrum of environments, ranging from deep-sea sediments and mangrove swamps to coral reefs, sponges, and open waters. Their distribution across such distinct niches reflects a high degree of metabolic flexibility and ecological resilience. These organisms are continuously exposed to environmental stressors such as elevated salinity, extreme hydrostatic pressure, fluctuating temperatures, and nutrient scarcity, all of which shape their survival strategies and evolutionary trajectory (Barka et al., 2016; Subramani & Aalbersberg, 2012). By adapting to these conditions, marine Actinobacteria not

Table 1 Key evolutionary innovations of marine Actinobacteria

S. no.	Adaptation	Description/Role	Representative example	References
1.	Osmoprotective mechanisms	Biosynthesis and accumulation of compatible solutes (e.g., glycine betaine and ectoine trehalose) to maintain cell integrity when facing high salt and osmotic stresses	Detected in *Streptomyces* and *Micromonospora* strains from marine sediments.	Subramani and Aalbersberg (2012) and Tiwari and Gupta (2012)
2.	Salt-tolerant enzymes & membrane modifications	Halotolerant enzymes and compositional adaptations of lipids that facilitate enzyme and membrane stability in saline environments	Marine *Nocardiopsis* spp. and *Streptomyces* spp.	Bull and Stach (2007)
3.	Enriched secondary metabolite BGCs	Genomes contain diverse arrays of BGC's, activated under marine stress conditions, that can also produce antibiotics and antifungal compounds	*Salinispora tropica* producing salinosporamide A.	Ziemert et al. (2014)
4.	Loss of sporulation (trade-off)	Monopolies of marine genera may lose sporulation capability, but may warrant enhanced secondary metabolism capacity and colonization	*Salinispora* spp.	Ziemert et al. (2014)
5.	Symbiotic associations	Develop symbiotic relationships with sponges, corals, and tunicates through producing bioactive compounds, allowing protection to host organisms	Sponge-associated *Micromonospora* spp.	Zhang et al. (2022a)
6.	Horizontal gene transfer (HGT)	Unique biosynthetic genes obtained via HGT, allowing the microbiome to yield increased metabolic diversity and ecologically based adaptability	Marine *Streptomyces* spp.	Zhang et al. (2022b)

Table 2 Novel marine Actinobacteria genera, species, and recent strains

Genus/species/strain	Source habitat/ host	Year	Notable features/ bioactivity	References
Salinispora spp. (*S. arenicola, S. tropica, S. pacifica*)	Marine sediments (obligate seawater)	2005–2006	First obligate marine actinomycete genus; producer of salinosporamide A	Maldonado et al. (2005) and Jensen and Mafnas (2006)
Marinispora spp.	Marine sediments	2010	Produces novel chemotypes	Maldonado et al. (2005)
Salinibacterium amurskyense	Marine sediments	2003	Halotolerant, non-spore-forming marine actinobacterium	Han et al. (2003)
Aeromicrobium marinum	Marine sediments	2003	Requires seawater for growth	Lee and Lee (2003)
Nocardioides nanhaiensis, N. antarcticus, N. litoris	South China Sea sediments	2014–2017	Novel Nocardioides species; potential secondary metabolites	Zhang et al. (2016)
Streptomonospora spp.	Marine sediments	2013	New genus of marine Actinobacteria	Pan et al. (2013)
Micromonospora maris (*Verrucosispora maris*)	Deep-sea sediments	2012	Produces abyssomicins and proximicins	Goodfellow et al. (2012)
Streptomyces corallincola sp. nov.	Coral (Favites pentagona)	2024	Coral-derived Streptomyces species with bioactive potential	Kanchanasin et al. (2024)
Kineosporia corallincola sp. nov.	Coral (Favites pentagona)	2024	Novel coral-associated Kineosporia species	Kanchanasin et al. (2024)
Micromonospora palythoicola sp. nov. (S2-005^{nT})	Zoanthid (Palythoa sp.)	2022	NRPS, PKS gene clusters; antimicrobial & cytotoxic activity	Zun et al. (2022)
Streptomyces poriticola sp. nov. (C6-003^{nT})	Coral (Porites sp.)	2022	Broad-spectrum antimicrobial & anticancer potential	Zun et al. (2022)
Actinomadura spongiicola sp. nov.	Sponge (Leucetta)	2022	Novel sponge-derived Actinomadura	Qin et al. (2022)
Brevibacterium spongiae sp. nov.	Sponge (Hymeniacidon)	2023	Produces secondary metabolites	Zhang et al. (2023)
Cytobacillus spongiae sp. nov.	Sponge (Diacarnus)	2023	Sponge-associated novel species	Gao et al. (2023)

(continued)

Table 2 (continued)

Genus/species/strain	Source habitat/host	Year	Notable features/bioactivity	References
Mycetocola spongiae sp. nov.	Deep-sea sponge	2022	Associated with Cacospongia mycofijiensis	Chen et al. (2022)
Nocardioides coralli sp. nov.	Coral (South China Sea)	2022	Coral-derived Nocardioides species	Shi et al. (2022)
Streptomyces marincola sp. nov.	Marine sediment	2022	Antimicrobial and anticancer metabolites	Fan et al. (2022)

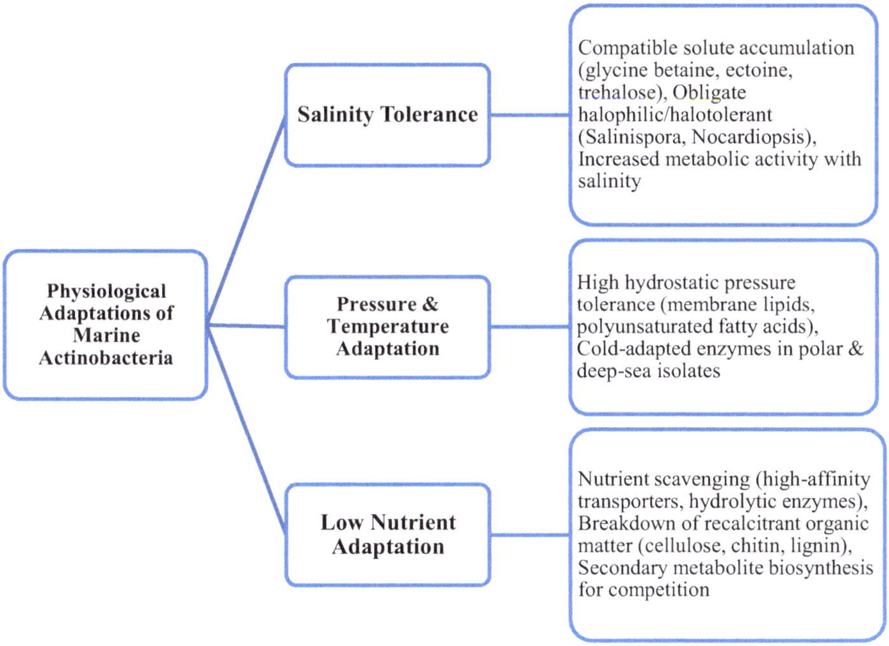

Fig. 2 Physiological adaptations to marine conditions

only sustain their own growth but also influence broader ecosystem functions and microbial community dynamics (Fig. 2).

3.1 Symbiotic Associations with Marine Fauna and Flora

A central ecological feature of marine Actinobacteria is their ability to establish symbiotic associations with marine fauna and flora. These partnerships are ecologically and evolutionarily significant because they provide reciprocal benefits: host organisms gain protection, defense, and nutrient advantages, while the bacteria secure access to a favorable ecological niche with reliable nutrient supplies.

3.1.1 Sponges (Porifera)

Marine sponges are particularly well-studied habitats for Actinobacteria. Their porous architecture and highly efficient water filtration systems promote dense microbial colonization. Genera such as *Streptomyces, Micromonospora, Nocardiopsis,* and *Salinispora* are common in sponge-associated communities. These bacteria are prolific producers of secondary metabolites with antimicrobial, anticancer, and antifouling properties, which support sponge survival against pathogenic microbes, colonization by fouling organisms, and predation. A well-known example is *Salinispora tropica*, isolated from *Haliclona* sp., which produces salinosporamide A, a potent proteasome inhibitor with clinical applications (Jensen et al., 2015).

3.1.2 Corals (Cnidaria)

Actinobacteria also occur within corals, especially reef-building species, where they are found in the surface mucus layer and skeletal framework. These bacteria play a protective role by producing antimicrobial agents that suppress pathogens such as Vibrio spp. In addition, they contribute to coral nutrition by fixing nitrogen and facilitating the recycling of organic matter. Species of Streptomyces have been reported to produce polyketides that enhance coral defense mechanisms, thereby helping maintain coral health and resilience in dynamic reef ecosystems (Raina et al., 2016).

3.1.3 Crustaceans (e.g., Crabs, Shrimps)

Crustaceans host Actinobacteria on their exoskeletons and gill tissues. These bacteria produce enzymes such as chitinases that degrade chitin and allow them to utilize crustacean exudates as nutrients. Furthermore, some strains secrete antimicrobial peptides that protect the host from bacterial and fungal infections. Such interactions enhance crustacean immunity and improve their survival in environments characterized by high microbial competition (Gao et al., 2019).

4 Isolation and Cultivation

The ability to isolate and cultivate marine Actinobacteria is fundamental for unlocking their potential in biotechnology and pharmaceutical development. Although their ecological significance and metabolic versatility are well recognized, a substantial proportion of marine Actinobacteria remain unexplored. This is largely due to the challenges of replicating the complex and variable conditions of the marine environment in laboratory settings. Many species are adapted to nutrient-limited habitats, and conventional cultivation methods often fail to support their growth. For this reason, the development of innovative strategies, integrating both culture-dependent and culture-independent approaches, is essential for accessing the hidden diversity of marine Actinobacteria and their novel bioactive metabolites (Fig. 3).

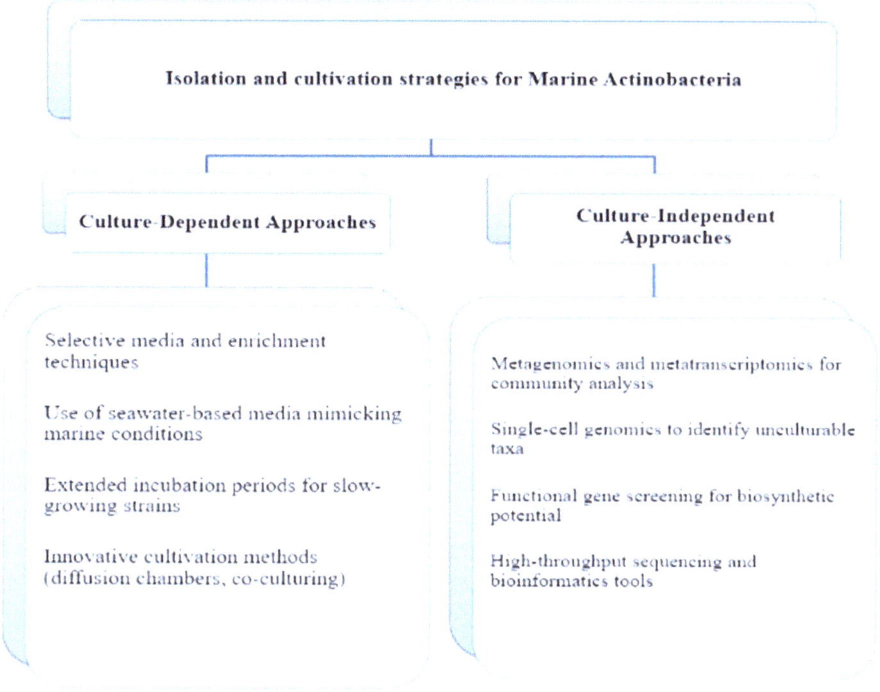

Fig. 3 Flowchart representation of culturing approach

Recent advances in microbial cultivation technologies have significantly improved the recovery of rare and previously unculturable Actinobacteria. One notable breakthrough is the use of in situ cultivation systems, including diffusion chambers and iChip devices, which permit microorganisms to grow in their natural habitats while remaining physically separated for later isolation. These methods maintain the environmental cues required for growth, thereby increasing the likelihood of recovering novel taxa.

Other innovative approaches include high-throughput cultivation platforms and microfluidic systems, which create miniature, controlled environments that replicate marine ecological conditions. Such systems are particularly effective for culturing slow-growing species that would otherwise be lost under standard laboratory conditions. Co-cultivation techniques have also emerged as valuable tools, allowing marine Actinobacteria to grow in association with helper microbes that provide essential nutrients, signaling molecules, or growth-promoting factors.

Beyond cultivation-based techniques, genome-guided and metabolomics-driven strategies are increasingly applied to direct isolation efforts. Genome mining enables researchers to predict metabolic pathways and identify optimal culture conditions for both primary metabolism and secondary metabolite production. When combined with metabolomics screening, these approaches not only improve cultivation success but also enhance the discovery of novel natural products with potential pharmaceutical value (Berdy et al., 2017). Together, these innovative methodologies are transforming the study of marine Actinobacteria, enabling access to rare microbial taxa and their bioactive compounds that were once thought to be beyond reach.

5 Biotechnological and Pharmaceutical Potential

Marine Actinobacteria have emerged as one of the most promising microbial resources for biotechnology and medicine. Their adaptation to the challenging conditions of the ocean, marked by high salinity, immense hydrostatic pressure, and limited nutrient availability, has driven the evolution of unique metabolic pathways. These adaptations enable them to synthesize structurally diverse metabolites that are often absent in terrestrial relatives. Many of these compounds possess strong pharmacological activities, with reported antimicrobial, anticancer, antifungal, antiviral, and anti-inflammatory effects (Fig. 4) (Manivasagan et al., 2014). Such diversity makes them invaluable for the discovery of novel therapeutic molecules and industrially relevant biomolecules.

Members of genera such as *Streptomyces, Salinispora, Micromonospora,* and *Nocardiopsis* are well-recognized for producing bioactive metabolites with distinct structural frameworks. Compared with terrestrial Actinobacteria, their marine counterparts often generate molecules with novel scaffolds, reducing the likelihood of rediscovering previously known compounds. This chemical novelty provides a foundation for the design and development of new classes of antibiotics, anticancer agents, and other therapeutic products. Halotolerant and halophilic marine

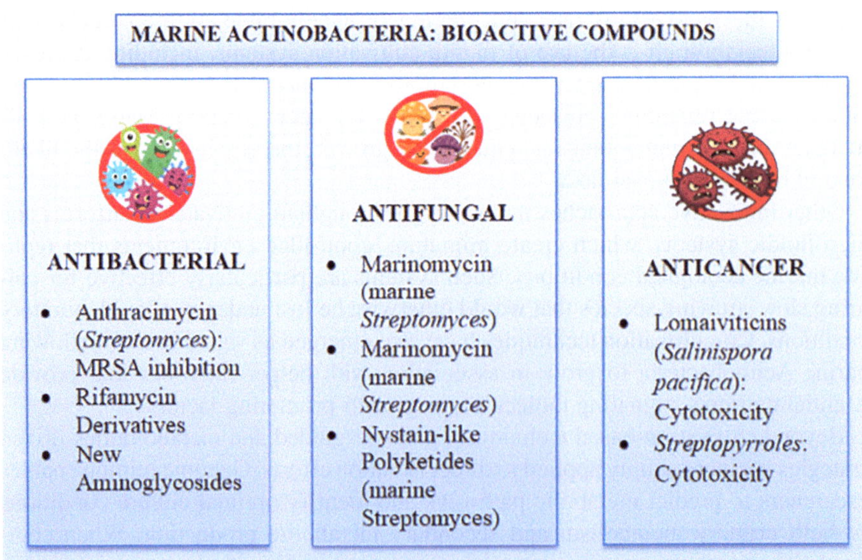

Fig. 4 Marine Actinobacteria: bioactive compounds

Actinobacteria produce enzymes that remain active under high salt concentrations, fluctuating temperatures, and extreme pH values. Their natural robustness makes these enzymes particularly attractive for industrial applications, ranging from food processing to biofuel production (Anand et al., 2021). Enzymes such as amylases, proteases, and cellulases derived from marine Actinobacteria are widely studied for their industrial potential. Halotolerant amylases, particularly from *Streptomyces* and *Micromonospora*, exhibit high thermostability and catalytic efficiency, making them useful in starch hydrolysis during baking processes. Proteases produced by marine strains are well-suited for detergents due to their ability to hydrolyze proteins in the presence of salt, while cellulases contribute significantly to biofuel production and biobleaching in the paper industry (Vasconcellos et al., 2019).

Marine Actinobacteria are also capable of breaking down complex organic pollutants, including hydrocarbons, plastics, and pesticides. Their enzymatic diversity enables them to degrade recalcitrant compounds that often persist in marine and industrial environments. Species belonging to *Nocardiopsis* and *Streptomyces* have demonstrated the ability to metabolize polyaromatic hydrocarbons in oil-contaminated sediments, making them valuable candidates for environmental clean-up and sustainable bioremediation. Another important trait of marine Actinobacteria is their ability to produce biosurfactants such as glycolipids and lipopeptides. These natural surfactants can reduce surface tension and promote emulsification. Their applications extend to enhanced oil recovery, bioremediation, food formulation, cosmetics, and pharmaceuticals. Importantly, biosurfactants produced by Streptomyces species show high emulsification activity and biodegradability, making them sustainable alternatives to chemically synthesized surfactants (Kiran et al., 2010). Taken together, these capacities highlight the potential of marine Actinobacteria to act as biofactories for eco-friendly industrial and environmental solutions.

6 Advances in Research and Methodologies

Research on marine Actinobacteria has expanded rapidly with the introduction of molecular biology, high-throughput sequencing, and synthetic biology. These tools have improved the discovery of bioactive compounds, helped in identifying biosynthetic gene clusters (BGCs), and enabled manipulation of metabolic pathways to unlock novel metabolites. Genome mining has proven essential in uncovering the hidden biosynthetic potential of marine Actinobacteria. Many of their metabolites are encoded by silent BGCs that remain unexpressed under laboratory conditions. Bioinformatics platforms such as antiSMASH and PRISM allow researchers to predict the chemical classes of metabolites associated with these clusters, including polyketides, non-ribosomal peptides, terpenes, and alkaloids.

Genera such as *Salinispora* have numerous cryptic clusters, offering untapped opportunities for novel drug discovery (Ziemert et al., 2016). Metagenomics combined with next-generation sequencing (NGS) has revolutionized the study of uncultivable marine Actinobacteria. By sequencing DNA directly from environmental samples, researchers can bypass cultivation barriers and gain insights into community structure and metabolic potential. Shotgun sequencing has also revealed gene clusters from marine sediments and symbiotic microbial consortia in sponges, corals, and algae. These approaches have enabled the identification of novel bioactive compounds with antimicrobial, anticancer, and anti-inflammatory activities (Fig. 4) (Charlop-Powers et al., 2015).

The integration of bioinformatics platforms has facilitated genome annotation, comparative genomics, and pathway reconstruction. Tools such as antiSMASH, IMG-ABC, and NaPDoS help in identifying BGCs, while BLAST, RAST, KEGG, and BioCyc support annotation and functional prediction. These resources have provided systematic insights into the genetic diversity of marine Actinobacteria and aided the identification of promising strains for natural product discovery (Medema et al., 2011). Synthetic biology approaches, including CRISPR-Cas genome editing, promoter engineering, and heterologous expression, are now applied to marine Actinobacteria to enhance secondary metabolite production. Such strategies allow activation of silent clusters and the creation of analogs with improved pharmacological potential. For example, engineering polyketide and peptide pathways can yield novel compounds with enhanced bioactivity, thereby expanding the chemical diversity of marine natural products (Zhou et al., 2021).

7 Emerging Biotechnological and Environmental Applications

Owing to their diverse metabolism and adaptability, marine actinobacteria are increasingly recognized as sustainable sources for bioactive molecules and enzymes with applications in pharmaceuticals, pollution control, and biotechnology.

7.1 Pharmaceutical Biotechnology

Marine Actinobacteria produce structurally distinct compounds with significant therapeutic value. *Salinispora tropica* produces salinosporamide A (Marizomib), a potent proteasome inhibitor under clinical trials for treating multiple myeloma and glioblastoma. Similarly, anthracimycin from marine Streptomyces has demonstrated strong activity against MRSA and *Bacillus anthracis*. Advances in genome mining and metagenomics continue to reveal cryptic clusters, underscoring the role of marine Actinobacteria in next-generation drug discovery (Berdy, 2022; Subramani & Aalbersberg, 2022).

7.2 Bioremediation and Pollution Mitigation

Several genera, including Rhodococcus and Nocardiopsis, are effective in degrading hydrocarbons and other pollutants, highlighting their role in oil spill management and marine pollution control. They also possess the ability to biosorb and bioaccumulate toxic heavy metals such as cadmium, lead, and mercury, reducing their impact on marine ecosystems. Genetic engineering has further enhanced their biodegradative capacity, positioning them as sustainable tools for green biotechnologies (Kumar et al., 2021; Goswami et al., 2023).

7.3 Blue Biotechnology and Aquaculture

In the realm of blue biotechnology, marine Actinobacteria are valued for producing enzymes, pigments, and bioactive molecules suitable for food, cosmetics, and renewable energy industries (Fig. 4). In aquaculture, they act as probiotics and biocontrol agents by inhibiting pathogens such as Vibrio and Aeromonas. They also promote immune responses in farmed marine organisms and improve water quality by degrading organic waste. These roles align with global initiatives seeking alternatives to antibiotics and chemical treatments in fisheries (Ghosh et al., 2021; Manivasagan et al., 2023).

8 Challenges and Knowledge Gaps

Marine Actinobacteria hold enormous promise as sources of novel bioactive compounds, yet several hurdles restrict their study and application. A major limitation lies in the difficulty of cultivating most strains under conventional laboratory conditions. Many of these organisms grow slowly or require extreme environmental factors such as high salinity, intense hydrostatic pressure, or nutrient-poor settings that are hard to replicate in vitro. In mixed cultures, rapidly growing microbes often dominate, suppressing the growth of rare or slow-growing Actinobacteria, which complicates isolation and population comparisons. Although advanced tools such as iChip technology, diffusion chambers, and co-culture systems have been introduced, these methods are still in early stages of development and lack standardization. Another challenge is the limited exploration of extreme marine habitats like hydrothermal vents, deep-sea sediments, polar oceans, and hypersaline waters. Research has historically focused on coastal sediments and invertebrate-associated species, leaving much of the actinobacterial diversity untapped. These unstudied environments are likely to harbor unique strains with biosynthetic gene clusters encoding unexplored secondary metabolites, but technological and logistical barriers, particularly deep-sea sampling and cultivation at high pressures, continue to restrict access.

Environmental threats and regulatory barriers further complicate progress in this field. Global climate change, rising ocean acidification, and pollution alter microbial communities, often reducing the abundance of sensitive actinobacterial taxa while favoring resistant strains. Runoff carrying heavy metals, plastics, and hydrocarbons disrupts microbial ecosystems and may lead to the permanent loss of species with valuable biosynthetic potential. Degradation of coral reefs and sponges, which act as important reservoirs of actinobacterial diversity, worsens this threat and underscores the need for urgent conservation. At the same time, marine bioprospecting is shaped by international legal and ethical frameworks. The Nagoya Protocol requires fair benefit-sharing from the use of genetic resources, but uneven implementation and unresolved questions of ownership—particularly in international waters—slow down collaborations and commercial development. Without coordinated global guidelines and without investment in conservation and advanced cultivation strategies, much of the biotechnological potential of marine Actinobacteria will remain beyond reach.

9 Conclusion

Marine Actinobacteria are a largely unexploited source of biodiversity with significant potential to address challenges in human health, environmental management, and ecosystem stability. These microorganisms have evolved unique adaptations that enable them to survive in extreme marine environments, from deep-sea vents to coral reefs, and produce secondary metabolites with diverse structures and valuable pharmaceutical properties. In addition to their potential in drug discovery, they offer applications in sustainable biotechnology, including bioremediation, enzyme production, and other marine-based biotechnologies. Although traditional cultivation methods and limited exploration of marine habitats have restricted access to their diversity, advances in genome mining, metagenomics, multi-omics approaches, and synthetic biology have expanded our ability to study their biosynthetic capabilities. Combining these advanced techniques with international collaborative research will help overcome existing limitations and unlock the sustainable and commercial potential of marine Actinobacteria. As environmental pressures grow and the need for new therapeutics increases, these microorganisms provide a promising pathway for natural product discovery and innovative, sustainable solutions.

References

Abdelmohsen, U. R., Yang, C., Horn, H., Hajjar, D., Ravasi, T., & Hentschel, U. (2014). Actinomycetes from Red Sea sponges: Sources for chemical and phylogenetic diversity. *Marine Drugs, 12*(5), 2771–2789.

Anand, S., Singh, A., Singh, V., Kaur, J., & Pandey, A. (2021). Halophilic and halotolerant actinobacteria: Biotechnological potential and future prospects. *World Journal of Microbiology and Biotechnology, 37*(6), 95. https://doi.org/10.1007/s11274-021-03055-3

Barka, E. A., Vatsa, P., Sanchez, L., Gavaut-Vaillant, N., Jacquard, C., Meier-Kolthoff, J. P., Clément, C., Ouhdouch, Y., & van Wezel, G. P. (2016). Taxonomy, physiology, and natural products of Actinobacteria. *Microbiology and Molecular Biology Reviews, 80*(1), 1–43.

Berdy, J. (2022). Bioactive microbial metabolites: A personal view. *Journal of Antibiotics, 75*(2), 71–91. https://doi.org/10.1038/s41429-022-00458-2

Berdy, B., Spoering, A. L., Ling, L. L., & Epstein, S. S. (2017). In situ cultivation of previously uncultivable microorganisms using the iChip. *Nature Protocols, 12*(11), 2232–2242. https://doi.org/10.1038/nprot.2017.074

Bull, A. T., & Stach, J. E. (2007). Marine actinobacteria: New opportunities for natural product search and discovery. *Trends in Microbiology, 15*(11), 491–499. https://doi.org/10.1016/j.tim.2007.10.004

Charlop-Powers, Z., Owen, J. G., Reddy, B. V., Ternei, M. A., & Brady, S. F. (2015). Chemical-biogeographic survey of secondary metabolism in soil. *Proceedings of the National Academy of Sciences, 111*(10), 3757–3762. https://doi.org/10.1073/pnas.1318021111

Chen, P., Zhang, L., Guo, X., Dai, X., Liu, L., Xi, L., Wang, J., Song, L., Wang, Y., Zhu, Y., Huang, L., & Huang, Y. (2016). Diversity, biogeography, and biodegradation potential of actinobacteria in the Deep-Sea sediments along the Southwest Indian Ridge. *Frontiers in Microbiology, 7*, 1340. https://doi.org/10.3389/fmicb.2016.01340

Chen, Y., Sang, J., Sun, W., Song, Q., & Li, Z. (2022). Mycetocola spongiae sp. nov., isolated from deep-sea sponge Cacospongia mycofijiensis. *International Journal of Systematic and Evolutionary Microbiology, 72*(3), Article 005291. https://doi.org/10.1099/ijsem.0.005291

Fan, L., Sun, S., Li, J., Sun, H., Xu, H., Zhang, X., & Wang, Y. (2022). Streptomyces marincola sp. nov. and its bioactive compounds. *Frontiers in Microbiology, 13*, 860308. https://doi.org/10.3389/fmicb.2022.860308

Fenical, W., Jensen, P. R., Palladino, M. A., Lam, K. S., Lloyd, G. K., & Potts, B. C. (2009). Discovery and development of the anticancer agent salinosporamide A (NPI-0052). *Bioorganic & Medicinal Chemistry, 17*(6), 2175–2180. https://doi.org/10.1016/j.bmc.2008.10.075

Frontiers in Microbiology. (2023). Actinobacteria from Arctic and Atlantic deep-sea sediments—Biodiversity and bioactive potential. *Frontiers in Microbiology, 14*, 1158441.

Gao, C., Sun, P., Chen, Y., & Wang, J. (2019). Actinobacteria in symbiosis with marine organisms: Diversity and bioactive potential. *Frontiers in Microbiology, 10*, 2177.

Gao, W., Chen, Y., Li, X., Zhang, J., Wang, Y., & Liu, H. (2023). *Cytobacillus spongiae* sp. nov., a sponge-associated actinobacterium. *Microbial Ecology, 85*(4), 1010–1020. https://doi.org/10.1007/s00248-022-02133-y

Ghosh, R., Rathore, R., Dubey, G., & Subudhi, E. (2021). Marine actinobacteria as probiotics in aquaculture: A review. *Aquaculture Reports, 19*, 100610. https://doi.org/10.1016/j.aqrep.2021.100610

Goodfellow, M., Brown, R., & Chun, J. (2012). Micromonospora maris sp. nov., isolated from deep-sea sediment. *International Journal of Systematic and Evolutionary Microbiology, 62*, 2037–2041. https://doi.org/10.1099/ijs.0.033332-0

Goswami, S., Thakur, S., Deka, R., Bora, S. S., & Bora, T. C. (2023). Marine actinobacteria in bioremediation and environmental sustainability. *Marine Pollution Bulletin, 188*, 114621. https://doi.org/10.1016/j.marpolbul.2023.114621

Han, S. K., Nedashkovskaya, O. I., Mikhailov, V. V., Kim, S. B., & Bae, K. S. (2003). Salinibacterium amurskyense gen. nov., sp. nov., a novel genus of the family Microbacteriaceae from marine environments. *International Journal of Systematic and Evolutionary Microbiology, 53*(6), 2061–2066. https://doi.org/10.1099/ijs.0.02783-0

Jensen, P. R., & Mafnas, C. (2006). Biogeography of the marine actinomycete *Salinispora*. *Environmental Microbiology, 8*(11), 1881–1888. https://doi.org/10.1111/j.1462-2920.2006.01093.x

Jensen, P. R., Mafnas, C., Mincer, T. J., & Fenical, W. (2015). Marine actinomycete diversity and natural product discovery. *Applied Microbiology and Biotechnology, 99*(19), 7983–7998.

Kanchanasin, P., Salahong, T., Sripreechasak, P., Suriyachadkun, C., Harunari, E., Igarashi, Y., Tanasupawat, S., Tawinwung, S., Vimolmangkang, S., Chaotham, C., & Phongsopitanun, W. (2024). Discovery of two new actinobacteria, Micromonospora palythoicola sp. nov. and Streptomyces poriticola sp. nov., isolated from marine invertebrates. *Scientific Reports, 14*(1), 22140. https://doi.org/10.1038/s41598-024-73040-4

Kiran, G. S., Selvin, J., & Lipton, A. P. (2010). Production of lipopeptide biosurfactant by a marine Nesterenkonia species and its application in food industry and environmental cleaning. *Bioresource Technology, 101*(22), 8953–8959. https://doi.org/10.1016/j.biortech.2010.07.078

Kumar, V., Naik, B., Gusain, O., Bisht, S., & Sharma, K. K. (2021). Actinobacteria in bioremediation: An overview of recent advances. *Frontiers in Environmental Science, 9*, 649232. https://doi.org/10.3389/fenvs.2021.649232

Lee, J., & Lee, J. (2003). Aeromicrobium marinum sp. nov., a marine actinobacterium. *International Journal of Systematic and Evolutionary Microbiology, 53*, 355–359. https://doi.org/10.1099/ijs.0.02350-0

López-García, P., Duperron, S., Philippot, P., Foriel, J., Susini, J., & Moreira, D. (2003). Bacterial diversity in hydrothermal sediment and epsilonproteobacterial dominance in experimental microcolonizers at the Mid-Atlantic Ridge. *Environmental Microbiology, 5*(10), 961–976.

Maldonado, L. A., Fenical, W., Jensen, P. R., Kauffman, C. A., Mincer, T. J., Ward, A. C., et al. (2005). Salinispora arenicola gen. nov., sp. nov. and Salinispora tropica sp. nov., obligate marine actinomycetes belonging to the family Micromonosporaceae. *International Journal of Systematic and Evolutionary Microbiology, 55*(5), 1759–1766.

Manivasagan, P., Venkatesan, J., Sivakumar, K., & Kim, S. K. (2014). Marine actinobacteria as a novel source of bioactive substances. *Enzyme and Microbial Technology, 67*, 26–41.

Manivasagan, P., Venkatesan, J., Kang, K. H., Sivakumar, K., Li-Chan, E. C. Y., & Kim, S. K. (2023). Blue biotechnology applications of marine actinobacteria. *Biotechnology Advances, 60*, 108029. https://doi.org/10.1016/j.biotechadv.2022.108029

Medema, M. H., Blin, K., Cimermancic, P., de Jager, V., Zakrzewski, P., Fischbach, M. A., et al. (2011). antiSMASH: Rapid identification, annotation and analysis of secondary metabolite biosynthesis gene clusters in bacterial and fungal genome sequences. *Nucleic Acids Research, 39*(suppl_2), W339–W346. https://doi.org/10.1093/nar/gkr466

Pan, H., Zhang, J., & Xu, L. (2013). Streptomonospora species isolated from marine sediment. *International Journal of Systematic and Evolutionary Microbiology, 63*, 3804–3809. https://doi.org/10.1099/ijs.0.051664-0

Pongen, Y. L., Thirumurugan, D., Ramasubburayan, R., & Prakash, S. (2023). Harnessing actinobacteria potential for cancer prevention and treatment. *Microbial Pathogenesis, 183*, 106324.

Qin, S., Li, J., Chen, H.-H., Chen, L.-Y., Zhang, Y.-J., Jiang, J.-H., et al. (2022). *Actinomadura spongiicola* sp. nov., from a marine sponge. *International Journal of Systematic and Evolutionary Microbiology, 72*(5), 005343. https://doi.org/10.1099/ijsem.0.005343

Raina, J. B., Tapiolas, D. M., Willis, B. L., & Bourne, D. G. (2016). Coral-associated bacteria and their role in the biogeochemical cycling of sulfur. *Applied and Environmental Microbiology, 82*(11), 3453–3462.

Shi, Y., Liu, Y., Wang, H., Zheng, T., & Zhang, W. (2022). Nocardioides coralli sp. nov., a coral-associated bacterium. *International Journal of Systematic and Evolutionary Microbiology, 72*(8), 005517. https://doi.org/10.1099/ijsem.0.005517

Subramani, R., & Aalbersberg, W. (2012). Marine actinomycetes: An ongoing source of novel bioactive metabolites. *Microbiological Research, 167*(10), 571–580. https://doi.org/10.1016/j.micres.2012.06.005

Subramani, R., & Aalbersberg, W. (2022). Marine actinobacteria as a source of novel bioactive metabolites. *Frontiers in Microbiology, 13*, 851144. https://doi.org/10.3389/fmicb.2022.851144

Tiwari, K., & Gupta, R. K. (2012). Rare actinomycetes: A potential storehouse for novel antibiotics. *Critical Reviews in Biotechnology, 32*(2), 108–132. https://doi.org/10.3109/07388551.2011.562482

Vasconcellos, S. P., Angolini, C. F. F., & García, G. D. (2019). Enzymes from marine-derived actinobacteria: Biotechnological applications. *Frontiers in Microbiology, 10*, 338. https://doi.org/10.3389/fmicb.2019.00338

Waksman, S. A., & Curtis, R. E. (1916). The actinomyces of the soil. *Soil Science, 1*(2), 99–134.

Zhang, L., Wang, Y., Li, X., & Goodfellow, M. (2016). Nocardioides species from marine sediments: Diversity and bioactive potential. *Marine Drugs, 14*(4), 56. https://doi.org/10.3390/md14040056

Zhang, H., Ahmed, I., Zhang, H., Wang, Y., Tian, X., & Zhang, G. (2022a). Evolutionary adaptation and secondary metabolite potential of marine actinobacteria. *Frontiers in Microbiology, 13*, 836543. https://doi.org/10.3389/fmicb.2022.836543

Zhang, W., Xie, Y., Wang, S., & Chen, J. (2022b). Advances in metagenomics of marine actinobacteria and their natural product discovery. *Frontiers in Microbiology, 13*, 868043. https://doi.org/10.3389/fmicb.2022.868043

Zhang, J., Chen, J., Wu, Y.-H., Yu, Z., & Xu, X.-W. (2023). *Brevibacterium spongiae* sp. nov., a new species from marine sponge. *MicrobiologyOpen, 12*(2), e1340. https://doi.org/10.1002/mbo3.1340

Zhou, Y., Li, Y., Zhang, R., & Chen, Y. (2021). Synthetic biology and metabolic engineering strategies for natural product discovery and production in *Streptomyces*. *Biotechnology Advances, 47*, 107697. https://doi.org/10.1016/j.biotechadv.2021.107697

Ziemert, N., Lechner, A., Wietz, M., Millán-Aguiñaga, N., Chavarria, K. L., & Jensen, P. R. (2014). Diversity and evolution of secondary metabolism in the marine actinomycete genus *Salinispora*. *Proceedings of the National Academy of Sciences, 111*(12), E1130–E1139. https://doi.org/10.1073/pnas.1324161111

Ziemert, N., Alanjary, M., & Weber, T. (2016). The evolution of genome mining in microbes—A review. *Natural Product Reports, 33*(8), 988–1005. https://doi.org/10.1039/c5np00125h

Zun, Z., Han, W., Liu, D., Wang, C., Liang, J., & Li, X. (2022). Diversity of actinobacteria from corals and their antimicrobial potential. *Frontiers in Microbiology, 13*, 860308. https://doi.org/10.3389/fmicb.2022.860308

Diversity of Marine Actinobacteria

A. Kowsalya and A. Sai Ramesh

Abstract

Marine actinobacteria are a highly diverse group of microorganisms recognized for their remarkable ability to generate novel bioactive compounds. These organisms hold immense promise in biotechnology and pharmaceuticals. Their adaptation to extreme marine environments has equipped them with distinctive metabolic pathways, enabling the synthesis of a wide range of secondary metabolites with antibacterial, antifungal, antiviral, anticancer, and anti-biofilm properties. Beyond their role in drug discovery, marine actinobacteria contribute to nanoparticle biosynthesis, enzyme production, and novel solutions to antimicrobial resistance, highlighting their versatile applications. However, significant challenges remain, including the difficulty of cultivating unculturable strains, the gap between genome mining and metabolite expression, and the threats posed by climate change and marine pollution. Emerging strategies such as integrative multi-omics, synthetic biology, and bioprospecting of unexplored marine niches hold great promise in unlocking their full potential. Establishing sustainable policies and frameworks for marine resource utilization will be crucial for advancing research while preserving ecological balance. Overall, marine actinobacteria stand at the forefront of modern biotechnology, offering transformative solutions for human health and industrial applications.

A. Kowsalya
Anna University, Chennai, Tamil Nadu, India

A. Sai Ramesh (✉)
Department of BioEngineering, Vels Institute of Science, Technology and Advanced Studies, Chennai, Tamil Nadu, India

K. Loganathan, Z. Li (eds.), *Marine Actinomycetes*,
https://doi.org/10.1007/978-981-95-9849-6_2

Graphical Abstract

Marine Actinobacteria's Potential

Bioactive Compounds

Rich source of novel compounds with diverse applications.

Diverse Activities

Wide range of activities including antibacterial and anticancer

Challenges

Difficulties in cultivation, metabolite expression and environmental threats

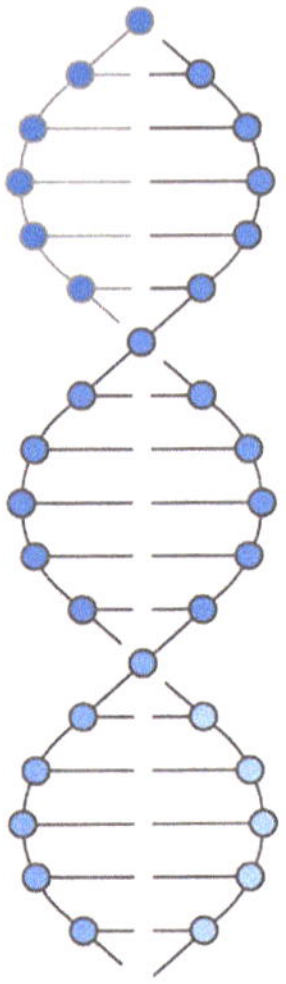

Metabolic Pathways

Unique pathways adapted from extreme marine environments.

Applications

Used in drug discovery, enzyme production, and resistance control.

Future Strategies

Multi-omics, synthetic biology, and sustainable policies for advancement.

Keywords

Marine actinobacteria · Drug discovery · Antimicrobial resistance · Synthetic biology · Bioactive compounds

1 Introduction

Marine actinobacteria are a highly diverse group of microorganisms recognized for their remarkable ability to generate novel bioactive compounds. Due to the useful chemicals that they produce, like antibiotics, antifungals, immunosuppressants, and many more, industrial microbiology and pharmaceutical biotechnology depend on them. Not only are actinobacteria useful for industry, but they are also important for recycling carbon, cleaning up waste, and helping plants grow more. They are very valuable microbes because they can survive in a variety of biological and metabolic conditions.

Marine environments are recently becoming part of the exploration process because they have the largest and least known types of microbes on Earth. Some of the biological problems that only happen in the marine environment are high salinity, changes in pressure, a lack of nutrients, and harsh temperatures. Because of selection pressure from these factors, marine actinobacteria have developed new metabolic pathways and naturally occurring products with different structures. These are things that you don't usually see in strains that live on land. A huge number of different species of actinobacteria live in coastal sands, mangroves, deep-sea sediments, hydrothermal vents, and other places in the ocean. The taxonomic and functional environment of the group has grown since new species and rare actinobacteria were found in these places (Qin et al., 2019; Rego et al., 2019). Interestingly, some of these marine isolates make secondary metabolites that can fight germs, cancer, free radicals, and inflammation. This shows how promising they are at solving important medical problems like drug resistance. Beyond making new medicines, the actinobacteria in the ocean are useful for other reasons. These animals help the aquaculture business and marine communities. They may also help protect the earth by getting rid of pollutants and recycling nutrients. Endophytes in mangroves affect how strong and productive plants are, and actinobacteria that live on sponges are very important for protecting their hosts. Actinobacteria play many important roles in marine environments, as shown by the wide range of biological activities they perform. Girão et al. (2022) say that these systems are becoming more and more known as great places for bioprospecting.

The present study reveals the different types of actinobacteria that live in the water, focusing on their functions, where they live in the environment, how they are classified, and their phylogeny. Habitat-specific factors affect the variety and metabolic skills of actinobacteria. These factors are looked at in this review. New marine species, evolutionary strategies for living in harsh environments, and possible biotechnological uses for animal waste are some of the other topics that are talked about. Recently, a lot of research has been done on actinobacteria that live in the ocean. This article is mostly about a study that was done in India (Puttaswamygowda et al., 2019).

2 Taxonomy and Phylogenetic Diversity of Marine Actinobacteria

The way marine actinobacteria are classified has been completely changed by molecular biology (Nafis et al., 2019). Systems based on genes and genomes have mostly taken the place of systems based on shape and function. The Actinobacteria phylum has a lot of different types of bacteria that have never been grown in a lab before (Narsing Rao & Li, 2022). *Streptomyces, Saccharopolyspora, Micromonospora, Nocardiopsis,* and *Salinispora* are some of the better-known names in this group (Table 1). Because it can make antibiotics and other chemicals, *Streptomyces* is still the group that scientists are studying the most. But sea strains make different chemicals and have very different genes from strains that live on land. It has also been very helpful to find natural products that can easily sort *Micromonospora* and *Nocardiopsis* from marine sediments and sponges. Salinosporamide A and other medicinal chemicals are made by *Salinispora* strains that can only grow in water. As an example, you can look into the connection between evolutionary biology and metabolite synthesis.

In addition, rare actinobacteria live in places near the sea. This group of organisms has a lot of promise that hasn't been used yet. In the genus *Actinokineospora, Verrucosispora, Dietzia, Nonomuraea,* and *Gordonia,* new natural chemical biosynthesis routes have been found (Subramani & Sipkema, 2019; Goel et al., 2021).

Table 1 Predominant diversity of marine actinobacteria and their ecological sources

Genus	Source (marine habitat)	Notable bioactive compounds/activities	References
Streptomyces	Sediments, sponges, seawater	Antibiotics (tetracyclines, streptomycin), anticancer agents	Manivasagan et al. (2014)
Micromonospora	Marine sediments, mangroves	Antibacterial and antitumor compounds	Subramani and Aalbersberg (2012)
Salinispora	Sediments (tropical/ subtropical regions)	Salinosporamide A (anticancer)	Jensen et al. (2007)
Nocardiopsis	Marine sediments, halophilic environments	Antimicrobial peptides, enzyme inhibitors	Goodfellow and Fiedler (2010)
Actinokineospora	Marine sponges, corals	Novel polyketides, antimicrobial metabolites	Abdelmohsen et al. (2014)
Dietzia	Deep-sea sediments	Hydrocarbon-degrading enzymes	Pathom-aree et al. (2006)

More and more Actinobacteria species are being found in strange places, like the polar regions, deep-sea sediments, and volcanic islands (Sottorff et al., 2019). Phylogenetic research shows that some actinobacteria that live in the ocean have evolved in very different ways from those that live on land. It's not clear how long these germs had been in the water.

The use of molecular technologies has completely changed how we think about the development and classification of actinobacteria. Multilocus Sequence Analysis (MLSA), Whole-genome sequencing, and 16S rRNA sequencing are the best ways to tell the difference between strains that are closely related and to find hidden variety. Recently, genome mining has led to some surprising findings, such as the discovery of sets of metabolic genes in marine actinobacteria that were not known before. It's possible that these genes could code for new chemicals that are useful. Comparative genomics says that biological and functional traits are linked to taxonomic groups. It has also shown how actinobacterial families have evolved together. These tools help with both accurate classification and the identification of taxa and molecules that have not been seen before.

A lot of what we know about marine actinobacterial variety comes from research that has been done at the regional level, especially in biodiverse countries like India. According to research done along the Gujarat coast, there is a lot of genetic and physiological variety among the actinobacteria isolates that were tested. These isolates were able to kill pathogens. Similarly, studying actinobacteria related to sponges off the coast of western India produced a wide range of phylogenetic trees (Gohain et al., 2020; Kurtböke et al., 2020). This showed how important marine invertebrates are as hosts for microbes whose lineages have not been known before (Baig et al., 2021). The Andaman and Nicobar Islands have a lot of different types of actinobacteria, which supports the idea that isolated marine areas are home to a wide range of microbes (Gobalakrishnan, 2022; López-Pérez et al., 2020). More study on marine ecosystems in India backs up these results, showing a lot of variety and possible uses in biotechnology (Sarkar & Suthindhiran, 2022). These specific results show that marine actinobacteria need to be studied in more depth in areas that haven't been studied much.

This last point is that marine actinobacteria are very different in terms of taxonomy and types of evolutionary trees. This group includes both recently found species and species that are well known. All of them have changed over time in order to play their biological parts and reach their full bioactive potential. Because molecular methods are getting better and regional case studies are being done, we are always learning more about how they are different. In the future, it will be easier to make new discoveries in science and medicine because of this.

3 Ecological Niches and Habitat-Specific Diversity

3.1 Marine Sediments and Coastal Zones

Because of the different amounts of salt, nutrients, and organic matter in marine sediments and coastal habitats, different actinobacteria can be found there (Gozari et al., 2019). Actinobacteria that can change their shape live in these places. They may be able to live and grow even in places with low oxygen, high tides, and nutrients that come from land waste that are hard to predict. Ribeiro et al. (2020) and Prins et al. (2024) said that marine sediments are home to many new groups and species. Some of these species can make bioactive chemicals that aren't found in terrestrial isolates because of the way their metabolism works. Marine environments depend on actinobacteria that come from sediment for energy flow and nutrient cycling because they can live in a lot of different conditions (Ibrahimi et al., 2020). Their ability to make enzymes that break down organic materials is what makes them useful. Actinobacteria can change in ways that make them more resistant to salt, hydrostatic pressure, and osmotic stress so that they can live in seaside areas. Microbes from both the ocean and the land come together in coastal areas, which is why this is the case. These changes mean that actinobacteria that live in sediment might be a good place to start looking for new bioactive compounds. These compounds might help fight the growing problem of antibiotic tolerance.

3.2 Sponge-Associated Actinobacteria

Actinobacteria do very well on sea sponges because they are the perfect hosts for their close relationships. By stopping these bacteria from growing, they protect sponges from diseases and biofouling species that are bad for them. Since sponges eat through filters, they are always moving through a lot of saltwater (Hazarika & Thakur, 2020). This lets a lot of different microbes live in their organs. Actinobacteria isolated from sponges make a lot of secondary chemicals. Some of these are

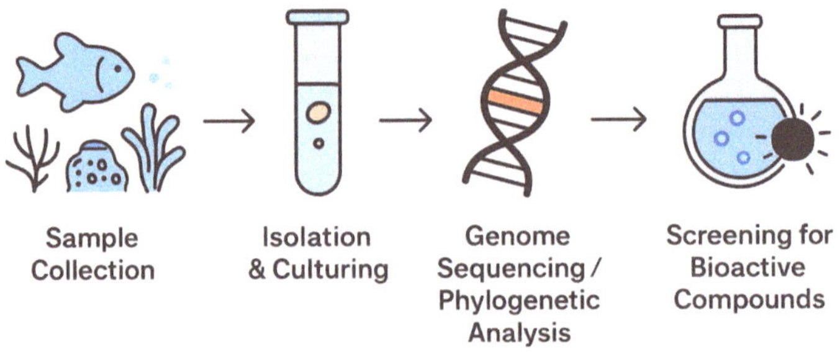

Fig. 1 Sponge-associated actinobacteria's complex metabolism

polyketides, alkaloids, and peptides that fight bacteria, fungi, and cancer. Sponge metabolism is known to be very complex (Liu et al., 2019; Baig et al., 2021) (Fig. 1). Actinobacteria get a stable climate and food sources in exchange for chemicals that the host sponge can use to protect itself. When compared to actinobacteria that live on their own, sponge-dwelling actinobacteria often make chemicals with different molecular structures. This symbiosis is very important for finding new natural products and for the pharmaceutical business in particular.

3.3 Mangroves and Rhizosphere

Mangrove settings are made up of salty sediments, organic matter, and sediments that don't have oxygen. These places are great for actinobacteria to live. Actinobacterial endophytes are very important for plants' life and health. This group gets more help from mangrove trees' rhizosphere than from any other plant. According to Jiang et al. (2018), Law et al. (2019), and Ye et al. (2023), these actinobacteria help plants take in nutrients, fix nitrogen, and dissolve phosphate. They also make chemicals that kill bacteria, which lowers the number of diseases that dirt can spread. Actinobacteria that were first found in mangroves have changed over time to deal with changing salt levels, higher temperatures, and redox changes. This is exactly what makes them able to keep stress-response genes and molecules. Endophytes from mangrove roots and rhizosphere soils have chemicals in them that kill fungi and protect cells, according to research. They not only help plants stay healthy, but they also break down complex organic substances like lignin and cellulose, which are an important part of biogeochemical cycling in mangrove forests. The actinobacteria that live in mangroves show how plants and microbes can live together in settings that are both hard and good for them.

3.4 Deep-Sea and Extreme Environments

Actinobacterial groups have changed over time to be able to survive in the harsh conditions of deep water, where there are little food and high hydrostatic pressure.

For getting food in oligotrophic environments, these living things have special metabolic and structural changes, like enzymes that can work in high pressure, metabolic pathways that can work in cold temperatures, and unique transport systems (Siro et al., 2022; Ribeiro et al., 2020). In the example given, actinobacteria that live in deep water can make chemicals that are very good at killing bacteria and cells. These chemicals are not found in actinobacteria that live in shallow water or on land (Millán-Aguiñaga et al., 2019). It is thought that evolutionary pressures in places with few resources, where chemical defense and competition are important for life, make deep-sea strains' metabolisms work harder. These extreme-environment actinobacteria are an area of bioprospecting that hasn't been studied enough, even though they could be used in biotechnological applications ranging from enzyme engineering to finding new drugs.

3.5 Marine Invertebrate-Associated Actinobacteria

A sponge, like many other sea creatures, can have actinobacterial symbionts live on it. Corals, tunicates, mollusks, and many more are all examples of these. Researchers Sheikh et al. (2018) and Kumar et al. (2020) say that these interactions protect invertebrate hosts from harmful organisms by making antibacterial or antifouling chemicals. Ecological niches and habitat-specific diversity of marine actinobacteria have a substantial role in the pharmaceutical industry (Table 2). Actinobacteria are very important to the biological processes of their hosts. They help break down nitrogen and get rid of harmful chemicals.

The medical value of invertebrate-associated actinobacteria has been emphasized by the fact that their byproducts are effective against cancer cells, viruses, and types of bacteria that are resistant to treatment. Actinobacteria have metabolic pathways that have changed to fit the chemical and physiological conditions of the plants that they live on. They are able to do this because of ecological specialization, which depends on how the host and microbe interact in certain situations. Invertebrates that live in water change their metabolisms to adapt to their environment. This makes more bioactive chemicals available, so these animals are a great source of possible new medicines.

4 Metabolic and Functional Diversity

Marine actinobacteria can change how much energy they use, which lets them make many different secondary chemicals and enzymes that are useful in biology, business, and healthcare. Their biosynthetic complexity and abundance are shown by the fact that they can make compounds with a wide range of structures. This part gives an overview of the metabolic byproducts and useful roles of marine actinobacteria, which include biotechnology, medicines, and environmental support (Fig. 2).

Table 2 Ecological niches and habitat-specific diversity of marine actinobacteria

Ecological niche/habitat	Representative genera	Key findings	References
Marine sediments	*Streptomyces, Micromonospora, Nocardiopsis*	Rich source of bioactive secondary metabolites; exhibits high species diversity due to nutrient-rich and stable conditions.	Narsing Rao and Li (2022) and Subramani and Sipkema (2019)
Marine sponges	*Salinispora, Actinokineospora, Verrucosispora*	Sponge–microbe symbiosis enhances production of novel antibiotics and cytotoxic compounds.	Sottorff et al. (2019) and Hazarika and Thakur (2020)
Coral reefs	*Streptomyces, Saccharopolyspora*	Contribute to coral health by producing antimicrobial agents that suppress pathogenic microbes.	Sarkar and Suthindhiran (2022)
Mangrove ecosystems	*Micromonospora, Nocardiopsis, Streptomyces*	Adapted to saline, anaerobic, and organic-rich environments; reported as reservoirs of novel enzymes and halotolerant metabolites.	Gohel et al. (2023) and Gobalakrishnan (2022)
Deep-Sea environments	*Salinispora, Streptomyces, Actinomadura*	Exhibit unique metabolic pathways for survival under high pressure and low temperature, a potential source of rare natural products.	Subramani and Sipkema (2019)
Marine algae-associated	*Streptomyces, Actinobacterium* sp.	Produce algicidal and antifouling metabolites, influencing algal health and marine ecological balance.	Narsing Rao and Li (2022)

4.1 Secondary Metabolites: Antibiotics, Anticancer, and Immunomodulators

Actinobacteria in the ocean make secondary chemicals that have different biological effects. In this group are antibiotics, which kill bacteria that have become resistant to many treatments, anticancer drugs, which stop the growth of tumor cells, and immunomodulators, which change how the host immune system responds (Girão et al., 2022; Al-Shaibani et al., 2021). The ocean is a great place for new biosynthetic pathways to form, which is why many of these metabolites are structurally different from their land-based cousins. Marine-derived glycosides, non-ribosomal peptides, and polyketides have a lot of promise in medicine and science. It is very important to stress the pharmacological importance of these substances because of the growing need for antibiotic resistance and new therapeutic drugs. Because their chemicals change the immune system, marine actinobacteria offer new and hopeful ways to treat cancer, inflammatory diseases, and autoimmune disorders.

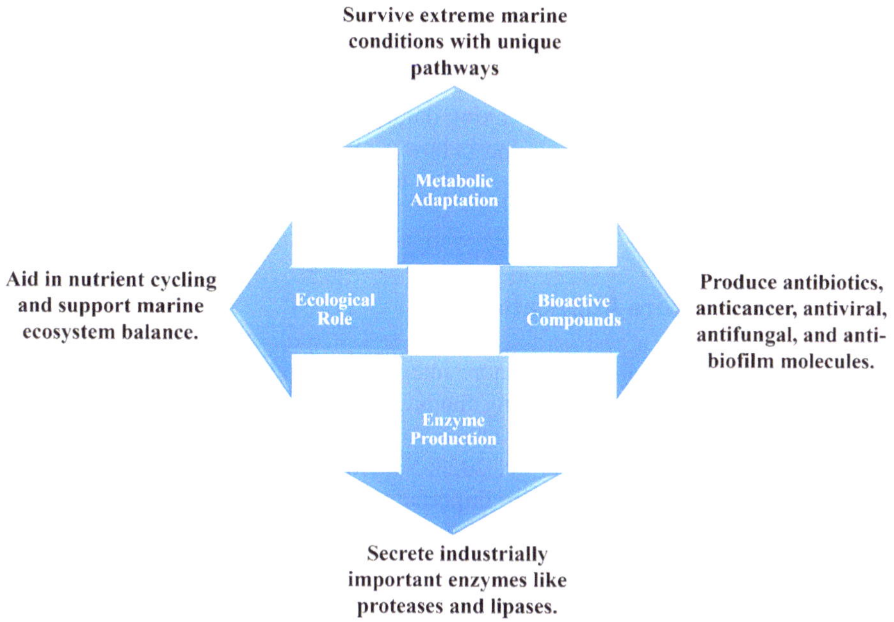

Fig. 2 Marine actinobacteria's metabolic and functional diversity

4.2 Halo-Metabolites, Flavonoids, and Alkaloids

Oceanic actinobacteria make many bioactive chemicals, some of which are well known as antibiotics and others that are not as well known, such as halometabolites, flavonoids, and alkaloids. Kasanah and Triyanto (2019) and Hozzein et al. (2021) studies back up what Singh and Dubey (2018) and Karthik (2022) already knew: halometabolites, which are made up of halogen atoms like chlorine and bromine, are more biologically active and have a faster metabolism. Drug research pipelines are interested in these compounds because they are very good at fighting cancer, fungus, and viruses. Flavonoids and alkaloids are two nutrients that marine actinobacteria make that help the defense system. Not only do they stop infectious diseases, but they also kill cancer cells. These metabolites give a lot of chemical information for medical studies. They show how a number of new chemicals have evolved in reaction to problems in the ocean's ecosystem.

4.3 Bioremediation Potential

Some data show that marine actinobacteria can be good for the environment and for people (Chauhan & Gohel, 2020; Thirumurugan et al., 2018). Heavy metals, hydrocarbons, and pesticides are all examples of complex organic pollutants. Living things can break these down, which can help clean them up (Rathore et al., 2020).

Marine ecosystems that are polluted can be cleaned up by enzymes that change harmful chemicals into less harmful ones. Actinobacteria that can break down PAHs live in sediments that have been contaminated with oil, for example. In this case, it shows how important they are for fixing damage that people have made. The fact that they can make biosurfactants shows how useful they are for biotechnology and how good they are for the earth. This makes it easier to break down fuels and clean up after oil spills.

4.4 Aquaculture Applications

Marine actinobacteria are getting more interest in aquaculture systems because it could be used as a disease-controlling and probiotic. There is proof that they can get rid of pathogens, improve the flow of nutrients, and help farmed fish and shellfish live faster and longer (James et al., 2023; Yaradoddi et al., 2022). Actinobacteria make bioactive compounds that are important in aquaculture because they help to reduce the need for pharmaceutical antibiotics and stop tolerance from forming. These compounds are used as antibiotics and siderophores. The ability to break down organic waste and improve water quality should be a strength for sustainable aquaculture businesses. These things not only increase output, but they also help meet the growing need for environmentally friendly farming around the world.

4.5 Novel Cultivation and OSMAC Approaches

Even though they have a lot of promise, many metabolic pathways involving marine actinobacteria still can't be reached in a normal lab setting. For example, Le Loarer et al. (2023) say that the One Strain-Many Compounds (OSMAC) method and new ways of growing plants were created to find this hidden chemical variety (Kasanah & Triyanto, 2019). Changes in the pH, salt, or nutrient amounts in the growth environment are some of these methods. Another way is to grow the bacteria in a mixed medium to bring out latent gene sets. These methods have made it possible for marine actinobacteria to use a lot more chemical space by helping them find new metabolites. It's exciting to think that OSMAC and new culture methods could increase the output of metabolites and help find new chemicals that could be used in medicine and industry. This is even more clear when you look at how genetics and synthetic biology have changed things.

5 Biotechnological and Pharmaceutical Applications

A lot of different bioactive chemicals are made by actinobacteria in the ocean. These compounds have opened up a lot of new ways to study and treat people. Their great work in finding new drugs led to the creation of many life-saving drugs, such as antibiotics, antivirals, and antifungals (Jagannathan et al., 2021; Siddharth, 2019).

Scientists are interested in them for more than just the fact that they might be able to fight biofilms. Because of this, new ways may come up to fight the worldwide problem of drug resistance (Goel et al., 2021; Ibrahimi et al., 2020).

Messaoudi et al. (2023) say that another interesting area of research is using marine actinobacteria as long-term biofactories to make metallic and non-metallic nanoparticles that could be used in medicine and business. These organisms can handle high temperatures and high pressures. Their enzymes can also be used in bioremediation, pharmaceutical processes, and industrial catalysis (Subash et al., 2024). All of these uses show how marine actinobacteria could be used to solve important medical and economic problems.

6 Challenges in Exploring Marine Actinobacteria

Even though it might be helpful, studying actinobacteria in the seas is very hard. It's hard to look at the metabolic profiles of many types of growth because they can't be grown in normal labs (Huang et al., 2021; Anteneh et al., 2022). Because of this, a lot of actinobacterial species are still unknown or don't have good descriptions, which makes the problem of secret diversity even worse. Genome sequencing doesn't solve the problem of how to get metabolites out of them because many groups of genes that are involved in biosynthesis don't work when tried in a controlled setting. Climate change, pollution, and ocean acidification are also threats to marine environments. These problems could wipe out important microbial resources before they are found. These problems show how important it is to find new ways to grow actinobacteria, use cutting-edge bioinformatics, and make long-term plans for protecting them so they can reach their full potential.

7 Future Perspectives and Research Directions

Future studies on marine actinobacteria will be shaped by methods that are kind to the environment and use cutting-edge technology. Integrating genetics, transcriptomics, proteomics, and metabolomics is a good way to learn more about how cells work and how their metabolic potential is controlled. It can also help find new metabolites. If bioprospecting attempts are stepped up, polar waters, deep-sea hydrothermal vents, and other ecological niches that haven't been explored yet could be the key to finding new strains with bioactivities that haven't been seen before (Majithiya & Gohel, 2022).

Besides that, there is hope that synthetic biology and genetic engineering can reveal hidden biosynthetic pathways. This would allow the production of many unique chemicals. For long-term benefits, it is also important to set up rules for healthy bioprospecting that balance scientific research with protecting the environment and sharing profits fairly. These ideas for the future could make it possible for marine actinobacteria to be used a lot more in biotechnology and medicine development around the world.

8 Conclusion

Actinobacteria from the ocean are one of the most useful microbes for bioengineering and drug development. Because they have a lot of different biochemical pathways, they are very helpful in the fight against antibiotic resistance, making enzymes, finding new drugs, and making nanoparticles. But because they are hard to grow, have a lot of different types, and are bad for the environment, their full potential has not been fully explored. To get around these problems, scientists should look into synthetic biology techniques, study the most remote sea habitats, and combine data from multiple omics. When it comes down to it, marine actinobacteria are the leaders in biotechnology and next-generation drug development. They are solving some of the biggest problems in business and healthcare today in the long run.

References

Abdelmohsen, U. R., Bayer, K., & Hentschel, U. (2014). Diversity, abundance and natural products of marine sponge-associated actinomycetes. *Natural Products Reports, 31*(3), 381–399. https://doi.org/10.1039/c3np70111e. PMID: 24496105.

Al-Shaibani, M. M., Radin Mohamed, R. M. S., Sidik, N. M., Enshasy, H. A. E., Al-Gheethi, A., Noman, E., et al. (2021). Biodiversity of secondary metabolites compounds isolated from phylum Actinobacteria and its therapeutic applications. *Molecules, 26*(15), 4504.

Anteneh, Y. S., Yang, Q., Brown, M. H., & Franco, C. M. (2022). Factors affecting the isolation and diversity of marine sponge-associated bacteria. *Applied Microbiology and Biotechnology, 106*(4), 1729–1744.

Baig, U., Dahanukar, N., Shintre, N., Holkar, K., Pund, A., Lele, U., et al. (2021). Phylogenetic diversity and activity screening of cultivable actinobacteria isolated from marine sponges and associated environments from the western coast of India. *Access Microbiology, 3*(9), 000242.

Chauhan, J. V., & Gohel, S. D. (2020). Molecular diversity and pharmaceutical applications of free-living and rhizospheric marine actinobacteria. In *Marine niche: Applications in pharmaceutical sciences: Translational research* (pp. 111–131). Springer Singapore.

Goodfellow, M., & Fiedler, H. P. (2010). A guide to successful bioprospecting: Informed by actinobacterial systematics. *Antonie Van Leeuwenhoek, 98*(2), 119–142. https://doi.org/10.1007/s10482-010-9460-2. Epub June 26, 2010. PMID: 20582471.

Girão, M., Ribeiro, I., & Carvalho, M. D. F. (2022). Actinobacteria from marine environments: A unique source of natural products. In *Natural products from actinomycetes: Diversity, ecology and drug discovery* (pp. 1–45). Springer Singapore.

Gobalakrishnan, R. (2022). Phylogenetic diversity of culturable marine actinobacteria isolated from the Havelock Island, the Andamans, India. *Ecological Genetics and Genomics, 23*, 100123.

Goel, N., Fatima, S. W., Kumar, S., Sinha, R., & Khare, S. K. (2021). Antimicrobial resistance in biofilms: Exploring marine actinobacteria as a potential source of antibiotics and biofilm inhibitors. *Biotechnology Reports, 30*, e00613.

Gohain, A., Manpoong, C., Saikia, R., & De Mandal, S. (2020). Actinobacteria: Diversity and biotechnological applications. In *Recent advancements in microbial diversity* (pp. 217–231). Academic.

Gohel, S. D., Majithiya, V. R., & Singh, S. P. (2023). Genetic and physiological diversity of marine actinobacteria from the Okha Coast, Gujarat, India. *Geomicrobiology Journal, 40*(6), 590–604.

Gozari, M., Bahador, N., Jassbi, A. R., Mortazavi, M. S., Hamzehei, S., & Eftekhar, E. (2019). Isolation, distribution and evaluation of cytotoxic and antioxidant activity of cultivable actinobacteria from the Oman Sea sediments. *Acta Oceanologica Sinica, 38*(12), 84–90.

Hazarika, S. N., & Thakur, D. (2020). Actinobacteria. In *Beneficial microbes in agro-ecology* (pp. 443–476). Academic.

Hozzein, W. N., Mohany, M., Alhawsawi, S. M., Zaky, M. Y., Al-Rejaie, S. S., & Alkhalifah, D. H. (2021). Flavonoids from marine-derived actinobacteria as anticancer drugs. *Current Pharmaceutical Design, 27*(4), 505–512.

Huang, Z., Mo, S., Yan, L., Wei, X., Huang, Y., Zhang, L., et al. (2021). A simple culture method enhances the recovery of culturable actinobacteria from coastal sediments. *Frontiers in Microbiology, 12*, 675048.

Ibrahimi, M., Korichi, W., Hafidi, M., Lemee, L., Ouhdouch, Y., & Loqman, S. (2020). Marine actinobacteria: Screening for predation leads to the discovery of potential new drugs against multidrug-resistant bacteria. *Antibiotics, 9*(2), 91.

Jagannathan, S. V., Manemann, E. M., Rowe, S. E., Callender, M. C., & Soto, W. (2021). Marine actinomycetes, new sources of biotechnological products. *Marine Drugs, 19*(7), 365.

James, G., Prasannan Geetha, P., Thavarool Puthiyedathu, S., & Vattringal Jayadradhan, R. K. (2023). Applications of Actinobacteria in aquaculture: Prospects and challenges. *3 Biotech, 13*(2), 42.

Jensen, P. R., Williams, P. G., Oh, D. C., Zeigler, L., & Fenical, W. (2007). Species-specific secondary metabolite production in marine actinomycetes of the genus Salinispora. *Applied and Environmental Microbiology, 73*(4), 1146–1152. https://doi.org/10.1128/AEM.01891-06. Epub December 8, 2006. PMID: 17158611; PMCID: PMC1828645.

Jiang, Z. K., Tuo, L., Huang, D. L., Osterman, I. A., Tyurin, A. P., Liu, S. W., et al. (2018). Diversity, novelty, and antimicrobial activity of endophytic actinobacteria from mangrove plants in Beilun Estuary National Nature Reserve of Guangxi, China. *Frontiers in Microbiology, 9*, 868.

Karthik, L. (2022). *Actinobacteria*. Springer Singapore.

Kasanah, N., & Triyanto, T. (2019). Bioactivities of halometabolites from marine actinobacteria. *Biomolecules, 9*(6), 225. https://doi.org/10.3390/biom9060225. PMID: 31212626; PMCID: PMC6627970.

Kumar, P. S., Ling, C. Y., Zhou, Z. B., Dong, Y. L., Sun, C. L., Song, Y. X., et al. (2020). Chemical diversity of metabolites and antibacterial potential of actinomycetes associated with marine invertebrates from intertidal regions of Daya Bay and Nansha Islands. *Microbiology, 89*(4), 483–492.

Kurtböke, D. İ., Okazaki, T., & Vobis, G. (2020). Actinobacteria in marine environments: From terrigenous origin to adapted functional diversity. *Encyclopedia of Marine Biotechnology, 3*, 1951–1978.

Law, J. W. F., Pusparajah, P., Ab Mutalib, N. S., Wong, S. H., Goh, B. H., & Lee, L. H. (2019). A review on mangrove actinobacterial diversity: The roles of Streptomyces and novel species discovery. *Progress in Microbes & Molecular Biology, 2*(1), a0000024.

Le Loarer, A., Dufossé, L., Bignon, J., Frédérich, M., Ledoux, A., Fouillaud, M., & Gauvin-Bialecki, A. (2023). OSMAC Method to assess impact of culture parameters on metabolomic diversity and biological activity of marine-derived Actinobacteria. *Marine Drugs, 22*(1), 23.

Liu, T., Wu, S., Zhang, R., Wang, D., Chen, J., & Zhao, J. (2019). Diversity and antimicrobial potential of Actinobacteria isolated from diverse marine sponges along the Beibu Gulf of the South China Sea. *FEMS Microbiology Ecology, 95*(7), fiz089.

López-Pérez, M., Haro-Moreno, J. M., Iranzo, J., & Rodriguez-Valera, F. (2020). Genomes of the "Candidatus Actinomarinales" order: Highly streamlined marine epipelagic Actinobacteria. *Msystems, 5*(6), 10–1128.

Majithiya, V. R., & Gohel, S. D. (2022). Marine invertebrates: Diversity and biological significance. In *Actinobacteria: Diversity, applications and medical aspects* (p. 23). IntechOpen.

Manivasagan, P., Venkatesan, J., Sivakumar, K., & Kim, S. K. (2014). Pharmaceutically active secondary metabolites of marine actinobacteria. *Microbiological Research, 169*(4), 262–278. https://doi.org/10.1016/j.micres.2013.07.014. Epub August 16, 2013. PMID: 23958059.

Messaoudi, O., Benamar, I., Azizi, A., Albukhaty, S., Khane, Y., Sulaiman, G. M., et al. (2023). Characterization of silver carbonate nanoparticles biosynthesized using marine actinobacteria and exploring of their antimicrobial and antibiofilm activity. *Marine Drugs, 21*(10), 536.

Millán-Aguiñaga, N., Soldatou, S., Brozio, S., Munnoch, J. T., Howe, J., Hoskisson, P. A., & Duncan, K. R. (2019). Awakening ancient polar Actinobacteria: Diversity, evolution, and specialized metabolite potential. *Microbiology, 165*(11), 1169–1180.

Nafis, A., Raklami, A., Bechtaoui, N., El Khalloufi, F., El Alaoui, A., Glick, B. R., et al. (2019). Actinobacteria from extreme niches in Morocco and their plant growth-promoting potentials. *Diversity, 11*(8), 139.

Narsing Rao, M. P., & Li, W. J. (2022). Diversity of actinobacteria in various habitats. In *Actinobacteria: Microbiology to synthetic biology* (pp. 37–58). Springer Nature Singapore.

Pathom-Aree, W., Stach, J. E., Ward, A. C., Horikoshi, K., Bull, A. T., & Goodfellow, M. (2006). Diversity of actinomycetes isolated from Challenger Deep sediment (10,898 m) from the Mariana Trench. *Extremophiles, 10*(3), 181–189. https://doi.org/10.1007/s00792-005-0482-z. Epub March 15, 2006. PMID: 16538400.

Prins, A., Dyani, S. S., Vreulink, J. M., Maldonado, L. A., & Le Roes-Hill, M. (2024). Actinobacteria diversity associated with marine sediments and a wetland system, Agulhas-South Africa. *Annals of Microbiology, 74*(1), 18.

Puttaswamygowda, G. H., Olakkaran, S., Antony, A., & Purayil, A. K. (2019). Present status and future perspectives of marine actinobacterial metabolites. In *Recent developments in applied microbiology and biochemistry* (pp. 307–319). https://doi.org/10.1016/B978-0-12-81 6328-3.00022-2

Qin, S., Li, W. J., Klenk, H. P., Hozzein, W. N., & Ahmed, I. (2019). Actinobacteria in special and extreme habitats: Diversity, function roles and environmental adaptations. *Frontiers in Microbiology, 10*, 944.

Rathore, D. S., Sheikh, M., & Singh, S. P. (2020). Marine Actinobacteria: New horizons in bioremediation. In *Recent developments in microbial technologies* (pp. 425–449). Springer Nature Singapore.

Rego, A., Raio, F., Martins, T. P., Ribeiro, H., Sousa, A. G., Séneca, J., et al. (2019). Actinobacteria and cyanobacteria diversity in terrestrial Antarctic microenvironments evaluated by culture-dependent and independent methods. *Frontiers in Microbiology, 10*, 1018.

Ribeiro, I., Girão, M., Alexandrino, D. A., Ribeiro, T., Santos, C., Pereira, F., et al. (2020). Diversity and bioactive potential of actinobacteria isolated from a coastal marine sediment in northern Portugal. *Microorganisms, 8*(11), 1691.

Sarkar, G., & Suthindhiran, K. (2022). Diversity and biotechnological potential of marine actinomycetes from India. *Indian Journal of Microbiology, 62*(4), 475–493.

Sheikh, M. A., Rathore, D. S., Gohel, S. D., & Singh, S. P. (2018). Marine actinobacteria associated with the invertebrate hosts: A rich source of bioactive compounds: A review. *Journal of Cell & Tissue Research, 18*(1), 6361–6374.

Siddharth, S. (2019). Isolation and characterization of bioactive compounds with antibacterial, antioxidant and enzyme inhibitory activities from marine-derived rare actinobacteria, Nocardiopsis sp. SCA21. *Microbial Pathogenesis, 137*, 103775.

Singh, R., & Dubey, A. K. (2018). Diversity and applications of endophytic actinobacteria of plants in special and other ecological niches. *Frontiers in Microbiology, 9*, 1767.

Siro, G., Donald, L., & Pipite, A. (2022). The diversity of deep-sea actinobacteria and their natural products: An epitome of curiosity and drug discovery. *Diversity, 15*(1), 30.

Sottorff, I., Wiese, J., & Imhoff, J. F. (2019). High diversity and novelty of Actinobacteria isolated from the coastal zone of the geographically remote young volcanic Easter Island, Chile. *International Microbiology, 22*(3), 377–390.

Subash, H. A., Santhosh, K., Kannan, K., & Pitchiah, S. (2024). Extraction of Keratin degrading enzyme from marine Actinobacteria of Nocardia sp and their antibacterial potential against oral pathogens. *Oral Oncology Reports, 9*, 100184.

Subramani, R., Aalbersberg, W. (2012). Marine actinomycetes: An ongoing source of novel bioactive metabolites. *Microbiological Research, 167*(10), 571–580. https://doi.org/10.1016/j.micres.2012.06.005. Epub July 15, 2012. PMID: 22796410.

Subramani, R., & Sipkema, D. (2019). Marine rare actinomycetes: A promising source of structurally diverse and unique novel natural products. *Marine Drugs, 17*(5), 249.

Thirumurugan, D., Vijayakumar, R., Vadivalagan, C., Karthika, P., & Khan, M. K. A. (2018). Isolation, structure elucidation and antibacterial activity of methyl-4, 8-dimethylundecanate from the marine actinobacterium *Streptomyces albogriseolus* ECR64. *Microbial Pathogenesis, 121*, 166–172.

Yaradoddi, J. S., Kontro, M. H., Ganachari, S. V., Banapurmath, N. R., Oli, A., Katti, A. S., & Sulochana, M. B. (2022). Actinobacteria in marine environments. In *Actinobacteria: Ecology, diversity, classification and extensive applications* (pp. 21–38). Springer Nature Singapore.

Ye, J. J., Zou, R. J., Zhou, D. D., Deng, X. L., Wu, N. L., Chen, D. D., & Xu, J. (2023). Insights into the phylogenetic diversity, biological activities, and biosynthetic potential of mangrove rhizosphere Actinobacteria from Hainan Island. *Frontiers in Microbiology, 14*, 1157601.

Isolation and Identification of Marine Actinobacteria

M. Kavin Raja, B. Madhumitha, Mithil Mohanraj, S. Taejashwini, Sohan Athmaraman, Kushi Poovaiah Chendira, and C. Subathra Devi

Abstract

Actinobacteria are some of the most productive and biotechnologically valuable bacterial phyla with the ability to produce bioactive compounds, industrial enzymes, and metabolites with agricultural and environmental applications. Actinobacteria were the key players in ecosystem processes as degraders of bulky organic substrates, while at the same time being humanity's main source of antibiotics and other therapeutic leads. Traditional cultivation methods access only 1–5% of actinobacterial diversity, and this shortfall has been referred to as the "great plate count anomaly." This chapter is a comprehensive overview of actinobacterial isolation techniques and provides an interface between historic methods and state-of-the-art techniques intended to realize the untapped potential hidden amongst non-culturable diversity. Ecological sources ranging from terrestrial, freshwater, and saltwater environments to extremophilic environments are surveyed with a focus on how adaptations to habitats result in biosynthetic performance. The approach to methods varies from selective and pretreatment techniques to state-of-the-art techniques, like ecological mimicry, microfluidics, and in situ cultivation and co-culture enrichment techniques. Molecular tools for identification, combining phylogenomics with morphological information, are outlined as powerful templates to characterize new taxa. The chapter concludes with future research trends like multi-omics techniques, AI-augmented cultivation techniques, and synthetic biology methods, while considering global bioprospecting's ethics and regulation ramifications. The present chapter serves as a scholar's reference and a hands-on guide to actinobacterial research advancement and biotechnology implementation using exhaustive protocols, case studies, and troubleshooting manuals.

M. Kavin Raja · B. Madhumitha · M. Mohanraj · S. Taejashwini · S. Athmaraman · K. P. Chendira · C. S. Devi (✉)
Department of Biotechnology, School of Bio Sciences and Technology, Vellore Institute of Technology, Vellore, Tamil Nadu, India
e-mail: csubathradevi@vit.ac.in

K. Loganathan, Z. Li (eds.), *Marine Actinomycetes*,
https://doi.org/10.1007/978-981-95-9849-6_3

Keywords

Actinobacteria · Isolation techniques · Uncultured diversity · Bioprospecting · Secondary metabolites · Biotechnology applications · Molecular identification

1 Introduction

Actinobacteria represent one of the most diverse and functionally versatile bacterial phyla, encompassing over 48 families and more than 379 genera, numbers that continue to expand with ongoing taxonomic revisions and the discovery of novel taxa. These Gram-positive bacteria possess exceptionally high G + C content (typically 55–75%) and exhibit distinctive morphological characteristics, including filamentous growth patterns, complex sporulation mechanisms, and aerial mycelium formation that often resembles fungal structures. The phylum is currently organized into six classes: *Actinobacteria, Acidimicrobiia, Coriobacteriia, Nitriliruptoria, Rubrobacteria, and Thermoleophilia.* The class Actinobacteria alone contains 16–18 orders and hosts the majority of biotechnologically important genera. The remarkable biosynthetic capabilities of actinobacteria have positioned them as the cornerstone of natural product discovery and biotechnology applications. These microorganisms are responsible for producing approximately two-thirds of all naturally derived antibiotics currently in clinical use, including landmark discoveries such as streptomycin from *Streptomyces griseus* (Nouioui et al., 2018; Schatz et al., 1944), as well as tetracyclines, macrolides, and glycopeptides (Bérdy, 2012). Beyond their pharmaceutical contributions, genera such as *Streptomyces, Micromonospora, Actinoplanes, and Rhodococcus* demonstrate distinctive biosynthetic repertoires that extend to anticancer agents, antifungals, immunosuppressants, and industrial enzymes. Recent genomic analyses reveal that many actinobacterial strains harbor complex regulatory networks controlling numerous silent biosynthetic gene clusters, representing an untapped reservoir of bioactive compounds (Barka et al., 2016; Augustijn et al., 2024; Mesquita et al., 2025; Hoyos et al., 2025; Mast & Stegmann, 2019). The ecological versatility of actinobacteria extends far beyond antibiotic production, encompassing critical roles in sustainable agriculture and environmental biotechnology. As plant growth-promoting rhizobacteria, biocontrol agents, and biofertilizers, they contribute significantly to crop productivity and soil health (Franco-Correa et al., 2010). Their enzymatic systems enable applications in bioremediation, heavy metal detoxification, and xenobiotic degradation (Amoroso et al., 2013), while their thermostable and pH-tolerant enzymes find widespread use in food processing, textiles, biofuels, and biopolymer industries (Genilloud et al., 2011; Mohammadipanah & Wink, 2016). Furthermore, actinobacteria demonstrate remarkable adaptability to extreme environments, from marine ecosystems where specialized genera like *Salinispora* have evolved unique seawater-dependent growth requirements, to oligotrophic desert environments that hold endemic species with novel metabolic capabilities (Krysenko, 2025). The isolation and cultivation of actinobacteria present unique challenges. Traditional isolation methods often favor the recovery of common Streptomyces species, while rare actinobacteria require specialized approaches and selective media to overcome competition from faster-growing microorganisms (Koech et al., 2025). Modern isolation strategies increasingly focus on accessing underexplored environments, including marine sediments, extreme habitats, and symbiotic associations with insects and plants, where novel actinobacterial diversity continues to be discovered. The development of sophisticated cultivation techniques, combined with genome-guided approaches for activating silent biosynthetic pathways, represents the current frontier in actinobacterial research, promising continued expansion of their biotechnological applications in the face of growing antimicrobial resistance and environmental challenges (Mohammadipanah & Wink, 2016).

2 The Cultivation Challenge and the Great Plate Count Anomaly

Though of tremendous value, the isolation and growth of actinobacteria are still a key limiting factor. Traditional techniques based on artificial culture media and fixed incubation conditions isolate only a small subset of actinobacterial diversity, usually fewer than 5% of the species found in nature (Staley & Konopka, 1985). This deviation, referred to as the "great plate count anomaly," has hindered entry to the immense biosynthetic potential of actinobacteria that are non-culturable (Girão et al., 2024; Epstein, 2013). Molecular methods, such as metagenomics and single-cell genomics, have uncovered a cryptic diversity of actinobacterial taxa and biosynthetic gene clusters, indicating an untapped pool of new natural products (Daniel, 2005; Ziemert et al., 2014). In addition, rediscovery of familiar compounds has stifled progress in natural product discovery programs, accelerating the demand for new isolation methods that can reach slow-growing or rare taxa with distinctive biosynthetic repertoires (Genilloud et al., 2011; Lewis, 2013). Exacerbating these issues, environmental degradation and habitat loss expose actinobacterial diversity to the risk of extinction, highlighting the need to devise new isolation frameworks for sustainable bioprospecting (Bull et al., 2016).

3 Marine Sample Collection and Processing

All sampling equipment must be properly sterilized before use. Glass containers, sampling tools, and other reusable equipment should be autoclaved at 121 °C for 15–20 min. For field sampling, sterile glass jars with screw caps were recommended as they prevent contamination and preserve sample integrity better than plastic containers. Equipment should be cleaned with a dilute solution of HCl and deionized water before autoclaving to prevent biofilm formation (Chifflet et al., 2019). Marine actinobacteria isolation requires careful site selection to maximize recovery of diverse strains. Deep-sea sediments, like Arctic and deep-sea environments, have shown exceptional actinobacterial diversity. Intertidal and subtidal sediments are the most productive environments for actinobacteria recovery. Sediment samples should be collected from multiple depth zones, such as for intertidal zones, where samples should be collected from the areas exposed during low tide in the upper 5–10 cm of the sediment layer. Subtidal zones are areas where the area is permanently submerged, typically at depths ranging from 4.5 to 450 m. Marine sediment samples should be collected using sterile core samplers or grab samplers to obtain undisturbed samples. The standard approach involves surface sediment collection, core sampling, and multiple site sampling (Jose & Jha, 2017).

3.1 Transport and Storage Considerations

Proper sample transport and storage are critical for maintaining the viability and diversity of actinobacterial populations. Soil samples should be transported at ambient temperature in breathable containers to prevent anaerobic conditions that may alter microbial community structure. Plastic bags, while convenient, can create moisture accumulation and anaerobic conditions that negatively impact actinobacterial survival. Storage temperature significantly affects actinobacterial viability, with refrigeration (4 °C) generally providing optimal conditions for short-term storage up to 1 week (Kurtböke, 2012). Freezing should be avoided for samples intended for cultivation-based studies, as freeze-thaw cycles can reduce viable cell counts by several orders of magnitude. Sample processing time represents a critical factor, with optimal results obtained when samples are processed within 24–48 h of collection (Hayakawa & Nonomura, 1987). Extended storage, even under optimal conditions, can result in selective survival of certain populations and loss of fastidious or stress-sensitive species.

4 Marine-Specific Pre-Treatment Methods

Marine actinobacteria isolation requires specific pre-treatment methods to overcome the challenges posed by competitive microorganisms and to selectively enrich for spore-forming actinobacterial strains. These methods are critical for the successful recovery of diverse actinobacterial species from marine environments.

4.1 Physical Pre-Treatment Approaches

4.1.1 Dry Heat Treatment Protocol

The heat shock method was the most widely employed pre-treatment for marine actinobacteria isolation. This technique specifically targets spore-forming bacteria while eliminating vegetative bacterial cells that may compete during isolation. Marine actinobacteria samples are processed by air-drying followed by heat treatment ranging from 60–65 °C for 15–30 min, which creates a selective pressure favoring heat-resistant spores. This method has been successfully applied to marine samples from the Arabian Sea, where dry heat treatment combined with pre-enrichment techniques enhanced actinobacteria recovery (Fang et al., 2017). The mechanism underlying this method's success lies in the inherent heat resistance of actinobacterial spores, which contain protective compounds such as dipicolinic acid and have robust cell wall structures that enable survival under thermal stress conditions that prove lethal to vegetative bacterial cells, thereby creating an enriched environment conducive to actinobacterial isolation and cultivation (Schneider et al., 2022).

4.2 Chemical Pre-Treatment Strategies

Chemical pre-treatment methods for marine actinobacteria isolation involve both chemical amendments and innovative techniques to selectively enrich the growth of actinobacteria. It exploits the inherent stress tolerances of actinobacterial spores and mycelia, allowing selective recovery of diverse genera such as *Streptomyces*, *Micromonospora*, and rare taxa (Wang et al., 2023).

4.2.1 Calcium Carbonate Pre-Treatment

Calcium carbonate ($CaCO_3$) pre-treatment is based on the natural association between actinobacteria and calcium-rich marine environments. Marine actinobacteria, particularly the genus *Brevibacterium*, are involved in calcium carbonate biomineralization processes. The calcium-dipicolinate complex found in actinobacterial spores serves as a stabilizing agent against environmental stress. This complex comprises 5–15% of the dry weight of actinobacterial spores and plays a crucial role in spore resistance to heat, desiccation, and other physical stresses. The presence of dipicolinic acid (DPA) chelated with calcium ions creates a stable molecular structure that maintains spore viability under harsh conditions, and supplementation with external calcium sources can enhance this protective mechanism during isolation procedures (Zhu et al., 2017). Research has demonstrated that among various calcium salts tested—including $CaCO_3$, $CaCl_2$, and calcium acetate [$(CH_3COO)_2Ca$] calcium carbonate consistently yields superior results for actinobacteria recovery. The optimal concentration has been established at 0.1% (w/v), which corresponds to approximately 10 mM calcium ions, providing significantly better colony-forming unit (CFU) counts compared to higher concentrations (1%) or the complete absence of calcium supplementation. This concentration appears to represent a critical threshold where calcium availability enhances spore germination and vegetative growth without creating osmotic stress or toxicity effects that might inhibit bacterial development. The calcium carbonate is incorporated directly into isolation media during sample processing, creating an environment that mimics the calcium-rich conditions naturally found in marine sediments, where actinobacteria thrive. Studies have consistently shown that $CaCO_3$ supplementation yields significantly more actinobacterial CFUs compared to other calcium salts (Fang et al., 2017).

4.2.2 pH Adjustment Pre-Treatment

A pH optimization strategy can enhance the recovery of actinobacteria. The isolation media are prepared at a series of pH levels to identify conditions that maximize CFU recovery. Studies have demonstrated that a neutral pH of 7.0 consistently yields the highest CFU counts, outperforming both alkaline and acidic conditions in comparative assays where pH values from 6.0 to 9.0 were evaluated. This preference for neutrality likely reflects the physiological advantage conferred by maintaining cytoplasmic pH near neutrality, which reduces the energetic burden of acid or alkaline stress on actinobacterial cells and supports optimal enzyme activity and membrane stability. Although natural marine environments are typically slightly alkaline (pH 8.0–8.5), adjusting isolation media to neutral pH has proven more effective for recovering diverse marine actinobacteria, providing a stable biochemical milieu that promotes spore germination and vegetative growth during the critical early phases of culture establishment (Fang et al., 2017).

4.2.3 Advanced Chemical Pre-Treatments

Phenol treatment (0.1–1.5%, 15–30 min) denatures proteins and disrupts membranes of many competitors while actinobacterial spores withstand short exposures, markedly enriching Streptomyces and non-streptomycete recovery from sediment and sponge samples. Anionic surfactants such as SDS (0.05–0.1%, 5–15 min) similarly lyse Gram-negative cells without impairing spore viability, especially in high-salinity suspensions. Alkali treatment (0.1 M NaOH, pH 10–1, 5–10 min) inactivates vegetative cells while preserving heat and alkali-tolerant actinobacteria, provided neutralization follows immediately to prevent carry-over toxicity. Quaternary ammonium compounds (e.g., Benzethonium chloride 0.01–0.05%) further suppress fungal and bacterial contaminants when applied in combination with mild heating. After chemical exposure, samples are centrifuged, washed thoroughly in sterile seawater or phosphate buffer (pH 7.2), and serially diluted before inoculation onto selective media (e.g., starch casein sea salt agar) supplemented with antibiotics such as nalidixic acid or nystatin to inhibit residual Gram-negatives and fungi. Plates are incubated at 28–30 °C for 2–4 weeks, during which chalky, slow-developing colonies are picked for 16S rDNA gene sequencing. By tailoring agent type, concentration, and exposure time to sample characteristics, researchers can maximize both the diversity and yield of marine actinobacteria for natural-product discovery and antimicrobial screening (Lim et al., 2017; Pisano et al., 1986).

5 Marine Cultivation Media and Conditions

5.1 Seawater-Based Media Development

The isolation of marine actinobacteria requires optimised media compositions that are similar to the physicochemical nature of marine environments. Pre-treatment methods used routinely for the isolation of slow-growing rare actinomycetes from marine sources usually involve dilution and the addition of sterile natural seawater,

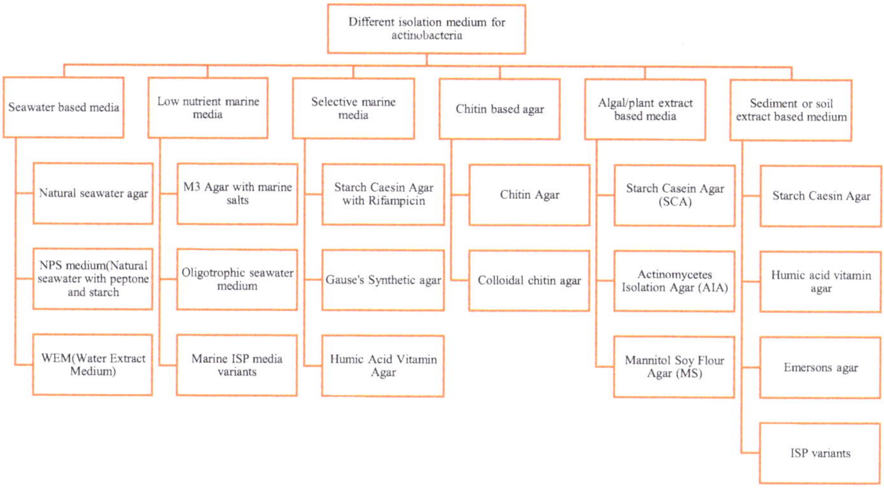

Fig. 1 Different isolation media for actinobacteria from different locations and entities (Joint et al., 2010; Abdelmohsen et al., 2010; Schneider et al., 2022; Meenakshi et al., 2024; Ibrahim et al., 2025)

artificial seawater, or deionised/distilled water supplemented with NaCl, multi-salts, and vitamin additives for maximum recovery (Subramani & Aalbersberg, 2012) (Fig. 1). Seawater-based media, including natural seawater agar and NPS medium (made by combining natural seawater with peptone and starch), have been shown to perform especially well during marine actinobacteria cultivation while maintaining their osmotic requirements (Jose & Jha, 2017). Poor-nutrient marine media, typified by the oligotrophic marine salts (M3) agar made with marine salts and oligotrophic seawater medium, play a significant role during the isolation of slow-growing marine actinobacteria adapted to impoverished marine environments. Media preparation must involve careful adjustment of salinity up to 3.5% for NaCl or the use of natural seawater, pH parameterization for marine regimes (typically in the pH range of 7.0–8.0), and the supplementation of marine trace elements for the inherent metabolic requirements of marine strains (Genilloud et al., 2011).

5.1.1 Artificial Seawater Formulations

The development of artificial seawater media has markedly enhanced both the accessibility and reproducibility of cultivating marine bacterioplankton. These formulations are successful in the expansion of many dominant species of marine life, and at the same time remove the variation and logistical concerns concerning the gathering and the localization of the two elements of the natural seawater (Henson et al., 2016). Studies have demonstrated that artificial seawater media can attain cultivation efficiencies of up to 45% when optimized for marine actinobacteria. A large-scale cultivation is also made possible by using standardized media compositions, resulting in higher reproducibility (Rodrigues & De Carvalho, 2022).

5.2 Novel Culture Media and Additives

Marine actinobacteria are recognized as valuable sources of novel bioactive molecules, yet a large fraction remains uncultivable on standard terrestrial formulations. Designing media and additives that mimic natural marine conditions has therefore become an important strategy to improve isolation success and secondary metabolite production. This section outlines current approaches to using marine-derived nutrients, natural extracts, co-culture, and micro-scale cultivation systems for marine actinobacteria.

5.2.1 Marine-Derived Nutrients

Replacement of freshwater with natural or artificial seawater provides needed salts (Na^{+}, Mg^{2+}, Ca^{2+}, K^{+}), trace metals, and osmotic balance present in marine ecosystems. Some authors have documented that seawater-based media can recover a wider variety of actinomycetes compared to freshwater-based media (Ibrahim et al., 2025). These ions stabilize cell envelopes and enzymatic systems of marine actinobacteria and prevent osmotic shock during isolation. Other supplements, such as sediment extracts, which are able to contribute complex organic carbon and micronutrients, or algal/seaweed extracts, which are usually rich in polysaccharides and signaling molecules, can enhance growth and induce some synthesis of pigments or metabolites in "silent" strains, those with biosynthetic gene clusters that are not expressed under standard laboratory conditions. A modified ISP-2 medium made with 75% natural seawater produced 30% more isolates from deep-sea sediments than the same formulation made with distilled water (Widada et al., 2023). Likewise, a high diversity of *Salinispora* species was produced in M1 medium containing 50% seawater and trace metals. Large-scale seawater or sediment extract collection and sterilization, however, can be labor-intensive and introduce batch variability. An alternative for promising consistency across labs can be achieved by the development of artificial seawater formulations with specific salts. The trace elements, such as strontium (Sr), lithium (Li), and boron (B), exist in seawater at varying concentrations of 8.1, 0.17, and 4.4 mg/L, respectively. The elements boost the metabolic activity of specific marine strains that have developed over a long time geologically (Roda-Gracia et al., 2023). The chelating character of seawater, which relies on the concentration of dissolved organic matter and carbonate species, substantially affects the bioavailability of metals, thereby leading to the alteration of the expression of biosynthetic routes that rely on metals in marine actinobacteria. The marine-specific trace elements such as strontium (Sr), lithium (Li), and boron (B) exist in seawater at varying concentrations of 8.1, 0.17, and 4.4 mg/L, respectively. The elements boost the metabolic activity of specific marine strains that have developed over a long time geologically.

5.2.2 Natural Marine Extracts

Sponge, coral, or other host organism extracts may include chemical cues that induce growth or biosynthesis of metabolites in associated actinobacteria. Incorporating such extracts into culture media can facilitate recovery of slow-growing or rare taxa and sometimes stimulate secondary metabolite expression. Siderophores (iron-chelating molecules), quorum-sensing molecules (cell-to-cell communication signals), and vitamins that resemble the actinobacteria's micro-niche are believed to be present in host extracts. Streptomyces strains that had not previously produced rifamycin-like compounds began to produce them when 1% sponge extract was added to starch casein agar (Kim et al., 2006). The availability of host material may restrict scale-up, and sterilizing crude extracts without eliminating heat-labile cues is still a significant challenge.

The sophistication of these extracts goes further than mere nutritional supplementation, as they contain species-specific signaling molecules such as N-acyl homoserine lactones, autoinducer molecules, and cyclic dipeptides capable of mobilizing latent biosynthetic potential (Santos et al., 2022). Extracts from marine sponges have proven promising, not least those from the *Haliclona* species with characteristic sulfated polysaccharides and alkaloids, potentially boosting the growth rates of actinobacteria by as much as 60% while simultaneously boosting the manufacture of antimicrobial agents. Concentration-based actions of these extracts need strict optimization, as concentrations of the extracts by the order of over 2% tend to demonstrate acts of inhibition due to natural antimicrobial products generated by the host organisms as a defense mechanism. Furthermore, extract composition fluctuations have been noted between seasons such that extracts harvested during the spring possessed a 25% higher induction potential for bioactivity compared to those harvested during the winter months, which would likely reflect higher metabolic activities and breeding cycles of the host organisms (Santos et al., 2019).

5.3 Specialized Incubation Conditions

The incubation of marine actinobacteria needs to be optimized due to varied heat adaptation and physiological characteristics. Generally, microorganisms fall into three classes based on their preferred temperatures, which are as follows: psychrophilic (−1 to +10 °C), mesophilic (+11 to +45 °C), and thermophilic (+46 to +75 °C) (Singh et al., 2009) (Table 1). For Arctic marine samples, it is necessary to create psychrophilic conditions with temperatures from 4 to 15 °C, whereas the vast majority of the temperate marine actinobacteria are mesophilic and flourish between 25 and 30 °C (Cheung et al., 2018). Salt tolerance evaluation by salinity gradient studies is indispensable for determining the osmotic adaptation potential of the isolated strains since marine actinobacteria exhibit a range of halotolerance. Slow-growing, rare actinobacteria, which often harbor the greatest potential for their valuable production of secondary metabolites due to their rarity, need extended incubation regimes spanning up to 4–8 weeks with multiple harvesting intervals.

Table 1 Specialized incubation and adaptation strategies for marine actinobacteria

Parameter	Condition	Purpose	References
Temperature	Psychrophilic: 4–10 °C	For Arctic/Antarctic isolates, 6–12 weeks incubation	Dhakal et al. (2017) and Ribeiro et al. (2023)
	Mesophilic: 25–30 °C	Optimal for most tropical/temperate marine isolates; 7–14 days	Dhakal et al. (2017) and Ribeiro et al. (2023)
	Deep-Sea stepwise: 4–8 °C and 15–20 °C	Mimics deep-sea habitat; acclimation improves rare isolate recovery	Yoo et al. (2023)
Salinity/osmotic adaptation	Gradient testing: 0.5–10% (w/v) NaCl in 0.5% increments	Determines halotolerance/halophily	Yoo et al. (2023)
	Progressive adaptation: 3.5%–1%	Allows lab adaptation while maintaining viability	Yoo et al. (2023)
Extended incubation	Long-term incubation: 4–8 weeks	Recovers slow-growing, rare actinobacteria	Dhakal et al. (2017) and Kamjam et al. (2017)
	Multiple harvests: 2–3, 4–5, 6–8 weeks	Captures both fast and slow-growing colonies	Liu et al. (2023)

6 Advanced Marine Cultivation Technologies

6.1 Co-Culture Techniques and Ecological Simulation

Culturing marine actinobacteria in the presence of other microbes induces two effects: competitive or symbiotic interaction that occurs in the natural environment, which likely results in the upregulation of specific biosynthetic gene clusters. The marine actinobacteria isolates, along with selected bacterial strains, resulted in either enhancing the yield of previously known bioactive compounds or new metabolite production (Aulia et al., 2023). The choice of the strain, ratio of inocula, and incubation conditions (static vs shaken) were important factors that will impact the outcome. For instance, a *Streptomyces* sp. cultivated with *Bacillus mycoides* in a medium with glycerol and yeast extract containing 75% seawater produced tryptamine derivatives and bacillamides that were absent or expressed only weakly in monoculture (Aulia et al., 2023). Diffusion-separated agar plugs to co-culture *Alteromonas* sp. and *Salinispora arenicola* have been the best alternative, and they noticed novel polyketide profiles (Tr, 2017). To quickly connect metabolites to gene clusters, such co-culture techniques can be used in combination with genomic or metabolomic screening. Purification challenges and contamination risks remain significant constraints in microbial studies, despite the advancements and optimizations.

6.2 High-Throughput and Miniaturized Systems

6.2.1 Dilution to Extinction

Dilution-to-extinction cultivation has become a potent high-throughput method to culture marine actinobacteria, which are generally not easily cultivable. The technique consists of serially diluted environmental samples and growth of such samples in 96-well/or greater microplates, and it allows isolation of formerly uncultivable microorganisms by recreating natural low-nutrient conditions (Benítez et al., 2021; Henson et al., 2016). Massive studies of DTE cultivation have effectively increased the phylogenetic and geographical diversity of cultured bacteriophages, such as actinobacteria in the sea. This method has been found especially useful in the isolation of novel clades like Actinobacteria acIV and other hitherto uncultivable actinobacterial clades. It has been demonstrated that DTE can be grown with cultivation efficiencies of 2–60 under optimized growth circumstances, with the technique being particularly effective when used with low-substrate media that are more reflective of natural seawater conditions (Henson et al., 2016). A notable advancement in DTE methodology involves the use of marine sponge tissue as direct inoculum, which has enabled the isolation of marine bacteria capable of producing bioactive compounds, including pederin-like compounds from Labrenzia species. This approach highlights the potential for integrating DTE with natural symbiotic systems to access unique microbial communities.

6.2.2 Microfluidics and Droplet-Based Cultivation

Microfluidic droplet culture is one of the most ingenious high-throughput methods of isolating marine actinobacteria. The technology allows entrapment of individual cells in droplets of picoliter size, thus providing a single microenvironment that will prevent nutrient competition and intercellular inhibition (Man et al., 2024). More recent inventions involve the establishment of a micron-scale Microbial Domestication Pod (μMD Pod), which is designed to be cultivated in situ to grow marine bacteria. These devices support the cell-to-cell interaction between vicinal species and retain isolation, thus overcoming the constraints faced by traditional methods of cell cultivation (Wheatley et al., 2022). The droplet-based systems have also exhibited an enhancement in performance compared to bulk-based cultivation strategies, reaching a degree of enrichment up to 13-fold with regard to bacterial species that are known to be difficult to cultivate. The technology can be used in automated sorting of growth-positive droplets and in a useful screening of particular traits, such as production of biosurfactants (Staskiewicz et al., 2024).

6.3 In Situ Cultivation Platforms

6.3.1 iChip Technology

The platform of the isolation chip (iChip) has revolutionized high-throughput in situ cultivation by creating hundreds of miniature chambers that could be deployed directly into marine environments. Each chamber is inoculated with one or several environmental cells and subsequently allows the diffusion of natural growth factors into the chamber while still maintaining isolation of individual cultures (Liu et al., 2020). Recent modifications to the design of the iChip have significantly improved its usage in marine environments, with the creation of FACS–iChip systems, which combine fluorescence-activated cell sorting with in situ culture systems. The platforms have already been able to isolate marine bacteria, which were previously uncultivable, thereby proving to be of higher recovery rates compared to the traditional cultivation methods.

6.3.2 Advanced Diffusion Chambers

Specialized systems of advanced diffusion chambers for marine sediment cultivation have been designed. These systems employ commercially available dialysis cassettes that are modified for environmental deployment, and are combined with microencapsulation techniques to enhance bacterial viability during in situ conservation (Pope et al., 2022). Microencapsulation is done in agarose after in situ microencapsulation. Studies have shown that microencapsulation markedly increases bacterial abundance and cell survival, especially in marine sediment bacteria. The method has made it possible to isolate new marine bacterial organisms that were never possible through the traditional process (Ding et al., 2022).

6.3.3 Microfluidic and High-Throughput Platforms

Miniaturized cultivation systems enable the simultaneous screening of an abundance of conditions. Microfluidic droplet systems encase cells in thousands of small chambers that vary in nutrients, pH, and salinity. In situ diffusion chambers and high-throughput microplate formats also expose isolates to a wide range of conditions and can also capture strains that fail to grow under standard conditions (Ibrahim et al., 2025). In order to isolate rare *Actinokineospora* species, instruments such as the iChip, which enables the direct in situ culture of microbes in their natural habitat, have recently been adapted for use with marine sediments (Santos et al., 2022). In order to find promising colonies prior to full-scale cultivation, future trends will involve combining microfluidics with automated image analysis and real-time metabolite sensors. These microfluidic system benefits go well beyond mere throughput and encompass the precise control of environmental gradients and the potential for the examination of single cells, which could provide insight into the heterogeneity of the actinobacterial populations (Mahler et al., 2018; Nauwynck et al., 2025). The microfluidic systems can replicate dynamic marine environments with changing salinity and nutrient supply, which would be hard for traditional culture systems to replicate. Droplet microfluidics also provides the potential for testing up to 10^6 conditions a day and every droplet as a single, freestanding bioreactor with volume from 1 to 1000 cells, from picolitres through nanoliters. The microfluidic systems include real-time monitoring capabilities, for example, fluorescence-based detection of the production of metabolites, pH sensors for acidification profiles, as well as optical density probes for growth kinetics (Aryal et al., 2024). Machine learning algorithms in combination with the platforms enable the computer-aided determination of promising growth regimens and the highlighting of aberrant metabolic profiles for further study. The development of marine-enabled microfluidic devices with pressure control for the reproducible establishment of deep-sea settings has opened up new possibilities for isolation of the previously enigmatic piezophilic actinobacteria from extreme marine ecosystems. Integration with mass spectrometry-on-a-chip technology could provide immediate metabolomic profiling of microfluidic cultures, which may revolutionize the marine natural products discovery pipeline by enabling the identification of the bioactive molecules in a few hours rather than culturing for a few weeks (Chen et al., 2021).

6.4 Robotic Systems and Automated Systems

These are entitled to High-Throughput screening platforms. The screening platforms have been automated to be able to screen thousands of marine bacterial cultures at the same time. These systems combine 96-well plate growing with a fast metabolomic analysis by using LC–MS/MS, which is capable of systematically characterizing the bacteria metabolomes and discovering promising bioactive compound manufacturers (Fiorini et al., 2022). Chemical screening techniques using plate-based 96-well bacterial culturing methods have been convenient in discovering marine microbial-produced invertebrate-related metabolites. The platforms have the potential to handle hundreds of isolates at the same time, keeping it sterile until real-time growth and metabolite generation can be studied (Berrue et al., 2011).

6.4.1 Microbioreactor Technology

More sophisticated microbioreactor systems have been developed in the areas of high-throughput cultivation and profiling of peptides in microbial research. These systems provide a means to strictly regulate the parameters of cultivation, like temperature, pH, dissolved oxygen, or nutrient delivery, as well as processing various cultures simultaneously (Khalil, 2025).

6.4.2 Single-Cell Isolation and Sorting

6.4.2.1 FACS-Based Approaches

Combined with high-throughput cultivation systems, fluorescence-activated cell sorting (FACS) has found use in order to select a specific mancouver cell in marine actinobacteria with high precision. Rates of thousands of processes can be done in these systems in a second, with viability and allowed downstream development (Xu et al., 2022). The development of FACS involving iChip technology (FACS–iChip) provides an effective means of mining the microbial dark matter, making it possible to isolate uncommon and hitherto unculturable marine actinobacteria (Liu et al., 2020).

6.4.2.2 Single-Cell Encapsulation

Techniques have been established in marine bacteria cultivation using highly sophisticated methods of encapsulation of single cells in agarose microbeads. These types of techniques provide the capability to isolate individual cells and allow the chemical communication of cells in the environment during incubating cells that are able to grow fastidious marine actinobacteria (Alkayyali et al., 2020).

6.4.2.3 Metabolomic-Guided Cultivation

Mobilizing high-throughput metabolomic studies has been combined with cultivation systems to systematically determine the bioactive capabilities of marine actinobacteria. Using this method will allow us to identify potential natural product producers by means of systematic metabolome profiling with LC–MS/MS and molecular networking. This method has been used to identify the strains with unique metabolic profiles in studies and has discovered some families of potentially novel compounds that would not have been observed otherwise through any of the other bioactivity screening methods alone (Fiorini et al., 2022).

6.5 Successful Marine Isolation Campaigns

6.5.1 Marine Sediment and Sponge Co-isolations

The research carried out a broad sampling of Andaman Sea intertidal sediments and sponge tissue, using a combined heat-shock (60–65 °C, 15 min) and starch casein sea salt agar method. This yielded 12 unique genera of *Micromonospora*, *Streptomyces*, and *Nonomuraea*, which exhibited antimicrobial, anticancer, antioxidant, and plant growth-promoting properties. Incorporation of sponge extracts at 1% (w/v) greatly improved recovery of slow-growing taxa by supplying host-borne signal molecules that initiated secondary metabolism.

6.5.2 Coral Tissue Isolations

Becerril-Espinosa et al. (2023) isolated coral tissues of *Porites astreoides* reefs in the Gulf of Mexico, employing dry-stamp plating with heat shock treatment. Species-specific actinobacterial patterns arose with predominant Streptomyces and *Micromonospora* at high prevalence, uniquely correlated with coral mucus. Isolates secreted new antibiotic compounds with activity against marine Vibrio pathogens, indicating symbiotic activities and bioprospecting value of coral-associated actinobacteria.

6.5.3 Deep-Sea Water Isolation

Ghosh et al. (2020) sampled water at 1200 m depth in the Bay of Bengal and used cold-adapted isolation protocols (4 °C incubation, 8 weeks). Routine psychrotolerant-directed media isolated eight genera with a predominance of Streptomyces spp. Some isolates were able to produce cold-active enzymes (lipases, proteases) and new antimicrobial compounds active at low temperatures, highlighting the deep-sea as a biotechnologically significant reservoir of actinobacteria.

6.6 Comparison of Methods and Performance

6.6.1 Conventional vs. DICA Strategies

Ahmad et al. (2024) compared conventional plating with the Diffusion-Based Integrative Cultivation Approach (DICA) on identical sediment samples. DICA obtained 58% new taxa compared with 12% with traditional methods, and isolated 12 bacterial classes with rare phyla. New isolates showed varying biosynthetic gene clusters, proving DICA superior in unlocking cryptic marine diversity.

6.6.2 Media Performance Comparisons

Jose and Jha (2017) compared natural seawater agar (NSA), starch-peptone sea salt (NPS), and artificial seawater media on different sediment samples. NSA and NPS yielded comparable genus richness (~10 genera), but NSA resulted in 20% greater CFU counts. Optimized artificial media realized 45% cultivation efficiency, with reproducible large-scale isolation.

6.6.3 Optimization of Recovery Rate

Huang et al. (2021) compared four media on coastal sediments, in which an enriched seawater-enriched medium, Williams' Medium E (WEM), enhanced actinobacterial recovery by 35% over ISP-2 with artificial seawater. Additional pretreatments with combined heat (65 °C) and $CaCO_3$ (0.1% w/v) further enhanced CFU counts by 40%, illustrating the potential of pretreatment-media optimization synergy.

6.7 Troubleshooting Common Challenges

6.7.1 Contamination Management

Sheikh et al. (2019) indicated seasonal variation in the recovery of morphotypes of Streptomyces from seawater samples; winter samples provided greater Streptomyces morphotypes. The addition of nalidixic acid (25 μg/mL) and nystatin (50 μg/mL) inhibited Gram-negative and fungal contaminants, respectively, enhancing actinobacterial purity from 60% to 85%.

6.7.2 Recovery of Slow-Growing Isolates

Maldonado et al. (2009) isolated close to 300 isolates from Gulf of California sediments through extended incubations (4–8 weeks) and gradual salinity adaptation (3.5 to 1% NaCl). This approach isolated rare taxa outside common genera by reducing osmotic shock and enabling delayed-appearing colonies to become mature.

6.7.3 Method-Specific Limitations

Rizescu (2022) reported that actinomycete densities were high but with predominantly Streptomyces and *Nocardiopsis* from heat-dilution pretreated polluted coastal sediments at 10–20 m depth. This skew was due to selective suppression of slow-growing non-streptomycete genera, and there is a requirement for other chemical pretreatments (e.g., SDS, quaternary ammonium compounds) to improve taxonomic recovery.

7 Marine Actinobacteria Characterization

7.1 Morphological and Physiological Assessment

The accurate identification and comprehensive characterization of microorganisms has undergone revolutionary changes over the past decade, with technological advances fundamentally transforming microbial systematics. Developing cutting-edge methods such as mass spectrometry, next-generation sequencing (NGS), and polymerase chain reaction (PCR) has dramatically increased microbial identification's precision, speed, and effectiveness. Modern microbiology employs an integrated polyphasic approach that combines traditional morphological and physiological methods with advanced molecular and genomic techniques to achieve robust taxonomic characterization (Prakash et al., 2007). This multifaceted strategy has become essential for understanding microbial diversity, ecological roles, and biotechnological applications.

7.1.1 Colonial Morphology Analysis

Despite the advent of molecular methods, morphological characterization remains a fundamental component of microbial identification, providing rapid preliminary information that guides subsequent analyses. Colonial morphology encompasses observable characteristics, including size, shape, elevation, margin definition, surface texture, opacity, and pigmentation patterns. Recent advances in digital imaging and automated colony analysis have standardized morphological assessments, reduced observer bias, and improved reproducibility. Colony size classifications have been refined through high-resolution imaging systems, enabling precise measurements that correlate with metabolic activity and growth rates. Micro-colonies (<0.5 mm), small colonies (0.5–2 mm), medium colonies (2–5 mm), and large colonies (>5 mm) represent distinct categories that often correlate with specific bacterial groups and environmental adaptations (Ferrari et al., 2016). Automated colony counting systems now integrate morphological analysis with image processing algorithms, providing objective measurements for taxonomic databases. Pigmentation analysis has benefited significantly from spectrophotometric and chromatographic advances. The interpretation and analysis of the spectral data should be coupled with other standard methods to ensure the reliability of FTIR spectroscopy analysis. Modern approaches include high-performance liquid chromatography (HPLC) for pigment separation, mass spectrometry for structural characterization, and UV-visible spectrophotometry for quantitative analysis (Jodłowska & Latała, 2011). Carotenoid profiling has emerged as a particularly valuable taxonomic tool, with specific carotenoid compositions serving as chemotaxonomic markers for bacterial groups. The standardization of growth conditions for pigment analysis has improved through controlled environment systems that regulate temperature, humidity, light exposure, and atmospheric composition. LED-based illumination systems provide consistent spectral output, while automated incubation systems maintain precise temperature control throughout the observation period. These advances have enhanced the reliability of pigmentation data for taxonomic applications. Digital imaging systems equipped with multispectral cameras enable the capture of pigmentation patterns across different wavelengths, revealing cryptic coloration not visible to the human eye. Machine learning algorithms trained on extensive image databases can now automatically classify pigmentation patterns, providing rapid preliminary identification results (Khan et al., 2021). These systems integrate seamlessly with laboratory information management systems (LIMS), creating comprehensive digital records for comparative analysis.

7.1.2 Microscopic Examination

Advanced microscopy techniques have revolutionized cellular morphology assessment, providing unprecedented detail about bacterial structure and specialized features. Super-resolution microscopy, including structured illumination microscopy (SIM) and photoactivated localization microscopy (PALM), now enables visualization of cellular structures below the diffraction limit (Fornasiero & Opazo, 2015). These techniques have revealed previously unknown structural details relevant to bacterial taxonomy and physiology. Automated microscopy systems incorporating artificial intelligence have transformed routine morphological analysis. Computer vision algorithms can automatically identify cell shapes, measure dimensions, assess arrangement patterns, and detect specialized structures such as flagella, pili, and inclusion bodies (Kotwal et al., 2021). These systems provide objective measurements with statistical analysis, reducing inter-observer variability and improving diagnostic accuracy. Live-cell imaging techniques enable real-time observation of bacterial growth, division, and morphological changes. Time-lapse microscopy systems with environmental control allow assessment of dynamic processes, including sporulation, germination, and biofilm formation (Jyothikumar et al., 2008). These approaches provide insights into bacterial behavior that complement static morphological assessments. Spore characterization has advanced through correlative microscopy approaches that combine light microscopy, electron microscopy, and spectroscopic techniques. Cryo-electron microscopy enables visualization of spore ultrastructure without fixation artifacts, while atomic force microscopy provides nanoscale surface characterization (Sexton & Tocheva, 2020). Chemical analysis of spore components using techniques such as Raman spectroscopy and infrared spectroscopy provides molecular-level information about spore composition and properties. Flow cytometry has emerged as a powerful tool for spore analysis, enabling rapid assessment of spore populations and individual spore characteristics. Multi-parameter flow cytometry can simultaneously measure spore size, optical density, and fluorescence properties, providing quantitative data for spore characterization (Chen et al., 2024). Fluorescent labeling techniques enable specific detection of spore components and metabolic activities.

7.1.3 Physiological Profiling

Automated growth monitoring systems have revolutionized physiological characterization by providing continuous, high-resolution growth data under precisely controlled conditions. Odin™ instruments are the all-in-one solution for cellular metabolic characterization, growth kinetics, and identification. Characterize phenotypes by monitoring growth curves and cell respiration kinetics for both individual microbes and microbial populations. These systems integrate turbidometric, respirometric, and colorimetric measurements to provide comprehensive growth profiles. Temperature tolerance assessment now employs gradient thermal cyclers and multi-zone incubation systems that enable simultaneous testing across wide temperature ranges. High-resolution thermal profiling reveals subtle differences in cardinal temperatures that distinguish closely related bacterial species (Bertaux et al., 2022). Thermostability testing protocols have been standardized to ensure reproducible results across laboratories. pH tolerance characterization utilizes automated pH control systems that maintain constant pH levels throughout extended incubation periods. Buffer capacity testing and organic acid tolerance assessments provide additional physiological information relevant to ecological adaptations and industrial applications (Velez-Suberbie et al., 2017). Real-time pH monitoring enables the detection of pH changes during growth, indicating specific metabolic activities. Osmotic stress response analysis employs defined synthetic media with controlled salt compositions, enabling precise assessment of halotolerance and osmoregulation mechanisms. Automated dilution systems provide gradual salt exposure protocols that simulate natural environmental variations (Janzen et al., 2019). Real-time oxygen monitoring during growth provides insights into respiratory activities and metabolic pathways.

7.2 Molecular Identification Methods

7.2.1 16S rRNA Gene Sequencing and Phylogenetic Analysis

16S rRNA sequencing was the primary technique used in bacterial strain identification and phylogenetic analysis. Full-length 16S rRNA gene sequencing is now made possible by third-generation sequencing technologies such as Oxford Nanopore and PacBio technologies, which cut out ambiguities inherent with short-read methodologies (Buetas et al., 2024; Yoon et al., 2017). Full-length sequences yield higher taxonomic resolution and facilitate the detection of evolutionarily informative secondary structure elements. Enhanced PCR primer pairs focus on highly conserved domains with broad phylogenetic coverage within bacterial domains. Optimal universal primer pairs like 27F/1492R and 8F/1541R have been developed with improved amplification efficiency and minimal bias (Lin et al., 2024). Primer validation experiments guarantee reproducible performance across a wide range of bacterial species, with special emphasis on the detection of emerging pathogens. Database curation activities have greatly enhanced the quality and completeness of 16S rRNA reference sequences. EzBioCloud database, SILVA, and Ribosomal Database Project (RDP) currently employ strict quality control procedures, such as chimera identification, sequence checking, and taxonomic validation (Quast et al., 2012; Molano et al., 2024). Updates include new species added through their recent description and rectification of taxonomic designations according to up-to-date nomenclature. Phylogenetic analysis has evolved through the application of advanced evolutionary models and computational methods. Maximum likelihood and Bayesian inference methods now include complicated substitution models that accommodate rate heterogeneity among sites and phylogenetic restraint (Quast et al., 2012). Web-based phylogenetic analysis tools offer accessible software for non-experts while applying stringent analytical rigor. The fIDBAC, using a 95% threshold, had an overall accuracy of 96.38%, lower than the top-hit scores given by fIDBAC (97.35%) but higher than the top-hit scores of 16S rRNA sequence alignment by 30.93%. This indicates the ongoing development of bacterial identification systems that incorporate multiple molecular strategies. Quality control programs for 16S rRNA gene sequencing have been harmonized using international collaborations and proficiency testing schemes. Automated quality check tools assess sequence length, base quality scores, chimera contamination, and database similarity scores. Uniform reporting formats enable data exchange and comparative analysis among research groups.

8 Recent Trends and Future Directions in Actinobacterial Isolation and Discovery

Conventional isolation strategies predominantly employed morphological and biochemical techniques, augmented by selective media tailored to preferentially support the growth of Streptomyces and closely related taxa. Despite this, the strategies employed did not adequately recognize the occurrence of rare actinobacteria and those variants tailored for particular ecological settings. This shortcoming resulted in the enhancement of pretreatment protocols (e.g., heat shock, chemical suppression) and also enhancements in media for each location-based isolation and identification of actinobacteria, which facilitated improved recovery rates from marine sediments and invertebrates (Saez et al., 2025). The advancements in molecular methodologies have significantly enhanced the field of actinobacterial classification, thereby resulting in taxonomical identification under which the identified actinobacterial classes fall. Since the 1990s, the application of 16S rRNA gene sequencing has allowed for taxonomic identification, unveiling cryptic diversity that is not identified through morphological examination alone (Narsing Rao & Li 2022; Chudzik et al., 2024). Collectively, these technological advancements have established integrated genomics–metabolomics platforms as the foundational elements of contemporary actinobacterial biodiscovery. Despite this, the strategies employed did not adequately recognize the occurrence of rare actinobacteria and those variants tailored for particular ecological settings. This shortcoming prompted the enhancement of pretreatment protocols (e.g., heat shock, chemical suppression) and the innovation of seawater-based media that facilitated improved recovery rates from marine sediments and invertebrates (Montaser & Luesch, 2011). The advancements in molecular techniques have notably improved the domain of actinobacterial classification. Currently, WGS and genome-centric taxonomy have yielded valuable insights into biosynthetic gene clusters (BGCs), while metabolomics techniques such as LC–MS/MS-based molecular networking have facilitated rapid dereplication and the discovery of novel bio-products (Tay et al., 2024; Medema et al., 2015). These technological advancements have established integrated genomics–metabolomics platforms as the foundation for actinobacterial discovery.

9 Conclusion

Isolation and culture of marine actinobacteria is an area of research that has developed from conventional culture-based methods retrieving less than 5% of environmental diversity to advanced high-throughput systems able to access previously uncultured taxa with unparalleled efficiency. This article has shown that it is necessary to take an integrated strategy consisting of marine-specific pretreatment strategies (heat shock, calcium carbonate addition, pH adjustment), niche seawater-based medium formulations, and cutting-edge cultivation technologies such as microfluidic systems, in situ cultivation platforms such as iChip technology, and co-culture methods mimicking natural microbial interactions. The distinctive physiological features of marine actinobacteria, represented by obligate marine genera like *Salinispora* and deep-sea *psychrophilic* specialized strains, require cultivation settings that reflect their indigenous marine habitats, optimal salinity gradients, trace element addition, and prolonged incubations. Recent isolation drives have isolated a variety of genera like *Micromonospora, Streptomyces,* and unusual taxa from extreme marine habitats, and these yielded new bioactive compounds with antimicrobial, anticancer, and industrial enzyme uses. The combination of computerized screening systems, metabolomics-assisted cultivation, and artificial intelligence-based optimization is the way forward for the discipline, while third-generation sequencing platforms and polyphasic characterization methods provide stable taxonomic determination and biotechnological evaluation. As worldwide challenges such as antimicrobial resistance and environmental degradation increase, methods for isolating marine actinobacteria are refining themselves towards the direction of sustainable bioprospecting strategies that reconcile conservation with the imperative to tap into the immense reservoir of bioactive potential in Earth's oceanic ecosystems, placing these extraordinary microorganisms at the center of pharmaceutical innovation, industrial biotechnology, and environmental rehabilitation endeavors.

Acknowledgments The authors acknowledge the Vellore Institute of Technology for providing all the necessary facilities to carry out the research.

References

Abdelmohsen, U. R., Pimentel-Elardo, S. M., Hanora, A., Radwan, M., Abou-El-Ela, S. H., Ahmed, S., & Hentschel, U. (2010). Isolation, phylogenetic analysis and anti-infective activity screening of marine sponge-associated actinomycetes. *Marine Drugs, 8*(3), 399–412. https://doi.org/10.3390/md8030399. PMID: 20411105; PMCID: PMC2857355.

Ahmad, T., Ishaq, S. E., Liang, L., Xie, R., Wang, Y., & Wang, F. (2024). A diffusion-based integrative approach for culturing previously uncultured bacteria from marine sediments. *Marine Life Science & Technology*. https://doi.org/10.1007/s42995-024-00240-2

Alkayyali, T., Pope, E., Wheatley, S. K., Cartmell, C., Haltli, B., Kerr, R. G., & Ahmadi, A. (2020). Development of a microbe domestication pod (MD Pod) for in situ cultivation of micro-encapsulated marine bacteria. *Biotechnology and Bioengineering, 118*(3), 1166–1176. https://doi.org/10.1002/bit.27633

Amoroso, M. J., Castro, G. R., Durán, A., Peraud, O., Hollmann, A., Dávila Costa, J. S., et al. (2013). The contribution of actinobacteria to the bioremediation of environmental pollutants. In *Actinobacteria-basics and biotechnological applications* (pp. 235–270). InTechOpen.

Aryal, P., Hefner, C., Martinez, B., & Henry, C. S. (2024). Microfluidics in environmental analysis: Advancements, challenges, and future prospects for rapid and efficient monitoring. *Lab on a Chip, 24*(5), 1175–1206. https://doi.org/10.1039/d3lc00871a

Augustijn, H. E., Roseboom, A. M., Medema, M. H., & Van Wezel, G. P. (2024). Harnessing regulatory networks in Actinobacteria for natural product discovery. *Journal of Industrial Microbiology & Biotechnology, 51*. https://doi.org/10.1093/jimb/kuae011

Aulia, G., Rachman, F., Untari, F., Rasyid, A., Hapsari, Y., Andriati, R., & Wibowo, J. T. (2023). Utilizing the co-culture method to improve the investigation of secondary metabolites of marine bacteria. *Journal of Applied Pharmaceutical Science*. https://doi.org/10.7324/japs.2024.136305

Barka, E. A., Vatsa, P., Sanchez, L., Gaveau-Vaillant, N., Jacquard, C., Meier-Kolthoff, J. P., et al. (2016). Taxonomy, physiology, and natural products of Actinobacteria. *Microbiology and Molecular Biology Reviews, 80*(1), 1–43. https://doi.org/10.1128/MMBR.00019-15

Becerril-Espinosa, A., Mateos-Salmón, C., Burgos, A., Rodríguez-Zaragoza, F. A., Meza-Canales, I. D., Juárez-Carrillo, E., Rios-Jara, E., & Ocampo-Alvarez, H. (2023). Dry stamping coral powder: An effective method for isolating coral symbiotic actinobacteria. *Microorganisms, 11*. https://doi.org/10.3390/microorganisms11122951

Benítez, X., Gonzalez, E. G., García, J., Zúñiga, P., de la Calle, F., & Cuevas, C. (2021). Detection of a pederin-like compound using a dilution-to-extinction-based platform for the isolation of marine bacteria in drug discovery strategies. *Microbial Biotechnology, 14*(1), 241–250. https://doi.org/10.1111/1751-7915.13679

Bérdy, J. (2012). Thoughts and facts about antibiotics: Where we are now and where we are heading. *The Journal of Antibiotics, 65*(8), 385–395.

Berrue, F., Withers, S. T., Haltli, B., Withers, J., & Kerr, R. G. (2011). Chemical screening method for the rapid identification of microbial sources of marine invertebrate-associated metabolites. *Marine Drugs, 9*(3), 369–381. https://doi.org/10.3390/md9030369

Bertaux, F., Sosa-Carrillo, S., Gross, V., Fraisse, A., Aditya, C., Furstenheim, M., & Batt, G. (2022). Enhancing bioreactor arrays for automated measurements and reactive control with ReacSight. *Nature Communications, 13*(1). https://doi.org/10.1038/s41467-022-31033-9

Buetas, E., Jordán-López, M., López-Roldán, A., D'Auria, G., Martínez-Priego, L., De Marco, G., Carda-Diéguez, M., & Mira, A. (2024). Full-length 16S rRNA gene sequencing by PacBio improves taxonomic resolution in human microbiome samples. *BMC Genomics, 25*(1). https://doi.org/10.1186/s12864-024-10213-5

Bull, A. T., Asenjo, J. A., Goodfellow, M., & Gómez-Silva, B. (2016). The Atacama Desert: Technical resources and the growing importance of novel microbial diversity. *Annual Review of Microbiology, 70*, 215–234.

Chen, L., Wang, X., & Zhang, Y. (2021). Machine learning approaches for automated bacterial colony classification from digital images. *Scientific Reports, 11*, 15416.

Chen, Z., Lu, Y., Cui, J., Feng, Y., Dong, H., Huang, X., Zhu, C., Xiong, X., Chen, H., Wang, Q., & Liu, G. (2024). Monitoring of Bacillus spore-forming dynamics through flow cytometry. *Frontiers in Microbiology, 15*. https://doi.org/10.3389/fmicb.2024.1450913

Cheung, M. K., Wong, C. K., Chu, K. H., & Kwan, H. S. (2018). Community structure, dynamics and interactions of bacteria, archaea and fungi in subtropical coastal wetland sediments. *Scientific Reports, 8*(1). https://doi.org/10.1038/s41598-018-32529-5

Chifflet, S., Quéméneur, M., Barani, A., Angeletti, B., Didry, M., Grégori, G., & Pradel, N. (2019). Impact of sterilization methods on dissolved trace metals concentrations in complex natural samples: Optimization of UV irradiation. *MethodsX, 6*, 1133–1146. https://doi.org/10.1016/j.mex.2019.04.020

Chudzik, A., Jalkanen, K., Täubel, M., Szponar, B., & Paściak, M. (2024). Identification of environmental Actinobacteria in buildings by means of chemotaxonomy, 16S rRNA sequencing, and MALDI-TOF MS. *Microbiology Spectrum, 12*(3). https://doi.org/10.1128/spectrum.03596-23

Daniel, R. (2005). The metagenomics of soil. *Nature Reviews Microbiology, 3*(6), 470–478.

Dhakal, D., Pokhrel, A. R., Shrestha, B., & Sohng, J. K. (2017). Marine rare actinobacteria: Isolation, characterization, and strategies for harnessing bioactive compounds. *Frontiers in Microbiology, 8*. https://doi.org/10.3389/fmicb.2017.01106

Ding, H., Liu, J., Yang, C., Guo, C., Ding, L., Jung, D., & Zhang, W. (2022). *Hanstruepera marina* sp. nov. and *Hanstruepera flava* sp. nov., two novel species in the family Flavobacteriaceae isolated by a modified in situ cultivation technique from marine sediment. *Frontiers in Microbiology, 13*. https://doi.org/10.3389/fmicb.2022.957397

Epstein, S. S. (2013). The phenomenon of microbial uncultivability. *Current Opinion in Microbiology, 16*(5), 636–642.

Fang, B. Z., Salam, N., Han, M. X., Jiao, J. Y., Cheng, J., Wei, D. Q., et al. (2017). Insights on the effects of heat pretreatment, pH, and calcium salts on isolation of rare Actinobacteria from karstic caves. *Frontiers in Microbiology, 8*, 1535.

Ferrari, A., Lombardi, S., & Signoroni, A. (2016). Bacterial colony counting with convolutional neural networks in digital microbiology imaging. *Pattern Recognition, 61*, 629–640. https://doi.org/10.1016/j.patcog.2016.07.016

Fiorini, F., Bajerski, F., Jeske, O., Lepleux, C., Overmann, J., & Brönstrup, M. (2022). A metabolomics-based toolbox to assess and compare the metabolic potential of unexplored, difficult-to-grow bacteria. *Marine Drugs, 20*(11), 713. https://doi.org/10.3390/md20110713

Fornasiero, E. F., & Opazo, F. (2015). Super-resolution imaging for cell biologists. *BioEssays, 37*(4), 436–451. https://doi.org/10.1002/bies.201400170

Franco-Correa, M., Quintana, A., Duque, C., Suarez, C., Rodríguez, M. X., & Barea, J. M. (2010). Evaluation of actinomycete strains for key traits related with plant growth promotion and mycorrhiza helping activities. *Applied Soil Ecology, 45*(3), 209–217.

Genilloud, O., González, I., Salazar, O., Martín, J., Tormo, J. R., & Vicente, F. (2011). Current approaches to exploit actinomycetes as a source of novel natural products. *Journal of Industrial Microbiology & Biotechnology, 38*(3), 375–389.

Ghosh, M., Gera, M., Singh, J., Prasad, R., & Pulicherla, K. K. (2020). A comprehensive investigation of potential novel marine psychrotolerant Actinomycetes sp. isolated from the Bay-of-Bengal. *Current Genomics, 21*(4), 271–282. https://doi.org/10.2174/1389202921666200330150642

Girão, M., Alexandrino, D. A. M., Cao, W., Costa, I., Jia, Z., & Carvalho, M. F. (2024). Unveiling the culturable and non-culturable actinobacterial diversity in two macroalgae species from the northern Portuguese coast. *Environmental Microbiology, 26*(4), e16620. https://doi.org/10.1111/1462-2920.16620

Hayakawa, M., & Nonomura, H. (1987). Humic acid-vitamin agar, a new medium for the selective isolation of soil actinomycetes. *Journal of Fermentation Technology, 65*(5), 501–509.

Henson, M. W., Pitre, D. M., Weckhorst, J. L., Lanclos, V. C., Webber, A. T., & Thrash, J. C. (2016). Artificial seawater media facilitate cultivating members of the microbial majority from the Gulf of Mexico. *mSphere, 1*(2). https://doi.org/10.1128/msphere.00028-16

Hoyos, H. A. V., Vargas, C. D. G., Giraldo, D. O., Monsalve, M. A. V., Rivera, J. C. A., & Vanegas, A. M. M. (2025). Actinobacteria: Source of antifungal secondary metabolites for agricultural sustainability. *Scientia Agropecuaria, 16*(2), 307–326. https://doi.org/10.17268/sci.agropecu.2025.023

Huang, Z., Mo, S., Yan, L., Wei, X., Huang, Y., Zhang, L., Zhang, S., Liu, J., Xiao, Q., Lin, H., & Guo, Y. (2021). A simple culture method enhances the recovery of culturable actinobacteria from coastal sediments. *Frontiers in Microbiology, 12*, 675048. https://doi.org/10.3389/FMICB.2021.675048

Ibrahim, J. A. A., Botcha, S., & Prattipati, S. D. (2025). Marine actinomycetes: A promising source of novel therapeutics and pharmaceutical bioactive compounds—A review. *The Microbe, 7*, 100383.

Janzen, N. H., Striedner, G., Jarmer, J., Voigtmann, M., Abad, S., & Reinisch, D. (2019). Implementation of a fully automated microbial cultivation platform for strain and process screening. *Biotechnology Journal, 14*(10). https://doi.org/10.1002/biot.201800625

Jodłowska, S., & Latała, A. (2011). The comparison of spectrophotometric method and high-performance liquid chromatography in photosynthetic pigments analysis. *OnLine Journal of Biological Sciences, 11*(2), 63–69. https://doi.org/10.3844/ojbsci.2011.63.69

Joint, I., Mühling, M., & Querellou, J. (2010). Culturing marine bacteria—An essential prerequisite for biodiscovery. *Microbial Biotechnology, 3*(5), 564–575. https://doi.org/10.1111/j.1751-7915.2010.00188.x. Epub 2010 Jun 21. PMID: 21255353; PMCID: PMC3815769.

Jose, P. A., & Jha, B. (2017). Intertidal marine sediment harbours Actinobacteria with promising bioactive and biosynthetic potential. *Scientific Reports, 7*(1). https://doi.org/10.1038/s41598-017-09672-6

Jyothikumar, V., Tilley, E. J., Wali, R., & Herron, P. R. (2008). Time-lapse microscopy of *Streptomyces coelicolor* growth and sporulation. *Applied and Environmental Microbiology, 74*(21), 6774–6781. https://doi.org/10.1128/aem.01233-08

Kamjam, M., Sivalingam, P., Deng, Z., & Hong, K. (2017). Deep sea actinomycetes and their secondary metabolites. *Frontiers in Microbiology, 8*. https://doi.org/10.3389/fmicb.2017.00760

Khalil, Z. G. (2025). Advancements in microbioreactor technology: High-throughput cultivation and peptide profiling in microbial research. *Methods in Molecular Biology*, 123–141. https://doi.org/10.1007/978-1-0716-4562-8_11

Khan, F. M., Gupta, R., & Sekhri, S. (2021). Automated bacteria colony counting on Agar plates using machine learning. *Journal of Environmental Engineering, 147*(12). https://doi.org/10.1061/(asce)ee.1943-7870.0001948

Kim, T. K., Hewavitharana, A. K., Shaw, P. N., & Fuerst, J. A. (2006). Discovery of a new source of rifamycin antibiotics in marine sponge actinobacteria by phylogenetic prediction. *Applied and Environmental Microbiology, 72*(3), 2118–2125. https://doi.org/10.1128/aem.72.3.2118-2125.2006

Koech, S. C., Plechatá, M., Pathom-Aree, W., Kamenik, Z., & Jaisi, A. (2025). Strategies for actinobacteria isolation, cultivation, and metabolite production that are biologically important. *ACS Omega, 10*(16), 15923–15934. https://doi.org/10.1021/acsomega.5c01344

Kotwal, S., Rani, P., Arif, T., Manhas, J., & Sharma, S. (2021). Automated bacterial classifications using machine learning based computational techniques: Architectures, challenges and open research issues. *Archives of Computational Methods in Engineering, 29*(4), 2469–2490. https://doi.org/10.1007/s11831-021-09660-0

Krysenko, S. (2025). Current approaches for genetic manipulation of Streptomyces spp.—Key bacteria for biotechnology and environment. *Biotech, 14*(1), 3. https://doi.org/10.3390/biotech14010003

Kurtböke, D. I. (2012). Biodiscovery from rare actinomycetes: An eco-taxonomical perspective. *Applied Microbiology and Biotechnology, 93*(5), 1843–1852.

Lewis, K. (2013). Platforms for antibiotic discovery. *Nature Reviews Drug Discovery, 12*(5), 371–387.

Lim, C. P., Hoon, K. C., & Cheah, Y. K. (2017). Actinobacteria from Greenwich Island and Dee Island: Isolation, diversity and distribution. *Life Sciences Medicine and Biomedicine, 1*(1). https://doi.org/10.28916/lsmb.1.1.2017.2

Lin, X., Waring, K., Ghezzi, H., Tropini, C., Tyson, J., & Ziels, R. M. (2024). High accuracy meets high throughput for near full-length 16S ribosomal RNA amplicon sequencing on the Nanopore platform. *PNAS Nexus, 3*(10). https://doi.org/10.1093/pnasnexus/pgae411

Liu, H., Xue, R., Wang, Y., Stirling, E., Ye, S., Xu, J., & Ma, B. (2020). FACS-iChip: A high-efficiency iChip system for microbial 'dark matter' mining. *Marine Life Science & Technology, 3*(2), 162–168. https://doi.org/10.1007/s42995-020-00067-7

Liu, S., Zhai, X., Liu, D., Liu, Y., Sui, L., Luo, K., Yang, Q., Li, F., Nikandrova, A. A., Imamutdinova, A. N., Lukianov, D. A., Osterman, I. A., Sergiev, P. V., Zhang, B., Zhang, D., Xue, C., & Sun, C. (2023). Bioprospecting of actinobacterial diversity and antibacterial secondary metabolites from the sediments of four saline lakes on the Northern Tibetan Plateau. *Microorganisms, 11*(10), 2475. https://doi.org/10.3390/microorganisms11102475

Mahler, L., Wink, K., Beulig, R. J., Scherlach, K., Tovar, M., Zang, E., Martin, K., Hertweck, C., Belder, D., & Roth, M. (2018). Detection of antibiotics synthetized in microfluidic picolitre-droplets by various actinobacteria. *Scientific Reports, 8*, 13087.

Maldonado, L. A., Fragoso-Yáñez, D., Pérez-García, A., Rosellón-Druker, J., & Quintana, E. T. (2009). Actinobacterial diversity from marine sediments collected in Mexico. *Antonie van Leeuwenhoek International Journal of General and Molecular Microbiology, 95*(2), 111–120. https://doi.org/10.1007/S10482-008-9294-3

Man, D. K. W., Hermans, S. M., Taubert, M., Garcia, S. L., Hengoju, S., Küsel, K., & Rosenbaum, M. A. (2024). Enrichment of different taxa of the enigmatic candidate phyla radiation bacteria using a novel picolitre droplet technique. ISME. *Communications, 4*(1). https://doi.org/10.1093/ismeco/ycae080

Mast, Y., & Stegmann, E. (2019). Actinomycetes: The antibiotics producers. *Antibiotics, 8*(3), 105. https://doi.org/10.3390/antibiotics8030105

Medema, M. H., Kottmann, R., Yilmaz, P., Cummings, M., Biggins, J. B., Blin, K., De Bruijn, I., Chooi, Y. H., Claesen, J., Coates, R. C., Cruz-Morales, P., Duddela, S., Düsterhus, S., Edwards, D. J., Fewer, D. P., Garg, N., Geiger, C., Gomez-Escribano, J. P., Greule, A., et al. (2015). Minimum information about a biosynthetic gene cluster. *Nature Chemical Biology, 11*(9), 625–631. https://doi.org/10.1038/nchembio.1890

Meenakshi, S., Hiremath, J., Meenakshi, M. H., & Shivaveerakumar, S. (2024). Actinomycetes: Isolation, cultivation and its active biomolecules. *Journal of Pure and Applied Microbiology, 18*(1), 118–143. https://doi.org/10.22207/JPAM.18.1.48

Mesquita, A., Cerqueira, D., Rocha, M., Silva, D., Martins, C., & Souza, B. (2025). A review on rare and symbiotic actinobacteria: Emerging biotechnological tools against antimicrobial resistance. *Journal of Basic Microbiology*. https://doi.org/10.1002/jobm.70036

Mohammadipanah, F., & Wink, J. (2016). Actinobacteria from arid and desert habitats: Diversity and biological activity. *Frontiers in Microbiology, 6*, 1541.

Molano, L. G., Vega-Abellaneda, S., & Manichanh, C. (2024). GSR-DB: A manually curated and optimized taxonomical database for 16S rRNA amplicon analysis. *mSystems, 9*(2). https://doi.org/10.1128/msystems.00950-23

Montaser, R., & Luesch, H. (2011). Marine natural products: A new wave of drugs? *Future Medicinal Chemistry, 3*(12), 1475–1489. https://doi.org/10.4155/fmc.11.118

Narsing Rao, M. P., & Li, W. J. (2022). Diversity of actinobacteria in various habitats. In *Actinobacteria: Microbiology to synthetic biology* (pp. 37–58). Springer Nature Singapore. https://doi.org/10.1007/978-981-16-5835-8_2

Nauwynck, W., Faust, K., & Boon, N. (2025). Droplet microfluidics for single-cell studies: A frontier in ecological understanding of microbiomes. *FEMS Microbiology Reviews*. https://doi.org/10.1093/femsre/fuaf032

Nouioui, I., Carro, L., García-López, M., Meier-Kolthoff, J. P., Woyke, T., Kyrpides, N. C., et al. (2018). Genome-based taxonomic classification of the phylum Actinobacteria. *Frontiers in Microbiology, 9*, 2007. https://doi.org/10.3389/fmicb.2018.02007

Pisano, M., Sommer, M., & Lopez, M. (1986). Application of pretreatments for the isolation of bioactive actinomycetes from marine sediments. *Applied Microbiology and Biotechnology, 25*(3). https://doi.org/10.1007/bf00253664

Pope, E., Cartmell, C., Haltli, B., Ahmadi, A., & Kerr, R. G. (2022). Microencapsulation and in situ incubation methodology for the cultivation of marine bacteria. *Frontiers in Microbiology, 13*. https://doi.org/10.3389/fmicb.2022.958660

Prakash, O., Verma, M., Sharma, P., Kumar, M., Kumari, K., Singh, A., Kumari, H., Jit, S., Gupta, S. K., Khanna, M., & Lal, R. (2007). Polyphasic approach of bacterial classification—An overview of recent advances. *Indian Journal of Microbiology, 47*(2), 98–108. https://doi.org/10.1007/s12088-007-0022-x

Quast, C., Pruesse, E., Yilmaz, P., Gerken, J., Schweer, T., Yarza, P., Peplies, J., & Glöckner, F. O. (2012). The SILVA ribosomal RNA gene database project: Improved data processing and web-based tools. *Nucleic Acids Research, 41*(D1), D590–D596. https://doi.org/10.1093/nar/gks1219

Ribeiro, I., Antunes, J. T., Alexandrino, D. A. M., Tomasino, M. P., Almeida, E., Hilário, A., Urbatzka, R., Leão, P. N., Mucha, A. P., & Carvalho, M. F. (2023). Actinobacteria from Arctic and Atlantic deep-sea sediments—Biodiversity and bioactive potential. *Frontiers in Microbiology, 14*. https://doi.org/10.3389/fmicb.2023.1158441

Rizescu, C. (2022). Selective isolation, antimicrobial screening and phylogenetic diversity of marine actinomycetes derived from the Coast of Bejaia City (Algeria), a polluted and microbiologically unexplored environment. *Journal of Applied Microbiology, 132*(4), 2870–2882. https://doi.org/10.1111/jam.15415

Roda-Garcia, J. J., Haro-Moreno, J. M., & López-Pérez, M. (2023). Evolutionary pathways for deep-sea adaptation in marine planktonic Actinobacteriota. *Frontiers in Microbiology, 14*. https://doi.org/10.3389/fmicb.2023.1159270

Rodrigues, C. J. C., & De Carvalho, C. C. C. R. (2022). Cultivating marine bacteria under laboratory conditions: Overcoming the "unculturable" dogma. *Frontiers in Bioengineering and Biotechnology, 10*. https://doi.org/10.3389/fbioe.2022.964589

Saez, J. M., Raimondo, E. E., Costa-Gutierrez, S. B., Aparicio, J. D., Angelucci, D. M., Donati, E., Polti, M. A., Tomei, M. C., & Benimeli, C. S. (2025). Enhancing environmental decontamination and sustainable production through synergistic and complementary interactions of actinobacteria and fungi. *Heliyon, 11*(3), e42135. https://doi.org/10.1016/j.heliyon.2025.e42135

Santos, J. D., Vitorino, I., De La Cruz, M., Díaz, C., Cautain, B., Annang, F., Pérez-Moreno, G., Martinez, I. G., Tormo, J. R., Martín, J. M., Urbatzka, R., Vicente, F. M., & Lage, O. M. (2019). Bioactivities and extract dereplication of actinomycetales isolated from marine sponges. *Frontiers in Microbiology, 10*. https://doi.org/10.3389/fmicb.2019.00727

Santos, J. D. N. D., João, S. A., Martín, J., Vicente, F., Reyes, F., & Lage, O. M. (2022). iChip-inspired isolation, bioactivities and dereplication of Actinomycetota from Portuguese beach sediments. *Microorganisms, 10*(7), 1471. https://doi.org/10.3390/microorganisms10071471

Schatz, A., Bugie, E., & Waksman, S. A. (1944). Streptomycin, a substance exhibiting antibiotic activity against gram-positive and gram-negative bacteria. *Proceedings of the Society for Experimental Biology and Medicine, 55*(1), 66–69.

Schneider, Y. K., Hagestad, O. C., Li, C., Hansen, E. H., & Andersen, J. H. (2022). Selective isolation of Arctic marine actinobacteria and a down-scaled fermentation and extraction strategy for identifying bioactive compounds. *Frontiers in Microbiology, 13*. https://doi.org/10.3389/fmicb.2022.1005625

Sexton, D. L., & Tocheva, E. I. (2020). Ultrastructure of exospore formation in streptomyces revealed by cryo-electron tomography. *Frontiers in Microbiology, 11*. https://doi.org/10.3389/fmicb.2020.581135

Sheikh, M., Rathore, D. S., Gohel, S. D., & Singh, S. P. (2019). Cultivation and characteristics of the Marine Actinobacteria from the Sea water of Alang, Bhavnagar. *Indian Journal of Geo-Marine Sciences, 48*(12), 1896–1901.

Singh, L., Mazumder, S., & Bora, T. (2009). Optimisation of process parameters for growth and bioactive metabolite produced by a salt-tolerant and alkaliphilic actinomycete, *Streptomyces tanashiensis* strain A2D. *Journal de Mycologie Médicale, 19*(4), 225–233. https://doi.org/10.1016/j.mycmed.2009.07.006

Staley, J. T., & Konopka, A. (1985). Measurement of in situ activities of nonphotosynthetic microorganisms in aquatic and terrestrial habitats. *Annual Review of Microbiology, 39*(1), 321–346.

Staskiewicz, K., Dabrowska-Zawada, M., Kozon, L., Olszewska, Z., Drewniak, L., & Kaminski, T. S. (2024). Droplet microfluidic system for high throughput and passive selection of bacteria producing biosurfactants. *Lab on a Chip, 24*(7), 1947–1956. https://doi.org/10.1039/d3lc00656e

Subramani, R., & Aalbersberg, W. (2012). Marine actinomycetes: An ongoing source of novel bioactive metabolites. *Microbiological Research, 167*(10), 571–580. https://doi.org/10.1016/j.micres.2012.06.005

Tay, D. W. P., Tan, L. L., Heng, E., Zulkarnain, N., Chin, E. J., Tan, Z. Y. Q., Leong, C. Y., Ng, V. W. P., Yang, L. K., Seow, D. C. S., Kanagasundaram, Y., Ng, S. B., Lim, Y. H., & Wong, F. T. (2024). Tandem mass spectral metabolic profiling of 54 actinobacterial strains and their 459 mutants. *Scientific Data, 11*(1). https://doi.org/10.1038/s41597-024-03833-9

Tr, K. (2017). Co-culture as the novel approach for drug discovery from marine environment. *Novel Approaches in Drug Designing & Development, 2*(4). https://doi.org/10.19080/napdd.2017.02.555594

Velez-Suberbie, M. L., Betts, J. P. J., Walker, K. L., Robinson, C., Zoro, B., & Keshavarz-Moore, E. (2017). High throughput automated microbial bioreactor system used for clone selection and rapid scale-down process optimization. *Biotechnology Progress, 34*(1), 58–68. https://doi.org/10.1002/btpr.2534

Wang, M., Zheng, N., Li, X., Zhao, K., & Xie, B.-B. (2023). Enrichment pretreatment expands the microbial diversity cultivated from marine sediments. *Microorganisms.* https://doi.org/10.3390/microorganisms11112771

Wheatley, S. K., Cartmell, C., Madadian, E., Badr, S., Haltli, B. A., Kerr, R. G., & Ahmadi, A. (2022). Microfabrication of a micron-scale microbial-domestication pod for in situ cultivation of marine bacteria. *RSC Advances, 12*(43), 28123–28127. https://doi.org/10.1039/d2ra05420e

Widada, J., Damayanti, E., Mustofa, M., Dinoto, A., Febriansah, R., & Hertiani, T. (2023). Marine-derived *Streptomyces sennicomposti* GMY01 with anti-plasmodial and anticancer activities: Genome analysis, in vitro bioassay, metabolite profiling, and molecular docking. *Microorganisms, 11*(8), 1930. https://doi.org/10.3390/microorganisms11081930030

Xu, Y., Jing, Y., Zhang, Y., Liu, Q., Xiu, J., Zhang, K., Jia, N., Zhou, M., Zhou, X., Huang, J., Nie, Y., & Wu, X. (2022). Single-cell-based high-throughput cultivation and functional characterization of biosurfactant-producing bacteria from soil and oilfield-produced water. *Microorganisms, 10*(11), 2216. https://doi.org/10.3390/microorganisms10112216

Yoo, Y., Lee, H., Lee, J., Khim, J. S., & Kim, J. (2023). Insights into saline adaptation strategies through a novel halophilic bacterium isolated from solar saltern of Yellow sea. *Frontiers in Marine Science, 10.* https://doi.org/10.3389/fmars.2023.1229444

Yoon, S., Ha, S., Lim, J., Kwon, S., & Chun, J. (2017). A large-scale evaluation of algorithms to calculate average nucleotide identity. *Antonie Van Leeuwenhoek, 110*(10), 1281–1286. https://doi.org/10.1007/s10482-017-0844-4

Zhu, Y., Ma, N., Jin, W., Wu, S., & Sun, C. (2017). Genomic and transcriptomic insights into calcium carbonate biomineralization by marine actinobacterium *Brevibacterium linens* BS258. *Frontiers in Microbiology, 8.* https://doi.org/10.3389/fmicb.2017.00602

Ziemert, N., Lechner, A., Wietz, M., Millán-Aguiñaga, N., Chavarria, K. L., & Jensen, P. R. (2014). Diversity and evolution of secondary metabolism in the marine actinomycete genus Salinispora. *Proceedings of the National Academy of Sciences, 111*(12), E1130–E1139.

Indigenous Marine Actinobacteria

A. Kowsalya and A. Sai Ramesh

Abstract

Indigenous constitute a highly diverse group of microorganisms inhabiting both coastal and deep-sea regions of the Indian Ocean. These gram-positive bacteria are notable for producing bioactive compounds with significant pharmaceutical and industrial potential. The distinctive conditions of the Indian marine environment, such as fluctuations in salinity, temperature, and pressure, have driven the development of unique metabolic pathways and secondary metabolites in these organisms. Recent research has emphasized the isolation and characterization of actinobacteria from sediments, sponges, coral reefs, and other marine habitats, leading to the discovery of new species and genera and broadening our understanding of microbial diversity in Indian waters. The bioactive molecules they produce exhibit strong antimicrobial, anticancer, and enzyme-inhibitory properties. Advancements in metabolomics, metagenomics, and genome mining have shed light on the biofuel potential of these organisms. New technologies such as synthetic biology, CRISPR–Cas genome editing, and artificial intelligence are transforming the identification and creation of new natural products. Moreover, Indian marine actinobacteria show promise in bioremediation and other biotechnological applications. This review highlights the potential indigenous actinobacteria as a valuable resource for natural product discovery and play an important role in maintaining marine ecosystem functions.

A. Kowsalya
Anna University, Chennai, Tamil Nadu, India

A. Sai Ramesh (✉)
Department of BioEngineering, Vels Institute of Science, Technology and Advanced Studies, Chennai, Tamil Nadu, India

K. Loganathan, Z. Li (eds.), *Marine Actinomycetes*,
https://doi.org/10.1007/978-981-95-9849-6_4

Graphical Abstract

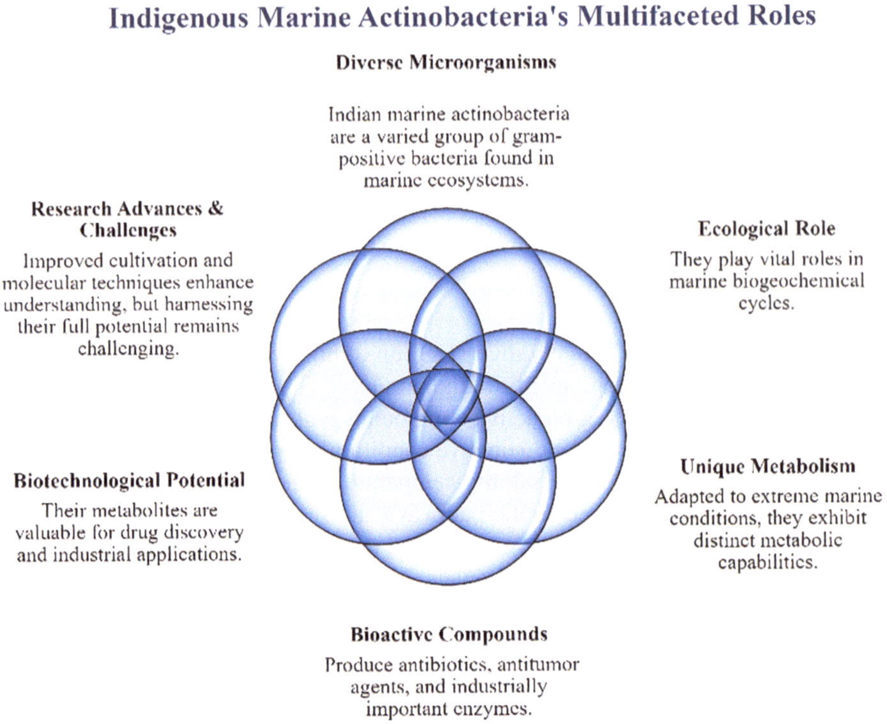

Keywords

Marine actinobacteria · Bioactive compounds · Indian marine ecosystem · CRISPR-Cas · Metabolomics · Climate change

1 Introduction

Global health issues, such as the spread of disease and antibiotic tolerance, mean that we need to find new ways to treat these conditions. Since the Earth's surface is mostly water, marine environments are full of bioactive substances and different types of life. A lot of research has been done on marine actinobacteria because they have a lot of different naturally occurring chemicals that are involved in both structure and function. The special stresses of marine settings have made these microorganisms stronger by giving them new ways to use energy. Chemicals that kill germs, cancer, and inflammation very well are made in these ways (Paul et al., 2021). Because they are so common and come in many types, marine actinomycetes are great for bioprospecting to find new prescription drugs and medicines. The growing number of diseases that can't be cured with antibiotics makes the need for other useful chemical sources even more clear. They make things that land animals don't make because of the stresses of the ocean (Rathinam et al., 2023) that we are going to look into.

These bacteria can make new secondary molecules as part of their many symbiotic relationships in the ocean. As a result, more ingredients can be used to study drugs. This review will talk about how techniques like growing and analyzing have helped recognize and classify these unique chemicals. It's important to study "rare" actinobacteria, like strains that aren't Streptomyces or are hard to separate and grow, says Batra et al. (2023). This is because these bacteria make a lot of biotechnologically important chemicals and can be hard to separate and grow. Instead of these issues, we need to use niche-specific medium that are like marine settings and co-cultivation to make the most of their biosynthetic potential.

This review will discuss how indigenous marine bacteria might fight drug resistance and enhance sustainable biotechnology by creating novel medications and industrial enzymes. Due to salt, temperature, and nutrient availability, indigenous actinobacterial strains in Indian maritime ecosystems may develop chemically unique and physiologically potent secondary metabolites. Besides the well-studied Streptomyces species, these unusual actinobacteria from these places must be isolated and characterized because they produce a large amount of market antibiotics. Genomic and phylogenetic studies demonstrate that deep-sea sediments include a varied array of Streptomyces lineages, a substantial but generally overlooked aspect of antibiotic-producing Streptomyces ecosystems. Streptomyces still has several bioactive chemicals.

2 Biodiversity, Ecology, and Distribution of Indian Marine Actinobacteria

The Table 1 lists marine actinobacteria from Indian coastal environments. These microbes' ecological and biotechnological roles show India's maritime ecosystems' richness and untapped potential. *Streptomyces* is known for its beneficial secondary metabolites. *Streptomyces* species were isolated from mangrove sediments, coral fragments, sponges, and coastal sediments. Marine environments use them for decomposition, symbiosis, competition, and secondary metabolite formation. Other famous genera include *Nocardiopsis, Salinispora, Micromonospora*, and *Actinomadura*. Each group has unique roles. Coastal sediment-dwelling *Salinispora* species produce anti-inflammatory and anti-cancer compounds. Indian actinobacteria inhabits the Arabian Sea, Bay of Bengal, Gulf of Mannar, Lakshadweep, and Andaman Sea. Based on its extensive distribution, India's maritime environment may include biotechnologically useful actinobacteria. Microorganisms produce enzymes and metabolites with several medical uses.

Table 1 List of indigenous marine actinobacteria

S.no	*Microorganism*	Typical marine sample source	Place	Ecological role	Application	Metabolites/enzymes	References
1	*Streptomyces griseus*	Coastal sediment	Gulf of Mannar, TN, India	Decomposer	Antibiotics	Streptomycin, proteases	Sarkar and Suthindhiran (2022)
2	*Streptomyces albus*	Sponge-associated	Lakshadweep, India	Symbiont, biofilm	Antibacterial	Polyketides, NRPS peptides	Paul et al. (2021)
3	*Streptomyces avermitilis*	Shelf sediments	Arabian Sea, India	Competitor	Antiparasitic	Avermectins	Rathinam et al. (2023)
4	*Streptomyces rochei*	Coral rubble	Gulf of Kachchh, Gujarat, India	Secondary metabolite producer	Antifungal	Macrocyclic lactones	Batra et al. (2023)
5	*Streptomyces nobilis*	Sandy sediment	Odisha coast, India	Decomposer	Antitumor leads	Noboritomycin-type PKs (polyketides)	Nair and Abraham (2020)
6	*Streptomyces indicus*	Mangrove sediment	Sundarbans, West Bengal, India	Organic matter turnover	Antibacterial	Indole alkaloids, chitinases	Jain and Tailor (2020)
7	*Streptomyces xinghaiensis*	Deep-sea sediment	Bay of Bengal, India	Oligotrophic survivor	Drug discovery	Polyketides, β-lactamases	Manikkam et al. (2019)
8	*Streptomyces* sp. (*lobophorin-producer*)	Sponge	Andaman & Nicobar, India	Symbiont	Antibacterial	Lobophorins	Wahaab and Subramaniam (2018)
9	*Salinispora tropica*	Marine sediment	Andaman Sea, India	Specialist in sediments	Anticancer	Salinosporamide A	Davies-Bolorunduro et al. (2021)
10	*Salinispora arenicola*	Continental shelf sediment	Arabian Sea, India	Nutrient recycling	Anti-inflammatory	Rifamycin-like PKs	Kannabiran (2024)
11	*Micromonospora marina*	Sediment	Vizhinjam, Kerala, India	Decomposer	Enzymes	Cellulases, micromonolactam	Mahapatra et al. (2020)
12	*Micromonospora chalcea*	Sponge	Rameswaram, TN, India	Symbiont	Antitumor	Anthracyclines	Godbole et al. (2023)

13	*Micromonospora* sp. (*abyssomicin-producer*)	Deep sediment	Central Indian Ocean	Decomposer	Antibacterial	Abyssomicins	Thangavel et al. (2019)
14	*Nocardiopsis dassonvillei*	Hypersaline sediment	Gulf of Khambhat, Gujarat, India	Halotolerant recycler	Antimicrobials	Peptidolactones, proteases	Shabari et al. (2022)
15	*Nocardiopsis alba*	Shelf sediment	Mangalore, Karnataka, India	Decomposer	Antibiotics	Phenazines	Izhar et al. (2024)
16	*Nocardiopsis synnemataformans*	Mangrove sediment	Kerala backwaters, India	Organic matter turnover	Enzymes	Chitinases, lipases	Kalimuthu et al. (2022)
17	*Actinomadura* sp.	Coral debris	Lakshadweep, India	Competitor	Antibacterials	Macrolides	Chauhan and Gohel (2020)
18	*Saccharopolyspora* sp.	Sediment	Chennai coast, TN, India	Decomposer	Antibiotics	PKS/NRPS (non-ribosomal peptides) hybrids	Manisha et al. (2019)
19	*Marinactinospora thermotolerans*	Deep-sea sediment	Bay of Bengal, India	Heat-tolerant decomposer	Novel antibiotics	Maritimycins	Dharumadurai (2022)
20	*Pseudonocardia* sp.	Sponge	Gulf of Mannar, India	Symbiont	Antifungal	Polyketides	Manikkam et al. (2024)
21	*Kocuria marina*	Seawater	Kochi, Kerala, India	Surface colonizer	Enzymes	Esterases, catalases	Muthusamy et al. (2025)
22	*Kocuria palustris*	Estuarine water	Mandovi–Zuari, Goa, India	Biofilm former	Antioxidants	Pigments, peroxidases	Kripa et al. (2020)
23	*Rhodococcus erythropolis*	Oil-impacted sediment	Ennore, Chennai, India	Hydrocarbon degrader	Bioremediation	Oxygenases, lipases	Jagannathan et al. (2021)
24	*Rhodococcus qingshengii*	Harbor sediment	Mumbai, Maharashtra, India	Pollutant degrader	Oil spill cleanup	Alkane hydroxylases	Nair and Abraham (2020)
25	*Dietzia maris*	Sediment	Paradip, Odisha, India	Hydrocarbon metabolism	Bioremediation	Biosurfactants, lipases	John et al. (2024)
26	*Gordonia* sp.	Oiled beach sand	Kakinada, AP, India	Hydrocarbon degrader	Bioremediation	Monooxygenases	Patil and Selvaraj (2025)

(continued)

Table 1 (continued)

S.no	*Microorganism*	Typical marine sample source	Place	Ecological role	Application	Metabolites/enzymes	References
27	*Arthrobacter* sp. (*Glutamicibacter*)	Beach sand	Digha, WB, India	Decomposer	Enzymes	Amylases, proteases	Quintero et al. (2018)
28	*Brevibacterium marinum*	Fish mucus	TN coast, India	Commensal	Food biotech	Proteases	Zothanpuia et al. (2018)
29	*Dermacoccus marinus*	Seawater	Lakshadweep, India	Surface colonizer	Antioxidants	Carotenoids	Girão et al. (2022)
30	*Saccharomonospora* sp.	Mangrove rhizosphere	Pichavaram, TN, India	Nutrient cycling	Enzymes	Xylanases	Hassan et al. (2019)
31	*Williamsia* sp.	Sediment	Cochin estuary, India	Organic matter turnover	Biocatalysis	Lipases	Kaur et al. (2020)
32	*Tsukamurella* sp.	Deep sediment	Bay of Bengal, India	Hydrocarbon metabolism	Bioremediation	Alkane oxygenases	Jayachandran et al. (2024)
33	*Corynebacterium marinum*	Seawater	Gujarat coast, India	Commensal	Enzymes	Lipases	Maghembe et al. (2020)
34	*Mycobacterium marinum*	Fish, seawater	TN & Kerala coasts, India	Pathogen	Disease model	Mycolic acids	Brahmam Pasumarthi et al. (2024)
35	*Nocardia* sp.	Sediment	Andaman Islands, India	Decomposer	Antibacterials	Polyketides	Tizabi (2022)
36	*Actinokineospora* sp.	Sponge	Gulf of Mannar, India	Symbiont	Antifungal	NRPS peptides	Pattnaik et al. (2022)
37	*Verrucosispora* sp. (*now Micromonospora*)	Deep sediment	Central Indian Ocean	Decomposer	Antibacterial	Abyssomicins, proximicins	Ruwandeepika et al. (2022)
38	*Kytococcus sedentarius*	Biofilm	Chennai Marina, India	Surface colonizer	Antimicrobials	Kytomycin	Mehetre et al. (2019)
39	*Janibacter* sp.	Seawater	Mumbai harbor, India	Biofilm, recycler	Enzymes	Esterases	Tamreihao et al. (2019)

40	*Nesterenkonia* sp.	Saltern brine	Tuticorin, TN, India	Halophile	Biocatalysis	Extremozymes	Rusyaev and Orlov (2024)
41	*Isoptericola* sp.	Mangrove sediment	Sundarbans, India	Organic matter turnover	Enzymes	Cellulases	Subramani and Sipkema (2022)
42	*Microbacterium maritypicum*	Seawater	Andaman Sea, India	Surface colonizer	Enzymes	β-glucosidase	Hussain et al. (2020)
43	*Brachybacterium* sp.	Beach sand	Goa, India	Decomposer	Enzymes	Proteases	Dixit et al. (2021)
44	*Agrococcus* sp.	Algal surface (seaweed)	Vizag, AP, India	Epiphyte	Biocontrol	Lytic enzymes	Dhandapani et al. (2022)
45	*Cellulomonas* sp.	Detritus	Chilika Lagoon, Odisha, India	Cellulose degrader	Waste valorization	Cellulases	Indira et al. (2021)
46	*Nocardioides* sp.	Sediment	Gulf of Mannar, India	Organic matter turnover	Biocatalysis	Oxygenases	Godbole et al. (2023)
47	*Kocuria flava*	Coral mucus	Lakshadweep, India	Symbiont	Antioxidants	Carotenoids	Meena et al. (2023)
48	*Streptomyces pactum*	Mangrove sediment	Sundarbans, India	Decomposer	Antibacterial	Actinomycins, chitinases	Sinha (2020)
49	*Streptomyces rubrolavendulae*	Sponge	Andaman & Nicobar, India	Symbiont	Antitumor leads	Anthracyclines	Shintre et al. (2020)
50	*Mariniluteicoccus* sp.	Deep-sea sediment	Indian Ocean ridge, global	Oligotrophic recycler	Novel metabolites	Terpenoids, PKs	Law et al. (2019)

3 Genomic and Metabolomic Insights

3.1 Genomic Features of Marine Actinobacteria

Indigenous marine actinobacteria's genetic variety helps to synthesize structurally distinct chemicals that survive harsh marine conditions. These creatures' genomes are 6–12 mega base pairs and have 70% GC, which helps them survive in salty conditions (Nair & Abraham, 2020). Many of these BGCs are "silent," however, genome mining has shown that marine actinobacteria have substantially higher metabolic potential. Their genomes contain genes for osmo-protectants like ectoine and trehalose, salt-tolerant ion transporters and chaperones, and barotolerant proteins to withstand deep-sea pressure. These organisms are useful for biotechnology because their genomes contain extremozymes, enzymes that can tolerate high temperatures, pressure, and salt (Jain & Tailor, 2020).

3.2 Comparative Genomics with Terrestrial Counterparts

Genomic comparisons show that land and ocean actinobacteria are genetically distinct. *Streptomyces* strains from soil are recognized for producing antibiotics, however marine isolates have unique PKS and NRPS clusters that adapt to marine biological niches. Marine *Salinispora* species encode salinosporamide gene clusters to produce powerful anticancer compounds, unlike land-based actinobacteria (Manikkam et al., 2019). Marine strains have increased stress tolerance, halophilicity, and hydrocarbon breakdown genes to adapt to the changing maritime environment. Marine actinobacteria are unique natural product repositories, and their genetic differences show how terrestrial and marine species evolved.

3.3 Metabolic Pathways and Adaptations

Marine actinobacteria use metabolic pathways to adapt to low-nutrient and uncertain salinity settings. Glycine betaine and ectoine are produced to fight osmotic stress. Bioremediation of oil-spoiled coastal waterways is possible because they metabolize petroleum molecules through hydrocarbon breakdown pathways (Wahaab & Subramaniam, 2018). Actinobacteria and sponges, corals, and algae often trade chemicals that protect them from dangerous germs. Symbiotic pathways awaken dormant BGCs, which produce novel secondary metabolites not expressed in free-living organisms.

3.4 Emerging Culture-Independent Approaches

Strategies without culture are needed since a large fraction of marine actinobacteria cannot be grown. Genomic sequencing of environmental DNA shows new BGCs and an unknown variety without culture. Single-cell genomics allows genome reconstruction from single cells without accounting for community complexity (Davies-Bolorunduro et al., 2021). Meta-transcriptomics can illuminate metabolic activity and real-time adaptation by examining transcribed genes. These approaches have revealed marine actinobacteria's biosynthesis pathways and bioactive compounds. By linking genetic potential to metabolite expression, these approaches enable novel bioprospecting prospects in Indian maritime habitats.

4 Bioactive Compounds and Biotechnological Applications

There are several applications of Indian marine actinobacteria in various fields of study that include protein engineering & design, molecular biology, marine ecology, biochemistry, drug discovery & development, medicine & therapeutics, synthetic biology, agriculture, vaccine design, neuroscience, microbiology, industrial & environmental biotechnology (Fig. 1).

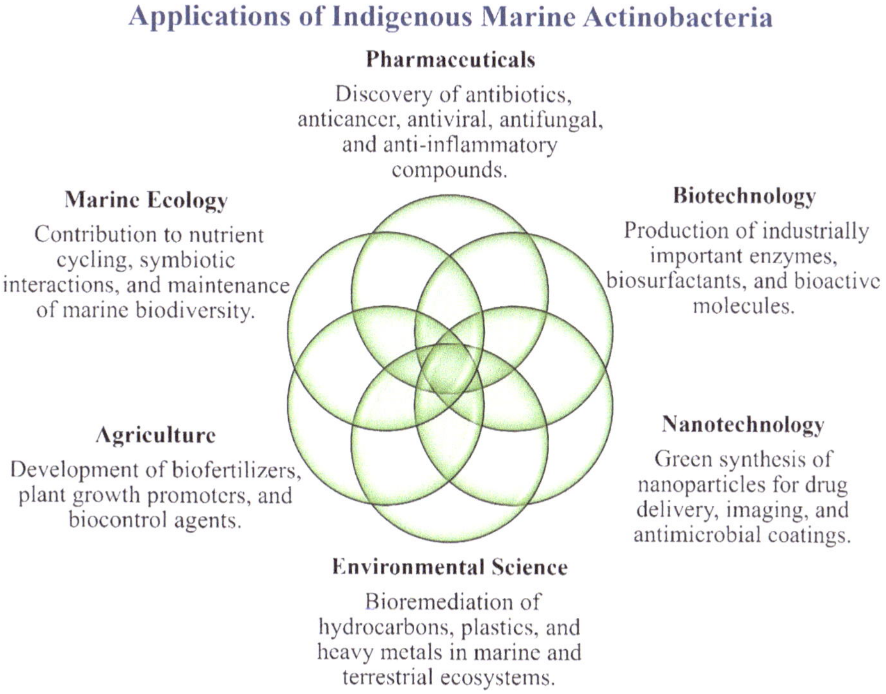

Fig. 1 Applications of indigenous marine actinobacteria

4.1 Secondary Metabolites and Their Structures

Water-borne actinobacteria produce several secondary metabolites, including various NRPS and PKS families. Repeated enzyme assembly lines give these compounds complex scaffolds and common modifications like glycosylation, halogenation, and cyclization. PKS and NRPS domains have been combined to create hybrid metabolites with different bioactivities (Mahapatra et al., 2020). Marine-specific chemicals, such as lobophorins, which are antibacterial and anticancer, salinosporamides, which suppress proteasomes, and abyssomicins, which may treat MRSA, demonstrate this diversity. Maritime surroundings encourage biosynthetic creativity and structural originality, as shown by these compounds.

4.2 Antimicrobial Activities

One huge use for marine actinobacterial waste is to look into how to fight illnesses that are not helped by antibiotics. Godbole et al. (2023) say that marine isolate scaffolds will be more useful as more "superbugs" like MRSA and MDR–TB show up. Microbes are killed by pyrroles and streptoketides in the ocean. Actinobacteria in the ocean also kill parasites and fungi, like *Plasmodium falciparum* and *Candida albicans*.

4.3 Anticancer and Other Therapeutic Potentials

Increasing data suggest marine actinobacteria compounds have significant anticancer properties. For instance, anthracyclines and macrolides kill cancer cell lines. Salinosporamide A's multiple myeloma clinical trial development is another (Thangavel et al., 2019). Kinase modulators and serine protease inhibitors from marine animals may help treat inflammatory, neurological, and metabolic diseases. These studies highlight marine metabolites' potential as direct cytotoxins and molecular probes for novel therapeutic targets.

4.4 Industrial and Environmental Applications

The extremozymes of marine actinobacteria can tolerate severe temperatures, pressures, and salinities, making them useful outside of medicine. Proteases, chitinases, and salt-tolerant cellulases are used more in bioprocessing, detergents, and biofuel production (Shabari et al., 2022). Marine strains are important in bioremediation, notably for oil spills and other hydrocarbon-containing environmental concerns, due to their hydrocarbon degradation pathways. Biotechnology-derived colours, antioxidants, and enzymes from actinobacteria are valuable in the food and nutraceutical industries.

4.5 Genome Mining for Biosynthetic Gene Clusters

With bioinformatics, genome mining of biosynthetic gene clusters (BGCs) has revealed new metabolites. AntiSMASH, PRISM, and BAGEL predict ribosomally synthesized peptides, cryptic routes, and PKS/NRPS clusters. Due to silent or cryptic BGCs, marine strains, especially those from the Indian Ocean, have tremendous promise for pharmaceutical leads (Izhar et al., 2024). Cryptic polyketide clusters in some Indian *Streptomyces* species can develop novel drug families when activated. Genome mining speeds drug discovery by avoiding screening and cultivation mistakes.

5 Conservation and Bioprospecting

5.1 Threats to Marine Actinobacterial Diversity

However, marine actinobacteria are fragile and very important. Their numbers are going down because of both natural and man-made causes. The microbiome is lost when heavy metals, oil spills, and microplastics pollute the environment. Marine resources that are used too much for biotechnological and pharmaceutical products have put stress on wild populations (Kalimuthu et al., 2022). These animals are in danger because climate change is making the oceans more acidic, salty, and warmer at the top. Loss of coral reefs, mangroves, and deep-sea habitats makes it harder for these bacteria to live in places where they can grow, which makes the problem worse.

5.2 Sustainable Exploitation Strategies

We need long-term plunder to fix these issues. Marine Protected Areas do a lot to protect places where a lot of different kinds of life live. They also have strict rules about bioprospecting. Biotechnology has made it so that large-scale wild gathering is no longer possible. The metabolic processes of marine actinobacteria are copied in different hosts by synthetic biologists to make life easier for local species (Chauhan & Gohel, 2020). It is possible to make a lot of important molecules in a lab or factory setting with little harm to the environment thanks to fermentation optimization. These plans try to prevent both of those problems at the same time so that people can always get new drugs.

5.3 Legal and Ethical Considerations in Bioprospecting

Bioprospecting in the ocean is regulated by both international and state laws to make sure it is fair and durable. The Nagoya Protocol on Access and Benefit Sharing (ABS) is a key part of CBD. Manisha et al. (2019) infers that this global deal controls how genetic resources are used and makes sure that native and local people get fair pay. The Indian Biodiversity Act of 2002 safeguards the country's marine genetic resources and makes sure that any discoveries made in the Indian seas are lawful and moral. In these models, it's significant to protect microbial species and do the right thing by partners and locals who use marine resources. Protecting marine actinobacteria and looking for new ones needs a diverse approach that includes taking care of the environment, making technology better, and making rules and laws (Dharumadurai, 2022). Strong rules and long-lasting habits can protect marine environments for future generations and make use of the scientific potential of these bacteria.

6 Future Perspectives and Challenges

6.1 Emerging Technologies for Studying Marine Actinobacteria

Modern computational and molecular methods have revolutionized marine actinobacteria research. CRISPR-Cas genome editing techniques allow researchers to fine-tune biosynthetic gene clusters (BGCs), which is promising. Genome editing can reveal latent processes in marine actinobacteria with cryptic or silent BGCs, potentially leading to new bioactive compounds (Manikkam et al., 2024). Restructuring and expressing complex marine actinobacteria gene clusters in heterologous microbial hosts is still possible in synthetic biology. Metabolites can be produced in great yields because the parent organisms are sluggish to develop or hard to cultivate. Synthetic biology is also enabling metabolic pathway rational engineering to create novel molecules with better pharmacological properties. Metabolomics and AI have accelerated chemical discovery. AI-driven dereplication algorithms may distinguish recognized metabolites from new structures, improving natural product screening (Muthusamy et al., 2025). Genomes, proteomics, and metabolomics datasets can be integrated using machine learning algorithms to predict bioactivity and guide targeted molecule separation.

6.2 Potential for Novel Drug Discovery

Marine actinobacteria may solve some of the world's biggest health issues. Due to the rise of antimicrobial resistance (AMR), marine actinobacteria may be a novel antibiotic. Previous research has identified drugs effective against MRSA, MDR–TB, and CRE. The prospect of novel β-lactamase inhibitors to combat resistant organisms is promising. Actinobacteria produce many antimicrobials and cancer-fighting chemicals (Kripa et al., 2020). Marine anthracyclines, macrolides, and peptides are strong cytotoxic against certain cancer cell lines. Apoptosis, cell signaling system disruption, and DNA replication suppression are some of their effects. Future research into marine actinobacterial metabolites as targeted enzyme inhibitors will target kinases and proteasomes, which are crucial to cancer growth. Neurological treatment also shows promise.

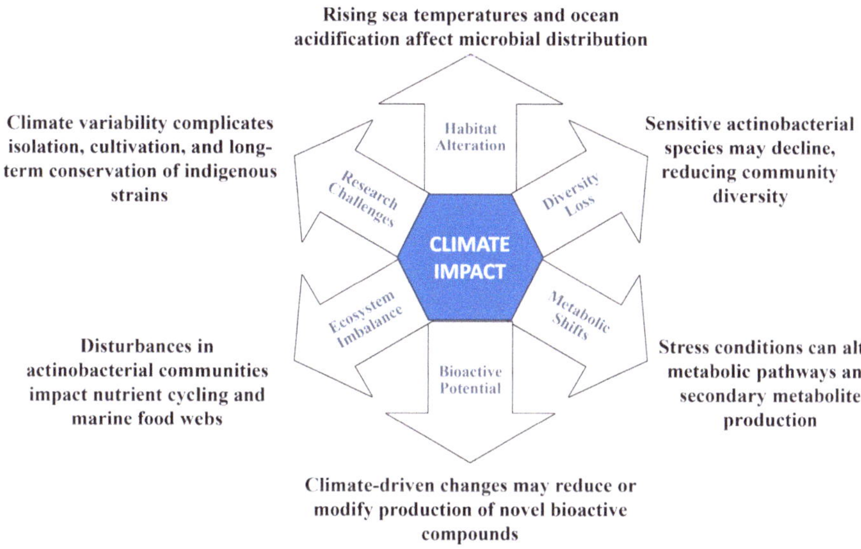

Fig. 2 Climate impacts on indigenous marine actinobacteria

6.3 Climate Change Impacts on Marine Actinobacterial Communities

Sudden climate changes cause a serious impact on Indian marine actinobacteria (Fig. 2). Technology and drugs may help marine actinobacteria, but climate change threatens their stability and variety. Ocean acidification alters marine pH and microbial metabolism, making it a major issue. Actinobacteria's proliferation, survival, and secondary metabolite production affect the availability of novel bioactive chemicals. Increasing ocean surface temperatures are another issue (Nair & Abraham, 2020). Global warming is projected to drive actinobacteria species to colder polar seas. If thermally sensitive strains that cannot adapt to changing surroundings go extinct due to this migration, unique and possibly irretrievable metabolites may be lost. Nutrient loading and climate change-induced ocean circulation shifts create marine hypoxia and deoxygenation, which threaten coastal and deep-sea ecosystems (Patil & Selvaraj, 2025).

7 Conclusion

Indigenous marine actinobacteria can yield bioactive compounds for medicine and biotechnology. The study highlighted marine actinobacteria's wide metabolic capacity, ecological and physiological importance, and the urgent need to conserve them. Metagenomics, genome mining, and metabolomics reveal bacteria' latent metabolic potential, supplementing cultivation-based approaches. India has abundant marine life and lengthy beaches, but actinobacteria have received little study.

These ecosystems have great potential for discovering new therapeutic compounds despite global issues, including antimicrobial resistance, cancer, and neurological diseases. Indigenous strains that adapted to specific environmental conditions may have genetic pathways for new medicinal chemicals. Conservation and ethical bioprospecting are needed to safeguard bacteria from pollution, overexploitation, and climate change. The Indian Biodiversity Act and Nagoya Protocol are needed for sustainable use and fair benefit distribution. Ecology, genetics, and biotechnology should be studied together. Synthetic biology for biochemical pathway reformation, CRISPR-based gene editing, and AI-driven metabolomics are needed to maximize marine actinobacteria's potential. Researchers, legislators, and business leaders must work together to protect marine ecosystems and benefit human health from marine microbial diversity.

References

Batra, B., Sharma, D., Bose, D., Parthasarthy, V., & Sarkar, A. (2023). Implications of bioprospecting marine diversity and sustainable production of bioactive compounds. In *Marine antioxidants* (pp. 27–43). Academic.

Brahmam Pasumarthi, R. B., Chahal, K., Kiranmai, T., & Jayarajan, D. (2024). Microbial biosynthesis of novel bioactive compounds. *Advances in Bioresearch, 15*(4), 309–321.

Chauhan, J. V., & Gohel, S. D. (2020). Molecular diversity and pharmaceutical applications of free-living and rhizospheric marine actinobacteria. In *Marine niche: Applications in pharmaceutical sciences: Translational research* (pp. 111–131). Springer Singapore.

Davies-Bolorunduro, O. F., Osuolale, O., Saibu, S., Adeleye, I. A., & Aminah, N. S. (2021). Bioprospecting marine actinomycetes for antileishmanial drugs: Current perspectives and future prospects. *Heliyon, 7*(8), e07710.

Dhandapani, R., Thangavelu, S., Ragunathan, L., Paramasivam, R., Velmurugan, P., & Muthupandian, S. (2022). Potential bioactive compounds from marine Streptomyces sp. and their in vitro antibiofilm and antibacterial activities against antimicrobial-resistant clinical pathogens. *Applied Biochemistry and Biotechnology, 194*(10), 4702–4723.

Dharumadurai, D. (Ed.). (2022). *Methods in actinobacteriology*. Humana Press.

Dixit, S., Shukla, A., Singh, V., & Upadhyay, S. K. (2021). Bioprospecting of natural compounds for industrial and medical applications: Current scenario and bottleneck. In *Bioprospecting of plant biodiversity for industrial molecules* (pp. 53–71). Wiley.

Girão, M., Ribeiro, I., & Carvalho, M. D. F. (2022). Actinobacteria from marine environments: A unique source of natural products. In *Natural products from actinomycetes: Diversity, ecology and drug discovery* (pp. 1–45). Springer Singapore.

Godbole, A. P., Wadetwar, R. N., Bundale, S. B., Nashikkar, N., & Kanojiya, P. S. (2023). Bioprospecting of actinomycetes from diverse ecosystems for antimicrobial activity. *Indian Journal of Pharmaceutical and Biological Research, 57*(3s), s599–s609.

Hassan, Q. P., Bhat, A. M., & Shah, A. M. (2019). Bioprospecting actinobacteria for bioactive secondary metabolites from untapped ecoregions of the Northwestern Himalayas. In *New and future developments in microbial biotechnology and bioengineering* (pp. 77–85). Elsevier.

Hussain, A., Hassan, Q. P., & Shouche, Y. S. (2020). New approaches for antituberculosis leads from Actinobacteria. *Drug Discovery Today, 25*(12), 2335–2342.

Indira, M., Venkateswarulu, T. C., Krupanidhi, S., & Peele, K. A. (2021). Microbial production of antimicrobial and anticancerous biomolecules. In *Bioprospecting of microorganism-based industrial molecules* (pp. 147–169). https://doi.org/10.1002/9781119717317.ch8

Izhar, S. K., Rizvi, S. F., Afaq, U., Fatima, F., & Siddiqui, S. (2024). Bioprospecting of metabolites from Actinomycetes and their applications. *Recent Patents on Biotechnology, 18*(4), 273–287.

Jagannathan, S. V., Manemann, E. M., Rowe, S. E., Callender, M. C., & Soto, W. (2021). Marine actinomycetes, new sources of biotechnological products. *Marine Drugs, 19*(7), 365.

Jain, A., & Tailor, V. (2020). Emerging trends of biotechnology in marine bioprospecting: A new vision. In *Marine niche: Applications in pharmaceutical sciences: Translational research* (pp. 1–36). https://doi.org/10.1007/978-981-15-5017-1_1

Jayachandran, K. V., Divya, L., & Oommen, O. V. (2024). Marine bioprospecting a promising future venture and precautionary measures to maintain its sustainability. In *Biodiversity and business: Bio prospecting and benefit sharing* (pp. 83–131). Springer.

John, G., Gopalakrishnan, A., Chakraborty, K., & Suresh, V. V. R. (2024). Marine biotechnology in India: Status and unfolding opportunities. In *Biotechnology in India—Reworking a strategy* (pp. 77–109). Springer Nature Singapore.

Kalimuthu, K., Kumaravel, M., Haldar, D., & Perumal, A. (2022). *Isolation and characterization of marine actinomycetes–bioprospecting for antimicrobial, antioxidant and anticancer properties*. Research Square.

Kannabiran, K. (2024). A review on bioprospecting of actinomycetes isolated from marine soil samples of India concerning their antimicrobial secondary metabolites. *Indian Journal of Natural Products and Resources (IJNPR) [Formerly Natural Product Radiance (NPR)], 15*(1), 13–20.

Kaur, S., Kalia, A., & Gangwar, M. (2020). Bioprospecting endophytic actinobacteria of medicinal plants as potential anticancer therapeutic agents. *Pollutants and Protectants, 197*. https://doi.org/10.54393/fbt.v5i4.208

Kripa, N. V., Reyhanath, P. V., Razia Beevi, M., & Ranjeet, K. (2020). Isolation of Streptomyces spp. with bioprospecting potential from Mangrove Regions of Ponnani, Kerala, India. *Journal of Aquatic Biology & Fisheries, 8*, 1–6.

Law, J. W. F., Pusparajah, P., Ab Mutalib, N. S., Wong, S. H., Goh, B. H., & Lee, L. H. (2019). A review on mangrove actinobacterial diversity: The roles of Streptomyces and novel species discovery. *Progress in Microbes & Molecular Biology, 2*(1), a0000024.

Maghembe, R., Damian, D., Makaranga, A., Nyandoro, S. S., Lyantagaye, S. L., Kusari, S., & Hatti-Kaul, R. (2020). Omics for bioprospecting and drug discovery from bacteria and microalgae. *Antibiotics, 9*(5), 229.

Mahapatra, G. P., Raman, S., Nayak, S., Gouda, S., Das, G., & Patra, J. K. (2020). Metagenomics approaches in discovery and development of new bioactive compounds from marine actinomycetes. *Current Microbiology, 77*(4), 645–656.

Manikkam, R., Pati, P., Thangavel, S., Venugopal, G., Joseph, J., Ramasamy, B., & Dastager, S. G. (2019). Distribution and bioprospecting potential of actinobacteria from Indian mangrove ecosystems. In *Microbial diversity in ecosystem sustainability and biotechnological applications: Volume 1. Microbial diversity in normal & extreme environments* (pp. 319–353). Springer Singapore.

Manikkam, R., Thenmozhli, G. S., Ranjani, S., Kishore Kumar, A., Azhagesan, S., Manigundan, K., et al. (2024). Translational research in microbial healthcare products with special reference to anticancer drug discovery from actinobacteria. In *Translational research in biomedical sciences: Recent progress and future prospects* (pp. 145–165). Springer Nature Singapore.

Manisha, M., Mithali, J., Vijayalakshmi, G., Gopikrishnan, V., Masilamaniselvam, M., & Radhakrishnan, M. (2019). Bioprospecting of actinobacteria from the Andaman marine ecosystem: Isolation, antagonistic potential, and taxonomy of potential strain. *Indian Journal Of Geo Marine Sciences, 48*(08), 1312–1318.

Meena, B., Anburajan, L., Aryamol, K., Vinithkumar, N. V., & Dharani, G. (2023). Studies on biodiversity and bioprospecting of active mud volcano associated bacteria and actinobacteria from Baratang, Andaman Islands, India. *Systems Microbiology and Biomanufacturing, 3*(2), 339–357.

Mehetre, G. T., Vinodh, J. S., Burkul, B. B., Desai, D., Santhakumari, B., Dharne, M. S., & Dastager, S. G. (2019). Bioactivities and molecular networking-based elucidation of metabolites of potent actinobacterial strains isolated from the Unkeshwar geothermal springs in India. *RSC Advances, 9*(17), 9850–9859.

Muthusamy, S., Manikkam, R., & Venugopal, G. (Eds.). (2025). *Mangrove microbiome: Diversity and bioprospecting*. Springer.

Nair, S., & Abraham, J. (2020). Natural products from actinobacteria for drug discovery. In *Advances in pharmaceutical biotechnology: Recent progress and future applications* (pp. 333–363). Springer Singapore.

Patil, A., & Selvaraj, S. (2025). Bioprospecting of bioactive compounds from marine fungi. In *Bioprospecting of multi-tasking fungi for therapeutic applications: Volume II* (pp. 1–27). Springer Nature Singapore.

Pattnaik, S., Imchen, M., Kumavath, R., Prasad, R., & Busi, S. (2022). Bioactive microbial metabolites in cancer therapeutics: Mining, repurposing, and their molecular targets. *Current Microbiology, 79*(10), 300.

Paul, S. I., Majumdar, B. C., Ehsan, R., Hasan, M., Baidya, A., & Bakky, M. A. H. (2021). Bioprospecting potential of marine microbial natural bioactive compounds. *Journal Of Applied Biotechnology Reports, 8*(2), 96–108.

Quintero, M., Velásquez, A., Jutinico, L. M., Jiménez-Vergara, E., Blandón, L. M., Martinez, K., et al. (2018). Bioprospecting from marine coastal sediments of Colombian Caribbean: Screening and study of antimicrobial activity. *Journal of Applied Microbiology, 125*(3), 753–765.

Rathinam, A. J., Santhaseelan, H., Dahms, H. U., Dinakaran, V. T., & Murugaiah, S. G. (2023). Bioprospecting of unexplored halophilic actinobacteria against human infectious pathogens. *3 Biotech, 13*(12), 398.

Rusyaev, S. M., & Orlov, A. M. (2024). The phenomenon of marine bioprospecting. *Biology Bulletin Reviews, 14*(1), 115–132.

Ruwandeepika, H. A. D., Fernando, G. C. P., & Jayaweera, T. S. P. (2022). An overview of biomedical, biotechnological, and industrial applications of actinomycetes. In *Natural products from actinomycetes: Diversity, ecology and drug discovery* (pp. 475–508). Springer.

Sarkar, G., & Suthindhiran, K. (2022). Diversity and biotechnological potential of marine actinomycetes from India. *Indian Journal of Microbiology, 62*(4), 475–493.

Shabari, G., Lokesh, R., & Kannabiran, K. (2022). Bioprospecting of actinomycetes for anti-bacterial potential: A review. *Research Journal of Biotechnology, 17*, 10.

Shintre, N. A., Tamhane, V. A., Baig, U. I., Pund, A. S., Patwardhan, R. B., & Deshpande, N. M. (2020). Diversity of culturable actinobacteria producing protease inhibitors isolated from the intertidal zones of Maharashtra, India. *Current Microbiology, 77*(11), 3555–3564.

Sinha, R. P. (Ed.). (2020). *Natural bioactive compounds: Technological advancements*. Academic.

Subramani, R., & Sipkema, D. (2022). Exploring uncharted sources of rare actinomycetes for the discovery of innovative antimicrobial compounds to combat multidrug-resistant bacterial infections. *Specialty Journal of Pharmacognosy, Phytochemistry, and Biotechnology, 2*(1–2022), 43–52.

Tamreihao, K., Mukherjee, S., Ningthoujam, D. S., & Roy, S. S. (2019). Prospects of Actinobacteria from underexplored ecosystems as anti-infective agents against *Mycobacterium tuberculosis*. In *Current perspectives on anti-infective agents* (pp. 1–20). Bentham Science Publishers.

Thangavel, S., Venugopal, G., Joseph, J., Ramasamy, B., & Dastager, S. G. (2019). Distribution and bioprospecting potential of actinobacteria from Indian mangrove. In *Microbial diversity in ecosystem sustainability and biotechnological applications: Volume 1. Microbial diversity in normal & extreme environments* (p. 319). Springer.

Tizabi, D. R. (2022). *Bioprospecting marine actinomycetes for novel anti-tuberculosis drugs*. Doctoral dissertation, University of Maryland, College Park.

Wahaab, F., & Subramaniam, K. (2018). Bioprospecting marine actinomycetes for multidrug-resistant pathogen control from Rameswaram coastal area, Tamil Nadu, India. *Archives of Microbiology, 200*(1), 57–71.

Zothanpuia, Passari, A. K., Leo, V. V., Chandra, P., Kumar, B., Nayak, C., et al. (2018). Bioprospection of actinobacteria derived from freshwater sediments for their potential to produce antimicrobial compounds. *Microbial Cell Factories, 17*(1), 68.

Bioactive Compounds from Marine Actinobacteria

Diksha Dubey, Dibyani Rout, Akshat Rajan Mehta, Sankalp Mandal, Rishita Jain, Prerna Kumari, Kothari Neeti Suresh, and C. Subathra Devi

Abstract

Marine actinobacteria signify a revolutionary frontier in the natural product discovery field by producing numerous secondary metabolites with unique therapeutic capabilities. Marine actinobacteria have been isolated from extreme marine environments and have evolved biosynthetic pathways that allow them to create different compounds. As terrestrial sources of natural products are exhausting and multidrug resistance is a major concern, marine actinobacteria represent a new and largely untapped chemical resource. Prominent classes of bioactive compounds produced by marine actinobacteria include polyketides, non-ribosomal peptides, alkaloids, and terpenes. Polyketides tend to have complex structures, such as salinosporamide A, a polyketide that has progressed to Phase I trials for cancer. Non-ribosomal peptides show significant antimicrobial and immunosuppressive potential (for example, a potent antibiotic, which is a non-ribosomal peptide (NRP). Alkaloids usually have broad-spectrum antibacterial properties. Terpenes also demonstrate anticancer, anti-inflammatory, and antiviral properties. In addition to antimicrobial applications, bioactive compounds produced by marine actinobacteria have also been developed by pharmaceutical companies for cancer therapy. The advancement of genomic mining, OSMAC methods, and co-cultivation has accelerated the discovery rate of natural products produced by actinobacteria. Marine actinobacteria are also contributing to the fields of biotechnology, biofuels, food processing, and bioremediation. Thus, marine actinobacteria are nature's chemists, adapted over millions of years to extreme environments and yield an extensive collections of bioactive molecules that will ultimately change the face of drug discovery and biotechnology.

D. Dubey · D. Rout · A. R. Mehta · S. Mandal · R. Jain · P. Kumari · K. N. Suresh · C. S. Devi (✉)
School of Bio Sciences and Technology, Vellore Institute of Technology, Vellore, Tamil Nadu, India
e-mail: csubathradevi@vit.ac.in

K. Loganathan, Z. Li (eds.), *Marine Actinomycetes*,
https://doi.org/10.1007/978-981-95-9849-6_5

Keywords

Marine actinobacteria · Bioactive compounds · Secondary metabolites · Drug discovery · Antibiotic resistance · Polyketides · Antimicrobial activity · Novel therapeutics

1 Introduction

Actinomycetes are one of the most interesting and important groups of microorganisms in nature. These Gram-positive bacteria have filamentous, branching structure which resembles fungi and have high guanine-cytosine (G + C) base pair content and generate the characteristic "earthy" smell of freshly turned soil. For almost 80 years, terrestrial actinomycetes, especially the genus *Streptomyces*, have been the major sources of antibiotics and have changed the practice of medicine. Actinomycetes represent the biological sources of nearly two-thirds of all antibiotics known today, including the earliest therapeutic agents streptomycin, tetracycline, and erythromycin. This remarkable biosynthetic capacity in actinomycetes has positioned them as the largest producers of secondary metabolites, which likely make up about 75% of all bioactive compounds produced by microbes (Ngema and Madoroba, 2024; Carroll et al., 2024).

It has an unsurpassed history of drug discovery based on terrestrial actinomycetes, occasionally called the golden age of antibiotic discovery, that has transformed medicine and saved lives all over the world. Throughout this period, a systematic screening of soil materials gave rise to several life-saving molecules, which eventually became the foundations of modern antibiotics being used today in clinical practice. Besides antibiotics, actinomycetes have also been used with success as life-saving antifungals (as the amphotericin B of *Streptomyces nodosus*), immunosuppressants (as tacrolimus of *Streptomyces tsukubaensis*), and anticancer agents (as either doxorubicin or Adriamycin of *Streptomyces peucetius*).

2 The Emergence of Drug Resistance and the Need for Novel Compounds

However, this magnificent source of discovery is beginning to give depressing indications of being depleted. The decades of extensive screening of earth-based microorganisms have led to a frustratingly exasperating state of diminishing returns as known structures are rediscovered more and more often, and new structures are discovered less and less often. This situation and the alarming trend in the world of the rise of multi-drug resistant (MDR) pathogens, also known as superbugs, due to their inability to respond to traditional anti-microbial therapies, have created a health crisis amongst humanity. The antimicrobial resistance has been termed by the World Health Organization as one of the top ten global health risks, and this necessitates the urgency of new chemical scaffolds having new mechanisms of action.

It is not only related to the cases of bacterial infections but also antifungal resistance, viral mutation, and continuous evolution of cancer cells to become resistant to the available chemotherapy options. The clinical fraternity has a known requirement for new molecular entities to get around the resistance mechanisms and offer a treatment option to the conditions that did not have a treatment choice in the past. The need to overcome resistance has directed scientific thought to environmental niches that have not been considered in the past, where microbial metabolic and biosynthetic potential can be explored.

3 Marine Environment: The Next Frontier in Natural Product Discovery

The oceans represent the final frontier in natural product discovery, covering over 70% of the Earth's surface and harboring an estimated 80% of all life forms. Marine environments present a vast, largely unexplored reservoir of biodiversity characterized by unique physicochemical conditions that drive the evolution of distinctive metabolic pathways. These extreme conditions include crushing hydrostatic pressures reaching over 1000 atmospheres in deep-sea environments, highly variable salinity levels, perpetual darkness in abyssal zones, limited oxygen availability, and complex chemical gradients.

To survive and thrive in such challenging conditions, marine organisms have evolved sophisticated biochemical strategies, producing secondary metabolites with chemical structures and bioactivities distinctly different from their terrestrial counterparts. The harsh marine environment creates intense evolutionary pressure that favors the development of novel biosynthetic pathways, resulting in structurally unique compounds with unprecedented mechanisms of action. Marine biodiversity surpasses terrestrial biodiversity at the phylum level, with more than 20 marine animal phyla having no terrestrial representatives, while few animal phyla exist solely on land (Pan et al., 2025).

4 Marine Actinomycetes: A Distinct Microbial Community

For years, actinomycetes identified from marine sediments were thought to simply be terrestrial spores that had entered the ocean through runoff and sedimentation. However, this was fundamentally challenged by early studies confirming that obligate marine actinomycetes exist, meaning they have not only evolved to tolerate but require seawater for maximal growth and metabolism. When the genus *Salinispora* was formally described taxonomically, it signified a major step forward in the study of marine microbiology as it demonstrated unequivocally that marine actinomycete lineages exist as discrete evolutionary lineages having evolved to ocean environments (Ngema and Madoroba, 2024; Carroll et al., 2024).

Recent research has identified numerous genera of marine actinomycetes outside of *Streptomyces*, such as *Salinispora*, *Marinispora*, *Actinomadura*, *Micromonospora*,

Nocardiopsis, and numerous rare and newly described taxa. These marine actinomycetes have been identified as particularly valuable resources of bioactive compounds derived from rare actinomycete bioactive compounds from marine substrata habitats such as deep-sea sediments, marine sponges, soft corals, and other invertebrate associations. Together, marine actinomycetes represent over 45% of bioactive metabolites derived from marine sources, with 83 species from 28 genera isolated from marine habitats, many of which are newly described and new to science (Ibrahim et al., 2025).

5 Structural Diversity and Bioactivity of Marine-Derived Compounds

Marine actinomycetes produce an extraordinary array of structurally diverse secondary metabolites that can be broadly classified into several major chemical families: polyketides, non-ribosomal peptides, alkaloids, terpenoids, macrolides, and quinones. These compounds demonstrate remarkable structural complexity and novel architectural features rarely encountered in terrestrial natural products. The unique marine environment has driven the evolution of biosynthetic machinery capable of incorporating unusual building blocks, including halogenated amino acids, rare sugars, and modified fatty acid chains, resulting in compounds with enhanced stability and bioactivity (De La Hoz-Romo et al., 2022).

The bioactive potential of marine actinomycete-derived compounds spans multiple therapeutic areas. These molecules demonstrate potent antimicrobial activity against drug-resistant pathogens, including methicillin-resistant *Staphylococcus aureus* (MRSA) and vancomycin-resistant *Enterococci* (VRE). Notable examples include abyssomicin C, which exhibits excellent activity against MRSA through a novel mechanism targeting chorismate biosynthesis, and salinosporamide A (marizomib), a proteasome inhibitor that has advanced to clinical trials for treating multiple myeloma and glioblastoma (Ngema and Madoroba, 2024; Carroll et al., 2024).

6 Polyketides

Polyketides constitute a highly diverse group of secondary metabolites that are produced by bacteria, specifically actinobacteria, through highly specialized multienzyme machines known as polyketide synthases (PKS), referred to as natural enzyme factories (Valliappan et al., 2014). They are characterized by complicated and diverse chemical structures that are based on analogous biosynthetic pathways to fatty acid synthesis, on which acetyl-CoA subunits are condensed and subsequently modified to yield aromatic, macrocyclic, and additional classes of compounds (Chinnathambi et al., 2023). It is this structural variance that directly causes the varied range of biological properties of polyketides, including antibiotic, anticancer, antifungal, and immunosuppressive activity, which makes them highly valued in medicinal chemistry, agriculture, and biotechnology (Shuikan et al., 2023).

7 Unique Production by Marine Actinobacteria

Marine actinobacteria are Gram-positive bacteria with a high GC ratio, which can live under extreme marine circumstances, such as high pressure, high salinity, and variable temperatures (Chinnathambi et al., 2023). The condition of living under these extreme conditions encourages evolution towards atypical metabolic pathways, resulting in the production of new secondary metabolites that are often different from those produced by actinobacteria from soil habitats (Shuikan et al., 2023). This is because marine actinobacteria will have stress-adapted biosynthetic gene clusters that contribute to their survival. Thus, marine actinobacteria can be expected to have a greater chance of being able to generate novel bioactive compounds than actinobacteria from soil habitat (Shuikan et al., 2023). According to the review by Shuikan et al. (2023), marine actinobacteria have shown themselves to be unique sources of antibiotics, anticancer compounds, and other drugs—particularly as multidrug resistance begins to expand. The genera that are most regularly found in marine habitats include *Streptomyces*, *Salinispora*, and *Micromonospora*, which produce the vast majority of metabolites that possess unique chemical moieties and various biological activities.

In their review from 2023, Chinnathambi et al. describe the extreme oceanic environment where marine actinobacteria live as a niche system for the evolution of unique enzyme systems and chemical processes. It is due to this unique evolutionary process that marine actinobacteria should be viewed as highly diverse, yet still underutilized, sources of drug discovery opportunity. Simply put, marine environments can represent a significant opportunity to discover new polyketides and similar secondary metabolites. Therefore, the systematic utilization of marine actinobacteria for drug discovery will greatly increase the opportunities for augmenting the chemical space drug screening to prioritize novel drugs and therapeutics.

8 Examples of Important Polyketides

The following section provides detailed descriptions of seven major polyketides derived from marine actinobacteria, each representing unique structural classes and bioactive properties that highlight the remarkable diversity and therapeutic potential of this compound family.

8.1 Salinosporamide A (Marizomib)

Salinosporamide A (Marizomib) is a naturally occurring product from obligate marine bacteria (*Salinispora tropica* and *Salinispora arenicola*) that were sourced from sediment sample collections from the ocean (Reed et al., 2007). Marizomib is a selective and irreversible proteasome inhibitor with an IC50 in the low to mid nanomolar range for all 3 catalytic moieties (β1, β2, β5) of the 20S proteasome. It is also noteworthy that Salinosporamide A was able to quickly progress into phase I clinical trials for the treatment of multiple myeloma, just 3 years after discovery in 2003 (Feling et al., 2003). Thus, Salinosporamide A is arguably one of, if not the quickest developed and evaluated marine drug. Part of the enthusiasm around Salinosporamide A lies in its ability to effectively penetrate the blood-brain barrier, which has garnered excitement over its potential to treat tumors of the central nervous system (CNS), namely glioblastoma. Initial clinical studies have revealed reasonable activity against brain tumors with tolerable side effects, including, but not limited to, fatigue, nausea, headache, and hallucinatory effects in certain individuals (Niewerth et al., 2014; Chinnathambi et al., 2023).

8.2 Arenimycin

Arenimycin is a quinone antibiotic from the marine-derived actinomycete *Salinispora arenicola* strain CNR-647, which belongs to the family of benzo [α] naphthacenes. Arenimycin displays significant antimicrobial activity against Gram-positive infections resistant to multiple drugs, including methicillin-resistant and rifampin-resistant strains of *Staphylococcus aureus*, as seen by 0.05–1.06 μg/mL minimum inhibitory concentration (MIC) values (Alam et al., 2022). Arenimycin is the first benzo[α] naphthacene quinone antibiotic isolated from marine actinobacteria, and its action against MRSA makes it a particularly exciting candidate. Arenimycin also demonstrates activity against coagulase-negative *Staphylococci*, *Enterococci*, and some mycobacterial strains, and shows some cytotoxicity towards eukaryotic cells, which may affect its therapeutic window (Asolkar et al., 2010).

8.3 Lomaiviticins A & B

Lomaiviticins A and B are complicated dimeric natural products derived from *Micromonospora* sp. and carry two unusual diazotetrahydrobenzo[b]fluorene (diazofluorene) functional groups. Lomaiviticin A demonstrates remarkable cytotoxicity in cultured human cancer cells at nanomolar to picomolar concentrations, and is therefore regarded as one of the most potent anticancer agents characterized to date. Lomaiviticin A induces DNA double-strand breaks by a radical-mediated mechanism, which involves both diazofluorene residues penetrating the DNA duplex and disrupting base pairing. This unusual mode of DNA binding results in the ejection of central AT bases while delivering both reactive centers proximal to each DNA strand. Lomaiviticin A is currently in preclinical evaluation, particularly for DNA repair-deficient tumor types, which have been described as more sensitive to its effects (He et al., 2001).

8.4 Marinomycins A–D

Marinomycins A–D are uncommon macrodiolides synthesized by *Marinispora* sp. CNQ-140 is a strain that was obtained from marine sediment collected off La Jolla, California. Marinomycins A–D have the structure of dimeric 2-hydroxy-6-alkenyl-benzoic acid lactones with conjugated tetraene-pentahydroxy polyketide chains, thus they represent a new structural class of polyene macrolides (Olano et al., 2009). Marinomycins are active against bacteria and cancer cell lines, with mean LC50 values ranging from 0.2 to 2.7 μM against a panel of 60 different cancer cell lines (from NCI). Marinomycin A (the most studied compound) displays remarkable tissue-type selectivity and is especially active against human melanoma cancer cell lines, with the most sensitive strain (melanoma SK-MEL-5) exhibiting a LC50 value of 5.0 nM (Olano et al., 2009). The large and GC-rich biosynthetic gene clusters, and dimeric structure make marinomycins a challenging target for heterologous expression. As such, a special cloning strategy was developed to facilitate top-down and bioprospecting projects aimed at producing marinomycins using the large GC-rich biosynthetic gene clusters (Kwon et al., 2006).

8.5 Pladienolides

Pladienolides are a type of large, ring-like chemical compound that were first isolated from the marine bacteria *Streptomyces* platensis Mer-11107. They inhibit a component of a key cellular splicing machine, the spliceosome, and specifically inhibit the SAP130 component of the U2 small nuclear ribonucleoprotein. Pladienolides thereby inhibit the function of the spliceosome, leading to the accumulation of unprocessed pre-messenger RNA and preventing cells from growing and dividing. A synthetic pladienolide called E7107 has been developed and has advanced to clinical trials as a cancer therapeutic. E7107 has shown some evidence of pharmacologic effect back to the blood cells in patients, but complete tumor shrinkage was not achieved, though the disease was stable in some patients (Ngema & Madoroba, 2024).

8.6 Lactonamycin

Lactonamycin is a natural product that has both antimicrobial and antitumor activity, which was isolated from marine *Streptomyces*. This compound is a polyketide antibiotic that exhibits dual bioactivity, as it displays antimicrobial activity against a variety of bacterial pathogens and antitumor activity against multiple cancer cell lines. Lactonamycin is a compound in the growing family of structurally related lactones that shows the potential for marine actinobacteria to produce bioactive secondary metabolites with other therapeutic targets. While there are ongoing studies unraveling lactonamycin's mode of action, it has been presented as a significant natural example of a marine-derived polyketide that can potentially treat both infectious disease and cancer (Adachi et al., 2013).

8.7 Griseorhodins

Griseorhodins are anthraquinone-centric polyketides arising from *Streptomyces* strains found in marine sediments. Griseorhodins are active against tumors. Anthraquinone compounds from marine sources are valuable additions to the growing list of biochemically-sourced natural products. Griseorhodins are grouped into the anthraquinone class, many of which demonstrate some level of antitumor activity and the ability to create stable DNA complexes that also produce DNA strand breaks. As these producing strains have a marine source, there may be distinctive structural alterations resulting in differences in bioactivity compared to terrestrial strains and species, as well as differences in altered bioactivity with the development of pathogen resistance, with impacts on the overall bioactivity of these compounds. These compounds also highlight the continued potential of actinobacteria from marine sediments as potential sources of new bioactive entities (Yunt et al., 2009) (Fig. 1 and Table 1).

Marine actinobacteria are an important source of polyketides, the most prominent and largest class of bioactive compounds. Polyketides are key sources of new compounds with a variety of pharmacological activities (Shuikan et al., 2023; Manivasagan et al., 2014). They can be important leads in the search and development of antibiotics, cancer therapeutics, and other therapeutics with a medical need for antibiotic resistance and cancer treatments (Ngema & Madoroba, 2024; Ibrahim et al., 2025). The metabolic capacity of marine actinobacteria will continue to generate new polyketides with potentially beneficial and active pharmaceutical activities, demonstrating the importance of continuing to investigate marine microbial diversity for drug discovery (Chinnathambi et al., 2023; Shuikan et al., 2023).

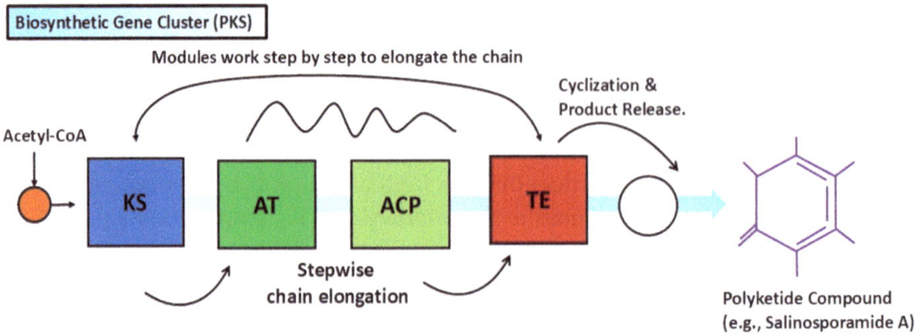

Fig. 1 Simplified polyketide synthase (PKS) pathway showing modular enzyme domains and the production of polyketide compounds in marine actinobacteria

Table 1 Major polyketides from marine actinobacteria

Compound	Source organism	Bioactivity	Special Notes	References
Salinosporamide A (Marizomib)	*Salinispora tropica*	Anticancer (proteasome inhibitor)	In clinical trials for glioblastoma and other cancers	Feling et al. (2003)
Arenimycin	*Salinispora arenicola*	Antibiotic	Effective against multidrug-resistant bacteria (MRSA)	Asolkar et al. (2010)
Lomaiviticins A & B	*Micromonospora* sp.	Anticancer, DNA cleavage	Highly potent, unusual diazofluoren structure	He et al. (2001) and Nicolaou (2004)
Marinomycins A-D	*Marinispora* sp.	Antibiotic, anticancer	Dimetric macrolides with strong bioactivity	Kwon et al. (2006) and Maloney (n.d.)
Pladienolides	*Streptomyces platensis* (marine isolate)	Anticancer (splicing inhibitor)	Basis for clinical drug E7107	Mizui et al. (2004) and Kotake et al. (2007)
Lactonamycin	Marine *Streptomyces* sp.	Antitumor, antimicrobial	Quinone-containing macrolide with strong cytotoxicity	Omura et al. (2001)
Griseorhodins	*Streptomyces* sp. (marine sediments)	Antibacterial, anticancer	Inhibits respiratory chain in pathogens	Maskey et al. (2008)

8.8 Non-ribosomal Peptides (NRPs)

From life-saving penicillin, an antibiotic, to potent immunosuppressants, nature has long been medicine's greatest chemist. One of the most fascinating classes of such natural products is the non-ribosomal peptides (NRPs), a group of microbial molecules with great therapeutic potential. Unlike the proteins and peptides that form the machinery of life, NRPs are not made by ribosomes using an mRNA template. Rather, they are constructed by gigantic enzymatic complexes known as Non-Ribosomal Peptide Synthetases (NRPSs), which work similarly to molecular-scale, programmable production lines. Their unusual biosynthetic origin is the basis of their staggering structural variety (Agrawal et al., 2017). Whereas ribosomal peptides are nearly universally linear chains of the 20 canonical L-amino acids, NRPs resist these limitations. Their topologies can be linear, cyclic, or branched and typically have a huge chemical repertoire of more than 500 distinct building blocks that include D-amino acids (the enantiomers of their natural counterparts), hydroxy fatty acids, and methyl- or hydroxyl-substituted amino acids that provide them with distinct chemical characteristics and elevated stability (Finking & Marahiel, 2004).

This structural versatility allows NRPs to perform an immense number of biological functions, such as useful antibiotics, antifungals, immunosuppressants, and even anti-cancer agents, and is therefore a cornerstone of drug discovery (Beyer et al., 2023).

9 The NRPS Assembly Line: A Molecular Factory

The biosynthesis of NRPs is a biochemical engineering achievement performed by a family of gigantic enzymes known as Non-Ribosomal Peptide Synthetases (NRPSs). They are real "megasynthases" which usually weigh over a million Daltons on the molecular scale. The brilliance of the NRPS system is that it is modular in design. It operates in a manner similar to a very sophisticated factory assembly line, with each module being a specialized workstation that is tasked with the selection, activation, and incorporation of one specific building block into the developing peptide chain (Strieker et al., 2010).

An average elongation module has three critical domains, each with a very specialized function:

Adenylation (A) Domain This is the assembly line's first "gatekeeper." It recognizes specifically one amino acid (or other monomer) from the pool in the cell and activates it by hydrolysis of ATP to produce an aminoacyl-AMP intermediate. The high specificity of the A-domain binding pocket is the main factor that determines what monomer ends up being added at a given position to the final peptide.

Thiolation (T) Domain Also referred to as the Peptidyl Carrier Protein (PCP), this is a "robotic swinging arm". Upon activation, the monomer is covalently linked to a flexible 4′-phosphopantetheine (Ppant) arm attached to the T-domain through a thioester bond. Through this covalent attachment, the T-domain can transfer the substrate from one of the other catalytic domains in the module to the next, preparing it for the next assembly step.

Condensation (C) Domain This domain is the "peptide bond forming" master. It is the catalyst of the central chain elongation chemical reaction. The C-domain choreographs the nucleophilic attack of the amino group of the amino acid attached to its own module's T-domain on the thioester bond of the elongating peptide chain in the grip of the T-domain of the adjacent module. This response lengthens the peptide by one residue and transfers the extended chain to the T-domain of the present module. This activation, tethering, and condensation is reiterated at every module on the NRPS, the peptide chain extending step by step.

Thioesterase (TE) Domain The last release station in this area is the release station. It cleft the peptide maturity of its covalent bond with the last T-domain. The TE-domain is a generic catalyst; it can be hydrolyzed to give a linear peptide or more usually, it is an intramolecular attack (macrolactamization or macrolactoniza-

Table 2 Marine actinobacteria major non-ribosomal proteins

Compound	Producer (Marine actinobacteria)	Type of NRP	Biological activity	References
Marinomycin A–D	*Streptomyces* sp. CNQ-418 (marine sediment)	Cyclic depsipeptide	Potent anti-cancer, anti-MRSA	Solanki et al. (2008)
Salinamides A & B	*Streptomyces* sp. CNB-091 (marine sediment, Bahamas)	Cyclic depsipeptide	Inhibitor of bacterial RNA polymerase	Solanki et al. (2008)
Arenamides A–C	*Salinispora arenicola* (marine sediment)	Cyclic depsipeptide	Anti-inflammatory (NF-κB inhibitors)	Solanki et al. (2008)
Cyclomarins A–F	*Streptomyces* sp. (marine sponge symbiont)	Cyclic heptapeptides	Antimalarial, anti-inflammatory	Finking and Marahiel (2004)
Lajollamides A–E	*Streptomyces nodosus* (marine sediment, La Jolla, USA)	Cyclic peptides	Antibacterial, cytotoxic	Strieker et al. (2010)
Streptochlorin	*Streptomyces* sp. (marine mud)	Halogenated cyclic peptide	Anti-angiogenic, antitumor	Solanki et al. (2008)
Gramicidin S-like peptides	Marine *Brevibacterium* sp.	Cyclic decapeptide	Antibacterial, antifungal	Solanki et al. (2008)
Nocardicin A	Marine *Nocardia* sp.	Monocyclic β-lactam NRP	Antibacterial (β-lactam antibiotic)	Walsh (2004)
Holomycin	Marine *Streptomyces* sp.	Dithiolopyrrolone peptide	Broad-spectrum antibiotic	Newman and Cragg (2020)

tion) to give a stable cyclic NRP, a characteristic common to this type of compound (Walsh, 2004) (Table 2).

10 Pharmacopeia of Biological Activities

The molecular key role of NRPs is the result of its structural complexity and uniqueness, which allows it to fit into the locks of key machinery within cells. The given specificity allows them to be powerful medicines that can target pathogens or remodel human biology.

Penicillin This is the old antibiotic, a 0-lactam NRP of the fungus Penicillium rubens. The strength of this specific drug is in the fact that it attacks the cell wall of bacteria, which is a hard outer shell that the bacteria cannot survive without. The second specific mechanism of action is penicillin, which inactivates the work of an enzyme called transpeptidase, which acts as a builder and binds the building material of the wall (peptidoglycan) in a cross-linkage. In the absence of this enzyme the cell wall becomes weaker and the bacterium bursts due to its own pressures (Iacovelli et al., 2021).

Vancomycin In case of bacterial resistance, we have the so-called last-resort-antibiotics as vancomycin. It is a large glycopeptide NRP that also targets cell wall assembly in a more devious way. Instead of interacting with the enzyme, it interacts with the building blocks themselves, i.e., the D-Ala-D-Ala tails of the peptidoglycan precursors. It closes these ends, thereby ensuring the construction workers (transpeptidase and transglycosylase enzymes) do not physically access the materials and stop the construction of walls. This will render it an important weapon against infections like MRSA.

11 Antifungal Agents

NRPs are also an effective source of antifungal compounds as they fight bacteria. Fungal cells are much more similar to human cells as compared to bacterial cells, hence are much harder to target. Fungi do possess a unique cell wall; however, like bacteria, that is a very good drug target (Le Govic et al., 2019).

Echinocandins This type of NRPs is a type of lipopeptide (caspofungin, a drug) and is an emerging treatment of invasive fungal infections. They do so by very specifically blocking an enzyme called 8—(1,3)-glucan synthase. It is an enzyme that is involved in the synthesis of glucan, which is the major structural polymer of the fungal cell walls. Echinocandins disrupt the structural integrity of the fungal cell by blocking the synthesis of glucan and thus lyse and die. The drug is highly effective and has no side effects since it is used without the need of a cell wall and without the presence of this enzyme in human cells.

12 Immunosuppressive Agents

The contribution of NRPs is even greater than the fight against microbes because it is thanks to them that we now have a powerful immune system that we can control. Organ transplantation and treatment of autoimmune diseases cannot be done without such a facility.

Cyclosporin A This is a metabolite of a fungus, Tolypocladium inflatum, that is a cyclic peptide that single-handedly transformed organ transplantation, turning it into a life-saving operation rather than an experimental procedure. It is brilliant in the sense that it can suppress the immune response to reject the organ. The access of cyclosporin A to the T-cells, the leaders of the immune force, is blocked, and the most important signaling protein, calcineurin. The calcineurin inhibition prevents the activation and the attack of the T-cells on the foreign organ transplanted in order to enable it to live in the new host.

13 Alkaloids

Alkaloids are naturally occurring organic compounds containing at least one atom of nitrogen. They are mostly found in plants, animals, fungi, and bacteria. The Alkaloid name originated from the Arabic word al-qali and the Greek suffixoid, which basically means "alkali-like," indicating basic properties of these compounds that come from the nitrogen atom present in them (Dey et al., 2020). Alkaloids derived from marine actinobacteria have been proven to possess various therapeutic properties like anti-tumor, anti-inflammatory, and anti-bacterial properties, which is why a lot of research focused on alkaloids is being done (De Rop et al., 2022). Hordenine was the first alkaloid to be isolated from marine sources (Pudi et al., 2016). The structure of alkaloids varies greatly, which is why a single classification system can't be used for it.

14 Classification of Alkaloids

Alkaloids can be classified based on biogenesis as true alkaloids, protoalkaloids, and pseudoalkaloids.

True alkaloids: True alkaloids are obtained from amino acids and share a heterocyclic ring containing nitrogen. They have high reactivity and potent biological activity. They are known for forming water-soluble salts, and a lot of them are crystalline in nature, so they conjugate with acids to form a salt. Most of them are solid and bitter, except nicotine (brown liquid). True alkaloids can be found in three forms: free state, as N-oxide, and as salts. Some examples include cocaine, morphine, and quinine (Dey et al., 2020).

Protoalkaloids: Protoalkaloids are derived from an amino acid without being part of the heterocyclic ring system. The main precursors of protoalkaloids are L-Tryptophan and L-tyrosine. They can be used for various medical disorders. Some examples include yohimbine, mescaline, and hordenine (Dey et al., 2020).

Pseudoalkaloids: Pseudoalkaloids are derived from amination or transamination reaction from the precursors or postcursors of amino acids instead of directly being derived from amino acids. They can be derived from non-amino acid precursors or from phenylalanine or acetate. Some examples are ephedrine, caffeine, and capsaicin (Dey et al., 2020) (Fig. 2).

Based on ring structure, alkaloids can be classified as tropane, pyrrolizidine, piperidine, pyridine, quinoline, isoquinoline, indole, steroidal, imidazole, purine, and pyrrolidine alkaloids.

Tropane alkaloids are distinguished by a tropane bicyclic ring that is generated from ornithine. They are mostly found in the Solanaceae family. Atropine, cocaine, and scopolamine are a few examples (Dey et al., 2020).

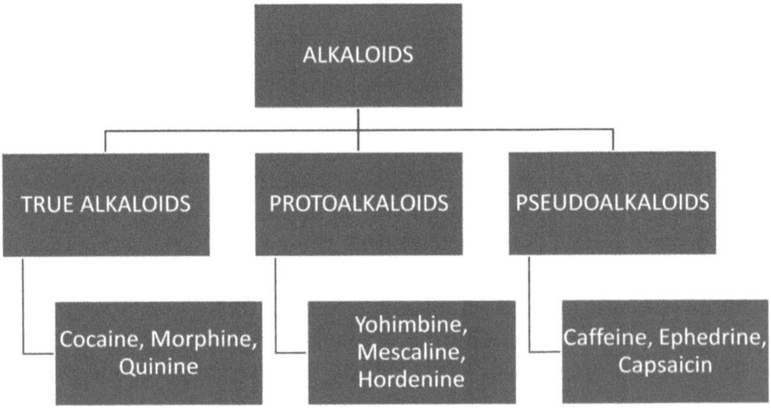

Fig. 2 Classification of alkaloids based on biogenesis

Pyrrolidine alkaloids are compounds that have a five-membered pyrrolidine ring. Many of them are isolated from plants. Examples are hygrine and cuscohygrine (Dey et al., 2020).

Pyrrolizidine alkaloids: These compounds are toxic and consist of fused rings with five and six members (pyrrolizidine nucleus). They are found in the Asteraceae and Fabaceae families. Senecionine is an example (Dey et al., 2020).

Piperidine alkaloids: Coniine and lobeline are important alkaloids that are based on a six-membered piperidine ring. They are derived from piperidine, which is already a lysine-derived alkaloid, but some are also derived from acetate and acetoacetate. They exhibit chronic neurotoxicity and have a characteristic odor (Dey et al., 2020).

Isoquinoline alkaloids: Alkaloids with an isoquinoline ring structure are known as isoquinoline alkaloids. They have various medicinal properties like antifungal, anticancer, antiviral, antispasmodic, antioxidant, and an enzyme inhibitor. They include papaverine, codeine, and morphine (Dey et al., 2020).

Quinoline alkaloids are based on the quinoline skeleton, which is benzene joined to pyridine. These include cinchonine, cinchonidine, quinine, and quinidine (Dey et al., 2020).

Indole alkaloids: This type of alkaloids are identified by an indole ring, which is benzene fused to pyrrole. Polyhalogenation is commonly seen in this type of alkaloid. Examples of indole alkaloids include ergot, strychnine, and reserpine (Dey et al., 2020).

Purine alkaloids: They are made up of a purine ring. These include theobromine, theophylline, and caffeine (Dey et al., 2020).

Steroidal alkaloids: Solanine and aconitine are well-researched; they have a steroid backbone joined to nitrogen. They are typically derived from plant species like Liliaceae, Apocynaceae, Solanaceae, and Buxaceae families, while some are also isolated from amphibians (Dey et al., 2020).

15 Therapeutic Applications of Alkaloids

15.1 Antibacterial Activity

Methicillin-resistant *Staphylococcus aureus*, or MRSA, is a deadly superbug that has developed resistance to most conventional antibiotics. Researchers have discovered that these alkaloids can efficiently battle MRSA.

These alkaloids are distinctive because of the way they combat bacteria. Marine actinomycete alkaloids frequently employ unique mechanisms, in contrast to many conventional antibiotics that bacteria have learnt to resist. Some function by breaking down the bacterial cell wall, while others obstruct vital cellular functions that are necessary for the survival and growth of bacteria. The MIC (Minimum Inhibitory Concentration), which is basically the minimal amount required to stop bacterial growth, is frequently used to determine effectiveness. Numerous marine alkaloids exhibit remarkably low minimum inhibitory concentrations (MICs), indicating their potency even at low dosages (De Rop et al., 2022).

15.2 Anti-cancer Property

Alkaloids combat a wide range of cancer cell types, such as blood, lung, and breast cancers. Many of them target the growth and division of the tumor cell. Certain alkaloids disrupt microscopic internal structures essential for cell division, known as microtubules. Cancer cells are unable to finish dividing when these microtubules are disturbed, so they die. This is especially important since cancer cells are more susceptible to this kind of attack because they divide far more quickly than healthy cells.

Studies have demonstrated that at extremely low concentrations, several alkaloids derived from marine actinomycetes can efficiently destroy tumor cells. For instance, when tested against breast cancer cell lines, several chemicals exhibit IC50 values (the concentration required to kill 50% of cancer cells) as low as 10–22 μg/mL. Since scientists often search for molecules with IC50 values below 30 μg/mL as prospective cancer medicines, this level of potency is regarded as extremely promising for drug development (Ravikumar et al., 2012). They may therefore be able to eradicate cancer cells with minimal harm to healthy, normal tissues, which is a significant advantage over conventional chemotherapy medications that frequently impact both diseased and healthy cells (De Rop et al., 2022).

15.3 Antifungal and Antiviral Activity

Marine actinomycete alkaloids exhibit strong antiviral and antifungal activity. Considering the increased prevalence of drug-resistant fungal infections, particularly in individuals with impaired immune systems, this is very important. Certain alkaloids have demonstrated efficacy against harmful fungus and Candida species, which cause annoying yeast infections.

Compared to antibacterial and anticancer actions, the antiviral capabilities have been explored less, but they nevertheless hold potential for future advancement. These substances may function by blocking the entry of viruses into host cells or by interfering with the mechanisms involved in viral replication (De Rop et al., 2022).

15.4 Anti-inflammatory and Neuroprotective Effects

Marine actinomycetes contain alkaloids with anti-inflammatory qualities that may be useful in the treatment of several inflammatory ailments (Liu et al., 2024). Additionally, several substances have demonstrated neuroprotective properties, which may provide novel treatments for neurodegenerative illnesses like Alzheimer's disease.

For example, it has been discovered that certain indolizidine alkaloids inhibit ROCK2 protein kinase, which is implicated in neurodegenerative diseases. This implies that these substances may aid in preventing damage to nerve cells and may even halt the course of illnesses affecting the brain and nervous system (De Rop et al., 2022) (Table 3).

15.5 Terpenes

Terpenes are the greatest class of natural products, biosynthesized from basic five-carbon isoprenoid building blocks (isopentenyl diphosphate and dimethylallyl diphosphate). Rare sesterterpenes, meroterpenes, and innovative hybrid compounds are among the many structurally diverse and bioactive terpene derivatives found in marine actinobacteria, particularly in the genera *Streptomyces*, *Salinispora*, and *Micromonospora*. Hemiterpenes (1 unit, C5), monoterpenes (2 units, C10), sesquiterpenes (3 units, C15), diterpenes (4 units, C20), sesterterpenes (5 units, C25), triterpenes (6 units, C30), tetraterpenes (8 units, C40), and polyterpenes (greater than 8 units) are hydrocarbons made up of isoprene units (C5H8). The two main pathways for biosynthesis are the cytosolic mevalonate (MVA) pathway and the plastidal methylerythritol phosphate (MEP) pathway. Both pathways produce the precursors isopentenyl diphosphate (IPP) and dimethylallyl diphosphate (DMAPP). Enzymes like geranyl pyrophosphate synthase (GPPS) for monoterpenes, farnesyl pyrophosphate synthase (FPPS) for sesquiterpenes, and geranylgeranyl pyrophosphate synthase (GGPPS) for diterpenes control the assembly of these precursors to extend chains. Structural changes also arise through cyclic formation, rearrangement, and modification from terpene synthases (TPS); TPS families vary according to species (e.g., TPS a for angiosperm sesquiterpenes, TPS d for gymnosperm sesquiterpenes). Terpenes may happen to possess antimicrobial, anticancer, anti-inflammatory, and antioxidant activities. For example, monoterpenes such as limonene exhibit antitumor activity, whereas sesquiterpenes such as artemisinin

Table 3 Examples of marine alkaloids

Compound	Source (Genus)	Structure/class	Biological activity	References
Lodopyridone	Marine Streptomyces	Pyridine alkaloid	Broad-spectrum antibacterial, cytotoxic	Pudi et al. (2016)
Indolocarbazoles	Marine Actinomycete	Indolocarbazole	Apoptotic, anticancer	Pudi et al. (2016)
Aporphine SSV	Marine Actinomycete	Aporphine alkaloid	Antimicrobial, anticancer	Alharbi (2016)
β-carbolines	Nocardia, others	β-carboline	Antibacterial, neuroactive	De La Hoz-Romo et al. (2025)

show antimalarial activity. Due to exclusive biosynthetic gene clusters, terpenes in marine actinobacteria are likely new, both encouraging ecological adaptation and economic interest (Selim et al., 2021; Shirai et al., 2010; Kim et al., 2021).

15.6 Sesterterpenes: Structure, Biosynthesis, and Activities

Sesterterpenes constitute the occasional subclass of terpenes with five isoprene units (C25), commonly with complicated cyclic or linear structures, and are mostly isolated from marine organisms, fungi, and certain actinobacteria. Their biosynthesis typically entails farnesyl geranyl pyrophosphate synthase (FGPPS) to condense precursors into farnesylgeranyl pyrophosphate (FGPP) that is cyclized by TPS; thereafter, in marine actinobacteria such as Streptomyces, by way of genome mining, specific clusters for the said compounds are revealed. Some of these include manoalide (linear sesterterpene from marine sponges but similar in actinobacterial pathways) that contains anti-inflammatory effects through phospholipase A2 inhibition, and the ophiobolins (cyclic, obtained from fungi and actinobacteria) that exhibit anti-T cell activity. In marine actinobacteria like *Streptomyces* MAR4 strains, sesterterpenes like stellatic acid, which are usually elicited by halogenation or hybridization, exhibit antioxidant and antidiabetic properties. Pharmacologically speaking, they have a wide range of properties, such as antifungal, antibacterial, and antitumor properties; for instance, derivatives of geranylfarnesol suppress inflammation and bacterial growth. Marine actinobacteria are important sources of novel sesterterpenes, for which genetic engineering research is used to increase yield (De La Hoz-Romo et al., 2022; Iobbi et al., 2025).

15.7 Sesterterpenes from Marine Actinobacteria

Sesterterpenes are C25 terpenoids representing one of the most chemically diverse and biologically promising subgroups of terpenoids. Marine actinobacteria produce several unique sesterterpenes with unprecedented structural features.

15.8 Marinoterpins A–C

Marine-derived *Streptomyces* species (CNQ-253 and AJS-327) produce marinoterpins A–C, which represent rare linear merosesterterpenoids. These compounds feature a unique combination of quinoline-N-oxides with linear sesterterpenoid side chains, along with rare bis-tetrahydrofuran ring motifs derived from cyclization of methyl group carbons to adjacent hydroxy-bearing methylene groups (Kim et al., 2021).

15.9 Salinosporamides

The salinosporamides, which are produced by *Salinispora tropica*, are the most clinically significant sesterterpenes from marine actinobacteria. Salinosporamide A (NPI-0052), a powerful inhibitor of the 20S proteasome that is presently undergoing Phase I clinical trials for the treatment of cancer, has a rare bicyclic β-lactone-γ-lactam ring system. This substance shows highly selective tumor cell growth inhibition and IC50 values of less than 2 ng/mL against HCT-116 human colon carcinoma cells (Reed et al., 2007).

16 Meroterpenes: Hybrid Nature and Pharmacological Potential

Meroterpenes are natural product hybrids of terpenoid moieties with non-terpenoid structures such as polyketides or quinones, to afford diversified scaffolds. Structurally divergent meroterpenes are constructed by marine actinobacteria by unique biosynthetic pathways. Prenyltransferases in marine actinobacteria, such as *Streptomyces* CNQ-253 or MAR4, attach terpene chains to aromatic peripheries to produce linear or cyclic hybrid structures. Biosynthesis usually consists of the MVA/MEP pathways in conjunction with polyketide synthases (PKS) or other proteins. The marinoterpins A–C from *Streptomyces*, which consist of quinoline-N-oxide and bis-tetrahydrofuran subunits in a linear sesterterpenoid chain derived from anthranilate-CoA ligase and tailoring genes, are just a few examples. The products exhibit strong cytotoxicity to the growth of cancer cells or antimicrobial activity against bacterial membranes. Because of their distinct halogenated structures, other meroterpenes, like *Streptomyces napyradiomycins*, exhibit antibacterial activity against MRSA. For higher production engineering, genome scanning identifies biosynthetic clusters (like the mrt cluster). These are produced by actinobacteria in response to stress in marine environments, providing us with novel pharmacological agents with antiviral or anti-inflammatory properties. They have the potential to be used in drug discovery because of their structural novelty, and because of their hybrid interactions, they typically have higher activities than single terpenes (Sweeney et al., 2024; Wang et al., 2024; Shirai et al., 2010).

17 Some Examples Are

17.1 Atolypene A

Atolypene A represents a bacterial meroterpenoid with a perhydrophenanthrene skeleton featuring unique methyl substitution patterns. Its biosynthesis follows a distinct sequence of epoxidation, prenylation, and cyclization, divergent from fungal meroterpenoid pathways. The compound exhibits moderate cytotoxicity against multiple human cancer cell lines (Wang et al., 2024).

17.2 Napyradiomycins and Related Compounds

MAR4 *Streptomyces* species produce napyradiomycins A–D, which are THN-based naphthoquinone meroterpenoids with demonstrated antibacterial and antiproliferative activities. These compounds utilize vanadium-dependent haloperoxidases (VHPOs) for their unique cyclization reactions (Sweeney et al., 2024).

17.3 Naphterpin Derivatives

Streptomyces sp. CNQ-509 produces naphterpin derivatives comprising naphthoquinone and geranyl moieties. These compounds exhibit radical scavenging activities and represent examples of hybrid isoprenoid natural products with potential therapeutic applications (Park & Kwon, 2018).

17.4 Flavonoids: Biosynthesis and Roles in Marine Actinobacteria

Flavonoids are polyphenolic secondary metabolites typically associated with plants but increasingly found in marine actinobacteria, from engineered or native biosynthetic pathways. Flavonoids are yielded by these microorganisms by classical fermentation, plant biosynthetic gene engineering, and de novo biosynthesis technologies. The categorization comprises flavones, flavonols, isoflavones, and anthocyanins based upon a C6-C3-C6 backbone derived by the phenylpropanoid pathway. In Streptomyces and other actinobacteria, chalcone synthase (CHS) catalyzes the union of malonyl-CoA and p-coumaroyl-CoA to yield chalcone, the precursor cyclized into the flavonoids; through genome sequencing, CHS-like genes enable that potential in non-plant microbes. Marine-derived strains, for instance, from the deep-sea terrains, give unique flavonoids like genistein analogues with stronger antioxidant and antimicrobial activities. Biologically, they exhibit anticancer (e.g., inducing apoptosis), antibacterial (membrane disruption), and anti-inflammatory actions; for instance, the actinobacterial flavonoids inhibit MRSA and exhibit stronger synthesis without a precursor over terrestrial microbes. Environmental regulation, for instance, through OSMAC methodology, boosts synthesis to indicate marine actinobacteria as under-exploited sources for drugs based upon flavonoids (Hozzein et al., 2021).

17.5 Macrolides: Structural Diversity and Therapeutic Applications

Macrolides are large macrocyclic lactone-containing molecules, typically 12–16–membered rings containing sugar moieties, and are produced abundantly by marine actinobacteria as antibiotics. Biosynthesis is of type I polyketide synthases (PKS-I) assembling polyketide chains from acetate/malonate units, followed by cyclization and glycosylation; in genera like *Streptomyces* or *Actinomadura*, gene clusters specify the enzymes to yield structurally rare variants. Some of these are erythromycin (antibacterial by ribosome inhibition) and salinomycin (anticancer ionophore) with marine isolates producing rare analogues like kebanmycins A–D from *Streptomyces* SCSIO 40068 showing potent anti-MRSA activity (MIC 0.125 μg/mL). Their pharmacologically distinct actions are bacterial protein synthesis, fungal growth, and cancer cells; novel marine macrolides possess antiprotozoal, insecticidal, and antiviral activities, often with reduced resistance profiles. Marine actinobacteria, under the pressure of harsh conditions, produce hybrid/or halogenated macrolides to enhance efficacy against anti-drug-resistant pathogen cultures. Sporolides A and B from *Salinispora* tropica are novel polyketide-derived macrolides of unprecedented structures. Whereas sporolide A shows unexpected biological activity, sporolide B shows predictable HIV-1 reverse transcriptase inhibition activity (De La Hoz-Romo et al., 2022; Alharbi, 2016; Dineshkumar et al., 2013).

17.6 Hybrid Molecules: Classification and Biological Effects

Marine hybrid molecules of actinobacteria combine structural elements of disparate biosynthetic pathways, for example, terpenoids with polyketides, alkaloids, or peptides to yield novel chimeras. Types encompass meroterpenoids (terpene-polyketide fusions) and hybrid isoprenoids (halogenated terpenes), including polyketide-nonribosomal peptide fusions. Biosynthesis entails combining separate gene clusters; for instance, in *Streptomyces* MAR4, sesterterpene synthases are combined by PKS or halogenases to yield products including indanopyrroles A–B containing pyrroloketoindane skeletons. These possess wide activity, such as indanopyrrole A, which has MIC values of 2–4 μg/mL for MRSA and VRE. Other fusions, for example, the marine *Streptomyces*-derived glycoabyssomicin A, enlist sugar groups for potential augmentation of antibacterial activity, although variants possess none at sub-threshold doses. Biologically, they possess anticancer, antimicrobial, and anti-inflammatory activity, and marine adaptation provides novel pharmacology (Park & Kwon, 2018; De La Hoz-Romo et al., 2022).

17.7 Marine Actinobacteria as Producers of Unusual and Novel Compounds

Marine actinobacteria, comprising novel-rare genera such as *Streptomyces*, *Nocardiopsis*, and *Salinispora*, are prolific producers of novel metabolites owing to their survival in harsh oceanic environments, resulting in novel biosynthetic gene clusters. They account for producing more than 70% of known microbial antibiotics, comprising terpenes, flavonoids, macrolides, and hybrid-type antibiotics, often of atypical structures such as halogenated or glycosylated analogues that are not found in terrestrial counterparts. An example is the finding of halogenated pyrroloketoindanes with broad-spectrum activity through genome mining in *Streptomyces* sp. CNY-716. The compounds solve the problem of antibiotic resistance, with activities against MRSA, VRE, and cancer cells; techniques such as OSMAC and genetic manipulation boost yields (Sweeney et al., 2024).

Marine actinobacteria demonstrate a remarkable capacity for producing unusual and novel compounds through several mechanisms:

1. *Environmental Adaptation:* The complex marine environment with extreme variations in pressure, salinity, light, and temperature influences their diverse chemical production (Miller et al., 2022).
2. *Genetic Diversity:* Marine actinobacteria harbor 20–50 biosynthetic gene clusters, allocating 0.8–3.0 Mb of their genomes to secondary metabolite production (Xie & Pathom-Aree, 2021).
3. *Co-cultivation Effects:* Bacterial interactions activate silent biosynthetic gene clusters and enhance production of compounds not observed in monoculture (Lee et al., 2020).
4. *Species-Specific Production:* Different Salinispora species produce distinct classes of secondary metabolites, with S. tropica producing salinosporamides and S. arenicola producing staurosporines (Jensen et al., 2006).

Key intermediates and enzymatic conversions are displayed in the biosynthesis pathways of various terpene classes involving endoplasmic reticulum and plastid enzymes (refer to Fig. 3). The complex terpene biosynthesis pathways of marine actinobacteria, which incorporate several enzymatic processes and regulatory systems to generate structurally varied molecules, serve as an example of the biosynthetic complexity of these organisms (Table 4).

18 Applications

18.1 Medical Applications

1. Antibiotics: The World Health Organization (WHO) has added the emergence of multidrug-resistant (MDR) bacterial pathogens to the list of most severe threats to human health and food security. The continuous dilemma renders the discovery of new antibiotics a pressing priority, since new compounds are critical in fighting resistant bacterial strains. Actinomycetes, in fact, continue to be one of the brightest areas of bioactive metabolites, with a broad diversity of antibiotics that have clinical significance. The antibiotics are grouped into a few categories that include β-lactams (such as penicillins, cephalosporins, carbapenems, monobactams, and PBP inhibitors), ansamycins, macrolides, tetracyclines, aminoglycosides, and peptide antibiotics. Actinomycetes have long been a major source of drug discovery. The antibiotics rifampicin and cycloserine, for instance, both come from actinomycetes and are prescribed to treat infections caused by Mycobacterium tuberculosis. Respiratory infections caused by Legionnaires'

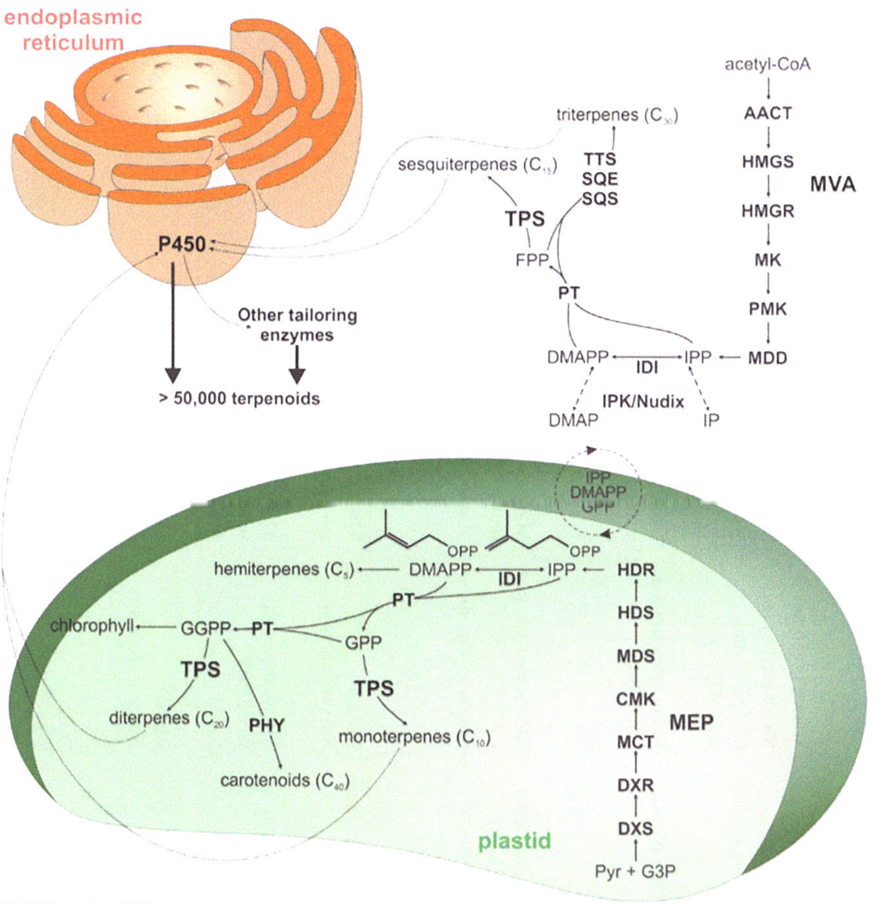

Fig. 3 Biosynthetic pathways of diverse terpene classes involving endoplasmic reticulum and plastid enzymes, showing key intermediates and enzymatic conversions (Karunanithi & Zerbe, 2019)

disease are often treated with erythromycin. Erythromycin is formed by Saccharopolyspora erythraea, an actinomycete member. Tetracycline is derived from *Streptomyces aureofaciens* and acts on bacterial ribosomes. Daptomycin, which is obtained from *Streptomyces roseosporus*, is employed to cure vancomycin-resistant Staphylococcus aureus (VRSA/MRSA) infections. Chloramphenicol, which is obtained from *Streptomyces* venezuelae, gentamicin, which is obtained from Micromonospora purpurea, and monensin, the antibiotic obtained from *Streptomyces cinnamonensis*, are some other examples that are popular in their field. All of them exhibit broad-spectrum activity against bacteria. Again, the examples given show the profound impact secondary metabolites, which originated from actinomycetes, had on the treatment of infectious diseases (Ngamcharungchit et al., 2023).

Table 4 Major terpenes from actinobacteria

Compound category	Compound name	Structure type	Source organism	Source location	Biological activity	References
Sesterterpenes	Marinoterpins A-C	Merosesterterpenoids (quinoline-N-oxides + linear sesterterpenoid)	Streptomyces sp. CNQ-253 and AJS-327	Marine sediment	Novel chemical motifs with rare tetrahydrofuran functionalities	Kim et al. (2021)
Sesterterpenes	Salinosporamide A (NPI-0052)	Î^2-lactone-Î^3-lactam bicyclic core	Salinispora tropica CNB-392	Ocean sediment	Potent 20S proteasome inhibitor, anticancer agent in Phase I clinical trials	Reed et al. (2007) and Fenical et al. (2008)
Sesterterpenes	Salinosporamide K	Bicyclic ring system (lacks chloro-ethyl side chain)	Salinispora pacifica CNT-133	Marine sediment	Proteasome inhibitor activity	Freel et al. (2011)
Meroterpenes	Atolypene A	Perhydrophenanthrene meroterpenoid	Streptomyces sp.	Marine environment	Moderate cytotoxicity against human cancer cell lines	Wang et al. (2024)
Meroterpenes	Napyradiomycins A-D	THN-based naphthoquinone meroterpenoids	MAR4 Streptomyces sp.	Marine environment	Antibacterial, antiproliferative activities	Sweeney et al. (2024)
Meroterpenes	Naphterpin derivatives	Naphthoquinone-geranyl meroterpenoids	Streptomyces sp. CNQ-509	Marine sediment, La Jolla, California	Radical scavenging activities	Park and Kwon (2018)
Meroterpenes	Marinones	THN-farnesyl meroterpenoids	MAR4 Streptomyces sp.	Marine environment	Antibiotic properties	Sweeney et al. (2024)

Flavonoids	Marine actinobacterial flavonoids	Polyphenolic compounds	Marine actinobacteria (various genera)	Marine environments	Anticancer, antimicrobial, antioxidant, and anti-inflammatory activities	Hozzein et al. (2021)
Macrolides	Sporolide A	Polyketide-derived macrolide	Salinispora tropica	Marine sediment	Unknown biological activity (novel structure)	Jensen et al. (2006) and Dineshkumar et al. (2013)
Macrolides	Sporolide B	Polycyclic macrolide	Salinispora tropica	Marine sediment	HIV-1 reverse transcriptase inhibition (predicted)	Dineshkumar et al. (2013)
Hybrid molecules	Marinophenazines	Phenazine-terpenoid hybrid	Streptomyces sp. CNQ-509	Marine sediment	Novel chemical scaffold	Park and Kwon (2018)
Hybrid molecules	Phenazine derivatives	Dimethyl phenazine-1,6-dicarboxylate, phencomycin	Micromonospora sp. UR56, Actinokinespora sp. EG49	Red marine sponge	Antibacterial and antibiofilm activity against S. aureus, P. aeruginosa	De La Hoz-Romo et al. (2022)
Hybrid molecules	Nitropyrrolins	Farnesyl-Î ± -nitropyrrole hybrid	Streptomyces sp. CNQ-509	Marine sediment	Novel farnesyl-nitropyrrole structure	Park and Kwon (2018)
Alkaloids	Staurosporines	Indolocarbazole alkaloids	Salinispora arenicola, S. pacifica	Marine sediment	Protein kinase inhibitors	Freel et al. (2011)
Other terpenes	Î^{2}-Eudesmol, Î^{2}-cedrene	Sesquiterpenes	Streptomyces hygroscopicus NRRL 15879	Marine-derived	Volatile bioactive compounds	Tarasova et al. (2023)
Other terpenes	5-Dimethylallylindole-3-carboxylic acid	Terpenoid-indole derivative	Streptomyces sp. MS239	Marine-derived	Novel terpenoid structure	Motohashi et al. (2008)

2. Anticancer drugs: In the search for new cancer-fighting drugs, scientists are increasingly turning to natural sources, especially tiny marine organisms called marine actinomycetes. These special bacteria are exciting because they produce a huge variety of natural chemicals that can fight tumors. Between 2007 and 2017, certain types of these bacteria, like Micromonospora, Salinispora, and Verrucosispora, were found to create many new compounds that show strong anti-cancer effects. These compounds usually belong to four main groups: polyketides, alkaloids, peptides, and quinones. One of the reasons these compounds are important is that they often cause fewer side effects than traditional chemotherapy drugs. For example, salinosporamide A is a promising anticancer compound that highlights this benefit.

   Some of these compounds are already used as cancer treatments. For instance, Adriamycin (also known as doxorubicin) is derived from the bacterium Streptomyces peucetius and works by inhibiting the replication of cancer cells' DNA, thereby preventing growth. Other well-known anti-cancer drugs from *Streptomyces* include actinomycin D, bleomycin, daunorubicin (an anthracycline), and mitomycin C. These medicines have been essential in treating various cancers and show how Marine actinomycetes have made a big impact on modern cancer therapy. In addition to these previously discussed drugs, marine actinomycetes continue to show a remarkable range of structurally distinctive compounds with antitumor activity. While these compounds have already undergone testing against different cancer cell lines and showed that there may be various types of cancers that they could potentially assist in fighting, major examples include streptochlorin, actinofuranones, aureoverticillactam, cyanosporasides, komodoquinones, thiocoraline, and marizomib.

   In a more recent discovery, *Streptomyces* from Nigeria yielded a couple of new substances: ULDF4 and ULDF5. These compounds showed cytotoxic activity against several human cancers, highlighting the immense potential of marine actinomycetes in the search for new cancer treatments, including blood cancers (acute myelocytic leukemia), cervical cancer, stomach cancer, and breast cancer. Structurally, these compounds appear like staurosporine and kigamicin, which are both well-characterized molecules associated with programmed cell death (apoptosis) and necrosis, respectively. Additionally, ketomycin, a marine compound, inhibits the migration and invasion of breast carcinoma cells by targeting the NF-κB signaling pathway relevant to tumor progression. The dual use of ketomycin as both a marine compound and an antibacterial compound, coupled with a relatively simple chemical structure, makes it a viable option for future drug development (Jagannathan et al., 2021; Ngamcharungchit et al., 2023)
3. Antivirals: Viruses such as influenza, SARS, MERS, and now COVID-19 have demonstrated how quickly an infection can spread and the severity of its impacts. COVID-19, especially, demonstrated how unprepared the world can be for new viral pathogens. This is why it is important to discover new, effective medicines to treat viral infections. Historically, nature (specifically microbes) has been one of the best-performing teachers of drug discovery. Natural product-based medicines approved by the FDA can be found in nearly half of all approved medi-

cines, and this medicinal benefit can also apply to antiviral drugs. Actinomycetes, a group of soil- and marine-living bacteria, have been one of the more valuable sources of drug discovery. Vidarabine (Ara-A), an antiviral agent, was one of the first antiviral drugs discovered and was derived from the actinomycetes known as *Streptomyces* antibioticus. Since then, many more antiviral compounds have been sourced from actinomycetes. For example, a structure related to Antimycin A from the strain *Streptomyces kaviengensis* was effective against the western equine encephalitis virus as an antiviral by disrupting energy and biosynthetic building blocks required by the virus. Additionally, the same actinomycete S. citricolor produces the antiviral aristeromycin that is an effective antiviral against influenza A, but has not been studied as a potential antiviral treatment for COVID-19. Further, another compound, pISC-2002, came from an actinobacterium isolate, *Streptomyces* chromofuscus, and also exhibited full antiviral activity against influenza (Kumar et al., 2024, Jakubiec-Krzesniak et al., 2018).

18.2 Industrial Applications

1. Biofuels: Actinomycetes are naturally adapted to degrade complex organic matter. Many actinomycete species, particularly species found in soil, are important drivers of carbon cycling as they degrade hydrocarbons and plant material. Their degradation abilities make them promising subjects for producing biofuels. Plant biomass degradation will yield simple sugars, the raw material for renewable energy in the form of bioethanol or biodiesel. Several actinomycetes, including Streptomyces, Rhodococcus, Nocardia, Corynebacterium, and Mycobacterium, have enzymatic systems capable of breaking down carbohydrates into simple sugars efficiently. Sugars can then be converted into energy-dense compounds that can serve as replacements for fossil fuels. *Streptomyces* has garnered interest due to its demonstrated ability to convert cellulose and other tough structural plant materials, making large-scale and renewable fuel production realistic. While most studies up to now have emphasized terrestrial species, marine actinomycetes are beginning to gain attention. Initial studies indicated that some marine strains produce carbohydrate-degrading enzymes such as cellulase and xylanase, both of which could be beneficial for the degradation of marine plant biomass for fuel production. This presents an exciting new aspect of biofuel production in marine environments that may yield unique advantages from marine actinomycete (Islam et al., 2025; Bharathi & Lee, 2024).
2. Food industry: Enzymes, which are industrially significant, like amylases, proteases, as well as cellulases, are already being used in food and beverage processing, like ameliorating bread structure, maintaining juice clarity, aiding brewing, and even participating in biofuel production. Marine enzymes are truly unique because they can be useful in destructive conditions, such as extreme salt content of water or extreme pH levels. Examples of such enzymes are lipases and chitinases, which are useful in our day-to-day lives. Lipases are used in producing biodiesel fuel, in making baked goods stronger by making the dough, and add

flavor to the cheese. On the other hand, chitinases offer ecologically friendly ways of preventing diseases in plants and convert the waste generated by seafood into consumer health products. In addition to the ability to make enzymes, marine actinomycetes can synthesize biosurfactants, natural pigments, and antioxidants, which are extremely sought-after in modern industries. Biosurfactants make food stability, cleaning of products, and even cleaning of oil spills possible. They are a more environmentally friendly alternative to hard chemical emulsifiers and detergents (Bruno et al., 2019).

18.3 Environmental Applications

1. Pesticide manufacturing: Marine actinomycetes are potentially good sources of natural pesticides. They do manufacture a variety of compounds that have insecticidal, antifungal, and antibacterial properties, which can protect crops against diseases and pests. They are particularly useful in the sense that they do not harm the environment as chemical pesticides do. The marine actinomycote pesticides assist in reducing chemical congealed on the soil or environment, enhance the health of the soil, and preventing possible resistance by pests. Certain of these products of marine actinomycetes prevent the proliferation of plant pathogens or pesticides that have a direct action on insects and are therefore essential in crop protection. As an example, a study carried out in soil of different locations in Mexico has found that different isolates of actinomycetes had a large inhibitory potential of phytopathogenic fungi, Rhizoctonia solani, Phytophthora capsici, or Fusarium oxysporum. Also, inoculated with the actinobacterial isolates, chili plants (Capsicum annuum) grew taller and heavier with fruit than the control, possibly because they produced growth promoters, such as indole acetic acid, gibberellins, or cytokinin-like compounds (Sánchez-García et al., 2019).
2. Bioremediation: Actinobacteria, especially actinomycetes, are also getting hype as natural partners in bioremediation, or the utilization of environmental friendly agents in cleaning of contaminants such as pesticides, heavy metals, and complex contaminations. Conventional ways of eliminating toxic chemicals in soil and water usually prove ineffective, especially with inorganic substances, which disintegrate to form even more toxic byproducts. Actinobacteria, on the other hand, are remarkable in their capacity to restore ecological balance by degrading organic and inorganic pollutants into harmless end products. Most species, including Streptomyces and Mycobacterium, can utilize pesticides as a carbon food source, which aids in the process of purification of polluted areas. As an example, tyrosinase of *Streptomyces espinosus* has been demonstrated to be an unusually active modifier of phenols, a toxic constituent of numerous pesticides, and is more efficient than other sources of such enzymes. Marine actinomycetes have also been highly active in heavy metal pollution, besides pesticides. They can withstand, absorb, accumulate, and even precipitate metals such as lead, cadmium, copper, and chromium. This renders them much more adaptable than most of the more conventional ways of cleanup that are often unable to deal with

such pollutants. It is even more difficult in the case of mixed pollution, i.e., pesticides, heavy metals, and other toxins can coexist. Actinomycetes, however, shine here as well. Their versatility and adaptability, particularly when used in mixed cultures, make them address several pollutants simultaneously. This capability to both decontaminate and absorb various pollutants makes them good prospects in environmental cleanup exercises in the future at large scales (Jagannathan et al., 2021).

19 Clinical and Commercial Success Stories

The clinical success stories on marine actinomycetes have shown the power of these natural products and demonstrated the therapeutic potential of marine actinomycetes. Salinosporamide A is perhaps the brightest example, as it has gone through its discovery, to Phase I clinical trials in only 3 years—an extraordinarily fast track in terms of drug development. This is a β-lactone-γ-lactam derivative synthesized by Salinispora tropica, a selective and irreversible proteasome inhibitor possessing special properties enabling it to pass the blood-brain barrier, and is especially useful when dealing with central nervous system malignancies (Ngema and Madoroba, 2024; Marine Natural products reviews, 2019–2023).

In addition to single compounds, entire classes of bioactive compounds have been obtained with marine actinomycetes, redefining drug discovery approaches. Recent studies have discovered 167 new bioactive substances produced by 58 other species of rare marine actinomycota in 24 genera, with these products having activities mostly of antibacterial, antifungal, antiparasitic, anticancer, and antimalarial nature. The novelty of their structures is commonly associated with novel mechanisms of action that can overcome the presence of resistance mechanisms (Ngema and Madoroba, 2024; Marine Natural Products reviews, 2019–2023).

20 Modern Discovery Approaches and Technological Advances

Both the development of highly sophisticated cultivation methods, molecular biology, and analysis chemistry have transformed the study of marine actinomycetes. The conventional plate-screen methodologies have been supplemented by advanced methodologies that are aimed at exploiting the complete biosynthetic capabilities of such organisms. One particularly powerful method is genome mining, which allows the researcher to discover cryptic or silent clusters of biosynthetic genes (BGCs) that encode the infrastructure to produce secondary metabolites not expressed in the standard laboratory conditions (Ngema and Madoroba, 2024; Marine Natural Products reviews, 2019–2023).

Many Compounds One Strain Approach (OSMAC) has been very successful in the case of. maximizing the chemical diversity that can be obtained using individual strains. This approach involves parameters of cultivation (media composition, temperature, pH, etc.) are also varied systematically. Aeration and salinity to induce the expression of the various biosynthetic pathways. Co-cultivation also encouraging results have been obtained with techniques that mimic natural microbial interactions. Competition between bacteria and signaling through communication stimulation of the synthesis of compounds that persist. Noisy in monoculture (Ngema and Madoroba, 2024; Marine Natural Products reviews, 2019–2023).

21 Future Perspectives and Research Directions

A new golden era of drug development is foreseen due to the systematic exploration of marine actinomycetes, which represents a paradigm shift for the pursuit of natural products. Far less than 1% of marine microbial diversity has been reported as adequately explored, as estimated by current research sources, indicating a vast volume of unrealized potential for new discoveries. Specifically, the deep-sea environments remain largely unmapped and may harbor entirely novel actinomycetes with hitherto unheard-of biochemical potential (Ngema and Madoroba, 2024; Carroll et al., 2024).

Novel approaches to unlocking the metabolic potential of marine microbes that cannot be cultured are being made possible by developments in metagenomics and synthetic biology. These technologies essentially bring the laboratory to organisms that cannot be cultivated using conventional methods by enabling researchers to clone and express biosynthetic gene clusters in heterologous hosts. By forecasting bioactive potential and improving cultivation conditions, the combination of artificial intelligence and machine learning techniques is speeding up the discovery process even further (Ngema and Madoroba, 2024; Carroll et al., 2024).

22 Conclusion

Marine actinomycetes have already been thought of as an attractive source of bioactive compounds due to their novel structure and high biological activity. Marine actinomycetes are capable of living in dynamic and in many ways extreme environments and have distinct metabolic pathways, which enable them to produce metabolites that are not like terrestrial microbes. The related metabolic diversity of marine actinomycetes is the foundation of their possible ability to generate such a variety of the library of secondary metabolites that have distinct therapeutic uses.

Two classes of secondary metabolites produced by marine actinomycetes, which are of particular interest, are polyketides and non-ribosomal peptides, which contain some of the most potent compounds to date in the clinic environment. These classes have diverse biological activities such as antimicrobial, antifungal, anticancer, and immunosuppressive activity. In addition to polyketides and non-ribosomal peptides, alkaloids have recently been of interest due to their antimicrobial and cytotoxic properties, with terpenes providing further chemical diversity with therapeutic implications of use in anticancer, anti-inflammatory, and anti-viral therapy.

Collectively, these compounds not only expand the chemical space on which drug discovery can be conducted but different approaches to the management of diseases that are poorly handled are availed.

Besides biomedical uses, marine actinomycete bioactive compounds may find varied use in agriculture and biotechnology. Actinomycetes, marine organisms, have been regarded as a promising source of bioactive compounds since they are structurally novel and show high biological activity. Marine actinomycetes survive in dynamic environments that are commonly harsh, and they have distinctive metabolic tracks in the capacity to generate metabolites that are not alike those of terrestrial microbes. We do not yet have much understanding of the underlying metabolic diversity of marine actinomycetes, but it is on this foundation that this large array of secondary metabolites with distinct therapeutic uses is possible.

References

Adachi, S., Watanabe, K., Iwata, Y., Kameda, S., Miyaoka, Y., Onozuka, M., Mitsui, R., Saikawa, Y., & Nakata, M. (2013). Total syntheses of lactonamycin and lactonamycin Z with late-stage A-ring formation and glycosylation. *Angewandte Chemie International Edition, 52*(7), 2087–2091. https://doi.org/10.1002/ANIE.201209205

Agrawal, S., Acharya, D., Adholeya, A., Barrow, C. J., & Deshmukh, S. K. (2017). Nonribosomal peptides from marine microbes and their antimicrobial and anticancer potential. *Frontiers in Pharmacology, 8*(NOV), 293247. https://doi.org/10.3389/FPHAR.2017.00828/FULL

Alam, K., Mazumder, A., Sikdar, S., Zhao, Y. M., Hao, J., Song, C., Wang, Y., Sarkar, R., Islam, S., Zhang, Y., & Li, A. (2022). Streptomyces: The biofactory of secondary metabolites. *Frontiers in Microbiology, 13*, 968053. https://doi.org/10.3389/FMICB.2022.968053/FULL

Alharbi, N. S. (2016). Novel bioactive molecules from marine actinomycetes. *Biosciences, Biotechnology Research Asia, 13*(4), 1905–1927. https://doi.org/10.13005/bbra/2346

Asolkar, R. N., Kirkland, T. N., Jensen, P. R., & Fenical, W. (2010). Arenimycin, an antibiotic effective against rifampin- and methicillin-resistant Staphylococcus aureus from the marine actinomycete *Salinispora arenicola*. *Journal of Antibiotics, 63*(1), 37–39. https://doi.org/10.1038/JA.2009.114;KWRD

Beyer, L. I., Schäfer, A. B., Undabarrena, A., Mattsby-Baltzer, I., Tietze, D., Svensson, E., Stubelius, A., Wenzel, M., Cámara, B., & Tietze, A. A. (2023). Mimicking nonribosomal peptides from the marine actinomycete streptomyces sp. H-KF8 leads to antimicrobial peptides. *ACS Infectious Diseases, 10*(1), 79–92. https://doi.org/10.1021/ACSINFECDIS.3C00206

Bharathi, D., & Lee, J. (2024). Recent advances in marine-derived compounds as potent antibacterial and antifungal agents: A comprehensive review. *Marine Drugs, 22*(8), 348. https://doi.org/10.3390/md22080348

Bruno, S., Coppola, D., Di Prisco, G., Giordano, D., & Verde, C. (2019). Enzymes from marine polar regions and their biotechnological applications. *Marine Drugs, 17*(10), 544. https://doi.org/10.3390/md17100544. PMID: 31547548; PMCID: PMC6835263.

Carroll, A. R., Copp, B. R., Grkovic, T., Keyzers, R. A., & Prinsep, M. R. (2024). Marine natural products. *Natural Product Reports, 41*(2), 162–207.

Chinnathambi, A., Salmen, S. H., Al-Garadi, M. A., Wainwright, M., & Ali Alharbi, S. (2023). Marine actinomycetes: An endless source of potentially therapeutic novel secondary metabolites and other bioactive compounds. *Journal of King Saud University, Science, 35*(9), 102931. https://doi.org/10.1016/J.JKSUS.2023.102931

De La Hoz-Romo, M. C., Díaz, L., & Villamil, L. (2022). Marine actinobacteria a new source of antibacterial metabolites to treat acne vulgaris disease—A systematic literature review. *Antibiotics, 11*(7), 965. https://doi.org/10.3390/antibiotics11070965

De La Hoz-Romo, M. C., Díaz, L., Gómez-León, J., Reyes, L. F., Quintero, M., & Villamil, L. (2025). Quorum quenching: a key biological activity of marine actinobacteria extracts for acne vulgaris control. *Biotechnology Reports, 47*, e00903.

De Rop, A.-S., Rombaut, J., Willems, T., De Graeve, M., Vanhaecke, L., Hulpiau, P., De Maeseneire, S. L., De Mol, M. L., & Soetaert, W. K. (2022). Novel alkaloids from marine actinobacteria: Discovery and characterization. *Marine Drugs, 20*(1), 6. https://doi.org/10.3390/md20010006

Dey, P., Kundu, A., Kumar, A., Gupta, M., Lee, B. M., Bhakta, T., Dash, S., & Kim, H. S. (2020). Analysis of alkaloids (indole alkaloids, isoquinoline alkaloids, tropane alkaloids). In *Recent advances in natural products analysis* (pp. 505–567). https://doi.org/10.1016/B978-0-12-816455-6.00015-9

Dineshkumar, K., Aparna, V., Madhuri, K. Z., & Hopper, W. (2013). Biological activity of Sporolides A and B from *Salinispora tropica*: In silico target prediction using ligand-based pharmacophore mapping and in vitro activity validation on HIV-1 reverse transcriptase. *Chemical Biology & Drug Design, 83*(3), 350–361. https://doi.org/10.1111/cbdd.12252

Feling, R. H., Buchanan, G. O., Mincer, T. J., Kauffman, C. A., Jensen, P. R., & Fenical, W. (2003). Salinosporamide A: A highly cytotoxic proteasome inhibitor from a novel microbial source, a marine bacterium of the new genus Salinospora. *Angewandte Chemie, International Edition, 42*(3), 355–357. https://doi.org/10.1002/ANIE.200390115

Fenical, W., Jensen, P. R., Palladino, M. A., Lam, K. S., Lloyd, G. K., & Potts, B. C. (2008). Discovery and development of the anticancer agent salinosporamide A (NPI-0052). *Bioorganic & Medicinal Chemistry, 17*(6), 2175–2180. https://doi.org/10.1016/j.bmc.2008.10.075

Finking, R., & Marahiel, M. A. (2004). Biosynthesis of nonribosomal peptides. *Annual Review of Microbiology, 58*, 453–488. https://doi.org/10.1146/annurev.micro.58.030603.123615

Freel, K. C., Nam, S., Fenical, W., & Jensen, P. R. (2011). Evolution of secondary metabolite genes in three closely related marine actinomycete species. *Applied and Environmental Microbiology, 77*(20), 7261–7270. https://doi.org/10.1128/aem.05943-11

He, H., Ding, W.-D., Bernan, V. S., Richardson, A. D., Ireland, C. M., Greenstein, M., Ellestad, G. A., & Carter, G. T. (2001). Lomaiviticins A and B, potent antitumor antibiotics from *Micromonospora lomaiWitiensis*. https://doi.org/10.1021/ja010129o

Hozzein, W. N., Mohany, M., Alhawsawi, S. M., Zaky, M. Y., Al-Rejaie, S. S., & Alkhalifah, D. H. (2021). Flavonoids from marine-derived actinobacteria as anticancer drugs. *Current Pharmaceutical Design, 27*(4), 505–512. https://doi.org/10.2174/1381612826666201216160154

Iacovelli, R., Bovenberg, R. A. L., & Driessen, A. J. M. (2021). Nonribosomal peptide synthetases and their biotechnological potential in *Penicillium rubens*. *Journal of Industrial Microbiology & Biotechnology, 48*(9–10), kuab045. https://doi.org/10.1093/JIMB/KUAB045

Ibrahim, J. A. A., Botcha, S., & Prattipati, S. D. (2025). Marine actinomycetes: A promising source of novel therapeutics and pharmaceutical bioactive compounds—A review. *The Microbe, 7*, 100383. https://doi.org/10.1016/J.MICROB.2025.100383

Iobbi, V., Parisi, V., Giacomini, M., De Riccardis, F., Brun, P., Núñez-Pons, L., & Bisio, A. (2025). Sesterterpenoids: Sources, structural diversity, biological activity, and data management. *Natural Product Reports, 42*(3), 443–481.

Islam, A., Teo, S. H., Islam, M. T., Mondal, A. H., Mahmud, H., Ahmed, S., Ibrahim, M., Taufiq-Yap, Y. H., Abdulkareem-Alsultan, G., Hossain, M. L., Sheikh, M. C., Rasee, A. I., Rehan, A. I., Waliullah, R. M., Awual, M. E., Hasan, M. M., Hossain, M. S., Kubra, K. T., Salman, M. S., et al. (2025). Harnessing visible light for sustainable biodiesel production with Ni/Si/MgO photocatalyst. *Renewable and Sustainable Energy Reviews, 208*, 115033. https://doi.org/10.1016/J.RSER.2024.115033

Jagannathan, S. V., Manemann, E. M., Rowe, S. E., Callender, M. C., & Soto, W. (2021). Marine actinomycetes, new sources of biotechnological products. *Marine Drugs, 19*(7), 365. https://doi.org/10.3390/md19070365

Jakubiec-Krzesniak, K., Rajnisz-Mateusiak, A., Guspiel, A., Ziemska, J., & Solecka, J. (2018). Secondary metabolites of actinomycetes and their antibacterial, antifungal and antiviral properties. *Polish Journal of Microbiology, 67*(3), 259–272. https://doi.org/10.21307/pjm-2018-048

Jensen, P. R., Williams, P. G., Oh, D., Zeigler, L., & Fenical, W. (2006). Species-specific secondary metabolite production in marine actinomycetes of the genus Salinispora. *Applied and Environmental Microbiology, 73*(4), 1146–1152. https://doi.org/10.1128/aem.01891-06

Karunanithi, P. S., & Zerbe, P. (2019). Terpene synthases as metabolic gatekeepers in the evolution of plant terpenoid chemical diversity. *Frontiers in Plant Science, 10*. https://doi.org/10.3389/fpls.2019.01166

Kim, M. C., Winter, J. M., Asolkar, R. N., Boonlarppradab, C., Cullum, R., & Fenical, W. (2021). Marinoterpins A–C: Rare linear merosesterterpenoids from marine-derived actinomycete bacteria of the family Streptomycetaceae. *The Journal of Organic Chemistry, 86*(16), 11140–11148. https://doi.org/10.1021/acs.joc.1c00262

Kotake, Y., Cao, R., Viatour, P., Sage, J., Zhang, Y., & Xiong, Y. (2007). *pRB family proteins are required for H3K27 trimethylation and Polycomb repression complexes binding to and silencing p16 INK4a tumor suppressor gene*. https://doi.org/10.1101/gad.1499407

Kumar, R., Kumar, S., Gupta, A. K., Singh, S., Kumar, G., Narsinghani, T., & Sindhu, R. K. (2024). Recent progress in bioactive discovery from marine microorganisms and future directions. *Biochemical and Cellular Archives, 24*(1).

Kwon, H. C., Kauffman, C. A., Jensen, P. R., & Fenical, W. (2006). Marinomycins A–D, antitumor-antibiotics of a new structure class from a marine actinomycete of the recently discovered genus "Marinispora". *Journal of the American Chemical Society, 128*(5), 1622–1632. https://doi.org/10.1021/JA0558948

Le Govic, Y., Papon, N., le Gal, S., Bouchara, J. P., & Vandeputte, P. (2019). Non-ribosomal peptide synthetase gene clusters in the human pathogenic fungus *Scedosporium apiospermum*. *Frontiers in Microbiology, 10*, 476594. https://doi.org/10.3389/FMICB.2019.02062/BIBTEX

Lee, L., Goh, B., & Chan, K. (2020). Editorial: Actinobacteria: Prolific producers of bioactive metabolites. *Frontiers in Microbiology, 11*. https://doi.org/10.3389/fmicb.2020.01612

Liu, Z., Sun, W., Hu, Z., Wang, W., & Zhang, H. (2024). Marine *Streptomyces*-derived novel alkaloids discovered in the past decade. *Marine Drugs, 22*(1), 51. https://doi.org/10.3390/md22010051

Maloney, E. D. (n.d.). *The moist static energy budget of a composite tropical intraseasonal oscillation in a climate model*. https://doi.org/10.1175/2008JCLI2542.1

Manivasagan, P., Kang, K. H., Sivakumar, K., Li-Chan, E. C. Y., Oh, H. M., & Kim, S. K. (2014). Marine actinobacteria: An important source of bioactive natural products. *Environmental Toxicology and Pharmacology, 38*(1), 172–188. https://doi.org/10.1016/J.ETAP.2014.05.014

Marine natural products—Natural Product Reports (RSC Publishing). (n.d.). https://doi.org/10.1039/D4NP00067F. Retrieved September 22, 2025, from https://pubs.rsc.org/en/content/articlehtml/2025/np/d4np00067f

Maskey, A. P., Basnyat, B., Thwaites, G. E., Campbell, J. I., Farrar, J. J., & Zimmerman, M. D. (2008). Emerging trends in enteric fever in Nepal: 9124 cases confirmed by blood culture 1993–2003. *Transactions of the Royal Society of Tropical Medicine and Hygiene, 102*(1), 91–95. https://doi.org/10.1016/J.TRSTMH.2007.10.003

Miller, T., Waturangi, D. E., & Yogiara, N. (2022). Antibiofilm properties of bioactive compounds from actinomycetes against foodborne and fish pathogens. *Scientific Reports, 12*(1). https://doi.org/10.1038/s41598-022-23455-8

Mizui, Y., Sakai, T., Iwata, M., Uenaka, T., Okamoto, K., Shimizu, H., & Asada, M. (2004). Pladienolides, new substances from culture of Streptomyces platensis Mer-11107 III. In vitro and in vivo antitumor activities. *The Journal of Antibiotics, 57*(3), 188–196.

Motohashi, K., Irie, K., Toda, T., Matsuo, Y., Kasai, H., Sue, M., Furihata, K., & Seto, H. (2008). Studies on terpenoids produced by actinomycetes. *The Journal of Antibiotics, 61*(2), 75–80. https://doi.org/10.1038/ja.2008.113

Newman, D. J., & Cragg, G. M. (2020). Natural products as sources of new drugs over the nearly four decades from 01/1981 to 09/2019. *Journal of Natural Products, 83*(3), 770–803. https://doi.org/10.1021/ACS.JNATPROD.9B01285
Ngamcharungchit, C., Chaimusik, N., Panbangred, W., Euanorasetr, J., & Intra, B. (2023). Bioactive metabolites from terrestrial and marine actinomycetes. *Molecules, 28*(15), 5915. https://doi.org/10.3390/molecules28155915
Ngema, S. S., & Madoroba, E. (2024). A mini-review of anti-listerial compounds from marine actinobacteria (1990–2023). *Antibiotics, 13*(4), 362. https://doi.org/10.3390/ANTIBIOTICS13040362
Nicolaou, A. I. (2004). Firm performance effects in relation to the implementation and use of enterprise resource planning systems. *Journal of Information Systems, 18*(2), 79–105. https://doi.org/10.2308/JIS.2004.18.2.79
Niewerth, D., Jansen, G., Riethoff, L. F. V., van Meerloo, J., Kale, A. J., Moore, B. S., Assaraf, Y. G., Anderl, J. L., Zweegman, S., Kaspers, G. J. L., & Cloos, J. (2014). Antileukemic activity and mechanism of drug resistance to the marine *Salinispora tropica* proteasome inhibitor salinosporamide A (Marizomib). *Molecular Pharmacology, 86*(1), 12–19. https://doi.org/10.1124/MOL.114.092114/-/DC1
Olano, C., Méndez, C., & Salas, J. A. (2009). Antitumor compounds from marine actinomycetes. *Marine Drugs, 7*(2), 210. https://doi.org/10.3390/MD7020210
Omura, M., Kobayashi, S., Mizukami, Y., Mogami, K., Todoroki-Ikeda, N., Miyake, T., & Matsuzaki, M. (2001). Eicosapentaenoic acid (EPA) induces Ca2+-independent activation and translocation of endothelial nitric oxide synthase and endothelium-dependent vasorelaxation. *FEBS Letters, 487*(3), 361–366. https://doi.org/10.1016/S0014-5793(00)02351-6
Pan, C., ul Hassan, S. S., Ishaq, M., Yan, S., & Jin, H. (2025). Marine actinomycetes: A hidden treasure trove for antibacterial discovery. *Frontiers in Marine Science, 12*, 1558320. https://doi.org/10.3389/FMARS.2025.1558320/BIBTEX
Park, J., & Kwon, H. (2018). New naphthoquinone terpenoids from marine actinobacterium, Streptomyces sp. CNQ-509. *Marine Drugs, 16*(3), 90. https://doi.org/10.3390/md16030090
Pudi, N., Varikuti, G., Badana, A., Gavara, M., Kumari, S., & Malla, R. (2016). Studies on optimization of growth parameters for enhanced production of antibiotic alkaloids by isolated marine actinomycetes. *Journal of Applied Pharmaceutical Science*, 181–188. https://doi.org/10.7324/japs.2016.601025
Ravikumar, S., Fredimoses, M., & Gnanadesigan, M. (2012). Anticancer property of sediment actinomycetes against MCF-7 and MDA-MB-231 cell lines. *Asian Pacific Journal of Tropical Biomedicine, 2*(2), 92–96. https://doi.org/10.1016/S2221-1691(11)60199-8
Reed, K. A., Manam, R. R., Mitchell, S. S., Xu, J., Teisan, S., Chao, T. H., Deyanat-Yazdi, G., Neuteboom, S. T. C., Lam, K. S., & Potts, B. C. M. (2007). Salinosporamides D-J from the marine actinomycete *Salinispora tropica*, bromosalinosporamide, and thioester derivatives are potent inhibitors of the 20S proteasome. *Journal of Natural Products, 70*(2), 269–276. https://doi.org/10.1021/NP0603471
Sánchez-García, B. M., Ramírez-Pimentel, J. G., Guevara-Acevedo, L. P., Raya-Pérez, J. C., Covarrubias-Prieto, J., Mora-Avilés, M. A. (2019). Actinobacteria with antagonistic potential in vitro to phytopathogenic fungi and promotion of growth in chili plants. *Revista Mexicana De Ciencias Agrícolas, 10*, 339–344. https://doi.org/10.29312/remexca.v0i23.2033.
Selim, M. S. M., Abdelhamid, S. A., & Mohamed, S. S. (2021). Secondary metabolites and biodiversity of actinomycetes. *Journal of Genetic Engineering and Biotechnology, 19*(1), 72. https://doi.org/10.1186/s43141-021-00156-9
Shirai, M., Okuda, M., Motohashi, K., Imoto, M., Furihata, K., Matsuo, Y., Katsuta, A., Shizuri, Y., & Seto, H. (2010). Terpenoids produced by actinomycetes: Isolation, structural elucidation and biosynthesis of new diterpenes, gifhornenolones A and B from *Verrucosispora gifhornensis* YM28-088. *The Journal of Antibiotics, 63*(5), 245–250. https://doi.org/10.1038/ja.2010.30
Shuikan, A. M., Alshuwaykan, R. M., Arif, I. A., Shuikan, A. M., Alshuwaykan, R. M., & Arif, I. A. (2023). The role of metagenomic approaches in the analysis of microbial community in

extreme environment. In *Life in extreme environments—Diversity, adaptability and valuable resources of bioactive molecules*. https://doi.org/10.5772/INTECHOPEN.108050

Solanki, R., Khanna, M., & Lal, R. (2008). Bioactive compounds from marine actinomycetes. *Indian Journal of Microbiology, 48*(4), 410–431. https://doi.org/10.1007/s12088-008-0052-z

Strieker, M., Tanović, A., & Marahiel, M. A. (2010). Nonribosomal peptide synthetases: Structures and dynamics. *Current Opinion in Structural Biology, 20*(2), 234–240. https://doi.org/10.1016/J.SBI.2010.01.009

Sweeney, D., Bogdanov, A., Chase, A. B., Castro-Falcón, G., Trinidad-Javier, A., Dahesh, S., Nizet, V., & Jensen, P. R. (2024). Pattern-based genome mining guides discovery of the antibiotic indanopyrrole a from a marine streptomycete. *Journal of Natural Products*. https://doi.org/10.1021/acs.jnatprod.4c00934

Tarasova, E. V., Luchnikova, N. A., Grishko, V. V., & Ivshina, I. B. (2023). Actinomycetes as producers of biologically active terpenoids: Current trends and patents. *Pharmaceuticals, 16*(6), 872. https://doi.org/10.3390/ph16060872

Valliappan, K., Sun, W., & Li, Z. (2014). Marine actinobacteria associated with marine organisms and their potentials in producing pharmaceutical natural products. *Applied Microbiology and Biotechnology, 98*(17), 7365–7377. https://doi.org/10.1007/S00253-014-5954-6/METRICS

Walsh, C. T. (2004). Polyketide and nonribosomal peptide antibiotics: Modularity and versatility. *Science, 303*(5665), 1805–1810. https://doi.org/10.1126/science.1094318

Wang, Z., Alsup, T. A., Pan, X., Li, L., Tian, J., Yang, Z., Lin, X., Xu, H., Rudolf, J. D., & Dong, L. (2024). Biosynthesis of a bacterial meroterpenoid reveals a non-canonical class II terpene synthase. *Chemical Science*. https://doi.org/10.1039/d4sc06010e

Xie, F., & Pathom-Aree, W. (2021). Actinobacteria from desert: Diversity and biotechnological applications. *Frontiers in Microbiology, 12*. https://doi.org/10.3389/fmicb.2021.765531

Yunt, Z., Reinhardt, K., Li, A., Engeser, M., Dahse, H.-M., Gü, M., Bruhn, T., Bringmann, G., & Rn Piel, J. (2009). *Cleavage of four carbon-carbon bonds during biosynthesis of the griseorhodin a spiroketal pharmacophore*. https://doi.org/10.1021/ja807827k

Melanin from Marine Actinobacteria and Its Medical Applications

Geetha Bharathi

Abstract

Melanin pigment is widely distributed in humans, animals, plants, and microorganisms and plays various protective and immune-regulatory roles. Among microbial pigments, melanin is considered very important due to its distinctive physical, chemical, and biological characteristics, including hydrophobicity, high molecular weight, and a negative charge. Actinomycetes from the marine ecosystems can produce pigments and secondary metabolites of pharmaceutical value. Members of the phylum Actinobacteria, chiefly Glycomyces and Streptomyces genera, can generate high volumes of bioactive compounds, contributing significantly to the pool of industrially and medically relevant metabolites. Actinobacteria produce nearly two-thirds of naturally occurring bioactive microbial products, which are effective against multidrug-resistant pathogens. Although research on marine actinomycetes is still an ongoing process, their potential as a major source of melanin and other secondary metabolites highlights their considerable biotechnological and pharmaceutical significance.

Keywords

Melanin · Marine actinomycetes · Actinobacteria · Microbial pigments · Secondary metabolites · Antioxidant activity · Pharmaceutical applications

1 Introduction

Melanin is a complex biological pigment found in diverse species, from humans to plants and bacteria. In human beings, it is found in the skin, hair follicles, and eyes. It protects against extreme sunlight and plays a major part in controlling the immune response in humans. Other functions of melanin include antioxidant, hypoglycemic, neuroprotective, anti-inflammatory, and control gastrointestinal tract damage, which

G. Bharathi (✉)
Independent Researcher (Medical Applications), Coimbatore, Tamil Nadu, India

K. Loganathan, Z. Li (eds.), *Marine Actinomycetes*,
https://doi.org/10.1007/978-981-95-9849-6_6

are well acknowledged and more researched. Among microbial pigments, melanin is receiving more attention for its wide range of applicability, encompassing biological, physical, and chemical dimensions (Tran-Ly et al., 2020). Melanin is a biological macromolecule found in various microorganisms, reported in many bacteria, the actinomycetes, and diverse fungal groups. So far, microbial pigments, including carotenoids, quinones, violaceins, prodigiosin, melanin, and phenazines, have been well studied. Pheomelanin and Eumelanin are present in humans and most mammalian skin. Since primeval times, skin color has been a much-debated topic. Until 1819, the pigment cells were termed chromatophores and studied in squid. In 1873, the term melanin was first used by Charles-Philippe Robin, and later studies reported that melanocytes synthesize melanin in the skin (Westerhof, 2006). Microbial melanin is naturally hydrophobic, of higher molecular mass, and negatively charged. Extraction of melanin from animals and plant tissue is feasible, chemical synthesis is high-cost and not environmentally-friendly (Fenical & Jensen, 2006).

The aquatic environment is an increasingly promising research area because it contains numerous pharmaceutically important compounds (Fenical & Jensen, 2006). Marine actinomycetes are found in water and marine sediments and are widely distributed across diverse marine ecosystems. Actinomycetes from the marine environment are biotechnologically significant, highly productive microorganisms, and the dyes are used in several industries, compared to their terrestrial counterparts. Melanin produced from microbial organisms is preferable to plant-based pigments (Venil & Lakshmana Perumalsamy, 2009), as pigment production is more prevalent in actinobacteria compared to any other bacteria (Sánchez-Marroquín & Zapata, 1954). Pigment production in actinomycetes is advantageous because these dyes are economically viable and grow well in a low-cost agro-industrial medium. It is a rapidly growing research domain with several industrial applications (Di Salvo et al., 2023). Actinomycetes belong to the Actinobacteria phylum, Actinomycetales order, comes under the Actinobacteria class, with the Actinobacteridae subclass.

Actinobacteria display characteristics of fungi, including hyphae and spores, and have a prokaryotic cell structure; these are considered bacteria in the kingdom Monera (Alenazi et al., 2023). Actinobacteria are gram-positive, belong to the Glycomyces genus, are morphologically indistinguishable from both fungi and Firmicutes, and contain a high level, about 70% Guanine and Cytosine in their genetic makeup.

In 1996, Clark and Norris defined the species *Acidimicrobium ferrooxidans* sp., of the *Acidimicrobiia* class. Stackebrandt and colleagues in 1997 proposed the novel family Acidimicrobiaceae within the Actinobacteria class, based on a study of 16S rRNA genes. In the biomedical domain, actinomycetes produce 70–80% of scientifically and therapeutically significant specialized metabolites, and nearly 60% of them are synthesized by Streptomyces species (Nayaka et al., 2020). Secondary metabolites of actinobacteria account for about 2/3 of all natural bioactive compounds for medicinal applications. So far, twenty-three thousand bioactive metabolites have been reported from microorganisms, and approximately ten thousand are derived from actinobacteria. Roughly seven thousand and six hundred actinobacterial compounds have been reported to combat multidrug-resistant pathogens and to exhibit other therapeutic potential, with Streptomyces alone contributing about five thousand of these compounds (Berdy, 2012). The exploration of marine actinomycetes as an origin of innovative natural products is currently in its initial stages (Chinnathambi et al., 2023).

2 Objective and Scope

Currently, identifying natural metabolites (secondary) from the marine environment has escalated due to the exceptional structures and novelty of these compounds. Overall, actinobacteria are acknowledged to synthesize bioactive natural products with valuable antibiotic properties that can inhibit multidrug-resistant pathogens (Das et al., 2018). This part explores the potential of melanin produced by marine actinobacteria for health-related applications. It examines the enzymes that regulate melanin synthesis, the environmental factors that influence its production, and the current challenges in the field. The biological activities of melanin, particularly its antioxidant, antimicrobial, and therapeutic properties, are emphasized. Finally, it suggests future research directions and possible clinical translation, aiming to contribute to marine biotechnology and natural product-based interventions.

3 Marine Ecosystems and Actinobacteria Diversity

Actinobacteria are found in both normal and extreme environments, including acidic and saline-alkaline soils, thermophilic, radiation-loaded environments, and minimal nutrient levels (Zenova et al., 2011). They grow in diverse ecosystems, from marine to desert, and even in polar regions. They also form symbiotic associations with plants, insects, animals, sponges, corals, and fish (Avila et al., 2007).

Reports from the late nineteenth century describe marine actinomycetes. Since 1980, research on marine microorganisms has reported their use in drug development and biotechnology (Khalifa et al., 2019). *Salinispora*, the primary obligate marine actinomycete, was defined in 2005 (Jensen et al., 2015). Other well-reported aquatic actinomycete genera contain the gram-positive *Dietzia* sp., *Rhodococcus*, *Streptomyces*, and *Micromonospora*.

4 Marine Actinomycetes

Actinobacteria are recognized within the Bacterial domain and constitute the largest taxonomic units across 18 primary lineages (Chandra & Chater, 2013). Actinobacteria are widely distributed across terrestrial and aquatic environments. Actinobacteria are classified based on morphological, genetic, chemotaxonomic, molecular, and secondary metabolite production criteria (Barka et al., 2015). Actinobacteria include a wide array of species, such as pathogens (*Corynebacterium* sp., *Mycobacterium*,

Nocardia, Propionibacterium, and *Tropheryma*), soil inhabitants (e.g., *Micromonospora* and *Streptomyces* sp.), plant symbionts (e.g., *Frankia* sp.), and gastrointestinal-dependent species. These classes are classically recognized by the gene sequencing of 16S ribosomal RNA, which has identified 39 families and 130 genera. Unlike other bacteria, actinobacteria are much older, and not possible to classify the phylogenetically nearest species to Actinobacteria with poise (Embley & Stackebrandt, 1994). Within the class Acidimicrobiia, the Genome Taxonomy Database (GTDB) explains 6 orders, and above fifty families (Parks et al., 2022). These families have alphanumeric names, and fewer than 5 genomes are available for each family. Hence, the unrefined members of the class Acidimicrobiia taxonomy demand further investigation.

Actinobacteria exhibit diverse morphologies, have a thin cell wall, and possess a linear or circular chromosome. They produce non-septate, slender mycelia, which include two regions: aerial mycelia and substrate mycelia. The aerial mycelia grow externally and are mainly involved in reproduction through binary fission, conidia, and sporulation, whereas substrate mycelia thrive in the agar/substrate medium, which is vital for the bacterium. Actinobacteria are either aerophilic, heterotrophic, or chemoautotrophic and thrive in a variety of nourishing environments, with several intricate polysaccharides (Lechevalier & Lechevalier, 1965; Zimmerman, 1990). Aquatic Actinobacteria are found in marine sediments (Wagner et al., 2014) and have evolved constitutional features to thrive under varying marine environments (Patin et al., 2016). Jose and Jha (2017) found that intertidal sediments are a fertile place for various Actinobacteria with substantial biosynthetic gene clusters.

Cultivation methods and metagenomic studies indicate that various invertebrates harbor actinobacterial species. Actinobacteria are spotted or identified from various marine organisms, including sponges, fish, seaweeds, sea grass, corals, mollusks, and ascidians (Valliappan et al., 2014). These actinobacterial species belong to the genera *Saccharopolyspora, Salinospora, Rhodococcus, Streptomyces, Nocardiaceae, Nonomurae,* and *Actinokinespora* (Hifnawy et al., 2020; Krishnamoorthy et al., 2020; Ramesh et al., 2021). Actinobacterial communities associated with corals can fix nitrogen and contribute to healthy corals (Lesser et al., 2018). Lee et al. (2018) reported that several actinobacterial species, including *Nocardiopsis Sinemet Forman's*, were isolated from squid in Jumunjin, Gangwon-do, Korea. You et al. (2005) confirmed the role of the bioactive metabolite from actinobacteria in shrimp farming owing to their antimicrobial abilities. Many studies have stated the spreading of Actinobacteria in places including Cairo, Egypt; Falmouth Town, Massachusetts; Karnataka State, India; Keffi Town, Nigeria; Presque Isle, Pennsylvania; and alkaline deserts in Wadi El Natrun and Wadi Araba, Egypt.

4.1 Melanin: Structure, Types, and Functions (www.chemspider.com)

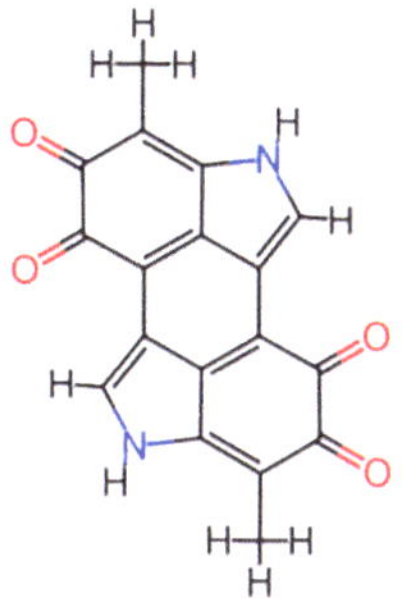

Molecular Formula: $C_{18}H_{10}N_2O_4$
Average Mass: 318.288
Monoisotopic Mass: 318.064057
ChemSpider ID: 4884931

Melanin is derived from indolic or phenolic compounds (Seelam et al., 2021). These pigments are extracted and purified for widespread applications (Gupta et al., 2022). Melanin is present in specialized cellular structures, namely melanosomes, within melanocytes. Melanin is the principal biomaterial reported to exhibit a free radical signal. Bacterial melanin synthesis is reported in Gram-positive and Gram-negative bacteria (Guo et al., 2023) and involves enzymatic pathways and metabolic directives.

Melanin is an amorphous biomaterial that is pervasive across all biological kingdoms on Earth. It is present in nature in single-celled organisms to multicellular animals and plants, and is linked with several functions, such as the synthesis of pigment, radical removal, defending against ultraviolet radiation, and thermal regulation. Melanin is a heterogeneous, complex polymer composed of repeating derivative units from phenolic or indole compounds. It contains 5,6-dihydroxyindole (DHI) and 5,6-dihydroxyindole 2-carboxylic acid (DHICA). These compounds produce different types of melanin. Melanin is principally dark brown to black in color, with red and yellow occasionally noted (El-Naggar & Saber, 2022). Melanin is resilient to light, acid concentrates, and reducing agents, while it is soluble in base/phenolic solutions (Jacobson, 2000). In nature, three types of melanin are identified (Seelam et al., 2021).

Mostly, melanin is grouped under five types based on the chemical precursors for the biosynthesis: (Fig. 1)

- Eumelanin (DOPA-o-dihydroxyphenylalanine-present in animals and plants),
- Pheomelanin (cysteinyl-Dopa, animals and plants)
- Neuromelanin (in animals)

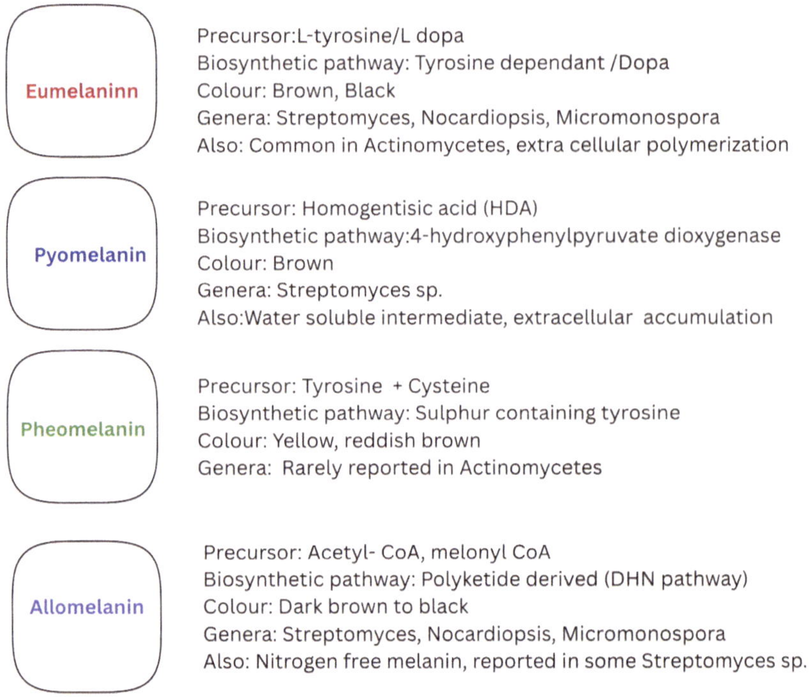

Fig. 1 Types of melanin: Actinomycetes produce structurally and biochemically distinct Melanin, depending on precursor availability and enzyme. These pigments have different biosynthetic pathways, physicochemical properties, and ecological roles

- Allomelanin (DHN-di-hydroxynaphthalene-in plants)
- Pyomelanin (in fungi and bacteria)

The precise chemical compounds of melanin are obscure, with an innate multiplicity and intricacy of chemical structure, and play several roles and properties that are yet to be identified.

The black or brown pigment eumelanin is most commonly found in animals. It is an extremely insoluble pigment produced by the melanocytes. Initially, tyrosinase hydroxylates tyrosine to produce the catecholamine 3,4-dihydroxyphenylalanine (DOPA), which further undergoes oxidation to produce 3,4-dihydroxyphenylalanine (dopaquinone). This is immediately cyclized and polymerized to form eumelanin, or couples with cysteine to form pheomelanin, which is red-yellow in color.

As eumelanin, pheomelanin is organically synthesized, excluding a sulfur-comprising precursor merged with it (El-Obeid et al., 2017). Eumelanin is typically detected in brown to dark brown skin and hair, whereas pheomelanin is noted in reddish and yellowish shades. Several living organisms synthesize a blend of these two melanins. Individuals with red hair typically have extra pheomelanin in their hair and skin. Research shows that individuals with higher pheomelanin levels are more prone to ultraviolet (UV) photodamage than those with higher eumelanin levels of the skin.

5 Melanin is an Enigmatic Pigment

In recent days, several studies have focused on acquiring pigments from natural sources, rather than synthetic ones. Natural pigments are extracted from plants, animals, and microbes, and the extraction of microbial pigments is a developing area of research with several industrial applications. In the field of biomedicine, it is an emerging therapeutic agent because of its biological compatibility and easy degradability (Ghattavi et al., 2022).

Microorganisms can produce several pigments, including β-carotene, prodigiosin, zeaxanthin, astaxanthin, lycopene, and melanin. Melanin synthesis occurs through several metabolic pathways that involve many intracellular organelles and the cytoplasm. This process requires anabolic and redox demands and is regulated by melanocytes (Snyman et al., 2024). Microbial melanin is produced through the malonyl-CoA (DHN pathway) or tyrosine conversion (DOPA pathway), through distinct enzymatic genomes. Several bacteria and *basidiomycete* fungi utilize the DOPA route to produce melanin. Ascomycetes and macroscopic fungi use the DHN route. The DOPA pathway is analogous to melanin synthesis in mammals, where tyrosine is transformed to dihydroxyphenylalanine, then by laccase to dihydroxyphenylalanine quinone. Activated DOPA quinone can be quickly oxidized and self-polymerized to generate melanin. Propylene glycol coenzyme A is a precursor of the DHN pathway. Decarboxylation and condensation of five malonyl-CoA molecules are catalysed by polyketo compound synthase to produce 1,3,6,8-tetrahydroxy naphthalene (THN). This undergoes several subsequent reactions to develop 1,8-dihydroxy naphthalene (DHN), which then crosslinks to produce dihydroxy naphthalene melanin. Phenolic hydroxyl, carboxyl, and amino functional groups of melanin absorb UV light. Because the biological action of melanin is strongly influenced by its composition and structure, functional properties have gained momentum and become the focus of melanin activity research.

Research shows that melanin protects against environmental stress. Melanin-producing bacteria exhibit antibacterial, anticancer, anti-inflammatory (Kurian et al., 2015), antivenom (Hung et al., 2004), antiviral, and liver-protective (Sava et al., 2003) properties. Melanin is used to treat neurodegenerative disabilities, including Alzheimer's, retinitis pigmentosa, schizophrenia, and dementia (Berliner et al., 1993). Synthetic soluble melanin has significant potential against human immunodeficiency virus, and is used for AIDS treatment (Montefiori & Zhou, 1991).

5.1 Biosynthesis of Melanin in Marine Actinobacteria

Melanin is synthesized by many bacterial species, and the *Streptomyces* group is the most intensively examined actinomycete for melanin production. Melanin derived from Streptomyces species exhibits UV light protection; it absorbs radiation across the entire UV wavelength (Tarangini & Mishra, 2014). Vasantha Bharathi et al. (2011) reported that an unpurified melanin preparation is active against *E. coli* and *L. vulgaris*. DOPA melanin from Streptomyces sp. MVCS6 was active against *P. aeruginosa RMMH7* and *V. parahaemolyticus RMMH12*. DOPA melanin's cytotoxicity against a cell line of cervical cancer exhibited a dose–dependent pattern, and the results show that eumelanin is an excellent bioactive compound with beneficial potential (Sivaperumal et al., 2015).

Melanin was extracted from the aquatic actinobacterium *Actinoalloteichus* sp. *MA-32* (Manivasagan et al., 2013), which exhibits activity against the bacteria and antioxidant activity. Melanin from the aquatic *S. cavourensis strain RD8* showed activity to inhibit the oxidation of other molecules and antibacterial activity (Dholakiya et al., 2017). Rare actinomycetes produce less melanin than *Streptomyces* spp., so the following marine actinomycetes, and *Nocardiopsis* sp., were collected from marine soil sediment samples from Kerala, India. These melanin pigments exhibited activity against biofilm formation, with a maximum of 64.20% and 65.99% against *Staphylococcus* sp., and human *Listeria monocytogenes* and *Bacillus* sp. (Kamarudheen et al., 2019).

Melanin, as an anti-quorum-sensing agent, was reported to reduce the spread of *Chromobacterium violaceum MTCC 2656* (Kamarudheen et al., 2019). From the Coastal region sediment of the Bay of Bengal, Streptomyces BI 244, a melanin-producing strain, was removed and exhibited activity against *Enterococcus faecalis, B. subtilis*, and *S. aureus*, and moderate activity against *S. typhi, Shigella* sp., *E. coli, K. pneumoniae*, and *C. albicans*. Melanin produced by Streptomyces sp. BI 244 expressed dose and time-based cytotoxicity to HeLa cells, and prevents the oxidation activity of other molecules.

5.2 Key Enzymes: Tyrosinase, Laccase, Polyketide Synthases

Melanin synthesis mainly depends upon enzymes belonging to the tyrosinase, laccase, and polyketide synthase families (Fig. 2).

Tyrosinases, a copper-containing metalloenzyme, facilitate the o-hydroxylation and two-electron oxidation of monophenols, such as tyrosine, to ortho-quinones, through reducing O_2 to water (Bijelic et al., 2015). The copper-enhanced melanin production in Frankia sp. was completed through an established copper-binding agent, potassium cyanide (KCN). This study highlighted the critical part of tyrosinase in the pathway of melanin biosynthesis. The effects of L-DOPA, another specific inhibitor of tyrosinases and an enhancer, have also been studied, with comparable results (Singh et al., 2021).

In 2006, the tyrosinase from Streptomyces castanet oglobisporus was identified initially, and the crystal structure was determined (Decker et al., 2006). In 2011, the tyrosinase domains of *B. megaterium* and the fungal *Agaricus bisporus* were extensively studied (Leiva et al., 2015). Tyrosinases aid in melanin synthesis in bacteria, *Streptomyces, Pseudomonas* sp., and *Bacillus megaterium* (Shuster & Fishman,

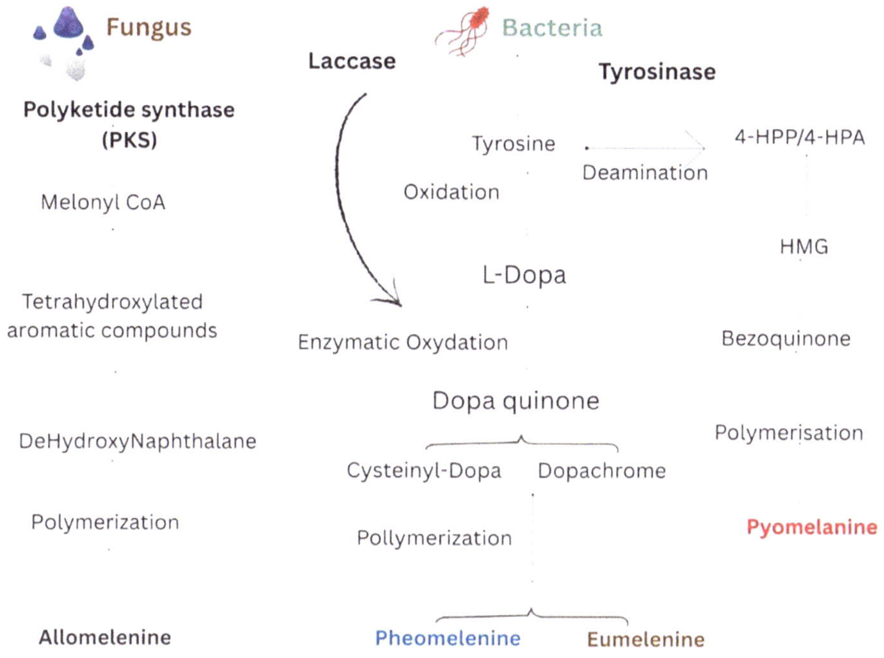

Fig. 2 Biosynthesis of microbial melanin. (Modified from Singh et al., 2021). A schematic illustration of the microbial melanin biosynthesis pathways mediated by tyrosinase, laccase, and polyketide synthase (PKS). Tyrosine or L-DOPA oxidation process through tyrosinase and laccase yields eumelanin, pheomelanin, and pyomelanin, whereas PKS catalyzes the pathway to DHN-melanin (allomelanin) from malonyl-CoA. This figure emphasizes enzymatic convergence toward quinone formation, which is followed by polymerization into various types of melanin

2009). Research on tyrosinases found in microbes has demonstrated that these organisms depend on tyrosinase for their endurance (Faccio et al., 2012).

5.2.1 Laccase

This is a copper-based glycoprotein, oxidoreductase enzyme, blue in color, that uses a wide array of substrates, including methoxy-replaced phenols, diamines, and polyphenols. Syringaldazine, namely 4-hydroxy-3,5-dimethoxybenzaldehyde azine, is designated as a laccase-deficient base. Laccases generate DHN-melanin (Langfelder et al., 2003) and DOPA melanin (Nosanchuk & Casadevall, 2003). The only variance between tyrosinase and laccase is substrate specificity; laccase catalyzes a single-phase oxidative process of dihydroxy phenols into quinones, whereas tyrosinase uses a two-stage reaction (Muniraj et al., 2021). Laccase activities are detected in nature, but in bacteria, Sondhi et al.(2023) and Pawlik et al. (2016) studied the melanin-producing *Sinorhizobium meliloti* strain. *Bacillus subtilis* synthesizes CotA, a multi-copper laccase enzyme that depends on copper to produce brown color spore pigments (Hullo et al., 2001). The human pathogen *Cryptococcus neoformans* yeast produces a laccase reported to synthesize melanin by oxidizing L and D-DOPA (Eisenman et al., 2007). Pavan et al. (2020) have examined the activity of tyrosinase and laccase, and reported that they act together in *Bacillus* sp., *Alteromonas*, and *Pseudomonas putida F6* strains.

5.2.2 Polyketide Synthase (PKS)

PKS enzymes are found in several microorganisms that synthesize secondary metabolites, for instance, toxins and dyes. PKS is linked with the pathways of DHN-melanin synthesis (Jia et al., 2025). The pathogenic fungus, *Penicillium marneffei*, with diverse PKS polyketide synthase genes (PKS) in its genome, allows the production of melanin biosynthesis. A study by Woo et al. was completed by generating PKS gene variations by removing the PKS gene (alb1), and the result showed failure to develop phenotypic attributes such as melanin synthesis, severity, and hydrogen peroxide resistance (Woo et al., 2010).

5.3 Different Classes of Melanin

5.3.1 Eumelanin

Eumelanins, which are black and brown, are produced from tyrosine via the L-DOPA pathway (Beeson et al., 2023) by the enzyme tyrosinase. In the absence of cysteine, L-DOPA converts to dopaquinone. Dopaquinone then cyclizes to form dopachrome, which is further converted to 5,6-dihydroxyindole-2-carboxylic acid (DHICA) and 5,6-dihydroxyindole (DHI). Continued oxidation and polymerization yield eumelanin or DOPA-melanin (Sugumaran et al., 2020).

5.3.2 Pheomelanin

Occasionally, with sulfur, the pathway of L-DOPA produces red and yellow color pheomelanins. During pheomelanogenesis, tyrosinase becomes less active, and the precursor tyrosine yields dopaquinone. Reacting with the thiol group, dopaquinone converts into pheomelanin, accompanied by cysteine. Around cysteine, dopaquinone synthesizes cysteinyldopa (CD isomer), which is oxidized, cyclized, and rearranged to yield 1,4-benzothiazines, which eventually polymerize to produce pheomelanin (Wakamatsu et al., 2021).

5.3.3 Allomelanin

DHN-melanins or allomelanins are being developed through the oxidation or polymerization of di-DHN (1,8-dihydroxynaphthalene) or THN (1,3,6,8-tetrahydroxy naphthalene). The primary transition in this pathway, THN, is produced by the enzyme polyketide synthase (Pfeifer & Khosla, 2001). Mutations in enzymatic genes in several microbes suggest that DHN melanin plays a vital role in their survival (Jia et al., 2025).

5.3.4 Pyomelanin

Pyomelanin was initially described in *P. aeruginosa* by Yabuuchi and Ohyama (1972) and Mendez et al. (2011), described it in *Burkholderia xenovorans LB400,* and by Steinert et al. in *Legionella pneumophila* (1995). During many environmental challenges, microorganisms are well supported by pyomelanins (Lorquin et al., 2022). The deaminated tyrosine, through the homogenized pathway, produces p-hydroxyphenyl pyruvate, which then undergoes various ring rearrangement, oxidative, and decarboxylation reactions, and at the end is oxidized to homogentisate. Homogentisate is then polymerized to form pyomelanin (Ahmad et al., 2017).

6 Environmental Factors

Environmental factors influence melanin production in actinobacteria. Significant factors include temperature, pH, humidity, nutrient availability, and UV radiation. These factors affect both the development of actinobacteria and the synthesis of melanin, which protects them from environmental stress.

1. *Temperature:*
   - Actinobacteria are characteristically mesophilic, flourishing in moderate temperatures (25–30 °C).
   - Though some thermophilic strains can endure higher temperatures (50–60 °C).
   - Temperature variations can affect melanin production, as some strains synthesize more melanin at higher temperatures as a protective mechanism.
2. *pH:*
   - Most actinobacteria desire neutral pH conditions (6–9), but some prefer acidic soils (pH 3.5).
   - As the ideal pH for melanin production varies between different strains, pH can impact the enzyme activity in melanin synthesis and affect melanin levels.
3. *Moisture:*
   - Actinobacteria favor drier conditions, as undue dampness may hamper their growth.
   - Some strains can produce more melanin under moisture conditions, as a protective mechanism against damage from submerged conditions.
4. *Nutrient Availability:*
   - Availability of carbon and nitrogen sources in the growth medium can disturb melanin production.
   - Absence of specific nutrients required for the biosynthesis of melanin can limit its production.
5. *UV Radiation:*
   - UV radiation stimulates melanogenesis in many organisms, including actinobacteria.

- Melanin, as a protective pigment against UV damage, produces more melanin UV exposure.

6. *Other Factors:*

   - Availability of oxygen and the existence of other microorganisms in the adjacent environment can concomitantly affect melanin synthesis in actinobacteria.

Melanin produced by microorganisms has proven to be an economical and environmentally friendly alternative to chemically synthesizing melanin (Julianti et al., 2022). The phenol group or indole structure of melanin, the physical and chemical characteristics, underlie a variety of biological actions. Melanins such as pyomelanin, synthesized by several Streptomyces species, including *S. fuscus*, and Nocardiopsis sp., exhibit cytotoxic, antioxidant (Mavridi-Printezi et al., 2023), bactericide, protein-binding, and antineoplastic properties (Rudrappa et al., 2023; El-Zawawy et al., 2024). These comprehensive bioactivities position pyomelanin as an ideal candidate for cancer treatment, antimicrobial and anti-aging therapy, protection from UV radiation, and oxidative stress.

Several studies have reported the functions of melanin in removing free radicals, chelating metal ions, and exhibiting antibacterial activity. These studies have also described that melanin is useful in agronomy and agro-food industries (Nair et al., 2017; Tran-Ly et al., 2020). Recently, biomedical applications have gained momentum in magnetic resonance imaging (MRI) contrast agents (Sun et al., 2018), melanin's anti-tumor (El-Naggar & El-Ewasy, 2017), immunomodulatory (Brenner & Hearing, 2008), and antimicrobial (Muthuraj et al., 2021; Nosanchuk & Casadevall, 2003) activities, radiation protection, photothermal properties (Kurian et al., 2015; Rudrappa et al., 2023), and other activities.

Melanin contains several irregular electrons and neutralizes free radicals using one-electron transfer, demonstrating its role in skin products that lessen tissue damage by various toxins. Melanin has a strong capacity and affinity to bind directly to metal ions, including Fe2+ (Li et al., 2016). The components and functional groups in melanin enable it to interact with various compounds (Solano, 2017). Melanin has anti-tumor properties, including inhibition of apoptosis and angiogenesis (Rózanowska et al., 1999). Research study stated that melanin nanoparticles, based on dosage, diminish cancer viability, and prolonged exposure showed harmfulness to the living cells (El-Zawawy et al., 2024). Cell lines tested for viability were unusually low when exposed to 0.5 μg/mL or higher levels of melanin. The authors concluded that melanin synthesized by *S. djakartensis NSS-3* sp. is biologically valuable as an antitumor agent. In the HEP2 carcinoma cell line, melanin inhibited cell viability by 53% at 60 μg concentration, reveals the dosage can play a vital role in cell viability (Arun et al., 2015).

Fu and colleagues (2013) studied the coral mucus *Streptomyces* sp. *OUCMDZ-1703* produced melanin under improved medium conditions, targeting the proteins PI3K, Akt, the mTOR pathway, and BcL2, which can be used to treat skin cancer. Melanin is an effective photothermal converter and captures light from the ultraviolet range to the near-infrared range. Hence, it is employed in multimodal imaging-guided photothermal therapy (PTT) to destroy tumor cells or tissues explicitly (Kim & Kim, 2024).

Melanin can either accept or donate an electron because it has an unpaired electron in its structure; hence, it can readily interact with free radicals and neutralize oxidative stress (Mavridi-Printezi et al., 2023). The high inhibitory oxidant activity of melanin is supported by its complex structure of hydroxyl and indole groups, which can scavenge the DPPH radical by donating a hydrogen atom. Li et al. (2018) reported the extracellular biological activity of melanin, including antiradiation and free radical scavenging activity, and its use in cosmetics, food, and pharmaceuticals as an antioxidant/UV protectant.

Shewanella colwelliana, V. cholerae, and *Alteromonas nigrifaciens* were the initial aquatic strains reported to synthesize melanin or melanin-alike pigments (Fuqua et al., 1991; Ruzafa et al., 1995). The pigment derived from *Vibrio cholerae* was described as an allomelanin, derived from homogentisic acid (Sánchez-Amat et al., 1998), synthesized due to tyrosine modification, without the tyrosinase-induced melanin synthetic pathway. *Cellulophaga tyrosinoxydans* has tyrosinase activity and yields a pigment of yellow color, which is proposed as a pheomelanin (Kahng et al., 2009). All Streptomyces strains are described as using tyrosinases to synthesize melanin pigments, namely black eumelanin from L-tyrosine (Kordjazi et al., 2024).

7 Antimicrobial Activities

Melanin is reported to hinder the development of several varieties of living organisms. Roy and Rhim (2021) reported that melanin resistance to pathogenic microbes in fungi, plants, and invertebrates is attributed to the host defense mechanism. The antibacterial properties of decontaminated melanin at different concentrations against multidrug-resistant (MDR) bacterial strains were quantitatively assessed (El-Zawawy et al., 2024). Melanin showed robust properties against all microorganisms. Mostly, *P. aeruginosa* (gram-) was affected by melanin, and even at low concentrations, *S. aureus (SA-04)* and *K. pneumoniae (KP-01)* were effectively inhibited. The minimal potential was reported against *E. coli (EC-03)*. The Minimum Inhibitory Concentrations and Minimum Bactericidal Concentration of melanin against infective microorganisms indicate its ability to suppress bacterial cell development, suggesting its use in the development of new anti-infective or germicides as adjuvant drugs.

Sivaperumal et al. (2015) reported the melanin pigment retrieved from Streptomyces sp. MVCS6, as dihydroxyphenylalanine melanin, revealed vigorous potential against several infective bacteria from hospital specimens, predominantly against *P. aeruginosa RMMH7* and Vibrio parahaemolyticus. Melanin is synthesized in marine actinobacteria, including *Nocardia* sp., *alkalophilic Nocardiopsis dassonvillei, Dietzia,* and *Marinispora* (Thirumurugan et al., 2018). Sundar and Sivaperumal (2022) reported melanin in *Nocardiopsis* sp. and noted its presence in

E. coli, related to two species. Sushma et al. (2025) suggested that melanin from *A. nidulans* is a potential Hooch remover and can be used in the cosmetic industry to produce various creams and lotions that can guard the skin from UV oxidative damage (de Cássia et al., 2005). Kotakonda et al. (2024) stated that melanin synthesized by *A. cyanogriseus isolate AU-RM-4* exhibits antibacterial potential against *S. aureus* and *K. pneumoniae.*

A high antioxidant activity of *S. cavourensis RD8* marine actinomycete that produces melanin, was stated by Dholakiya et al. (2017). Srinivasan et al. (2017) described that *S. bellus MSA1* produces melanin as an intracellular product from marine sediments. El-Naggar and El-Ewasy (2017), described that melanin produced by the actinobacterium *Streptomyces glaucescens*, a eumelanin, demonstrated antioxidant activity, anti-inflammatory potential, wound-healing applications, and UV protection.

Current challenges in marine natural product-based medicine include the need for improved methods for isolating, cultivating, and screening marine microorganisms, as well as identifying the chemical diversity and expanding the range of effective and safe clinical applications. Future research should emphasize linking advanced technologies such as metagenomics, genomics, and synthetic biology to identify new compounds and improve their production. In addition, the structure-activity relationships of marine-derived molecules should be clearly understood to simplify their clinical translation and the development of new therapeutic drugs.

8 Conclusion

Melanin is a biological macromolecule broadly used in cosmetic, medical, and biotechnology industries. Melanin has great potential in the agricultural sector, as its biological properties enable pathogen control and UV protection. Actinobacteria are recognized as economically important prokaryotes and an abundant source of novel compounds with diverse therapeutic applications. The importance of aquatic actinomycetes will continue to rise, supported by advances in genetic engineering, microbial experimental evolution, and advanced techniques such as cloning genes. Melanin from marine actinomycetes has comprehensive usage in medicine and environmental protection, and is presently being explored for several industrial purposes. Microbial melanin presents promising opportunities in skincare, biomedical, agricultural, and environmental sectors. Despite promising indicators, it remains in the process of achieving its full potential. Continued research is warranted to enhance its production methods to accomplish the next frontiers in bio-based production.

References

Ahmad, S., Lee, S. Y., Khan, R., Kong, H. G., Son, G. J., Roy, N., Choi, K., & Lee, S. (2017). Identification of a gene involved in the negative regulation of pyomelanin production in *Ralstonia solanacearum*. *Journal of Microbiology and Biotechnology, 27*(9), 1692–1700. https://doi.org/10.4014/jmb.1705.05049

Alenazi, A., Anwar, Y., Aboaba, S., & Bataweel, N. (2023). A review on Actinomycetes distribution, isolation, and their medical applications. *Novel Research in Microbiology Journal., 7*, 1918–1931. https://doi.org/10.21608/nrmj.2023.294180

Arun, G., Eyini, M., & Gunasekaran, P. (2015). Characterization and biological activities of extracellular melanin produced by Schizophyllum commune (Fries). *Indian Journal of Experimental Biology, 53*(6), 380–387. PMID:26155678.

Ávila, E., Carballo, J. L., & Cruz-Barraza, J. A. (2007). Symbiotic relationships between sponges and other organisms from the Mar de Cortés (Mexican Pacific Coast): Same problems, same solutions. In M. R. Custódio, G. Lôbo-Hajdu, E. Hajdu, & G. Muricy (Eds.), *Porifera research: Biodiversity, innovation and sustainability* (pp. 147–156). Museu Nacional.

Barka, E. A., Vatsa, P., Sanchez, L., Gaveau-Vaillant, N., Jacquard, C., Meier-Kolthoff, J. P., Klenk, H. P., Clément, C., Outdoes, Y., & van Wezel, G. P. (2015). Taxonomy, physiology, and natural products of actinobacteria. *Microbiology and Molecular Biology Reviews: MMBR, 80*(1), 1–43. https://doi.org/10.1128/MMBR.00019-15

Beeson, W., Gabriel, K., & Cornelison, C. (2023). Fungi as a source of eumelanin: Current understanding and prospects. *Journal of Industrial Microbiology & Biotechnology, 50*(1), kuad014. https://doi.org/10.1093/jimb/kuad014

Bérdy, J. (2012). Thoughts and facts about antibiotics: Where we are now and where we are heading. *The Journal of Antibiotics, 65*(8), 385–395. https://doi.org/10.1038/ja.2012.27

Berliner, D. L., Erwin, R. L., McGee, D. R., & Inventors. (1993). *Methods of treating Parkinson's disease using melanin*. United States Patent No. 5,210,076.

Bijelic, A., Pretzler, M., Molitor, C., Zekiri, F., & Rompel, A. (2015). The structure of a plant tyrosinase from walnut leaves reveals the importance of "substrate-guiding residues" for enzymatic specificity. *Angewandte Chemie (International Ed. in English), 54*(49), 14677–14680. https://doi.org/10.1002/anie.201506994

Brenner, M., & Hearing, V. J. (2008). The protective role of melanin against UV damage in human skin. *Photochemistry and Photobiology, 84*(3), 539–549. https://doi.org/10.1111/j.1751-1097.2007.00226.x

Chandra, G., & Chater, K. F. (2013). Developmental biology of Streptomyces from the perspective of 100 actinobacterial genome sequences. *FEMS Microbiology Reviews, 38*(3), 345–379. https://doi.org/10.1111/1574-6976.12047

Chinnathambi, A., Salmen, S. H., Al-Garadi, M. A., Wainwright, M., & Alharbi, S. A. (2023). Marine actinomycetes: An endless source of potentially therapeutic novel secondary metabolites and other bioactive compounds. *Journal of King Saud University, Science, 35*(9), 102931. https://doi.org/10.1016/j.jksus.2023.102931

Clark, D. A., & Norris, P. R. (1996). *Acidimicrobium ferrooxidans* gen. nov., sp. nov.: Mixed-culture ferrous iron oxidation with Sulfobacillus species. *Microbiology (Reading, England), 142*(4), 785–790. https://doi.org/10.1099/00221287-142-4-785

Das, R., Romi, W., Das, R., Sharma, H. K., & Thakur, D. (2018). Antimicrobial potentiality of actinobacteria isolated from two microbiologically unexplored forest ecosystems of Northeast India. *BMC Microbiology, 18*(1), 71. https://doi.org/10.1186/s12866-018-1215-7

de Cássia, R., Goncalves, R., & Pombeiro-Sponchiado, S. R. (2005). Antioxidant activity of the melanin pigment extracted from *Aspergillus nidulans*. *Biological & Pharmaceutical Bulletin, 28*(6), 1129–1131. https://doi.org/10.1248/bpb.28.1129

Decker, H., Schweikardt, T., & Tuczek, F. (2006). The first crystal structure of tyrosinase: All questions answered? *Angewandte Chemie (International Ed. in English), 45*(28), 4546–4550. https://doi.org/10.1002/anie.200601255

Dholakiya, R. N., Kumar, M. A., & Mody, K. H. (2017). Production and characterization of melanin from *Streptomyces cavourensis* strain RD8 using response surface optimization. *Environmental Pollution and Protection, 2*, 168–178. https://doi.org/10.22606/epp.2017.24002

Di Salvo, E., Lo Vecchio, G., De Pasquale, R., De Maria, L., Tardugno, R., Vadalà, R., & Cicero, N. (2023). Natural pigments production and their application in food, health, and other industries. *Nutrients, 15*(8), 1923. https://doi.org/10.3390/nu15081923

Eisenman, H. C., Mues, M., Weber, S. E., Frases, S., Chaskes, S., Gerfen, G., & Casadevall, A. (2007). Cryptococcus neoformans laccase catalyses melanin synthesis from both D- and L-DOPA. *Microbiology, 153*, 3954–3962.

El-Naggar, N. E. A., & El-Ewasy, S. M. (2017). Bioproduction, characterization, anticancer and antioxidant activities of extracellular melanin pigment produced by newly isolated microbial cell factories *Streptomyces glaucescens* NEAE-H. *Scientific Reports, 7*, 42129. https://doi.org/10.1038/srep42129

El-Naggar, N., & Saber, W. (2022). Natural melanin: Current trends, and future approaches, with especial reference to microbial source. *Polymers, 14*(7), 1339. https://doi.org/10.3390/polym14071339

El-Obeid, A. S., Kamal-Eldin, A., Abdelhalim, M. A. K., & Haseeb, A. M. (2017). Pharmacological properties of melanin and its function in health. *Basic & Clinical Pharmacology & Toxicology, 120*(6), 515–522. https://doi.org/10.1111/bcpt.12748

El-Zawawy, N. A., Kenawy, E. R., Ahmed, S., et al. (2024). Bioproduction and optimization of newly characterized melanin pigment from *Streptomyces djakartensis* NSS-3 with its anticancer, antimicrobial, and radioprotective properties. *Microbial Cell Factories, 23*, 23. https://doi.org/10.1186/s12934-023-02276-y

Embley, T. M., & Stackebrandt, E. (1994). The molecular phylogeny and systematics of the actinomycetes. *Annual Review of Microbiology, 48*, 257–289. https://doi.org/10.1146/annurev.mi.48.100194.001353

Faccio, G., Kruus, K., Saloheimo, M., & Thöny-Meyer, L. (2012). Bacterial tyrosinases and their applications. *Process Biochemistry, 47*, 1749–1760. https://doi.org/10.1016/j.procbio.2012.08.018

Fenical, W., & Jensen, P. (2006). Developing a new resource for drug discovery: Marine actinomycete bacteria. *Nature Chemical Biology, 2*, 666–673. https://doi.org/10.1038/nchembio841. (PDF) Review on Melanin from Marine Actinomycetes. Available from: https://www.researchgate.net/publication/343809590_Review_on_Melanin_from_Marine_Actinomycetes. Accessed 23 July 2025

Fu, P., Kong, F., Wang, Y., Wang, Y., Liu, P., Zuo, G.-Y., & Zhu, W. (2013). Antibiotic metabolites from the coral-associated actinomycete Streptomyces sp. OUCMDZ-1703. *Chinese Journal of Chemistry, 31*, 100–104. https://doi.org/10.1002/cjoc.201201062

Fuqua, W. C., Coyne, V. E., Stein, D. C., Lin, C.-M., & Weiner, R. M. (1991). Characterization of *mel*A: A gene encoding melanin biosynthesis from the marine bacterium *Shewanella colwelliana. Gene, 109*(1), 131–136. https://doi.org/10.1016/0378-1119(91)90598-6

Ghattavi, K., Homaei, A., Kamrani, E., & Kim, S.-K. (2022). Melanin pigment derived from marine organisms and its industrial applications. *Dyes and Pigments, 201*, 110214. https://doi.org/10.1016/j.dyepig.2022.110214

Guo, L., Li, W., Gu, Z., Wang, L., Guo, L., Ma, S., Li, C., Sun, J., Han, B., & Chang, J. (2023). Recent advances and progress on melanin: From source to application. *International Journal of Molecular Sciences, 24*(5), 4360. https://doi.org/10.3390/ijms24054360

Gupta, P., Imchen, M., & Kumavath, R. (2022). Exploration and characterization of melanin pigment produced by actinomycetes. In D. Dharumadurai (Ed.), *Methods in actinobacteriology* (Springer protocols handbooks). Humana. https://doi.org/10.1007/978-1-0716-1728-1_99

Hifnawy, M., Hassan, H. M., Mohammed, R., Fouda, M., Sayed, A. M., Hamed, A., et al. (2020). Induction of antibacterial metabolites by co-cultivation of two Red-Sea-sponge-associated actinomycetes Micromonospora sp. UR56 and Actinokinespora *sp. EG49. Marine Drugs, 18*(5), 243. https://doi.org/10.3390/md18050243

Hullo, M. F., Moszer, I., Danchin, A., & Martin-Verstraete, I. (2001). CotA of *Bacillus subtilis* is a copper-dependent laccase. *Journal of Bacteriology, 183*(18), 5426–5430. https://doi.org/10.1128/jb.183.18.5426-5430.2001

Hung, Y. C., Sava, V., Hong, M. Y., & Huang, G. S. (2004). Inhibitory effects on phospholipase A2 and antivenin activity of melanin extracted from *Thea sinensis* Linn. *Life Sciences, 74*(16), 2037–2047. https://doi.org/10.1016/j.lfs.2003.09.048

Jacobson, E. S. (2000). Pathogenic roles for fungal melanins. *Clinical Microbiology Reviews, 13*(4), 708–717. https://doi.org/10.1128/CMR.13.4.708

Jensen, P. R., Moore, B. S., & Fenical, W. (2015). The marine actinomycete genus Salinispora: A model organism for secondary metabolite discovery. *Natural Product Reports, 32*(5), 738–751. https://doi.org/10.1039/c4np00167b

Jia, H., Liu, N., Zhang, L., Li, P., Meng, Y., Yuan, W., Li, H., Tantai, D., Qu, Q., Cao, Z., & Dong, J. (2025). Fungal melanin in plant pathogens: Complex biosynthesis pathways and diverse biological functions. *Plants (Basel, Switzerland), 14*(14), 2121. https://doi.org/10.3390/plants14142121

Jose, P. A., & Jha, B. (2017). Intertidal marine sediment harbours Actinobacteria with promising bioactive and biosynthetic potential. *Scientific Reports, 7*, 10041. https://doi.org/10.1038/s41598-017-09672-6

Julianti, E., Abrian, I. A., Wibowo, M. S., Azhari, M., Tsurayya, N., Izzati, F., Juanssilfero, A. B., Bayu, A., Rahmawati, S. I., & Putra, M. Y. (2022). Secondary metabolites from marine-derived fungi and actinobacteria as potential sources of novel colorectal cancer drugs. *Marine Drugs, 20*(1), 67. https://doi.org/10.3390/md20010067

Kahng, H. Y., Chung, B. S., Lee, D. H., Jung, J. S., Park, J. H., & Jeon, C. O. (2009). *Cellulophaga tyrosinoxydans* sp. nov., a tyrosinase-producing bacterium isolated from seawater. *International Journal of Systematic and Evolutionary Microbiology, 59*(4), 654–657. https://doi.org/10.1099/ijs.0.003210-0

Kamarudheen, N., Naushad, T., & Rao, B. K. V. (2019). Biosynthesis, characterization and antagonistic applications of extracellular melanin pigment from marine Nocardiopsis Sps. *The Indian Journal of Pharmaceutical Education and Research, 53*(2S), s112–s120.

Khalifa, S. A. M., Elias, N., Farag, M. A., Chen, L., Saeed, A., Hegazy, M. F., & El-Seedi, H. R. (2019). Marine natural products: A source of novel anticancer drugs. *Marine Drugs, 17*(9), 491. https://doi.org/10.3390/md17090491

Kim, D., & Kim, H. (2024). Effectiveness of photothermal therapy using various noble-metal photothermal agents. *International Journal of Thermal Sciences, 200*, 108998. https://doi.org/10.1016/j.ijthermalsci.2024.108998

Kordjazi, T., Mariniello, L., Giosafatto, C. V. L., Porta, R., & Restaino, O. F. (2024). Streptomycetes as microbial cell factories for the biotechnological production of melanin. *International Journal of Molecular Sciences, 25*(5), 3013. https://doi.org/10.3390/ijms25053013

Kotakonda, M., Marappan, M., & Biswas, B. (2024). Characterization and antibacterial activity of melanin pigment from marine bacterium *Actinoalloteichus cyanogriseus*. *Letters in Drug Design & Discovery, 21*(5), 938–947. https://doi.org/10.2174/1570180820666230131093932

Krishnamoorthy, M., Dharmaraj, D., Rajendran, K., Karuppiah, K., Balasubramanian, M., & Ethiraj, K. (2020). Pharmacological activities of coral reef associated actinomycetes, *Saccharopolyspora* sp. IMA1. *Biocatalysis and Agricultural Biotechnology, 28*, 101748. https://doi.org/10.1016/j.bcab.2020.101748

Kurian, N. K., Nair, H. P., & Bhat, S. G. (2015). Evaluation of the anti-inflammatory property of melanin from marine Bacillus spp. BTCZ31. *Asian Journal of Pharmaceutical and Clinical Research, 8*(1), 251–255.

Langfelder, K., Streibel, M., Jahn, B., Haase, G., & Brakhage, A. A. (2003). Biosynthesis of fungal melanins and their importance for human pathogenic fungi. *Fungal Genetics and Biology, 38*(2), 143–158. https://doi.org/10.1016/S1087-1845(02)00526-1

Lechevalier, H., & Lechevalier, M. P. (1965). Classification des actinomycètes aérobies basée sur leur morphologie et leur composition chimique [Classification of aerobic actinomycetes based on their morphology and their chemical composition]. *Annales de l'Institut Pasteur, 108*(5), 662–673.

Lee, N. Y., Cha, C. J., Im, W. T., Kim, S. B., Seong, C. N., Bae, J. W., et al. (2018). A report of 42 unrecorded actinobacterial species in Korea. *Journal of Species Research, 7*(1), 36–49. https://doi.org/10.12651/JSR.2018.7.1.036

Leiva, F. J., Saenz-Díez, J. C., Martínez, E., Jiménez, E., & Blanco, J. (2015). Environmental impact of *Agaricus bisporus* mycelium production. *Agricultural Systems, 138*, 38–45. https://doi.org/10.1016/j.agsy.2015.05.003

Lesser, M. P., Morrow, K. M., Pankey, S. M., & Noonan, S. H. (2018). Diazotroph diversity and nitrogen fixation in the coral *Stylophora pistillata* from the Great Barrier Reef. *The ISME Journal, 12*(3), 813–824. https://doi.org/10.1038/s41396-017-0008-6

Li, Y. W., Xie, Y. J., Wang, Z., et al. (2016). Structure and function of iron-loaded synthetic melanin. *ACS Nano, 10*(11), 10186–10194. https://doi.org/10.1021/acsnano.6b05502

Li, C., Ji, C., & Tang, B. (2018). Purification, characterisation and biological activity of melanin from Streptomyces sp. *FEMS Microbiology Letters, 365*(19), fny077. https://doi.org/10.1093/femsle/fny077

Lorquin, F., Piccerelle, P., Orneto, C., Robin, M., & Lorquin, J. (2022). New insights and advances on pyomelanin production: From microbial synthesis to applications. *Journal of Industrial Microbiology & Biotechnology, 49*(4), kuac013. https://doi.org/10.1093/jimb/kuac013

Manivasagan, P., Venkatesan, J., Sivakumar, K., & Kim, S. K. (2013). Actinobacterial melanins: Current status and perspective for the future. *World Journal of Microbiology and Biotechnology, 29*(10), 1737–1750. https://doi.org/10.1007/s11274-013-1352-y

Mavridi-Printezi, A., Menichetti, A., Mordini, D., Amorati, R., & Montalti, M. (2023). Recent applications of melanin-like nanoparticles as antioxidant agents. *Antioxidants, 12*(4), 863. https://doi.org/10.3390/antiox12040863

Mendez, V., Agullo, L., Gonzalez, M., & Seeger, M. (2011). The homogentisate and homoprotocatechuate central pathways are involved in 3- and 4-hydroxyphenylacetate degradation by *Burkholderia xenovorans* LB400. *PLoS One, 6*(3), e17583. https://doi.org/10.1371/journal.pone.0017583

Montefiori, D. C., & Zhou, J. Y. (1991). Selective antiviral activity of synthetic soluble L-tyrosine and L-dopa melanins against human immunodeficiency virus in vitro. *Antiviral Research, 15*(1), 11–25. https://doi.org/10.1016/0166-3542(91)90037-r

Muniraj, I., Shameer, S., & Uthandi, S. (2021). Tyrosinase and laccase-producing *Bacillus aryabhattai* TFG5 and its role in the polymerization of phenols. *BMC Microbiology, 21*, 187. https://doi.org/10.1186/s12866-021-02258-3

Muthuraj, R., Meghashyama, P. B., Bhat, S. B., Dhanyakumara, H. M., Halaswamy, K. N., Shashiraj Bidhayak Chakraborty, S. S., & Nayaka, P. S. (2021). Isolation, identification and characterization of antimicrobial activity exhibiting actinomycete *Streptomyces paradoxus* strain KUASN-7 from soil. *International Journal of Current Microbiology and Applied Sciences, 10*(8), 164–176. https://doi.org/10.20546/ijcmas.2021.1008.021

Nair, A. S., Kumar, B. P., & Geo, J. A. (2017). Microbial production of textile grade pigments. *African Journal of Microbiology Research, 11*(42), 1532–1537. https://doi.org/10.5897/AJMR2017.8205

Nayaka, S., Muthuraj, R., Chakraborty, B., Bhat, M. P., Pallavi, S. S., Shashiraj, K. N., et al. (2020). A potential bioactive secondary metabolites and antimicrobial efficacy of *Streptomyces thermocarboxydus* strain KSA-2, isolated from Kali River, Karwar. *Current Research in Microbiology and Infection (CRMI), 1*(1), 5–13. https://doi.org/10.31559/CRMI2020.1.1.2

Nosanchuk, J. D., & Casadevall, A. (2003). The contribution of melanin to microbial pathogenesis. *Cellular Microbiology, 5*(4), 203–223. https://doi.org/10.1046/j.1462-5814.2003.00268.x

Parks, D. H., Chuvochina, M., Rinke, C., Mussig, A. J., Chaumeil, P. A., & Hugenholtz, P. (2022). GTDB: An ongoing census of bacterial and archaeal diversity through a phylogenetically consistent, rank normalized and complete genome-based taxonomy. *Nucleic Acids Research, 50*(D1), D785–D794. https://doi.org/10.1093/nar/gkab776

Patin, N. V., Duncan, K. R., Dorrestein, P. C., & Jensen, P. R. (2016). Competitive strategies differentiate closely related species of marine actinobacteria. *The ISME Journal, 10*(2), 478–490. https://doi.org/10.1038/ismej.2015.128

Pavan, M. E., López, N. I., & Pettinari, M. J. (2020). Melanin biosynthesis in bacteria: Regulation and production perspectives. *Applied Microbiology and Biotechnology, 104*, 1357–1370. https://doi.org/10.1007/s00253-019-10257-0

Pawlik, A., Wójcik, M., Rułka, K., Motyl-Gorzel, K., Osińska-Jaroszuk, M., Wielbo, J., et al. (2016). Purification and characterization of laccase from *Sinorhizobium meliloti* and analysis of the lacc *gene. International Journal of Biological Macromolecules, 92*, 138–147. https://doi.org/10.1016/j.ijbiomac.2016.07.012

Pfeifer, B. A., & Khosla, C. (2001). Biosynthesis of polyketides in heterologous hosts. *Microbiology and Molecular Biology Reviews: MMBR, 65*(1), 106–118. https://doi.org/10.1128/MMBR.65.1.106-118.2001

Ramesh, C., Tulasi, B. R., Raju, M., Thakur, N., & Dufossé, L. (2021). Marine natural products from tunicates and their associated microbes. *Marine Drugs, 19*(6), 308. https://doi.org/10.3390/md19060308

Roy, S., & Rhim, J. W. (2021). New insight into melanin for food packaging and biotechnology applications. *Critical Reviews in Food Science and Nutrition*, 1–27. https://doi.org/10.1080/10408398.2021.1878097

Rózanowska, M., Sarna, T., Land, E. J., & Truscott, T. G. (1999). Free radical scavenging properties of melanin interaction of eu- and pheo-melanin models with reducing and oxidising radicals. *Free Radical Biology & Medicine, 26*(5–6), 518–525. https://doi.org/10.1016/s0891-5849(98)00234-2

Rudrappa, M., Nayaka, S., & Kumar, R. S. (2023). In silico molecular docking approach of melanin against melanoma-causing MITF proteins and anticancer, oxidation-reduction, photoprotection, and drug-binding affinity properties of extracted melanin from Streptomyces sp. strain MR28. *Applied Biochemistry and Biotechnology, 195*(7), 4368–4386. https://doi.org/10.1007/s12010-023-04358-4

Ruzafa, C., Sanchez-Amat, A., & Solano, F. (1995). Characterization of the melanogenic system in *Vibrio cholerae,* ATCC 14035. *Pigment Cell Research, 8*(3), 147–152. https://doi.org/10.1111/j.1600-0749.1995.tb00656.x

Sanchez-Amat, A., Rufaza, C., & Solano, F. (1998). Comparative tyrosine degradation in *Vibrio cholera* strains. The strain ATCC 14035 as a prokaryotic melanogenic model of homogentisate-releasing cell. *Comparative Biochemistry and Physiology Part B, 119*, 557–562. https://doi.org/10.1016/s0305-0491(98)00028-5

Sánchez-Marroquín, A., & Zapata, M. (1954). Observations on the pigment of *Streptomyces coelicolor. Applied Microbiology, 2*(2), 102–107. https://doi.org/10.1128/am.2.2.102-107.1954

Sava, V. M., Hung, Y. C., Blagodarsky, V. A., Hong, M. Y., & Huang, G. S. (2003). The liver-protecting activity of melanin-like pigment derived from black tea. *Food Research International, 36*(5), 505–511. https://doi.org/10.1016/S0963-9969(02)00199-0

Seelam, S. D., Agsar, D., Halmuthur, M. S. K., Reddy Shetty, P., Vemireddy, S., Reddy, K. M., Umesh, M. K., & Rajitha, C. H. (2021). Characterization and photoprotective potentiality of lime dwelling Pseudomonas mediated melanin as sunscreen agent against UV-B radiations. *Journal of Photochemistry and Photobiology. B, Biology, 216*, 112126. https://doi.org/10.1016/j.jphotobiol.2021.112126

Shuster, V., & Fishman, A. (2009). Isolation, cloning and characterization of a tyrosinase with improved activity in organic solvents from *Bacillus megaterium. Journal of Molecular Microbiology and Biotechnology, 17*(4), 188–200. https://doi.org/10.1159/000233506

Singh, S., Nimse, S. B., Mathew, D. E., Dhimmar, A., Sahastrabudhe, H., Gajjar, A., Ghadge, V. A., Kumar, P., & Shinde, P. B. (2021). Microbial melanin: Recent advances in biosynthesis, extraction, characterization, and applications. *Biotechnology Advances, 53*, 107773. https://doi.org/10.1016/j.biotechadv.2021.107773

Sivaperumal, K., Kamala, R., & Rajaram, R. (2015). DOPA melanin isolated and characterised from a marine actinobacterium, Streptomyces sp. MVCS6 from the Versova coast. *Natural Product Research, 29*(17), 2117–2121. https://doi.org/10.1080/14786419.2014.988712

Snyman, M., Walsdorf, R. E., Wix, S. N., & Gill, J. G. (2024). The metabolism of melanin synthesis—From melanocytes to melanoma. *Pigment Cell & Melanoma Research, 37*(3), e13165. https://doi.org/10.1111/pcmr.13165

Solano, F. (2017). Melanin and melanin-related polymers as materials with biomedical and biotechnological applications—Cuttlefish ink and mussel foot proteins as inspired biomolecules. *International Journal of Molecular Sciences, 18*(7), 1561. https://doi.org/10.3390/ijms18071561

Sondhi, S., Chopra, N. K., Kumar, A., & Gupta, N. (2023). Laccase: A green solution for environmental problems. *Advances in Environmental Engineering Research, 4*(2), 030. https://doi.org/10.21926/aeer.2302030

Srinivasan, M., Keziah, S. M., Hemalatha, M., & Devi, C. S. (2017, November). Pigment from *Streptomyces bellus* MSA1 isolated from marine sediments. In *IOP conference series: Materials science and engineering* (Vol. 263, No. 2, p. 022049). IOP Publishing.

Stackebrandt, E., Rainey, F. A., & Ward-Rainey, N. L. (1997). Proposal for a new hierarchic classification system, actinobacteria classis nov. *International Journal of Systematic and Evolutionary Microbiology, 47*(2), 479–491. https://doi.org/10.1099/00207713-47-2-479

Steinert, M., Engelhard, H., Flugel, M., Wintermeyer, E., & Hacker, J. (1995). The LLY protein protects *Legionella pneumophila* from light but does not directly influence its intracellular survival in *Hartmannella vermiformis*. *Applied and Environmental Microbiology, 61*(6), 2428–2430. https://doi.org/10.1128/aem.61.6.2428-2430.1995

Sugumaran, M., Evans, J., Ito, S., & Wakamatsu, K. (2020). Nonenzymatic spontaneous oxidative transformation of 5,6-dihydroxyindole. *International Journal of Molecular Sciences, 21*(19), 7321. https://doi.org/10.3390/ijms21197321

Sun, W., Liu, C., Zhang, F., Zhao, M., & Li, Z. (2018). Comparative genomics provides insights into the marine adaptation in sponge-derived Kocuriaflava S43. *Frontiers in Microbiology, 9*, 1257. https://doi.org/10.3389/fmicb.2015.01048

Sundar, R., & Sivaperumal, P. (2022). Melanin pigments from sediment-associated Nocardiopsis sp. marine actinobacterium and antibacterial potential. *Journal of Advanced Pharmaceutical Technology & Research, 13*(Suppl 1), S88–S92. https://doi.org/10.4103/japtr.japtr_339_22

Sushma, B., Arivarasu, L., & Sivaperumal, P. (2025). Antioxidant potential from melanin pigment produced by marine actinobacterium of Nocardiopsis species. *Cuestiones de Fisioterapia, 54*(3), 4327–4341. https://doi.org/10.48047/w6pqe447

Tarangini, K., & Mishra, S. (2014). Production of melanin by soil microbial isolate on fruit waste extract: Two step optimization of key parameters. *Biotechnology Reports, 4*, 139–146. https://doi.org/10.1016/j.btre.2014.10.001

Thirumurugan, D., Vijayakumar, R., Vadivalagan, C., Karthika, P., & Alam Khan, M. K. (2018). Isolation, structure elucidation and antibacterial activity of methyl-4,8-dimethylundecanate from the marine actinobacterium *Streptomyces albogriseolus* ECR64. *Microbial Pathogenesis, 121*, 166–172. https://doi.org/10.1016/j.micpath.2018.05.025

Tran-Ly, A., Reyes, C., Schwarze, F., & Ribera, J. (2020). Microbial production of melanin and its various applications. *World Journal of Microbiology and Biotechnology, 36*, 170. https://doi.org/10.1007/s11274-020-02941-z

Valliappan, K., Sun, W., & Li, Z.-Y. (2014). Marine actinobacteria associated with marine organisms and their potential in producing pharmaceutical natural products. *Applied Microbiology and Biotechnology, 98*(17), 7365–7377. https://doi.org/10.1007/s00253-014-5954-6

Vasanthabharathi, V., Lakshminarayanan, R., & Jayalakshmi, S. (2011). Melanin production from marine Streptomyces. *African Journal of Biotechnology, 10*(54), 11224–11234. https://doi.org/10.5897/AJB11.296

Venil, C. K., & Lakshmanaperumalsamy, P. (2009). An insightful overview on microbial pigment, prodigiosin. *Electronic Journal of Biology, 5*(3), 49–61.

Wagner, M., Abdel-Mageed, W. M., Ebel, R., Bull, A. T., Goodfellow, M., Fiedler, H. P., & Jaspars, M. (2014). Dermacozines H-J isolated from a deep-sea strain of *Dermacoccus abyssi* from Mariana Trench sediments. *Journal of Natural Products, 77*(2), 416–420. https://doi.org/10.1021/np400952d

Wakamatsu, K., Zippin, J. H., & Ito, S. (2021). Chemical and biochemical control of skin pigmentation with special emphasis on mixed melanogenesis. *Pigment Cell & Melanoma Research, 34*(4), 730–747. https://doi.org/10.1111/pcmr.12970

Westerhof, W. (2006). The discovery of the human melanocyte. *Pigment Cell Research, 19*(3), 183–193. https://doi.org/10.1111/j.1600-0749.2006.00313.x

Woo, P. C. Y., Tam, E. W. T., Chong, K. T. K., Cai, J. J., Tung, E. T. K., Ngan, A. H. Y., Lau, S. K. P., & Yuen, K. Y. (2010). High diversity of polyketide synthase genes and the melanin biosynthesis gene cluster in *Penicillium marneffei. FEBS Journal, 277*(16), 3750–3758. https://doi.org/10.1111/j.1742-4658.2010.07776.x

Yabuuchi, E., & Ohyama, A. (1972). Characterization of "pyomelanin"-producing strains of *Pseudomonas aeruginosa. International Journal of Systematic Bacteriology, 22*(2), 53–64. https://doi.org/10.1099/00207713-22-2-53

You, J. L., Cao, L. X., Liu, G. F., Zhou, S. N., Tan, H. M., & Lin, Y. C. (2005). Isolation and characterization of actinomycetes antagonistic to pathogenic Vibrio spp. from nearshore marine sediments. *World Journal of Microbiology and Biotechnology, 21*(5), 679–682. https://doi.org/10.1007/s11274-004-3851-3. https://www.chemspider.com/Chemical-Structure.4884931.html

Zenova, G. M., Manucharova, N. A., & Zvyagintsev, D. G. (2011). Extremophilic and extremotolerant actinomycetes in different soil types. *Eurasian Soil Science, 44*, 417–436. https://doi.org/10.1134/S1064229311040132

Zimmermann, W. (1990). Degradation of lignin by bacteria. *Journal of Biotechnology, 13*(2–3), 119–130. https://doi.org/10.1016/0168-1656(90)90098-V

Chemical Additives from Marine Actinobacteria

Harishkumar Rajendran, Vinodhini Shanmugam, Kaviyarasi Renu, Anne Sahithi Somavarapu Thomas, Swarnkumar Reddy, and Saran Krishnamoorthy

Abstract

Actinobacteria constitute a diverse and widely distributed group of bacteria prevalent in both freshwater and marine ecosystems, where they play crucial roles in ecological processes such as nutrient cycling and symbiotic interactions. Freshwater actinobacteria, including genera such as Streptomyces and Nocardiopsis, are more abundant in lake sediments than in dynamic river environments, where they produce antimicrobial compounds effective against pathogens. Marine actinobacteria occupy a range of habitats, from surface waters to deep-sea sediments, including extreme environments, and exhibit extensive genetic and metabolic diversity, resulting in unique bioactive metabolites with pharmaceutical potential.

These compounds are promising candidates for the development of antibiofilm, antibacterial, and anticancer therapies. Key challenges in this field include difficulties in culturing, activation of silent biosynthetic gene clusters, low yields, complex purification processes, and genomic complexities. Emerging solutions include integrated omics, genome mining, synthetic biology, co-cultivation, and heterologous expression to unlock novel metabolites. Cost-effective fermenta-

H. Rajendran (✉) · V. Shanmugam · S. Reddy
Department of Biotechnology, Sri Sankara College of Arts and Science, Kanchipuram, Tamil Nadu, India
e-mail: rharishkumar@sankaracollege.edu.in; svinodhini@sankaracollege.edu.in; bswarnkumar@sankaracollege.edu.in

K. Renu · S. Krishnamoorthy
Center of Molecular Medicine and Diagnostics (COMManD), Department of Biochemistry, Saveetha Dental College and Hospitals, Saveetha Institute of Medical and Technical Sciences, Saveetha University, Chennai, Tamil Nadu, India
e-mail: kaviyarasir.sdc@saveetha.com; 162511001.sdc@saveetha.com

A. S. S. Thomas
Centre for Global Health Research, Saveetha Medical College and Hospital, Chennai, Tamil Nadu, India

K. Loganathan, Z. Li (eds.), *Marine Actinomycetes*,
https://doi.org/10.1007/978-981-95-9849-6_7

tion, scalable bioprocesses, and regulatory pathways are essential for industrial applications. Marine and freshwater actinobacteria are promising sources of novel therapeutic agents and biotechnological innovations, underscoring the need for continued research and innovative cultivation strategies to maximize their potential applications.

Keywords

Actinobacteria · Sources of marine actinobacteria · Secondary metabolites · Biosynthesis · Applications

1 Introduction

Actinobacteria are a diverse and widely distributed group of bacteria found in numerous ecosystems, including aquatic environments such as freshwater lakes, rivers, and marine areas (Zothanpuia et al., 2018). Although traditionally regarded as soil bacteria, their presence in aquatic ecosystems has been increasingly acknowledged. It is currently understood that actinobacteria are cosmopolitan, prevalent in both freshwater and marine sediments, and associated with various hosts, playing a crucial ecological role in producing biologically active compounds valuable for biotechnology and pharmaceuticals (Lewin et al., 2016). In freshwater environments, actinobacteria have been isolated from sediments in lakes and rivers, with lakes often exhibiting higher abundance owing to the stability of sediments compared to the ever-changing conditions of rivers. These actinobacteria include common species such as Streptomyces and rare species such as *Nocardiopsis* and *Rhodococcus*. They generate antimicrobial substances effective against various pathogenic microorganisms, making freshwater actinobacteria a promising source of new biologically active compounds (Zothanpuia et al., 2018).

Marine actinobacteria inhabit a wide range of environments, from surface waters to deep-sea sediments, including extreme habitats such as hydrothermal vents (Manivasagan et al., 2013). These microorganisms exhibit significant genetic and metabolic diversity and frequently produce unique biologically active metabolites that are distinct from those of their terrestrial counterparts. Various cultured actinobacteria, including *Streptomyces, Brachybacterium, and Microbacterium*, reside in marine sediments, with some potentially representing novel species (Claverías et al., 2015). The detection and distribution of actinobacteria in aquatic systems underscore their ecological significance in nutrient cycling, symbiotic interactions, and as a source of biologically active secondary metabolites with potential biotechnological applications. Despite existing advancements, the diversity and functional roles of these organisms in many aquatic environments remain only partially understood, highlighting the importance of further research in this area (Lewin et al., 2016).

1.1 Ecological Niches of Actinomycetes

Marine actinobacteria occupy a range of significant ecological niches within the ocean, including sediments, seawater, and associations with marine organisms, such as sponges, corals, shellfish, and algae. These bacteria are particularly abundant in marine sediments, where they contribute to the decomposition and recycling of complex organic compounds, such as chitin, and play a crucial role in marine biogeochemical cycles (Ribeiro et al., 2023). They establish stable communities across various marine ecosystems, including intertidal zones, coastal regions, and deep-sea environments, by adapting to conditions such as high salinity, fluctuating oxygen levels, pressure, and low temperatures (Manivasagan et al., 2013). Marine actinobacteria are present in both planktonic (free-floating) environments and biofilms and exhibit metabolic activity in their natural habitats, underscoring their adaptation to diverse marine ecosystems. Beyond their ecological importance, marine actinobacteria are recognized for synthesizing novel biologically active compounds with antimicrobial, antitumor, antioxidant, and other pharmacological properties. They represent a promising avenue for the discovery of natural products, as many marine species exhibit genetic and metabolic differences from their terrestrial counterparts, frequently yielding unique secondary metabolites.

Furthermore, marine actinobacteria have been isolated from extreme and less-explored marine environments, such as deep ocean sediments and hydrothermal vents, thereby expanding our understanding of their biological diversity and ecological functions (Jose & Jha, 2017).

1.2 The Importance of Chemical Additives from Marine Actinobacteria in Biotechnology and Industry

Marine actinobacteria are of significant importance in biotechnology and industry, primarily because of their capacity to synthesize a diverse array of valuable chemical additives, including antibiotics, enzymes, biologically active compounds, and biosurfactants. These metabolites have been extensively applied in various fields, including pharmaceuticals, aquaculture, environmental bioremediation, and other industrial processes.

- *Source of New Biologically Active Compounds*: Marine actinobacteria are prolific producers of secondary metabolites with unique structures and activities that often differ from those of their terrestrial counterparts because of the distinct characteristics of the marine environment. These metabolites include antibiotics, anticancer agents, and antioxidants. Notably, approximately two-thirds of natural antibiotics are derived from actinobacteria, and marine strains possess novel and unique molecules that are crucial for antibiotic discovery and drug development (Manivasagan et al., 2013).
- *Industrial Enzymes and Additives*: Marine actinobacteria synthesize enzymes and biochemical compounds of significant industrial relevance, which are uti-

lized in the textile, food, detergent, and chemical industries. These enzymes frequently exhibit distinctive characteristics, such as resistance to extreme conditions, including high salinity, pressure, and temperature, which are typical of the marine environment (James et al., 2023).

- *Environmental and Aquaculture Applications*: The metabolic versatility of these organisms enables them to decompose organic and inorganic waste, rendering them valuable for bioremediation and waste management in marine and aquaculture systems. Actinobacteria can also produce biosurfactants that facilitate hydrocarbon decomposition, enhance aquaculture water quality, and purify pollutants.
- *Ecological Role*: Actinobacteria play a crucial role in the nutrient cycle and decomposition of organic matter in marine ecosystems, thereby contributing to their health and sustainability. This ecological function is closely linked to enzymatic and metabolic adaptations in the marine environment.
- *Biotechnological Potential*: Recent advancements in genetic engineering and experimental evolution of microorganisms have significantly broadened the scope of utilizing marine actinobacteria for the production of novel chemical compounds and biotechnological products. This progress has spurred innovation in medicine, agriculture, and industry. Consequently, marine actinobacteria-derived compounds are of considerable interest for sustainable and novel biotechnological applications, underscoring their essential role as sources of chemical diversity with substantial industrial and therapeutic potential (James et al., 2023).

1.3 Rationale for Marine Sources as Alternative Reservoirs of Novel Compounds

The distinctive adaptations of marine actinobacteria justify the utilization of marine sources as alternative reservoirs for novel compounds, given the extreme and diverse marine environments that differ markedly from terrestrial habitats of these organisms. These environmental variations result in marine actinobacteria producing a wide array of secondary metabolites, often characterized by novel structures and unique biological activities that are absent in their terrestrial counterparts. Actinobacteria are prolific producers of biologically active secondary metabolites with extensive therapeutic applications, including antibiotics, antitumor agents, antifungal drugs, immunosuppressants, and enzymes. Although terrestrial actinobacteria have been extensively studied, marine actinobacteria remain largely unexplored, offering substantial chemical diversity and untapped genetic potential for drug discovery. This makes them attractive candidates for the discovery of new biologically active compounds, including those with antimicrobial, antitumor, and anti-inflammatory properties. The extreme conditions of the marine environment, such as high salinity, pressure, temperature fluctuations, and unique nutrient availability, compel marine actinobacteria to develop specialized metabolic pathways, leading to the biosynthesis of rare and structurally unique compounds. These include novel antibiotics such as salinosporamide and ilamycin, as well as other secondary metabolites with significant industrial and pharmaceutical potential (Ghosh et al., 2022).

2 Diversity and Ecology of Marine Actinobacteria

Exploring the diversity and ecology of marine actinobacteria has revealed considerable biodiversity, including numerous rare and novel taxa. This underscores the importance of these bacteria in marine ecosystems and their potential for pharmaceutical and biotechnological applications. These microorganisms constitute a stable and persistent population across diverse ocean habitats, highlighting their indigenous nature and ecological significance in marine environments. Marine actinobacteria are a highly diverse and ecologically significant group of microorganisms that are widely distributed in various marine environments, including sediments, seawater, sponges, corals, and seaweeds. Unlike their terrestrial counterparts, marine actinobacteria have adapted to extreme and diverse marine conditions, ranging from high pressure and low temperatures at great depths to acidic environments near hydrothermal vents, endowing them with unique genetic and metabolic characteristics (Claverías et al., 2015) (Fig. 1).

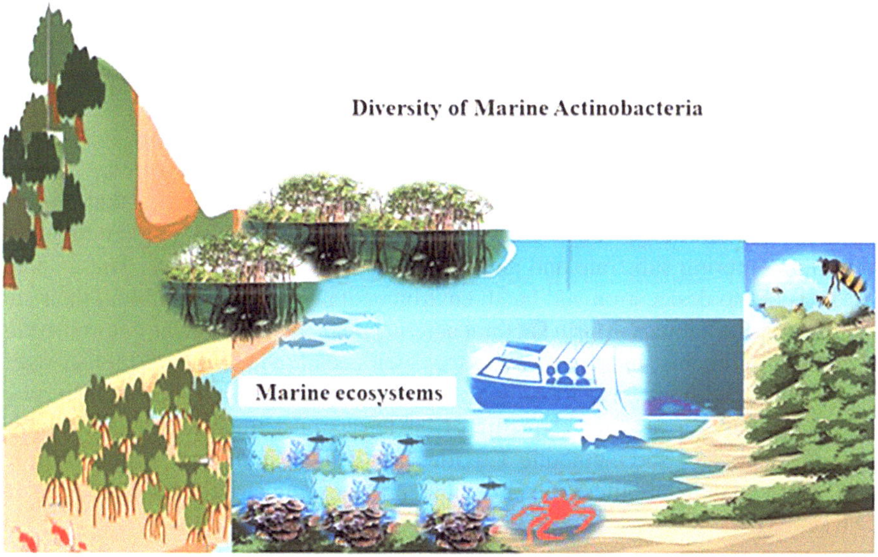

Fig. 1 Diversity of marine actinobacteria

2.1 Taxonomic Diversity

Marine actinobacteria represent a phylogenetically and taxonomically diverse group within the phylum Actinobacteria, also known as the "*Actinomycota.*" These microorganisms inhabit a broad spectrum of marine environments, including open oceans, deep-sea sediments, coral reefs, hydrothermal vents, and symbiotic niches within marine organisms. Recent advancements in both culture-dependent and independent methodologies have revealed significantly greater taxonomic diversity than previously recognized. *Phylum:* Actinobacteria includes a wide range of gram-positive bacteria with high levels of guanylate cyclases (G + C) in their DNA. Several orders and families have been consistently identified in the marine environment. *Order: Acidimicrobiales* and *Micrococcales*: Common in marine sediments and seawater, and are often identified by metagenomics. *Streptomycetales*: The well-known genus *Streptomyces* is present in both shallow and deep marine ecosystems (Jagannathan et al., 2021).

The diversity of these organisms is evident in their richness at both the genus and species levels, featuring notable genera such as *Streptomyces, Salinispora, and Nocardiopsis*, some of which are endemic to marine environments. From an ecological standpoint, marine actinobacteria are integral members of marine microbial communities and play crucial roles in the degradation of organic matter, nutrient cycling, and maintenance of microbial community dynamics through their antagonistic activity. Their ecological contribution is vital for the functioning of ecosystems in marine habitats, such as sediments, coral reefs, and sponges. Marine actinobacteria are a rich source of novel bioactive secondary metabolites, including antibiotics, anticancer agents, and enzymes, with significant potential for biotechnological and pharmaceutical applications in the future.

2.2 General Features of Marine Actinobacteria

Marine actinobacteria are a group of gram-positive aerobic bacteria that are predominantly non-motile. They can form filamentous structures similar to fungal mycelia, featuring substrate and aerial mycelia that produce spores. These spores are resistant to desiccation and harsh environmental conditions, enabling their survival in marine habitats. Marine actinobacteria inhabit diverse marine environments, including sediments, mangroves, and the open ocean, where they establish stable populations and contribute to ecological functions, such as the decomposition of organic matter (e.g., chitin). They differ from terrestrial actinobacteria by exhibiting unique phenotypes and metabolic activities, particularly in the production of bioactive secondary metabolites with potential pharmaceutical applications (Adler & Dücker, 2018).

2.3 Important Features of Marine Actinobacteria

Gram-positive filamentous bacteria, distinguished by their branched growth pattern, possess DNA with a high GC content (70–80%), indicative of genetic stability. Actinobacteria are aerobic, predominantly non-motile, and capable of forming spores. These spores are resistant to desiccation, facilitating their long-term survival. Morphologically, they resemble fungi, featuring both substrate and aerial mycelium, and demonstrate diverse metabolic capabilities, including the decomposition of refractory organic materials such as chitin. They inhabit various marine environments, including sediments, marine snow, mangroves, and deep-sea habitats. Actinobacteria produce a wide array of bioactive secondary metabolites, including antibiotics, anticancer drugs, and antifungal, antiviral, and immunosuppressive compounds. Among marine species, unique genera such as *Salinispora*, which are obligate marine actinobacteria, are included (Jagannathan et al., 2021).

2.3.1 Actinobacteria Habitat

Actinobacteria inhabit a wide range of environments, including terrestrial, aquatic, and extreme conditions. The ability to adapt to various ecological niches is a key factor in their success and diversity.

2.3.1.1 Soil

The primary habitat of many actinobacteria, especially genera such as *Streptomyces*, *Micromonospora*, and Nocardia, plays an important role in the decomposition of organic matter, including cellulose and lignin, and their presence is responsible for the earthy odor of soil (due to geosmin produced by *Streptomyces*).

2.3.1.2 Aquatic Environment

They inhabit both freshwater and marine ecosystems, and marine actinobacteria, such as *Salinispora*, have developed unique metabolic pathways and produce novel bioactive compounds. Some freshwater species play a role in nutrient cycling and degradation.

2.3.1.3 Extreme Environments

It is present in acidic, alkaline, saline, and thermophilic habitats such as hot springs, deserts, and salt lakes. Extremophilic actinobacteria are a source of new enzymes used in biotechnology owing to their stability under harsh environments.

2.3.2 Plant-Associated Habitats

Rhizosphere Many actinobacteria form beneficial associations with plant roots, promoting plant growth through nitrogen fixation and phytohormone production.

Endophytes Some species live inside plant tissues, providing protection against pathogens and improving nutrient uptake.

2.3.3 Human and Animal Microbiota

Actinobacteria, such as *Bifidobacterium*, are integral components of the gut microbiota, particularly in infants. They contribute significantly to digestion, immune system development, and defense against pathogenic organisms (Cui et al., 2022; Qin et al., 2019).

2.3.3.1 Marine Sediments

Actinobacteria are a substantial group of gram-positive bacteria known for their exceptional capacity to synthesize a diverse array of bioactive secondary metabolites. These compounds have been extensively used in medicine, particularly as antibiotics, antifungals, anticancer agents, and immunosuppressive agents. While terrestrial ecosystems have traditionally served as the primary source for the discovery of Actinobacteria, recent decades have witnessed increasing interest in marine environments, especially marine sediments, as reservoirs of untapped microbial diversity. Marine sediments provide unique physicochemical conditions that promote microbial adaptation and metabolic innovation, leading to the identification of novel Actinobacteria taxa and bioactive compounds. The subsequent sections elaborate on the diversity of Actinobacteria and the biotechnological potential of the four major marine sediment types (Fig. 2).

2.3.3.2 Deep-Sea Sediments

The deep-sea environment, characterized by high hydrostatic pressure, low temperatures, limited light, and nutrient deficiencies, represents one of the most extreme habitats on earth. These harsh conditions necessitate the development of unique survival strategies by microorganisms, including the production of specialized secondary metabolites. Recent investigations of dormant sulfide hills and other deep-sea sites have led to the isolation of several new genera of actinobacteria, including *Actinotalea* and *Dietzia*. Many of these deep-sea isolates exhibit potent antimicrobial and anticancer properties, underscoring the pharmaceutical potential of deep-sea actinobacteria. In addition to their metabolic potential, deep-sea actinobacteria often exhibit unique morphological and physiological features that distinguish them from terrestrial counterparts. This underscores the importance of continued bioprospecting of deep-sea marine sediments (Jose et al., 2021).

2.3.3.3 Intertidal Sediments

Intertidal zones are dynamic environments situated between high and low tide levels, where sediments alternately submerge and emerge at the surface. These zones experience frequent fluctuations in temperature, salinity, oxygen availability, and nutrient levels, which support the development of robust and metabolically diverse microbial communities. Investigations of actinobacteria from intertidal sediments have identified several genera with known biosynthetic capabilities, including *Streptomyces*, *Micromonospora*, and *Nocardiopsis* sps. Various isolates from these zones have demonstrated the capacity to produce potent antibacterial compounds, some of which are active against multidrug-resistant pathogens. The relative paucity

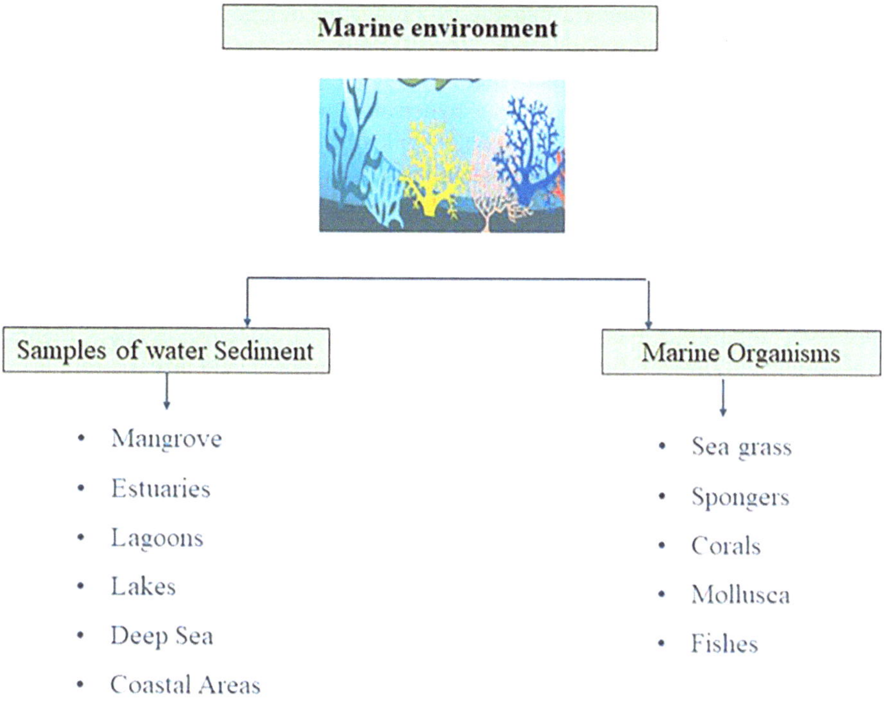

Fig. 2 Different sources of marine actinobacteria

of knowledge regarding intertidal sediments renders them promising sources for future microbial and natural product studies (Ribeiro et al., 2023).

2.3.3.4 Mangrove Sediments

Mangrove ecosystems are distinctive coastal habitats found in tropical and subtropical regions. These environments are characterized by varying salinity levels, high organic matter content, and dense root systems, which create microhabitats conducive to the growth and interactions of microorganisms.

Actinomycetes isolated from mangrove sediments, particularly those belonging to the genus Streptomyces, have been shown to produce diverse bioactive metabolites. Many of these compounds exhibit strong antibacterial and antifungal properties, suggesting their potential therapeutic efficacy. Beyond their medicinal applications, mangrove actinomycetes play crucial ecological roles in nutrient cycling and organic matter decomposition, further underscoring their significance.

2.3.3.5 Hypersaline Sediments

High-salinity environments, such as salt lakes, salt lagoons, and coastal salt marshes, are extreme ecosystems that experience high osmotic stress. Such conditions exert selective pressure on microbial communities, favoring the growth of halotolerant and halophilic actinobacteria capable of producing unique osmoprotectants and stress-responsive compounds.

One of the most important examples is *Streptomyces chilikensis*, a metal-tolerant strain of actinobacteria isolated from hypersaline water. This organism produces novel secondary metabolites that are not typically found in actinobacteria in less extreme environments. The unique metabolism of hypersaline actinobacteria makes them valuable candidates for the development of new drugs and industrial enzymes adapted to harsh processing conditions (Ribeiro et al., 2023).

Marine sediments are a diverse but largely unexplored resource for identifying novel actinomycetes and their bioactive compounds. Each sediment type, including deep-sea, intertidal, mangrove, and hypersaline environments, imposes distinct environmental stresses that promote microbial innovation. Ongoing exploration and bioprospecting in these settings hold significant potential for developing new antibiotics and other therapeutic agents, particularly considering the escalating challenge of antibiotic resistance (Rahman et al., 2019).

2.3.4 Marine Animals and Their Symbiotic Actinobacteria

In symbiotic relationships, actinomycetes often produce defensive compounds that protect host animals from pathogens, fouling organisms, and other environmental threats. In return, actinomycetes benefit from stable habitats and nutrient access. These symbiotic relationships are increasingly recognized for their ecological and medical potential.

2.3.4.1 Sponges (Phylum: Porifera)

Sponges are one of the most abundant sources of actinomycetes in the ocean. These simple filter-feeding organisms can harbor dense microbial communities, with symbiotic bacteria comprising up to 40% of the sponge biomass. In this microenvironment, actinomycetes play important roles in both host defense and nutrient cycling. Numerous strains of actinomycetes isolated from sponges have been shown to produce a variety of bioactive compounds, including antibacterial, antifungal, antiparasitic, and anticancer agents, among others. These metabolites may help sponges resist microbial invasion and surface pollution and may also represent promising resources for pharmaceutical development (Chen et al., 2021; Sun et al., 2015).

2.3.4.2 Corals (Cnidaria)

Corals, particularly reef-building corals, host complex microbial communities within their mucus, tissues, and calcium carbonate skeletons. These communities include actinomycetes, such as Jiangella and Micromonospora, which demonstrate potent antibacterial properties. These bacteria are thought to promote coral health by producing compounds that inhibit the growth of pathogens that cause coral bleaching and disease. Coral reefs are threatened by climate change and pollution, and understanding and harnessing these microbial interactions may be critical for their conservation and restoration of coral reefs (Siro et al., 2022).

2.3.4.3 Mollusks and Ascidians (Tunicates)

Mollusks (e.g., snails, bivalves, and octopuses) and ascidians (marine tunicates) are important hosts for various actinomycetes. In particular, ascidians have attracted attention because of their ability to produce potent anticancer compounds and enzyme inhibitors. For example, several actinomycete strains isolated from ascidians produce protease inhibitors that regulate host-pathogen interactions and have potential biomedical applications. Such invertebrates often possess chemical defense systems derived from symbiotic microorganisms, providing them with an advantage in densely populated, chemically competitive marine environments (Wang et al., 2020).

2.3.4.4 Fish

Marine fish primarily live on the mucous membranes of their intestines, gills, and skin. These microorganisms produce extracellular enzymes such as amylase, protease, and lipase, which aid in digestion and nutrient absorption. Additionally, some actinomycetes produce antibacterial compounds that potentially support fish health by suppressing pathogens in their microbiomes. Actinomycetes found in fish may also play a role in immune regulation and pathogen resistance; however, research in this area is still in its early stages and requires further investigation. Using these bacteria as probiotics and biocontrol agents, particularly in aquaculture, can improve fish health and sustainability (Manivasagan et al., 2014a; Valliappan et al., 2014).

2.3.5 Marine Plants and Their Symbiotic Actinobacteria

While much attention has been paid to the symbiotic relationships between animals and microorganisms, seaweeds and macroalgae also host diverse groups of actinomycetes that contribute to host defense, environmental sustainability, and the availability of natural resources.

2.3.5.1 Seaweeds (Macroalgae)

Brown and red seaweeds host diverse epiphytic and endophytic actinomycetes. These microorganisms often produce antibacterial secondary metabolites that may enhance algal resistance to adhesion and infections. Researchers have isolated diverse actinomycete strains from kelp (*Laminaria ochroleuca*), including those from the genera *Streptomyces, Rhodococcus, Nonomuraea, Microbispora, Isoptericola, Nocardiopsis,* and *Microbacterium.*

Notably, approximately 50% of the isolates showed antifungal activity (e.g., against *Candida albicans*), and several strains showed activity against *Staphylococcus aureus.* A larger study on seaweeds identified at least 22 genera of actinomycetes in brown, green, and red algae, including *Streptomyces, Micromonospora, Nocardiopsis, Rhodococcus, Amycolatopsis, Kocuria,* and *Salinispora.* In addition, actinomycetes from brown seaweed (Padina pavonica) have shown antibacterial activity against various human and animal pathogens, suggesting that they may be capable of chemical defense.

2.3.5.2 Seagrass (Angiosperms)

Seagrasses contain a rich microbiota, including actinomycetes, which are thought to play a role in pathogen suppression. A seaweed-derived strain of Streptomyces argenteolus TMA13 exhibited potent antibacterial activity against fish pathogens, including *Aeromonas hydrophila*, *Edwardsiella tarda*, *Vibrio harveyi*, and *Vibrio anguillarum*, and was resistant to biofilm formation. Seaweeds, such as *Syringodium isoetifolium*, an actinomycete isolate closely related to Streptomyces, yield potent anthraquinone compounds with broad-spectrum antibacterial activity against fish pathogens. A review of seaweed-associated actinomycetes has confirmed their antibacterial activity and called for further genome-wide studies to identify novel bioactive compounds in these bacteria.

2.3.5.3 Seawater and Marine Snow

Although not as abundant as host-associated niches, marine suspended media support actinomycetes with promising properties. Although the open ocean is not a dense habitat for actinomycetes, species such as Streptomyces, Salinispora, and Marinispora have been isolated from water samples, thereby increasing our knowledge of free-living marine actinomycetes.

Marine snow is an accumulation of organic particles that settle in the water column and serve as a nutrient medium for microorganisms. Actinomycetes cultured from these particles exhibit antagonistic activity toward other microorganisms, highlighting their ecological importance in marine microbial dynamics.

2.3.6 The Ecological Role of Actinobacteria

Actinobacteria perform essential ecological functions that support ecosystem stability, nutrient cycling, and biodiversity. They are habituated in various environments, as mentioned in Table 1.

- *Decomposition and nutrient cycling:* They decompose complex organic polymers, such as cellulose, chitin, keratin, and lignin. They participate in the carbon, nitrogen, and sulfur cycles, contributing to soil fertility and ecosystem productivity.

Table 1 Actinobacteria across environments

Environment type	Typical roles & adaptations
Soil (Terrestrial)	Organic matter decomposition, antibiotic production, and soil health
Freshwater & marine	Nutrient cycling, diverse phylogenetic groups, involved in aquatic ecosystems
Deserts/extreme environments	Stress-resistant, novel species, potential sources of unique secondary metabolites
Plant & animal symbioses	Growth promotion, disease protection, mutualistic interactions (e.g., ants, rhizospheres)
Caves & cold regions	Survival in darkness and low nutrients, cold-resistance, and reproduction via spores

- *Antibiotic production:* They produce a wide range of antimicrobial compounds that inhibit the growth of competing microorganisms, and this function not only shapes microbial communities but is also used for pharmaceutical purposes (e.g., streptomycin and tetracycline).
- *Symbiotic relationships:* Form mutualistic relationships with plants (e.g., Frankia in root nodules of actinorhizal plants for nitrogen fixation) and affect plant health, resilience, and nutrient uptake.
- *Bioremediation:* Some species can degrade pollutants, such as hydrocarbons, pesticides, and heavy metals, and are used to clean up contaminated soils and water using microbial remediation methods.
- *Industrial applications:* Their metabolic versatility makes them valuable for the production of enzymes, biosurfactants, and bioconversion processes. They serve as sources of new genes and bioactive molecules for biotechnology and drug discovery (Chen et al., 2016).

2.3.7 Biotechnological Potential

The great diversity of marine actinobacteria has made them prime targets for bioprospecting, especially as sources of new natural products and bioactive compounds. Actinobacteria produce a wide range of secondary metabolites with diverse biological activities, including *antimicrobials,* which are critical for combating drug-resistant pathogens. *Anticancer/antineoplastic agents:* Marine-derived actinobacteria have yielded novel cytotoxic compounds effective against various human cancer cell lines. *Enzymes:* Marine strains produce potent enzymes, such as amylases, proteases, and chitinases, which function over a wide range of temperatures and pH levels, making them valuable for industrial applications. *Antioxidants and biosurfactants:* These compounds have potential applications in the food, cosmetics, and pharmaceutical industries. *Bioremediation agents:* Some marine actinobacteria can degrade pollutants and remove heavy metals, such as uranium, from the environment (Sarkar & Suthindhiran, 2022) (Fig. 3).

2.3.8 Challenges and Prospects

Despite their potential, research on marine actinobacteria diversity faces several challenges. *Cultivation:* Many marine actinobacteria are rare and difficult to culture using traditional laboratory methods. Therefore, developing new methods is essential for accessing this untapped diversity. *Molecular methods:* Combining traditional cultivation with advanced molecular methods, such as metagenomics, provides a more complete picture of the true diversity and biosynthetic potential of these organisms. *Conservation:* The conservation of marine biodiversity, particularly in poorly understood or fragile ecosystems such as hydrothermal vents, is critical for future bioprospecting and drug discovery research (Claverías et al., 2015).

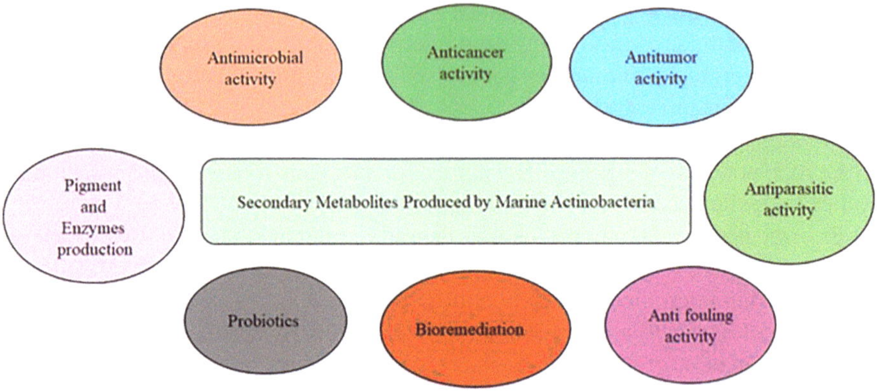

Fig. 3 Importance of secondary metabolites produced by marine actinobacteria

3 Biosynthesis of Chemical Additives

Marine actinobacteria are prolific producers of secondary metabolites with diverse chemical structures and functions. They belong to a diverse class of gram-negative bacteria that thrive in ocean sediments, corals, sponges, and other aquatic niches (Srinivasan et al., 2021). Marine actinobacteria are notable for their rich secondary metabolic profiles and potential for adaptation to dynamic marine environments. These secondary metabolites include antibiotics, pigments, surfactants, enzyme inhibitors, enzymes, and signaling molecules, collectively referred to as chemical additives. Unlocking the metabolic pathways involved in the synthesis of these additives enhances the production of potential secondary metabolites for sustainable application. This section discusses the biosynthesis of secondary metabolites by marine actinobacteria (Carretero-Molina, 2019).

Genetic richness is a key potential of marine actinobacteria, with each strain harboring 15–25 biosynthetic gene clusters (BGCs). Many of these traits remain dormant under standard culture conditions, limiting the synthesis of secondary metabolites (Dinesh et al., 2017). Induction of these gene clusters through molecular inducers or environmental cues expands the range of secondary metabolites, including antimicrobial peptides and enzyme inhibitors. With the increase in global demand for eco-friendly additives, knowledge of the biosynthesis of chemical additives enables genetic engineering to enhance yield and plays a vital role in drug discovery (van Bergeijk et al., 2020).

3.1 Biosynthetic Pathways in Marine Actinobacteria

Sophisticated enzymatic pathways are the heart of chemical additive production in marine actinobacteria, which assemble complex molecules from simple precursors to produce these bioactive compounds. Non-ribosomal peptide synthetases (NRPS)

and polyketide synthases (PKS) are the two primary systems that dominate the synthesis of these compounds (Martín et al., 2021). NPRS pathways synthesize peptide-based additives through a non-ribosomal mechanism, incorporating unusual amino acids and creating cyclic or branched peptides. Nisin, a polycyclic antibacterial peptide isolated from *Streptomyces* sp., is predominantly used as a food preservative to prevent spoilage. The PKS pathway is differentiated into three types by the iterative condensation of acyl-CoA units, resulting in polyketide antibiotics and antifouling agents. Type I modular PKSs from *Salinispora* and *Streptomyces* synthesize metabolites such as salinosporamide A, which acts as a potent proteasome inhibitor (Amoutzias et al., 2016).

Hybrid PKS–NRPS biosynthetic systems represent a convergence of polyketide and non-ribosomal peptide assembly lines, resulting in structurally diverse molecules that often exhibit pronounced bioactivity (Paulus, 2018). In marine actinobacteria, these pathways harness the modularity of polyketide synthases (PKS), which catalyze the sequential condensation of simple acyl-CoA building blocks, and the versatility of non-ribosomal peptide synthetases (NRPS), which can incorporate various amino acid residues, including non-proteinogenic types, through adenylation, thiolation, and condensation (Finking & Marahiel, 2004).

The interplay between PKS and NRPS modules enables the construction of complex hybrid metabolites, such as thiocoraline, which is synthesized by *Micromonospora* isolated from marine sediments (Perez Baz et al., 1997).

Thiocoraline features both polyketide- and peptide-derived structural elements, endowing it with notable antitumor activity through DNA bisintercalation and inhibition of replication enzymes, a mechanism supported by its unique macrocyclic and aromatic frameworks (Erba et al., 1999).

In addition to hybrid pathways, marine actinobacteria leverage terpene synthases to build isoprenoid scaffolds. These enzymes catalyze the cyclization and rearrangement of prenyl diphosphate precursors to produce molecules with diverse ring systems and stereochemical properties (De Rop et al., 2021). For instance, marine-derived terpenoids exhibit anti-inflammatory, cytotoxic, and antifouling activities, reflecting their adaptation to competitive aquatic environments (De La Hoz-Romo et al., 2025).

Moreover, alkaloid biosynthetic routes utilize specialized enzymes, including aminotransferases and cyclases, to construct nitrogen-containing heterocycles. Pyrrolizidine alkaloids, noted for antiviral and anti-inflammatory properties, exemplify this class. Their biosynthesis involves the formation of bicyclic structures through intramolecular cyclization and subsequent functional modifications, which are tightly regulated at genetic and enzymatic levels in marine actinobacteria (Tarasova et al., 2023; Morgan et al., 2023).

Collectively, these biosynthetic strategies highlight the genetic and enzymatic adaptability of marine actinobacteria, enabling the generation of structurally and functionally diverse chemical additives with considerable pharmacological and ecological significance.

3.2 Regulation and Diversification of Biosynthesis

3.2.1 Environmental Modulation

Environmental modulation plays a decisive role in determining the secondary metabolite profiles of marine Actinobacteria. Recent research has shown that physical and chemical factors, such as salinity, temperature, hydrostatic pressure, nutrient limitation, and the presence of competing microorganisms, directly influence the expression of biosynthetic gene clusters (BGCs). For instance, co-cultivation experiments, in which actinobacteria are grown together with other marine microbes, increase metabolite diversity by simulating natural ecological competition and activating previously silent BGCs.

Metabolomic profiling and transcriptomic analyses have revealed that nutrient stress, such as phosphate or iron limitation, can induce the production of novel polyketides and peptides by upregulating specific genes involved in secondary metabolism. Manipulation of osmotic conditions, particularly salinity, has also been shown to alter biosynthetic output, promoting the synthesis of distinctive glycosylated compounds and halogenated metabolites that contribute to chemical defenses and ecological fitness in marine organisms.

Advanced approaches now combine these environmental triggers with genome mining and systems biology to optimize product yields and discover novel chemical entities. For example, recent studies have demonstrated that targeted changes in pH or the addition of specific signaling molecules, such as quorum-sensing autoinducers, can unlock cryptic pathways and enhance metabolic flux toward desirable chemical additives. This emergent understanding underscores the dynamic interplay between marine microhabitats and the genetic regulation of specialized metabolite production, offering new avenues for bioprospecting and industrial biotechnological applications.

3.2.2 Mutasynthesis and Precursor Engineering

Mutasynthesis and precursor engineering are increasingly influential strategies for diversifying and enhancing the biosynthesis of natural products in marine actinobacteria, with several recent advances demonstrating their utility (Manivasagan et al., 2014a). In mutasynthesis, the targeted inactivation of key biosynthetic genes creates blockages in native metabolic pathways, enabling the incorporation of externally supplied substrate analogues into the metabolic pathway. For example, studies on marine actinobacterial genera, such as *Salinispora* and *Streptomyces*, have utilized mutant strains incapable of producing certain polyketide intermediates, which, upon feeding with structurally modified precursors, yield novel derivatives with improved pharmacological properties or bioactivity profiles (Hu et al., 2022).

Precursor engineering extends this principle by modulating the supply and structure of metabolic precursors, such as acetate, propionate, and diverse amino acids, thereby expanding the chemical diversity of the final product. Recent studies have employed genome mining and metabolic profiling to identify bottlenecks in biosynthetic gene clusters (BGCs) using mutasynthesis to introduce rare or synthetic

Table 2 Methods of chemical synthesis of chemical additives

Method	Key approach	Example additives
Conventional fermentation	Natural fermentation and extraction	Antibiotics, enzymes, alkaloids
Metal nanoparticle biosynthesis	CFE-mediated reduction of metal salts	AgNPs, AuNPs, ZnO NPs
Metabolic & synthetic engineering	Genetic pathway enhancement or transfer	Novel polyketides, peptides, and tailored surfactants
Precursor-directed/ combinatorial	Feeding analogues or combining biosynthetic genes	New antibiotics, anticancer agents
Mutasynthesis/ semi-synthesis	Targeted genetic block + precursor supplementation	Unique derivatives of commercial additives

building blocks in situ. For instance, this technique has led to the generation of aberrant polyketides and peptides, highlighted by the creation of abyssomicin C from *Salinispora arenicola*, wherein the addition of non-native precursors resulted in improved antibacterial efficacy against resistant pathogens (Selim et al., 2021),

High-throughput screening methods and integrative metabolic characterization tools further support these strategies by enabling the parallel assessment of thousands of engineered actinobacterial strains and their metabolic outputs. Such approaches also utilize antisense gene silencing and pathway refactoring to optimize flux toward the desired additives while minimizing competitive or unwanted side products, effectively balancing chemical innovation with biosynthetic feasibility (Table 2) (Kumar et al., 2025).

Horizontal gene transfer and BGC rearrangement, which are well-documented in marine actinobacteria, create additional opportunities for mutasynthesis in these organisms. By swapping domains or modules within PKS and NRPS clusters, researchers can design custom biosynthetic routes that respond to precursor changes, yielding metabolites with enhanced or altered bioactivity. These advances increasingly position mutasynthesis and precursor engineering as central tools in marine drug discovery and biotechnological innovation, enabling the targeted biosynthesis of previously inaccessible chemical additives with therapeutic and industrial relevance to humans (Table 3).

3.2.3 Genome Mining and Synthetic Biology

Recent advances in genome mining and synthetic biology have revolutionized the exploration and exploitation of marine actinobacteria for natural product biosynthesis. Genome mining, the computational identification and analysis of biosynthetic gene clusters (BGCs) within sequenced genomes, has revealed that marine actinobacteria harbor an unexpectedly large and diverse repertoire of gene clusters encoding enzymes responsible for producing complex secondary metabolites (Gontang et al., 2010). For instance, recent *in-silico* studies on various *Streptomyces* and *Nocardiopsis* species isolated from marine sponges demonstrated the presence of 17–28 distinct BGCs per strain, many of which exhibit low similarity to known clusters, suggesting a high potential for the discovery of novel bioactive compounds.

Table 3 Examples of chemical additives and pathways

Compound/class	Biosynthetic pathway	Marine source	Application
Thiocoraline	PKS	*Micromonospora sp.*	Cytotoxic agent
Salinosporamide Derivatives	PKS-NRPS Hybrid	*Salinispora tropica*	Anti-cancer, proteasome inhibitor
Naseseazine B	NRPS	*Streptomyces sp.*	Antimicrobial
Isocoumarins	PKS	*Streptomyces sp.*	Anti-inflammatory
Silver/zinc nanoparticles	Reductive biosynthesis	*Nocardiopsis/ Streptomyces*	Antimicrobial, catalyst

This genomic richness is complemented by insights into adaptive genetic features, such as enhanced systems for protein turnover, posttranslational modification, and signal transduction, which facilitate secondary metabolism under the unique environmental pressures of marine habitats. By combining genome mining with functional genomics and metabolomics, researchers can link specific BGCs to their chemical products and ecological roles, accelerating natural product characterization and expanding the chemical space accessible from marine actinobacteria (Selim et al., 2021).

Synthetic biology approaches have further empowered this field by enabling the refactoring and heterologous expression of BGCs in more tractable host organisms, thereby overcoming the challenges posed by difficult-to-culture marine strains and cryptic cluster activation (Gomez-Escribano & Bibb, 2014). Techniques such as CRISPR–Cas9–mediated genome editing, pathway reconstitution, and modular cloning have been employed to manipulate gene clusters to increase the yield, structural diversification, and production of novel derivatives. These synthetic strategies also facilitate the rational design of natural products, in which combinations of PKS, NRPS, and tailoring enzymes can be engineered to produce molecules with tailored functionalities and enhanced bioactivities (Gomez-Escribano & Bibb, 2014).

Moreover, advances in systems biology have enabled the integrated modelling of metabolic fluxes within marine actinobacteria, helping to optimize precursor supply and cofactor availability to maximize biosynthetic efficiency. High-throughput sequencing and machine learning algorithms coupled with genome mining to predict and prioritize BGCs have the greatest biotechnological promise and streamline experimental efforts.

3.2.3.1 Factors Influencing Biosynthesis

Various factors influence the biosynthesis of chemical additives in marine actinobacteria, which are known for their ability to produce diverse and valuable secondary metabolites. These variables span genetic, environmental, nutritional, and cultivation aspects, affecting both the quantity and spectrum of the compounds produced (Table 4).

The biosynthetic potential of marine actinobacteria is dictated by the presence and expression of biosynthetic gene clusters (BGCs) that encode polyketides, nonribosomal peptides, terpenes, and other metabolites. Comparative genomic studies

Table 4 Factors and impacts on biosynthesis

Factor	Impact on biosynthesis
Genetic composition	Diversity/activation of biosynthetic gene clusters
Salinity/temperature	Metabolite diversity and quantity
Nutrient sources	Enhanced production yield and spectrum
Co-culture	Activation of silent BGCs, novel compound yields
Molecular methods	Diversification, structure-activity relationship

have revealed that strains adapted to marine environments, such as *Streptomyces albidoflavus* VIP-1, possess stress-response genes, anti-stress mechanisms, and expanded BGCs, contributing to elevated biosynthetic diversity relative to their terrestrial relatives (Dejong et al., 2016). The activation of cryptic or silent BGCs, often triggered by specific conditions or pathway engineering, leads to the emergence of novel metabolites.

Key environmental parameters, such as temperature, salinity, pressure, and nutrient availability, profoundly shape the metabolite profiles. Marine actinobacteria are exposed to fluctuating salinities, which can induce the biosynthesis of protective and bioactive additives (Dhakal et al., 2017). Temperature changes alter metabolic pathways and enzyme activities, potentially increasing or decreasing metabolite concentrations and diversity. Unique marine conditions, including interactions with other microorganisms and exposure to stressors, also serve as powerful triggers of secondary metabolite production (Ibrahim et al., 2025).

The composition of the culture medium, particularly the carbon and nitrogen sources, is one of the most significant factors influencing the biosynthesis of chemical additives in the fermentation industry.

Optimized sources, such as starch and yeast extract, in precise combinations and concentrations, have resulted in marked increases (up to 218%) in the production of anti-infective metabolites (Manivasagan et al., 2014b). The presence of seawater, correct salt concentration, and supplementation with rare earth elements such as lanthanum chloride can further stimulate biosynthetic activity. Agitation speed, pH, calcium carbonate levels, and the incorporation of elicitors (e.g., soybean meal and sodium chloride) also play vital roles in maximizing metabolite production.

Co-culturing marine actinobacteria with other microbes activates silent gene clusters and boosts the yield of chemical additives during fermentation. These strategies mimic natural microbial competition and signaling, unlocking the production of compounds not observed in monoculture settings. Advanced techniques, such as mutasynthesis, precursor-directed biosynthesis, and combinatorial biosynthesis, are instrumental in diversifying and optimizing the types of additives produced. The introduction of specific precursors or analogues, genetic modifications, and pathway recombination enables rational management of metabolic outputs. Synthetic biology approaches further facilitate the tailored expression of desired metabolites (Koech et al., 2025).

4 Major Classes of Chemical Additives from Marine Actinobacteria

Actinobacteria are one of the largest taxonomic units among the major groups of bacteria. Actinobacteria genomes sequenced to date belong to species that are important in biotechnology, ecology, medicine, and veterinary science (Jagannathan et al., 2021). The observed genomic variability of these organisms reflects their biodiversity. The vast majority of actinobacteria are free-living organisms that inhabit both aquatic and terrestrial environments. Although many species of bacteria and fungi naturally produce bioactive substances, actinobacteria, or more specifically actinomycetes, are the most sought-after group of microorganisms capable of producing these secondary metabolites (Elnahas et al., 2021). The importance of this order is due to its ability to produce different classes of antibiotics with distinct chemical structures and mechanisms of action. According to recent studies, *Actinomycetes* are a valuable resource for the discovery of new natural products because of their special sets of enzymes that allow them to produce molecules with potential use in a variety of fields (De Simeis & Serra, 2021). The fact that different genera and species of actinomycetes can produce the same class of antibiotics and, in rare cases, the same chemical molecule, further suggests that metabolic pathways within the order are highly conserved. Actinomycetes also synthesize several types of antibiotics. Among them are aminoglycosides, ansamycins, epoxides, peptides, β-lactams, macrolides, tetracyclines, lincosamides, and aminocoumarins (Stincone & Brandelli, 2020). During the search for actinomycete secondary metabolites, compounds have been discovered that can be utilized for various industrial purposes, such as the production of plastics (Al-Shaibani et al., 2021). Phenolic acid decarboxylase from another species has been introduced into *S. lividans* to enable the conversion of cellulose to 4-vinylphenol, which is an additive for plastics (Noda et al., 2015). In another study, the synthesis of benzoate in Streptomyces maritimus was improved, allowing the direct extraction of benzoic acid from cellulose using a new technology. Benzoic acid is widely used in pharmaceuticals, food preservation, cosmetics, and personal care products. Although this is not a comprehensive enumeration of all potential applications of actinomycetes, it illustrates their adaptability to industrial environments (Silva et al., 2022). We anticipate that the unique characteristics of marine actinomycetes will be progressively utilized in these industries as they are identified and cultivated. Various marine actinomycetes have been used to isolate these biologically active chemicals (Table 5).

Table 5 Various marine actinomycetes have been used to isolate biologically active chemicals

Chemical family	Bioactive compounds	Chemical process	Marine source	References
Bisindole pyrrole	Lynamicins, spiroindimicins	Deep-sea marine sediment	*Streptomyces* sp.	Zhang et al. (2012)
Dechlorinated lipopeptide	Taromycin A	Moderately effective against MDR infections	*Saccharomonospora* sp.	Reynolds et al. (2018)
Steroidal alkaloids	Anandins	Marine sediments from the mangrove zone	*Streptomyces anandii*	Zhang et al. (2017)
Carotenoid	Sioxanthin	Siderophore	*Salinispora* sp.	Richter et al. (2015)
Alkaloid	Antimycin	Marine sediments from the mangrove zone	*Streptomyces* sp.	Li et al. (2022)
Pentangular polyphenols	Hexaricins A-C	–	*Streptosporangium* sp.	Tian et al. (2016)
Paulomycin derivatives	Paulomycin G	Deep-sea marine sediment	*Micromonospora matsumotoense*	Sarmiento-Vizcaíno et al. (2017)
Quinomycin-like depsipeptide	Retimycin A	Cytotoxicity against HCT-116	*Salinispora* sp.	Duncan et al. (2015)
Dipeptide	Petrocidin A	Symbiont to sponge, *Petrosia ficiformis*	*Streptomyces* sp.	Lee et al. (2021)
Polyene macrolactams	Lobosamides A-C	Anti-protozoan parasite, *Trypanosoma brucei*	*Micromonospora* sp.	Schulze et al. (2015)
Peptide	Bonnevillamides	Sediment	*Streptomyces* sp.	Wu et al. (2017)
Peptide	3-Isopropylhexahydro1H-pyridol [1,2-a] pyrazine-1,4(6H)-dione	Marine samples	*Rhodococcus* sp.	Gopi et al. (2014)
Benzofluorene	Nenestatin A	Antibacterial activity	*Micromonospora echinospora*	Jiang et al. (2017)
Alkaloid	Manzamine A	Symbiont to sponge Acanthostrongylophoraigens	*Micromonospora* sp.	Kim et al. (2017)
Quinones	N-acetyl-N-demethyl-mayamycin and Streptoanthraquinone A	Marine sediments	*Streptomyces* sp.	Liang et al. (2016)

(continued)

Table 5 (continued)

Chemical family	Bioactive compounds	Chemical process	Marine source	References
Triene polyol	Succinilenes	Sediment	*Streptomyces* sp.	Bae et al. (2017)
Glycosidic spirotetronates	Tetrocarcin N and O	Modest antibacterial activity	*Micromonospora*	Gong et al. (2018)
Napyradiomycin	MDN-0170	Symbiont to Ascidian	*Streptomyces* sp.	Mayer et al. (2021)
Ether	Nesterenkoniane	Deep-sea water sample	*Nesterenkonia flava*	Subramani and Sipkema (2019)
Ester	Dinactin	Marine sediments	*Streptomyces* sp.	Tuttle et al. (2019)
Peptide	N-(4-minocyclooctyl)-3,5-dinitrobenzamide	Sediment sample from the mangrove ecosystem	*Pseudonocardia endophytica*	Mangamuri et al. (2016)
Spirotetronate	Lobophorin K	Symbiont to deep-sea coral *Lophelia pertusa*	*Streptomyces* sp.	Braña et al. (2017)
Rifsaliniketal	Saliniketal	–	*Salinispora* sp.	Feng et al. (2016)

4.1 Antimicrobial Compounds

Marine actinomycetes are a major source of marine natural products because of their unique chemical structures and diverse biological activities. A statistical study of marine microbial natural products from 2010 to 2013 showed that actinomycetes of marine origin accounted for 28% (253 of 895) of the newly discovered marine microbial natural products (Zhao Chengying, 2013). Antimicrobial substances have diverse chemical structures, including sterols, polyketides, terpenoids, and nitrogen-containing molecules. Most of these substances are natural antibacterial products, accounting for 87% of all recently discovered natural marine products derived from actinomycetes. Additionally, *Streptomyces* sp. Ni-80 yielded urauchimycins, which exhibited antifungal activity against Candida albicans at 10 μg/mL. *Streptomyces* sp. HB202 yielded eight novel antibacterial streptophenazines, A–H (3–10) (Mitova et al., 2008). The MIC values for these drugs ranged from 15.6 to 62.5 μg/mL, indicating a wide range of inhibitory activities against bacterial species. *Streptomyces* sp. SCSGAA 0027 isolated four naguanoic acids, namely B–E, which exhibited weak antibiofilm activity against *Shewanella onedensis* MR-1 biofilms (Nong et al., 2016). In another study, a culture of *Streptomyces* sp. SCSIO 41399 yielded isotyrandamycin B (26), which exhibited antibacterial activity against *Streptococcus agalactiae*, with an MIC value of 11.5 μM (Cong et al., 2019). By the end of 2024, 353 new antibacterial compounds derived from marine actinomycetes have been reported to date. In addition to 2007 and 2009, more secondary metabolites have been identified in marine actinomycetes since 2016.

4.2 Enzymes and Biocatalysts

Microorganisms have been reported to produce approximately 23,000 different antibiotics. Ten thousand antibiotics have been identified in actinobacteria. Streptomyces can produce a broad range of primary and secondary metabolites with the potential to be bioactive, including antibiotics (Salwan & Sharma, 2020). Multidomain polyketide synthases (PKSs) synthesize a class of polyketide molecules with diverse structures. Most polyketide chemicals produced by marine microbes are industrially and biologically important. It exhibits antimicrobial, cytotoxic, antioxidant, antiviral, and anticancer properties are among its biological properties. In addition, marine microalgae are used in biotechnological applications as test organisms in ecotoxicology and as biocatalysts in the biotransformation of marine natural products and the creation of secondary metabolites (Gamal & Shreadah, 2024). Research on the production of marine biocatalysts using various bioengineering techniques is of interest. Various bioprocess strategies adopted for the cultivation of marine organisms for enzyme production have been described, including laboratory- and pilot-scale bioreactors for the production of protease, chitinase, agarase, and peroxidase from marine bacteria or fungi using three main operating modes: batch, fed-batch, and continuous. A remarkable example is Pat-G, a practical and versatile biocatalyst for peptide cyclization. Studies on epoxide hydrolases with special stereochemistry have also shown significant advantages for marine enzymes (Hecko et al., 2023).

4.3 Biosurfactants

Biosurfactants can separate at the interface between different liquid phases, such as water, oil, and air, owing to their amphiphilic nature. Different microorganisms produce different types of biosurfactants, such as rhamnolipids, glycolipids, and sophorolipids. Low-molecular-weight biosurfactants, such as glycolipids and lipopeptides, effectively reduce interfacial and surface tension (Das et al., 2013). Examples of high-molecular-weight substances include extracellular polysaccharides, lipopolysaccharides, proteins, and lipoproteins, which exhibit a high degree of surface binding. Compared to commercial surfactants such as sodium lauryl sulphate, biosurfactants offer unique advantages such as non-toxicity, increased foaming, increased biodegradability, high selectivity, and special activity under extreme salinity, pH, and temperature conditions. Therefore, unlike synthetic surfactants, biosurfactants have a higher degree of biodegradability, are environmentally friendly, and possess highly desirable surface characteristics (Jahan et al., 2020). To make these compounds economical on a commercial scale, despite their potential for several applications, manufacturing expenses must be reduced, and high-yield mutant strains must be created. The potential uses of Biosurfactant-based products have a wide range of applications, including lotions, facial cosmetics, cleansers, and hair care products. These products have special qualities, such as anti-radical properties, stimulation of dermal fibroblast metabolism, and hygroscopic characteristics, which promote healthy skin physiology. Currently, much research is being conducted on the production and analysis of biosurfactants that can replace chemical surfactants in cosmetic formulations (Karnwal et al., 2023). Bacterial lipases are another type of biosurfactant that exhibits enzymatic activity. Bacterial lipases are enzymes with important physiological and industrial applications. According to the latest market analysis, the market for enzymes used in cosmetics is expected to grow at a compound annual growth rate of 5% in 2015, driven by advances in technology and consumer awareness of the effective properties of enzymes (Kumar et al., 2014). Toothpaste is typically formulated with complex mixtures of surfactants, humectants, abrasives, tartar control chemicals, anticaries agents, pH buffers, humectants, and binders.

4.4 Antioxidants and Anti-Inflammatory Agents

Bioactive chemicals derived from *Streptomyces* exhibit antibacterial, anticancer, antitumor, cytotoxic, immunosuppressive, antihypertensive, antioxidant, plant growth-promoting, insecticidal, and herbicidal properties. CRISPR–Cas12a and CRISPR–Cas9 stimulate dormant biosynthetic gene clusters, revealing novel antibacterial and anticancer agents. These methods have been successfully applied to genera such as *Micromonospora* and *Actinoplanes* to activate dormant gene clusters, enhance metabolite production, and probe the activity of specific genes (Heng et al., 2021). Several strains of *Streptomyces* and other strains, such as *Salmonella typhimurium* KCTC1925 and *Staphylococcus aureus* KCTC1927, have demonstrated antioxidant, anti-inflammatory, and antibacterial activities (Ribeiro et al., 2023). Additionally, recent studies have isolated actinomycetes from major marine environments and found that their exopolysaccharides have immunomodulatory activity. The isolated compounds demonstrated anti-inflammatory activity, as confirmed in lipopolysaccharide-activated macrophages, by reducing the production of tumor necrosis factor, nitric oxide, and interleukin levels from *Streptomyces* sp. 13G036 (Gao et al., 2022).

4.5 Metal Chelators and Biocides

Members of the genus *Streptomyces* and the CMNR group (*Corynebacterium, Mycobacterium, Nocardia*, and *Rhodococcus*) are actinobacteria with large genomes and numerous linear and circular megaplasmids, which help explain their remarkable metabolic capabilities and resistance to toxic metals (Presentato et al., 2020). Different species of the Actinobacteria phylum have been reported to carry out metal(loid) sorption methods based on the metal(loid) diversity found in a particular ecological niche and its physicochemical properties. The structural chemistry of the cell wall and membrane surface, that is, the charge and orientation of the metal(loid)-binding groups, determines the yield, affinity, and specificity of the bioprocess. In addition, reactive functional groups that can mediate the sorption and deposition of metals on the bacterial surface, such as hydroxyl [OH−], carbonyl [CO], carboxyl [COO−], phosphate [PO43−], sulfate [SO42−], and amino [NH2+/NH3+], are present. For example, when cells were exposed to Zn2+ and Cd2+, *S. zinciresistens* exhibited a bald, non-spore-forming phenotype with short and thin mycelium, which was most likely a result of the interaction between OH−, CO, and NH, resulting in the formation of dense granules that can be observed on the cell surface using scanning electron microscopy (SEM) (Oyegoke & Adnan, 2024). In contrast, extracellular polymeric substances (EPS) play an important role as biosorbents of these chemicals, preventing their penetration into bacterial cells, owing to their exceptional ability to bind and capture metalloid ions from the extracellular environment. For example, *Amycolatopsis* sp. (Yang et al., 2023). AB0-generated EPSs are primarily composed of polysaccharides with approximately 20 repeating glucose units that can bind Cu2+. Upon the addition of Cu2+ and Pb2+, *K. rizophila* (formerly *Micrococcus luteus*) DE2008 cells secret more EPS containing more carbohydrates and uronic acids than EPS obtained without metal addition, resulting in increased metal immobilization efficiency (Presentato et al., 2020). Thus, a deeper understanding of these biotic processes is critical for developing effective bioremediation strategies that utilize pure cultures of actinobacteria to clean metalloid-contaminated environments in the context of the circular economy and green chemistry.

5 Case Studies and Notable Discoveries

5.1 Environmental and Industrial Additives

Bioremediation Agents Certain marine actinobacteria, including species such as *Streptomyces* and *Mycobacterium*, can break down pesticides and various organic pollutants, making them ideal for purifying contaminated marine and terrestrial environments. These enzymes, such as tyrosinase, are highly effective in eliminating phenolic compounds from polluted water, surpassing the efficiency of traditional methods (Jagannathan et al., 2021; Roy et al., 2014).

Plastic Industry Additives Genetically modified strains of *Streptomyces lividans* can transform cellulose into 4-vinylphenol, which is a significant component of plastic production. Similarly, Streptomyces maritimus has been altered to effectively synthesize benzoic acid from cellulose, a compound extensively utilized in the cosmetic industry, as a food preservative, and in pharmaceuticals (Jagannathan et al., 2021; Roy et al., 2014). Research has demonstrated that Streptomyces species, including *S. lividans*, can absorb heavy metals such as Cu (II) and Cd (II) (Amoroso et al., 1998).

5.2 Agricultural and Food Applications

Enzyme Production Enzymes such as amylase, lipase, and protease produced by marine actinobacteria are beneficial in industries such as food processing, textiles, and detergent production.

Aquaculture Probiotics Incorporating planktonic *Streptomyces* into shrimp diets improves digestion and boosts disease resistance, serving as a safe probiotic option in aquaculture (Roy et al., 2014).

5.3 Novel Bioactive Compounds

Antimicrobial Agents Numerous secondary metabolites derived from marine actinobacteria exhibit strong antimicrobial, anticancer, and quorum-quenching properties, offering potential leads for the development of new pharmaceutical and biotechnological products (Wang et al., 2020; Manivasagan et al., 2014b; Betancur et al., 2017).

Phytopathogen Control Research indicates that marine actinobacteria sourced from coral reefs can suppress plant pathogens and interfere with microbial communication, positioning them as promising biocontrol agents in agriculture (Jagannathan et al., 2021; Betancur et al., 2017).

5.4 Industrial Relevance

Structural Diversity Marine actinobacteria generate a diverse array of chemical structures, including polyketides, peptides, and terpenoids, making them an abundant source for the discovery of new industrial additives.

Biotechnological Potential Industries can exploit their ability to thrive in harsh environments by utilizing their distinctive metabolic processes to create new or more effective chemical additives, biopolymers, and agents for environmental remediation (Jagannathan et al., 2021; Amoroso et al., 1998).

Significant discoveries of chemical additives derived from marine actinobacteria (actinomycetes) include novel antibiotics, alkaloids, polyketides, peptides, macrolactams, and specialized metabolites, all of which have applications in industrial, pharmaceutical, and environmental contexts (Manivasagan et al., 2014b).

(a) In 2024, 45 novel antibacterial compounds were identified from marine actinomycetes, predominantly comprising polyketides (such as salinipostins and marinomycins), alkaloids, macrolactams, and peptides. These compounds exhibit significant activities against multidrug-resistant pathogens. Of particular note is corynetoxin U17a, which exhibits strong efficacy against *Staphylococcus aureus* at remarkably low concentrations (Pan et al., 2025).
(b) Between 2017 and 2021, 77 novel alkaloids were identified from marine actinobacteria, encompassing a range of structural classes, including indoles, diketopiperazines, glutarimides, indolizidines, and pyrroles. Notable examples include strepchazolins A and B, which are piperidines derived from *Streptomyces chartreusis*, and novel quinolines, specifically marinoterpins A–C, which exhibit unique biosynthetic origins (De Rop et al., 2021).
(c) Guanahanolide A, a meroterpenoid isolated from coral-derived Streptomyces species, exerts cytotoxic effects against various cancer cell lines. Furthermore, novel antimicins obtained from sponge-derived Streptomyces strains exhibit significant antifungal activity (Kanchanasin et al., 2024).
(d) Over 100 novel bioactive compounds have been isolated from 38 species of actinomycetes across 15 genera, including *Nocardiopsis, Salinispora, Marinispora, Micromonospora,* and *Streptosporangium*, as reported in various studies. Many of these compounds are currently undergoing preclinical and clinical development as antibiotics or other bioactive agents (Ibrahim et al., 2025).
(e) Marine actinomycetes possess unique biosynthetic pathways, cryptic gene clusters, and extrachromosomal plasmids owing to the challenging and competitive

nature of their environment. These genetic elements encode a diverse array of secondary metabolites, many of which are activated only under specific environmental stresses or microbial interactions, resulting in the production of structurally novel compounds (Jagannathan et al., 2021).

(f) The industrial significance of these findings is evident in several sectors. In bioremediation, enzymes such as tyrosinase play a crucial role, and the plastics industry benefits from the production of compounds such as benzoic acid. Therefore, the application of unique enzymes in food processing is crucial. Additionally, metabolites that inhibit phytopathogens are of particular importance in agricultural biocontrol (Jagannathan et al., 2021; Dhakal et al., 2017).

6 Technological Advances in Discovery and Production

Technological advances have revolutionized the discovery and production of chemical additives from marine actinobacteria (actinomycetes), enabling the identification of novel bioactive compounds and the efficient development of their industrial and therapeutic applications.

6.1 Advances in Discovery

High-throughput sequencing has revealed a substantial diversity of biosynthetic gene clusters (BGCs) in marine actinomycetes, many of which were previously considered "cryptic," meaning they were not expressed under laboratory conditions. Genome mining platforms facilitate the prediction and prioritization of novel natural product scaffolds for targeted isolation and characterization (Jagannathan et al., 2021; Dhakal et al., 2017; Koech et al., 2025).

(a) *Metagenomics:* The use of methodologies to examine environmental DNA from uncultivable marine actinomycetes has significantly expanded the potential for the discovery of previously inaccessible biosynthetic pathways.
(b) *Bioassay-guided fractionation & high-throughput screening:* By integrating cell-based phenotypic assays with advanced analytical chemistry techniques, such as LC-MS/MS and NMR, the rapid identification and structural elucidation of potent compounds, including novel antivirals and antibiotics, is significantly enhanced (Dhakal et al., 2017; Kanchanasin et al., 2024).
(c) *Bioinformatics and molecular fingerprinting:* Genes encoding polyketide synthases and non-ribosomal peptide synthetases are vital markers. Molecular fingerprinting enables the identification of actinobacteria with notable biosynthetic potential (Jose & Jha, 2017).

6.2 Advances in Production

(a) *Metabolic engineering & synthetic biology:* The yield and variety of chemical additives can be enhanced through targeted modifications of biosynthetic pathways using CRISPR, gene knockouts, and enzyme engineering. For instance, by engineering *S. lividans*, it became possible to produce 4-vinylphenol, a plastic additive, from cellulose (Jagannathan et al., 2021).
(b) *Fermentation and bioprocess optimization:* Advancements in bioreactor design and scale-up techniques have become more sophisticated, enabling the economical large-scale production of bioactive substances, enzymes, and chemical precursors that are important for the food, cosmetics, pharmaceutical, and plastics industries (Dhakal et al., 2017; Zhao et al., 2016).
(c) *Chemoenzymatic and semi-synthetic methods:* Combining biological production with chemical synthesis methods, such as mutasynthesis and total enzymatic synthesis, enables the development of new analogues with enhanced functional characteristics that may be more effective than the natural compounds. For instance, the bioinspired synthesis of salinosporamides and their derivatives aims to increase their potency (Dhakal et al., 2017).

Functional Diversification: Through chemical modification and strategic design, new functional molecules, such as altered benzoic acid or innovative detergents, have been developed with tailored properties for industrial applications (Jagannathan et al., 2021).

6.3 Multi-strategy Screening & One Strain Many Compounds (Osmac) of Marine Actinobacteria

Marine actinobacteria include genera such as *Streptomyces, Verrucosispora, Salinispora,* and *Micromonospora.* Many studies have used multi-strategy screening pipelines that combine traditional bioassay-guided discovery with contemporary omics-based methods, genome mining, combinatorial biosynthesis, and synthetic biology to uncover hidden chemical diversity (Subramani & Aalbersberg, 2012; Breitling et al., 2021). The OSMAC approach has been successfully applied to the study of marine actinobacteria. This affects salinity, pH, aeration, and nutrient composition. In addition, inorganic salts, such as NaBr or NaCl, activate silent BGCs and produce new metabolites. Strain prioritization by metabolomics (SPM), which profiles the chemical space of marine actinobacteria and identifies those with distinct metabolite fingerprints, uses LC-MS/MS or NMR to further prioritize strains for study (Shaaban et al., 2021).

6.4 Artificial Intelligence (AI) and Machine Learning (ML)

Marine actinobacteria, especially those belonging to the genera *Streptomyces, Salinispora, Micromonospora*, and *Verrucosispora,* produce many bioactive secondary metabolites. A large amount of their biosynthetic potential remains unrealized because of the difficulties presented by their large, high-GC-content genomes and the prevalence of cryptic or silent BGCs (León-Zayas et al., 2017; Jensen et al., 2015). Despite their value, traditional genome mining and culture-based methods frequently fall short of revealing their complete chemical diversity. By combining genomic, transcriptomic, and metabolomic data, AI and ML tools have begun to overcome these restrictions and more accurately identify, categorize, and rank BGCs (Navarro-Muñoz et al., 2020). AI-driven analyses combined with metabolomics platforms, such as GNPS, enable targeted experimental validation by highlighting distinct molecular features, predicting bioactivity profiles, and facilitating the rapid dereplication of known compounds (Wang et al., 2016). From the above findings, we can predict that many studies will expedite the discovery of novel marine actinobacteria metabolites with pharmaceutical relevance by optimizing culture conditions and systematically activating silent pathways by coordinating computational predictions, such as metagenomic exploration or the OSMAC approach.

6.5 High-Throughput Screening

High-throughput screening (HTS) has become a crucial method for accelerating the discovery of bioactive compounds from marine actinobacteria by allowing the rapid assessment of numerous strains or extracts using automated miniature assays. A total of 142 morphologically unique predatory actinobacterial strains were isolated from the Moroccan Atlantic coastal sediments. Rapid antimicrobial screening was used to identify isolates that exhibited strong antimicrobial activity against multidrug-resistant pathogens, including methicillin-resistant *Staphylococcus aureus* (MRSA) (Ibrahimi et al., 2020). Antarctic biomolecules from marine actinobacteria can be used for environmentally friendly industrial and medicinal applications.

Fridmanniella sp. SPF8's antifreeze proteins (AFP) are strong emulsifiers, cryoprotectants, and antioxidants with low cytotoxicity, making them attractive options for use in industrial formulations, cryopreservation, and biomedicine (Kannan & Sivaperumal, 2025). High-throughput microplate-based bioassays have been used to screen the extracts of deep-sea *Streptomyces* and related actinobacteria against *S. aureus,* MRSA, *Candida albicans,* and *Acinetobacter baumannii* after cultivating them in small-scale fermentation (Xu et al., 2018).

6.6 Biomedical Applications

There are various biomedical applications that include nanoparticles used for dental caries. The marine-derived bacterium, *Streptomyces plicatus,* synthesized with the help of zinc oxide nanoparticles (ZnONPs), has significant antibacterial and antibiofilm activities against *Streptococcus mutans*, which is associated with dental caries. It has also been used to demonstrate dose-dependent effects against OSCC (Soni et al., 2024).

The study on actinobacteria highlights the promising path of marine actinobacteria for finding new therapeutic molecules from unusual sources, as highlighted by the Actinobacterium SRV02 study, which promotes sustainable resource use and medical innovation. This is in line with the larger goal of minimizing our environmental footprint while utilizing nature's power to address pressing global issues (Parthasarathy et al., 2024).

7 Challenges and Future Prospects

7.1 Cultivation and Access to Silent BGCS

Marine microorganisms remain unculturable using current laboratory techniques, limiting the discovery of novel metabolites. The primary challenges in harnessing marine actinobacteria are the optimization of chemical additives for cultivation and unlocking the silent or cryptic biosynthetic capabilities of natural diversity (Ahamefule et al., 2020). Marine microorganism culture conditions were tested for BGCs. This often remains silent and underexplores novel chemical structures (Rui et al., 2025).

OSMAC is one of the strategies in which progress has been made for genome mining, co-cultivation to activate the silent clusters, and heterologous expression (Rui et al., 2025). Although systematic, these strategies for marine-derived actinobacteria have a success rate that varies considerably with labor-intensive tasks. The integration of metabolomics, bioinformatics, and transcriptomics helps prioritize BCGs and study the conditions under which they regulate metabolite generation (Sharma et al., 2018).

7.2 Low Yields and Purification

In marine actinobacteria, active compounds are often produced at low yields, either through fermentation or directly from marine-isolated samples. Furthermore, it requires structural characterization, which hampers purification and application development. Scaling up its production often requires heterologous hosts or synthetic biology to reconstruct pathways or employ heterologous hosts. These approaches may alter the activities or structures of the compounds. Marine actinobacteria produce complex molecules that are difficult to separate from culture media matrices. This requires chromatographic and analytical tools, which are not readily accessible (Rui et al., 2025).

7.3 Genomic and Bioinformatic Challenges

Recent advancements in next-generation sequencing and genome mining have empowered translational computational predictions, and the discovery of biosynthetic potential into actual compounds remains notorious. Linking specific BGCs to their metabolites and genetic action by activating sophisticated multi-tiered data integration (Sharma et al., 2018). Actinobacterial genomes are large and contain repetitive sequences, which complicate accurate assembly. Functional gene annotation remains a major bottleneck, especially if it is a unique or poorly characterized biosynthetic enzyme. Recent advancements in genetic engineering and heterologous expression systems have brought hope, but they must be paired with domain experts and tool development.

7.4 Environmental and Ecological Barriers

The marine ecosystem is complex, and in specific symbiotic or environmental contexts, it increases the production of metabolites. Replicating these conditions is difficult. Combined with other microorganisms, acting like its host or simulating stressors, may help with secondary metabolism, which is difficult to standardize (Bull et al., 2005). Biogeographic factors influence metabolite novelty through distinct biosynthetic reports (Kumar & Rao, 2012).

7.5 Commercial and Industrial Translation Hurdles

Scaling up fermentation, formulation, regulatory approval, and cost efficiency affects commercial viability (Aladejana, 2023). Marine actinobacteria enzymes, such as extemozymes (e.g., cellulase), often have advantageous traits, such as salt or temperature tolerance (Suriya et al., 2016).

8 Future Prospects

8.1 Integrated Omics and Synthetic Biology

Understanding regulatory networks and pathway activation can be improved by thoroughly integrating data from the genome, transcriptome, metabolome, and proteome. When combined with synthetic biology, this could make it possible to produce novel compounds in tractable hosts in a customizable manner (Sharma et al., 2018).

8.2 Cultivation Innovation

Co-cultivation methods, high-throughput microculture, and marine microhabitat simulation can aid in eliciting silent pathways. Improved environmental mimicry might be possible with new culture platforms, such as in situ cultivation equipment or microfluidics (Sharma et al., 2018).

8.3 Heterologous Expression and Pathway Refactoring

Advanced heterologous platforms, built-in model actinomycetes, or chassis organisms, can express marine BGCs under controllable promoters. Refactoring pathways for improved yields or altered structures can widen the chemical space (Sharma et al., 2018).

8.4 Development of Industrial Bioprocesses

Developing scalable bioprocesses is essential for converting discoveries into valuable products. Cost-effective production models (such as biorefineries), early engagement for regulatory clarity, and co-development with industry partners are essential (Sharma et al., 2018).

9 Conclusion

Marine actinobacteria are a significant source of chemical additives, with numerous instances illustrating their effective application in bioremediation, plastics, agriculture, food, and pharmaceuticals. Their ability to produce a wide range of structurally diverse and potent secondary metabolites offers substantial potential for future industrial and environmental advancements (Jagannathan et al., 2021; Dhakal et al., 2017). These developments emphasize the marine environment as a repository of unique chemical compounds derived from actinomycetes. Ongoing discoveries continue to highlight their importance in the development of novel antibiotics, anticancer agents, and industrial biotechnological products (Jagannathan et al., 2021; De Rop et al., 2021; Dhakal et al., 2017; Ibrahim et al., 2025; Roy et al., 2014; Pan et al., 2025).

References

Adler, A., & Dücker, E. (2018). When Pasteurian science went to sea: The birth of marine microbiology. *Journal of the History of Biology, 51*(1), 107–133.

Ahamefule, C. S., et al. (2020). Marine bioactive compounds against *Aspergillus fumigatus*: Challenges and future prospects. *Antibiotics, G*(11), 813.

Aladejana, E. B. (2023). Biological properties of polyherbal formulations: A review of their antimicrobial, anti-inflammatory, antioxidant, and toxicological activities. *Pharmacognosy Journal, 15*(5), 933–963.

Al-Shaibani, M. M., et al. (2021). Biodiversity of secondary metabolites compounds isolated from phylum actinobacteria and its therapeutic applications. *Molecules, 26*(15), 4504.

Amoroso, M. J., et al. (1998). Screening of heavy metal-tolerant actinomycetes isolated from the Sali River. *The Journal of General and Applied Microbiology, 44*(2), 129–132.

Amoutzias, G., Chaliotis, A., & Mossialos, D. (2016). Discovery strategies of bioactive compounds synthesized by nonribosomal peptide synthetases and type-I polyketide synthases derived from marine microbiomes. *Marine Drugs, 14*(4), 80.

Bae, M., et al. (2017). ϘM-HiFSA-aided structure determination of Succinilenes A–D, new triene polyols from a marine-derived Streptomyces sp. *Marine Drugs, 15*(2), 38.

Betancur, L. A., et al. (2017). Marine Actinobacteria as a source of compounds for phytopathogen control: An integrative metabolic-profiling/bioactivity and taxonomical approach. *PLoS One, 12*(2), e0170148.

Braña, A., et al. (2017). Lobophorin K, a new natural product with cytotoxic activity produced by Streptomyces sp. M-207 associated with the deep-sea coral *Lophelia pertusa*. *Marine Drugs, 15*(5), 144.

Breitling, R., et al. (2021). Synthetic biology approaches to actinomycete strain improvement. *FEMS Microbiology Letters, 368*(10), fnab060.

Bull, A. T., et al. (2005). Marine actinobacteria: Perspectives, challenges, future directions. *Antonie Van Leeuwenhoek, 87*(1), 65–79.

Carretero-Molina, D., Ortiz-López, F. J., Martín, J., Oves-Costales, D., Díaz, C., de la Cruz, M., & Reyes, F. (2019). New napyradiomycin analogues from Streptomyces sp. strain CA-271078. *Marine Drugs, 18*(1), 22.

Chen, P., et al. (2016). Diversity, biogeography, and biodegradation potential of actinobacteria in the deep-sea sediments along the Southwest Indian Ridge. *Frontiers in Microbiology, 7*, 1340.

Chen, J., et al. (2021). Natural products from actinomycetes associated with marine organisms. *Marine Drugs, 1G*(11), 629.

Claverías, F. P., et al. (2015). Culturable diversity and antimicrobial activity of Actinobacteria from marine sediments in Valparaíso bay, Chile. *Frontiers in Microbiology, 6*, 737.

Cong, Z., et al. (2019). Cytotoxic anthracycline and antibacterial tirandamycin analogues from a marine-derived Streptomyces sp. SCSIO 413SS. *The Journal of Antibiotics, 72*(1), 45–49.

Cui, P., et al. (2022). Diversity and antibacterial potential of the Actinobacteria associated with *Apis mellifera ligustica*. *Frontiers in Microbiology, 13*, 1056176.

Das, I., et al. (2013). Biosurfactant from marine actinobacteria and its application in cosmetic formulation of toothpaste. *Der Pharmacia Lettre, 5*(5), 1–6.

De La Hoz-Romo, M. C., et al. (2025). Marine actinobacteria metabolites: Unlocking new treatments for acne vulgaris. *Frontiers in Microbiology, 15*, 1501951.

De Rop, A.-S., et al. (2021). Novel alkaloids from marine actinobacteria: Discovery and characterization. *Marine Drugs, 20*(1), 6.

De Simeis, D., & Serra, S. (2021). Actinomycetes: A never-ending source of bioactive compounds—An overview on antibiotics production. *Antibiotics, 10*(5), 483.

Dejong, C. A., et al. (2016). Polyketide and nonribosomal peptide retro-biosynthesis and global gene cluster matching. *Nature Chemical Biology, 12*(12), 1007–1014.

Dhakal, D., et al. (2017). Marine rare actinobacteria: Isolation, characterization, and strategies for harnessing bioactive compounds. *Frontiers in Microbiology, 8*, 1106.

Dinesh, R., et al. (2017). Endophytic actinobacteria: Diversity, secondary metabolism and mechanisms to unsilence biosynthetic gene clusters. *Critical Reviews in Microbiology, 43*(5), 546–566.

Duncan, K. R., et al. (2015). Molecular networking and pattern-based genome mining improves discovery of biosynthetic gene clusters and their products from Salinispora species. *Chemistry C Biology, 22*(4), 460–471.

Elnahas, M., Elkhateeb, W., & Daba, G. (2021). Marine actinomycetes the past, the present and the future. *Pharmaceutical Research, 5*(2), 000241.
Erba, E., et al. (1999). Mode of action of thiocoraline, a natural marine compound with anti-tumour activity. *British Journal of Cancer, 80*(7), 971–980.
Feng, Y., et al. (2016). Rifamycin biosynthetic congeners: Isolation and total synthesis of rif-saliniketal and total synthesis of salinisporamycin and saliniketals A and B. *Journal of the American Chemical Society, 138*(22), 7130–7142.
Finking, R., & Marahiel, M. A. (2004). Biosynthesis of nonribosomal peptides1. *Annual Review of Microbiology, 58*, 453–488.
Gamal, R., & Shreadah, M. A. (2024). Marine microalgae and their industrial biotechnological applications: A review. *Journal of Genetic Engineering and Biotechnology, 22*(4), 100407.
Gao, M., et al. (2022). Anti-inflammatory butenolides from a marine-derived Streptomyces sp. 13G03C. *Applied Sciences, 12*(9), 4510.
Ghosh, S., et al. (2022). Novel bioactive compounds from marine sources as a tool for functional food development. *Frontiers in Marine Science, G*, 832957.
Gomez-Escribano, J. P., & Bibb, M. J. (2014). Heterologous expression of natural product biosynthetic gene clusters in *Streptomyces coelicolor*: From genome mining to manipulation of biosynthetic pathways. *Journal of Industrial Microbiology and Biotechnology, 41*(2), 425–431.
Gong, T., et al. (2018). Tetrocarcin Q, a new spirotetronate with a unique glycosyl group from a marine-derived actinomycete *Micromonospora carbonacea* LS27C. *Marine Drugs, 16*(2), 74.
Gontang, E. A., et al. (2010). Sequence-based analysis of secondary-metabolite biosynthesis in marine actinobacteria. *Applied and Environmental Microbiology, 76*(8), 2487–2499.
Gopi, M., et al. (2014). Marine natural product, Pyrrolo [-a] pyrazine–dione, hexahydro-(C7H10N2O2) of antioxidant properties from Bacillus species at Lakshadweep archipelago. *Journal of Coastal Life Medicine (JCLM), 2*, 632–637.
Hecko, S., et al. (2023). Enlightening the path to protein engineering: Chemoselective turn-on probes for high-throughput screening of enzymatic activity. *Chemical Reviews, 123*(6), 2832–2901.
Heng, E., et al. (2021). CRISPR-Cas strategies for natural product discovery and engineering in actinomycetes. *Process Biochemistry, 102*, 261–268.
Hu, D., et al. (2022). Exploration of secondary metabolite production potential in actinobacteria isolated from *Kandelia candel* mangrove plant. *Frontiers in Marine Science, G*, 700685.
Ibrahim, J. A. A., Botcha, S., & Prattipati, S. D. (2025). Marine actinomycetes: A promising source of novel therapeutics and pharmaceutical bioactive compounds—A review. *The Microbe, 7*, 100383.
Ibrahimi, M., et al. (2020). Marine actinobacteria: Screening for predation leads to the discovery of potential new drugs against multidrug-resistant bacteria. *Antibiotics, G*(2), 91.
Jagannathan, S. V., et al. (2021). Marine actinomycetes, new sources of biotechnological products. *Marine Drugs, 1G*(7), 365.
Jahan, R., et al. (2020). Biosurfactants, natural alternatives to synthetic surfactants: Physicochemical properties and applications. *Advances in Colloid and Interface Science, 275*, 102061.
James, G., et al. (2023). Applications of Actinobacteria in aquaculture: Prospects and challenges. *3 Biotech, 13*(2), 42.
Jensen, P. R., Moore, B. S., & Fenical, W. (2015). The marine actinomycete genus Salinispora: A model organism for secondary metabolite discovery. *Natural Product Reports, 32*(5), 738–751.
Jiang, X., et al. (2017). Isolation, structure elucidation and biosynthesis of benzo[b]ffuorene nenestatin A from deep-sea derived *Micromonospora echinospora* SCSIO 0408S. *Tetrahedron, 73*(26), 3585–3590.
Jose, P. A., & Jha, B. (2017). Intertidal marine sediment harbours Actinobacteria with promising bioactive and biosynthetic potential. *Scientific Reports, 7*(1), 10041.
Jose, P. A., Maharshi, A., & Jha, B. (2021). Actinobacteria in natural products research: Progress and prospects. *Microbiological Research, 246*, 126708.

Kanchanasin, P., et al. (2024). Discovery of two new actinobacteria, *Micromonospora palythoicola* sp. nov. and *Streptomyces poriticola* sp. nov., isolated from marine invertebrates. *Scientific Reports, 14*(1), 22140.

Kannan, K., & Sivaperumal, P. (2025). Bioprospecting Antarctic actinobacteria: Discovery of functional cryoprotective proteins for innovations in food sciences and biomedical applications. *Food Bioscience, 66*, 106134.

Karnwal, A., et al. (2023). Microbial biosurfactant as an alternate to chemical surfactants for application in cosmetics industries in personal and skin care products: A critical review. *BioMed Research International, 2023*(1), 1–21.

Kim, C.-K., et al. (2017). Manzamine alkaloids from an Acanthostrongylophora sp. Sponge. *Journal of Natural Products, 80*(5), 1575–1583.

Koech, S. C., et al. (2025). Strategies for actinobacteria isolation, cultivation, and metabolite production that are biologically important. *ACS Omega, 10*(16), 15923–15934.

Kumar, S. R. S., & Rao, K. V. B. (2012). In-vitro antimicrobial activity of marine actinobacteria against multidrug resistance *Staphylococcus aureus*. *Asian Pacific Journal of Tropical Biomedicine, 2*(10), 787–792.

Kumar, V., et al. (2014). Global scenario of industrial enzyme market. In *Industrial enzymes: Trends, scope and relevance* (pp. 176–196). Nova Science Publishers.

Kumar, H., et al. (2025). In silico exploration of biosynthetic gene clusters in marine Streptomyces sp. and Nocardiopsis sp. from the western coast of India: Genome-based profiling using whole genome sequencing. *Journal of Genetic Engineering and Biotechnology, 23*(2), 100483.

Lee, Y.-J., Cho, Y., & Tran, H. N. K. (2021). Secondary metabolites from the marine sponges of the genus Petrosia: A literature review of 43 years of research. *Marine Drugs, 1G*(3), 122.

León-Zayas, R., et al. (2017). The metabolic potential of the single cell genomes obtained from the Challenger Deep, Mariana Trench within the candidate superphylum Parcubacteria (OD 1). *Environmental Microbiology, 1G*(7), 2769–2784.

Lewin, G. R., et al. (2016). Evolution and ecology of actinobacteria and their bioenergy applications. *Annual Review of Microbiology, 70*(1), 235–254.

Li, K., et al. (2022). Natural products from mangrove sediments-derived microbes: Structural diversity, bioactivities, biosynthesis, and total synthesis. *European Journal of Medicinal Chemistry, 230*, 114117.

Liang, Y., et al. (2016). Bioactive polycyclic quinones from marine Streptomyces sp. 182SMLY. *Marine Drugs, 14*(1), 10.

Mangamuri, U. K., et al. (2016). Isolation and biological evaluation of N-(4-aminocyclooctyl)-3, 5-dinitrobenzamide, a new semisynthetic derivative from the Mangrove-associated actinomycete *Pseudonocardia endophytica* VUK-10. *3 Biotech, 6*(2), 158.

Manivasagan, P., Venkatesan, J., & Kim, S.-K. (2013). Introduction to marine actinobacteria. In *Marine microbiology* (pp. 1–19). Wiley.

Manivasagan, P., et al. (2014a). Marine actinobacteria: An important source of bioactive natural products. *Environmental Toxicology and Pharmacology, 38*(1), 172–188.

Manivasagan, P., et al. (2014b). Pharmaceutically active secondary metabolites of marine actinobacteria. *Microbiological Research, 16G*(4), 262–278.

Martín, J. F., Liras, P., & Sánchez, S. (2021). Modulation of gene expression in actinobacteria by translational modification of transcriptional factors and secondary metabolite biosynthetic enzymes. *Frontiers in Microbiology, 12*, 630694.

Mayer, A. M. S., et al. (2021). Marine pharmacology in 201C–2017: Marine compounds with antibacterial, antidiabetic, antifungal, anti-inflammatory, antiprotozoal, antituberculosis and antiviral activities; affecting the immune and nervous systems, and other miscellaneous mechanisms of A. *Marine Drugs, 1G*(2), 49.

Mitova, M. I., et al. (2008). Subinhibitory concentrations of antibiotics induce phenazine production in a marine Streptomyces sp. *Journal of Natural Products, 71*(5), 824–827.

Morgan, R. N., et al. (2023). New insights on biological activities, chemical compositions, and classifications of marine actinomycetes antifouling agents. *Microorganisms, 11*(10), 2444.

Navarro-Muñoz, J. C., et al. (2020). A computational framework to explore large-scale biosynthetic diversity. *Nature Chemical Biology, 16*(1), 60–68.

Noda, S., et al. (2015). 4-Vinylphenol biosynthesis from cellulose as the sole carbon source using phenolic acid decarboxylase- and tyrosine ammonia lyase-expressing *Streptomyces lividans. Bioresource Technology, 180*, 59–65.

Nong, X.-H., et al. (2016). Nahuoic acids B–E, polyhydroxy polyketides from the marine-derived Streptomyces sp. SCSGAA 0027. *Journal of Natural Products, 7G*(1), 141–148.

Qin, S., et al. (2019). Editorial: Actinobacteria in special and extreme habitats: Diversity, function roles and environmental adaptations, second edition. *Frontiers in Microbiology, 10*, 944.

Oyegoke, T., & Adnan, A. (2024). Impact of selected adsorbent functional groups on chromium sorption capacities in an effluent treatment: A DFT study. *The Journal of Engineering Research [TJER], 21*(1), 59–70.

Pan, C., et al. (2025). Marine actinomycetes: A hidden treasure trove for antibacterial discovery. *Frontiers in Marine Science, 12*, 1558320.

Parthasarathy, S., et al. (2024). Bio-valorization of secondary metabolites produced by actinobacterium SRV02, isolated from shell waste environment combats clinical bacterial pathogens and ameliorates anticancer effects against lung cancer cells. *Waste and Biomass Valorization, 15*(12), 7043–7056.

Paulus, C. (2018). *Metabolic and genomic profiling of actinobacteria strains: From new natural products to biosynthetic pathways*. Dissertation zu Erlangung des Grades.

Perez Baz, J., et al. (1997). Thiocoraline, a novel depsipeptide with antitumor activity produced by a marine Micromonospora. II. Physico-chemical properties and structure determination. *The Journal of Antibiotics, 50*(9), 738–741.

Presentato, A., et al. (2020). Processing of metals and metalloids by actinobacteria: Cell resistance mechanisms and synthesis of metal(loid)-based nanostructures. *Microorganisms, 8*(12), 2027.

Rahman, M. S., et al. (2019). Source of metal contamination in sediment, their ecological risk, and phytoremediation ability of the studied mangrove plants in ship breaking area, Bangladesh. *Marine Pollution Bulletin, 141*, 137–146.

Reynolds, K. A., et al. (2018). Isolation and structure elucidation of lipopeptide antibiotic taromycin B from the activated taromycin biosynthetic gene cluster. *The Journal of Antibiotics, 71*(2), 333–338.

Ribeiro, I., et al. (2023). Actinobacteria from Arctic and Atlantic deep-sea sediments—Biodiversity and bioactive potential. *Frontiers in Microbiology, 14*, 1158441.

Richter, T. K. S., Hughes, C. C., & Moore, B. S. (2015). Sioxanthin, a novel glycosylated carotenoid, reveals an unusual subclustered biosynthetic pathway. *Environmental Microbiology, 17*(6), 2158–2171.

Roy, S., et al. (2014). Isolation and characterization of tyrosinase produced by marine actinobacteria and its application in the removal of phenol from aqueous environment. *Frontiers in Biology, G*(4), 306–316.

Rui, S., et al. (2025). Biological activity of secondary metabolites of actinomycetes and their potential sources as antineoplastic drugs: A review. *Frontiers in Microbiology, 16*, 1550516.

Salwan, R., & Sharma, V. (2020). Molecular and biotechnological aspects of secondary metabolites in actinobacteria. *Microbiological Research, 231*, 126374.

Sarkar, G., & Suthindhiran, K. (2022). Diversity and biotechnological potential of marine actinomycetes from India. *Indian Journal of Microbiology, 62*(4), 475–493.

Sarmiento-Vizcaíno, A., et al. (2017). Paulomycin G, a new natural product with cytotoxic activity against tumor cell lines produced by deep-sea sediment derived *Micromonospora matsumotoense* M-412 from the Avilés Canyon in the Cantabrian Sea. *Marine Drugs, 15*(9), 271.

Schulze, C. J., et al. (2015). Genome-directed lead discovery: Biosynthesis, structure elucidation, and biological evaluation of two families of polyene macrolactams against *Trypanosoma brucei. ACS Chemical Biology, 10*(10), 2373–2381.

Selim, M. S. M., Abdelhamid, S. A., & Mohamed, S. S. (2021). Secondary metabolites and biodiversity of actinomycetes. *Journal of Genetic Engineering and Biotechnology, 1G*(1), 72.

Shaaban, K. A., et al. (2021). Boshramycinones AC: New anthracyclinones produced by a marine-derived Streptomyces sp.: Isolation, structure elucidation and biological activities. *Natural Product Research, 35*(8), 1281–1291.
Sharma, P., Dutta, J., & Thakur, D. (2018). Future prospects of actinobacteria in health and industry. In *New and future developments in microbial biotechnology and bioengineering* (pp. 305–324). Elsevier.
Silva, G. D. C., et al. (2022). The potential use of actinomycetes as microbial inoculants and biopesticides in agriculture. *Frontiers in Soil Science, 2*, 833181.
Siro, G., et al. (2022). Marine actinomycetes associated with stony corals: A potential hotspot for specialized metabolites. *Microorganisms, 10*(7), 1349.
Soni, J., et al. (2024). Bioinspired green synthesis of ZnO nanoparticles by marine-derived *Streptomyces plicatus* and its multifaceted biomedicinal properties. *Microbial Pathogenesis, 1G3*, 106758.
Srinivasan, R., et al. (2021). Marine bacterial secondary metabolites: A treasure house for structurally unique and effective antimicrobial compounds. *Marine Drugs, 1G*(10), 530.
Stincone, P., & Brandelli, A. (2020). Marine bacteria as source of antimicrobial compounds. *Critical Reviews in Biotechnology, 40*(3), 306–319.
Subramani, R., & Aalbersberg, W. (2012). Marine actinomycetes: An ongoing source of novel bioactive metabolites. *Microbiological Research, 167*(10), 571–580.
Subramani, R., & Sipkema, D. (2019). Marine rare actinomycetes: A promising source of structurally diverse and unique novel natural products. *Marine Drugs, 17*(5), 249.
Sun, W., et al. (2015). Actinomycetes from the South China Sea sponges: Isolation, diversity, and potential for aromatic polyketides discovery. *Frontiers in Microbiology, 6*, 1048.
Suriya, J., et al. (2016). Extremozymes from marine actinobacteria. *Advances in Food and Nutrition Research, 7G*, 43–66.
Tarasova, E. V., et al. (2023). Actinomycetes as producers of biologically active terpenoids: Current trends and patents. *Pharmaceuticals, 16*(6), 872.
Tian, J., et al. (2016). Discovery of pentangular polyphenols hexaricins A–C from marine Streptosporangium sp. CGMCC 4.730S by genome mining. *Applied Microbiology and Biotechnology, 100*(9), 4189–4199.
Tuttle, R. N., et al. (2019). Detection of natural products and their producers in ocean sediments. *Applied and Environmental Microbiology, 85*(8), e02830-18.
Valliappan, K., Sun, W., & Li, Z. (2014). Marine actinobacteria associated with marine organisms and their potentials in producing pharmaceutical natural products. *Applied Microbiology and Biotechnology, G8*(17), 7365–7377.
van Bergeijk, D. A., et al. (2020). Ecology and genomics of Actinobacteria: New concepts for natural product discovery. *Nature Reviews Microbiology, 18*(10), 546–558.
Wang, M., et al. (2016). Sharing and community curation of mass spectrometry data with Global Natural Products Social Molecular Networking. *Nature Biotechnology, 34*(8), 828–837.
Wang, C., Lu, Y., & Cao, S. (2020). Antimicrobial compounds from marine actinomycetes. *Archives of Pharmacal Research, 43*(7), 677–704.
Wu, G., et al. (2017). Bonnevillamides, linear heptapeptides isolated from a great salt lake-derived Streptomyces sp. *Marine Drugs, 15*(7), 195.
Xu, D., et al. (2018). Bioprospecting deep-sea actinobacteria for novel anti-infective natural products. *Frontiers in Microbiology, G*, 787.
Yang, L., et al. (2023). Hyperproduction of extracellular polymeric substance in *Pseudomonas fluorescens* for efficient chromium (VI) absorption. *Bioresources and Bioprocessing, 10*(1), 17.
Zhang, W., et al. (2012). Spiroindimicins A–D: New bisindole alkaloids from a deep-sea-derived actinomycete. *Organic Letters, 14*(13), 3364–3367.
Zhang, Y.-M., et al. (2017). Anandins A and B, two rare steroidal alkaloids from a marine *Streptomyces anandii* H41-5S. *Marine Drugs, 15*(11), 355.
Zhao Chengying, Z. T., & Weiming, Z. (2013). New marine natural products of microbial origin from 2010 to 2013. *Chinese Journal of Organic Chemistry, 33*(06), 1195–1234.

Zhao, X. Q., Xu, X. N., & Chen, L. Y. (2016). Chapter seven—Production of enzymes from marine actinobacteria. In S.-K. Kim & F. Toldrá (Eds.), *Advances in food and nutrition research* (pp. 137–151). Academic.

Zothanpuia, et al. (2018). Bioprospection of actinobacteria derived from freshwater sediments for their potential to produce antimicrobial compounds. *Microbial Cell Factories, 17*(1), 68.

Exploring the Biosynthetic Potential of Marine Actinobacteria Through Genome Mining

G. Sainath, M. A. Jayasri, and K. Suthindhiran

Abstract

Actinobacteria isolated from marine environments are currently gaining substantial attention due to their ability to produce a wide range of bioactive compounds with diverse biological activities. Structurally unique and emerging bioactive compounds extracted from marine actinobacteria have been found to have profound biotechnological applications. Recent advancements in bioinformatics and high-throughput sequencing have facilitated the emergence of genome mining as a prominent strategy for uncovering the biosynthetic potential of marine actinobacteria, surpassing traditional culturing methods. The diverse range of Biosynthetic Gene Clusters (BGCs) identified in actinobacterial genomes, which includes non-ribosomal peptide synthetases (NRPSs), polyketide synthases (PKSs), terpenoids and RiPPs and their functions will be discussed. Moreover, the chapter examines genome mining tools, such as antiSMASH, PRISM, BAGEL, and ClusterFinder, used to detect the secondary metabolites produced by marine actinobacteria. Further, this chapter will explore challenges and future prospects of combining genome mining with metabolomics and synthetic biology to enhance the discovery and production of novel secondary metabolites from rare and unculturable marine microorganisms.

Keywords

Actinobacteria · Bioactive compounds · Genome mining · Biosynthetic gene clusters

G. Sainath · M. A. Jayasri · K. Suthindhiran (✉)
Marine Biotechnology and Bioproducts Laboratory, Department of Biomedical Sciences, School of Bio Sciences and Technology, Vellore Institute of Technology, Vellore, Tamil Nadu, India
e-mail: ksuthindhiran@vit.ac.in

K. Loganathan, Z. Li (eds.), *Marine Actinomycetes*,
https://doi.org/10.1007/978-981-95-9849-6_8

1 Introduction

The marine environment remains an underexplored biological reservoir, housing diverse organisms that serve as an excellent source of bioactive compounds. Hence, the aquatic ecosystem is regarded as a treasure trove of bioactive compounds with interesting features, as they withstand extreme temperatures, varying pH levels, high pressure, salinity, and low nutrient contents (Yang et al., 2020; Ghosh et al., 2022). Nowadays, bioactive compounds and their derivatives are gaining attention for their profound biological activities, including anticancer, antibacterial, and anti-cholesterol effects. Notably, marine-derived natural products from bacteria and sponges have significant advantages over plant derivatives and synthetic drugs due to their unique diversity and novel mechanisms of action (Albarano et al., 2020; Saeed et al., 2021). Marine actinobacteria have emerged as abundant secondary metabolite producers and are known for their remarkable biotechnological applications (Manivasagan et al., 2014). Since 2020, approximately 70% of the essential pharmaceutical compounds derived from marine microorganisms are produced by actinobacteria, while fungi account for 20%, the genus Bacillus contributes 7%, and the remaining 2% comes from other marine microorganisms (Saeed et al., 2021). From previous studies, it is evident that actinobacteria produce a wide variety of structurally diverse secondary metabolites with varied morphologies and metabolic capabilities (Ngamcharungchit et al., 2023). Following that, the clinically significant natural antibiotics were found to be extracted from secondary metabolites of actinobacteria (Bhattacharjee et al., 2023). Especially belonging to the genus *Streptomyces*, it was found to contain bioactive compounds that show potential applications in agriculture, drug development, and industrial biotechnology. Traditional culture-based techniques have several limitations, including being time-consuming, prone to contamination, and difficult to culture marine actinobacteria under in vitro laboratory conditions (Ahmad et al., 2024). Currently, the advancements in machine learning, bioinformatic tools, and genome sequencing technologies help to explore the secondary metabolites from the aquatic environment (Wang et al., 2024). Genome mining is a bioinformatic approach for exploring marine bioactive natural products through interpreting genomic data to identify the biosynthetic gene clusters (BGCs) that encode for secondary metabolites or novel biomolecules (Maurya et al., 2019). Through genome mining strategies, researchers can also predict the presence of cryptic BGCs that are not expressed under standard culture conditions, which further leads to the activation of the biosynthetic pathway for producing emerging secondary metabolites (Hur et al., 2023). The integration of genome sequencing, genome mining, and biosynthetic approaches unlocks the vast reservoir of unexplored secondary metabolites and novel compounds with significant biotechnological applications.

2 Actinobacteria

Actinobacteria are gram-positive, filamentous aerobic bacteria with the ability to form spores. They have a DNA with a rich G + C content of about 70% and exhibit radial colony morphology and mycelial structure resembling fungi (Prem Anand et al., 2024). Actinomycetes remain an important area of research for discovering novel compounds that act as a key source of new secondary metabolites, offering a wide range of research because of their remarkable resources and applications (Fig. 1) in the medical field, etc. (Xu et al., 2023). They can survive in various habitats and are widely distributed in both aquatic and terrestrial environments, especially being found to be abundant in salt lakes, freshwater, alkaline soils and extreme conditions such as deep-sea sediments and polar regions (Barka et al., 2016). *Streptomyces* is the most isolated genus of the order Actinomycetales for its ability to contribute substantial resources to the biotechnology industry. They are responsible for generating approximately 50% of the explored secondary metabolites, including enzymes, antibiotics, anticancer, and anti-inflammatory agents (Selim et al., 2021). Majorly, the genera such as *Mycobacterium* and *Streptomyces* are well-studied terrestrial organisms already been sequenced hundreds of genomes and well-discovered compared to marine actinobacteria (Schorn et al., 2016). The marine environment is known for its high salinity, pressure and low temperature; hence, compared to terrestrial actinomycetes, isolated from the sea exhibit a diverse range of secondary metabolites with unique structures and biological activities (Xu et al., 2023). A previous study revealed that secondary metabolites isolated from *Nocardiopsis* sp. and *Streptomyces* sp., dominant genera of actinobacteria, had the potential to develop into novel colorectal cancer drugs (Julianti et al., 2022).

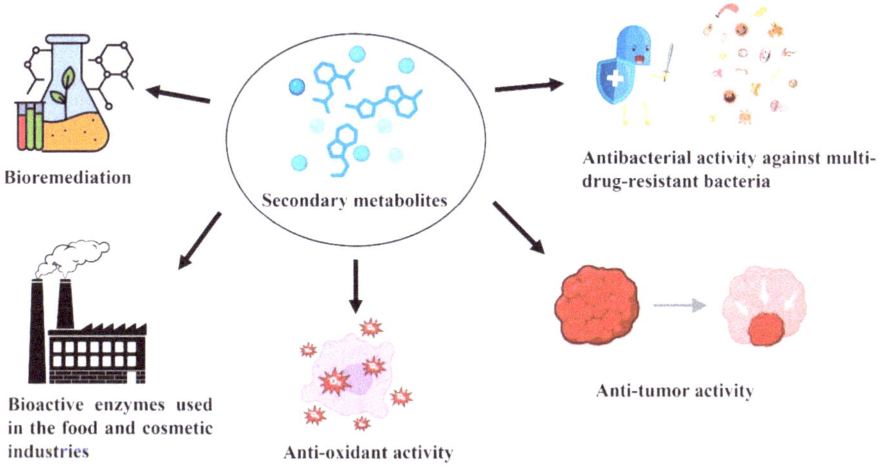

Fig. 1 Applications of marine actinobacteria-derived secondary metabolites

2.1 Marine Actinobacteria as a Source of Bioactive Compounds

Marine actinobacteria are among the richest and most promising sources of bioactive compounds, possessing significant pharmaceutical potential. Genera such as *Streptomyces*, *Salinispora*, *Micromonospora*, and *Actinomadura* have developed metabolic pathways that enable them to produce a diverse range of secondary metabolites. These compounds exhibit various biological activities such as antibacterial, antiviral, antiparasitic, and immunomodulatory properties (Table 1). As a result, they play a crucial role in drug discovery (Sarkar & Suthindhiran, 2022; Chinnathambi et al., 2023). Actinobacteria have been discovered to be a well-known antimicrobial agent that effectively inhibits multidrug-resistant pathogens. In addition to this, bioactive enzymes, such as chitinases isolated from *Streptomyces*, have demonstrated prominent antifungal activity and are applied in agriculture, cosmetics, and biomedical fields. According to the study (Shi et al., 2022a), 37 biosynthetic gene clusters were identified, including eight new clusters in the genomic sequence

Table 1 Marine Actinobacteria-derived metabolites and associated bioactivities

S. no	Organisms	Compounds	Activity	References
1	*Streptomyces longisporoflavus*	Myristic acid triglyceride, oleic acid, palmitic acid	Antibacterial and antifungal	Mothana et al. (2022)
2	*Streptomyces glaucescens NEAE-H*	Melanin pigment	Anticancer and antioxidant	Sarmiento-Tovar et al. (2022)
3	*Kocuria palustris*	Thiopeptides	Antibacterial activity against MRSA and fungi	Martín et al. (2013)
4	*Gordonia* sp. *WA8–44*	Collismycin A, actinomycin D, and actinomycin X2	Antifungal activity against *Candida albicans*	Liu et al. (2023)
5	*Streptomyces* sp. *MBTG13*	2-alkyl-4-hydroxyquinoline derivatives	Antifungal: inhibits hyphal transition in *Candida albicans*, antimalarial	Kim et al. (2019)
6	*Actinoalloteichus caeruleus*	Fibrinolytic enzymes	Treat thrombosis	Suresh and Devi (2025)
7	*Brachybacterium paraconglomeratum*	Polyhydroxybutyric acid (PHB)	Bioplastics, enhanced drug delivery	Mandragutti and Sudhakar (2023)
8	*Nocardiopsis* sp.	Macrolides, borrelidins c–e	Antimicrobial and anti-angiogenic	Shi et al. (2022b)
9	*Micrococcus roseus*	Beta-carotene, canthaxanthin	Antioxidant, antitumor, anti-inflammatory	Tizabi and Hill (2023)
10	*Actinomadhura sp.*	Isotrichodermin (IB-00208)	Antitumor activity	Meenakshi et al. (2024)

of *Streptomyces marincola*, which are responsible for the production of secondary metabolites. Consists of non-ribosomal peptide synthase (NRPS), siderophore, polyketide synthase (PKS), terpene, indole, thiopeptide, and lanthipeptide. *Micromonosporas* and *Streptomycyces* were found to have significant growth inhibitor compounds that promote antitumor activity (Khalifa et al., 2019). The anticancer effects of Salinosporamide A, isolated from the actinobacterium *Salinospora tropica* CNB-392, were studied in human colon carcinoma cell lines, specifically HCT-116. Additionally, it has demonstrated efficacy against a variety of other tumors, including pancreatic carcinoma, multiple myeloma, and chronic lymphocytic leukemia (CLL) (Udwary et al., 2007). It is evident that marine actinobacteria, particularly species of *Streptomyces*, produce a diverse range of halogenated compounds that exhibit a broad spectrum of biological activities. It has been reported that halogenated compounds derived from marine actinomycetes possess prominent antibacterial and anticancer effects. For instance, the natural substance chlorizidine, isolated from *Streptomyces*, exhibited cytotoxic effects on the HCT-116 human colorectal cancer cell line due to the presence of a chlorinated 5H-pyrrolo [2,1-a]isoindol-5-one ring, a novel structure (Alvarez-Mico et al., 2013). Moreover, Actinomycetes play a significant role in biogeochemical cycles and the production of potent bacteriocins. Most antibiotics used in aquaculture, such as tetracyclines, aminoglycosides, and macrolides, are derived from Actinobacteria. Also, they exhibit bioremediation agent by responding to various sources of environmental pollution, ultimately leading to the degradation and transformation of pollutants into harmless compounds. *Streptomyces gedanensis* strain LK-3 was isolated from marine sediments in the Nicobar Islands and was determined to produce biosurfactants. These microorganisms are known to generate biosurfactants that can enhance the removal of oil from polluted water surfaces. Additionally, these biosurfactants are used in the cosmetics, food, and pharmaceutical industries (Sarkar & Suthindhiran, 2022). Furthermore, it is fascinating that actinomyces from the marine environment have the ability to secrete metabolites such as antifouling agents fit in various chemical classes, including polyketides, peptides, sterols, isoprenoids and phenazines. Likewise, *Streptomyces* sp. is known to produce fatty acids, quorum-sensing inhibitors, steroids and terpenoids. Prior research has emphasized that the *Streptomyces parvulus* strain HY026 actively reduces biofilm formation of waste culture media consisting of *Staphylococcus aureus, Micrococcus luteus, and Pseudomonas aeruginosa* PAO1 through a potential compound, Actinomycin D (Morgan et al., 2023). Apart from this, actinobacteria are a promising source of enzymes, which are industrially important with various functions, as they are isolated from extreme environmental conditions. Enzymes derived from marine actinobacteria, such as amylases, lipases, proteases, keratinases and xylanases, are utilized in food processing for various purposes. These enzymes play a key role in protein hydrolysis, starch breakdown, fat modification, and fiber degradation. Hence, they enhance food texture, flavor, and shelf life, while also aiding in biofuel production from biomass. Keratinases derived from marine actinobacteria are essential for recycling waste from poultry (Mukhtar et al., 2017; Zhao et al., 2016).

2.2 Genes Responsible for the Biosynthesis of Natural Bioactive Products

The genomic sequence of bacteria contains the outline of potential bioactive molecules that the strain can produce. Genes responsible for producing natural bioactive compounds in marine actinobacteria are typically organized into clusters known as biosynthetic gene clusters (BGCs). These BGCs encode various enzymatic pathways that facilitate the production of diverse secondary metabolites, including polyketides (PKS), non-ribosomal peptides (NRPS), Terpenoids, Ribosomally synthesized and post-translationally modified peptides (RiPPs), siderophores, and bacteriocins. These gene clusters exist within the genomic islands, which indicates that the genome is attained through Horizontal gene transfer (HGT) (Singh et al., 2021). Marine actinobacterial genomes allocate a significant portion, about 10% of their genome, to secondary metabolite-producing biosynthetic gene clusters (BGCs), highlighting their potential for biosynthesis (Paulus et al., 2017).

2.2.1 Polyketides (PKS)

The PKS gene encodes multifunctional enzymes that are essential for biosynthesizing polyketides, a large and structurally varied class of secondary metabolites with significant biological activities. Polyketides are a significant family of natural compounds that encompasses a diverse range of chemical entities, including polyphenols, polyenes, macrolides, polyethers, enediynes, and complex lipids. These compounds exhibit important biological and pharmacological activities, acting as antifungal antibiotics, immunosuppressants, antitumor, and antiparasitic agents. Due to their diverse structural organization, polyketides are classified into three types: I, II, and III (Sabatini et al., 2018). These enzymes, which successively add simple building blocks—acetate (2 carbon units) or propionate (3 carbon units), operate as assembly-line enzymes to form complex polyketide chains via decarboxylative Claisen condensation reactions (Robbins et al., 2016). The PKS module has distinct key domains, such as ketosynthase (KS), acyltransferase (AT), and acyl carrier protein (ACP), and tailoring domains such as ketoreductase (KR), dehydratase (DH), and enoyl reductase (ER), which modify the chain by reducing keto groups and the Thioesterase (TE) domain. The structure and diverse activity of the produced polyketides are determined by the order and combination of these domains (Sabatini et al., 2018), which ultimately lead to pharmacologically important natural products, such as antibiotics, immunosuppressants, and anticancer agents (Bloudoff & Schmeing, 2017).

2.2.2 Non-ribosomal Peptides (NRPS)

Non-ribosomal peptide synthetase (NRPS) genes play a crucial role in marine actinobacteria by encoding the enzymatic machinery required for the biosynthesis of non-ribosomal peptides. Strains with NRPS genes typically produce a greater variety and higher quantities of bioactive compounds compared to those without these genes (Bloudoff & Schmeing, 2017). Compared to other genes, NRPS is abundant and is known for its effective production of secondary metabolites. The natural compounds, notable for their biological activities like actinomycin D (anticancer), vancomycin (antibacterial), and compounds, are renowned for their novel modes of action used for drug development (Salwan & Sharma, 2020). NRPS gene consist of catalytic domains such as condensation, adenylation, thiolation and Thioesterase domains which facilitates the synthesis of peptides with potential bioactivity (Huang et al., 2024). As the name suggest, non-ribosomal peptides are synthesized independent of ribosomes and mRNA hence this unique way of NRP synthesis creates structurally diverse and novel marine bioactive compounds with anticancer and immunosuppressant activities (Gokulan et al., 2014).

2.2.3 Terpenoids

Terpene synthetase genes catalyze the formation of terpenoids, resulting in the synthesis of structurally diverse natural compounds. These genes convert multiple precursors like geranylgeranyl diphosphate (GGPP), geranyl diphosphate (GPP), and farnesyl diphosphate (FPP) into diterpenes, sesquiterpenes, monoterpenes, etc. Terpene synthetase enzymes exhibit a wide range of catalytic versatility and substrate specificity, thereby generating various terpenoid products from the same substrate (Zhou et al., 2020). They are classified into medium-sized multiple families, such as TSP-a, TSP-b, and TSP-c. Terpenoids such as found to have promising bioactivities such as plant growth regulation, defense and commercially valuable compounds with fragrances, and bioactive therapeutic compounds (Zhou et al., 2020).

2.2.4 Ribosomally Synthesized and Post-translationally Modified Peptides (RiPPs)

RiPPs are originated from precursor peptides, which are initially derived from the ribosome through mRNA. It consists of a leader peptide and a core peptide. The leader peptide recognizes the biosynthesis enzymes, whereas the core peptide undergoes post-translational modifications and further it is converted to structurally diverse bioactive peptides (Han & Won, 2024). RiPPs have genes that are smaller compared to other BGCs yet code for structurally sophisticated molecules. During post-translational modification, the enzymes modify the core peptide through various processes such as methylation, dehydration, cyclization, prenylation and transformation into mature RiPPs (Fu et al., 2023). They are classified into different classes namely thiopeptides, cyanobactins and lantipeptides. (RiPPs) are regarded as an important class of natural products found in all domains of life, from bacteria to humans. Due to their bioactive properties specificity and stability, RiPPs have gained interest from various industries having potential therapeutic ability such as antivirals, anticancer agents and antibiotics (Pfeiffer et al., 2024).

3 Genome Mining—An Effective Tool to Unlock Silent Biosynthetic Gene Clusters

Traditional culture-dependent methods, which include isolation, extraction, antimicrobial activity testing, and purification of secondary metabolites, are found to have limitations, such as time-consuming and labor-intensive processes. To overcome this, genome mining is currently an effective in-silico strategy for identifying gene clusters responsible for producing novel bioactive secondary metabolites that cannot be expressed under laboratory conditions (Biermann et al., 2022). The recent advancements in genome sequencing technology, along with bioinformatics, have helped reveal a tremendous amount of genome sequence data that aids in recognizing interesting novel biosynthetic genes and detecting cryptic BGC genes. Strategies such as epigenetic modification, heterologous expression, and genetic manipulation are employed to activate the silent BGCs, thereby unlocking novel therapeutic natural products with potential anticancer and antibacterial activities that were previously undiscovered through traditional methods (Scherlach & Hertweck, 2021). Genome mining includes a series of steps (Fig. 2): (1) Genome sequencing-Retrieving the whole-genome sequence of the target microorganism through NCBI blastn; (2) Gene cluster identification using bioinformatic tools, such as antiSMASH, ClustScan, and PRISM, is employed to detect novel biosynthetic gene clusters with accuracy and predict potential secondary metabolites. (3) Prioritizing novel BGCs encoding for effective natural metabolites by analyzing comparative genomics and domain architecture prediction to ensure that the gene encodes for structurally diverse secondary metabolites with significant applications; (4) Experimental methods to activate cryptic BGCs, through genomic and cultivation-based methods. Genomic methods include Heterologous expression, which involves

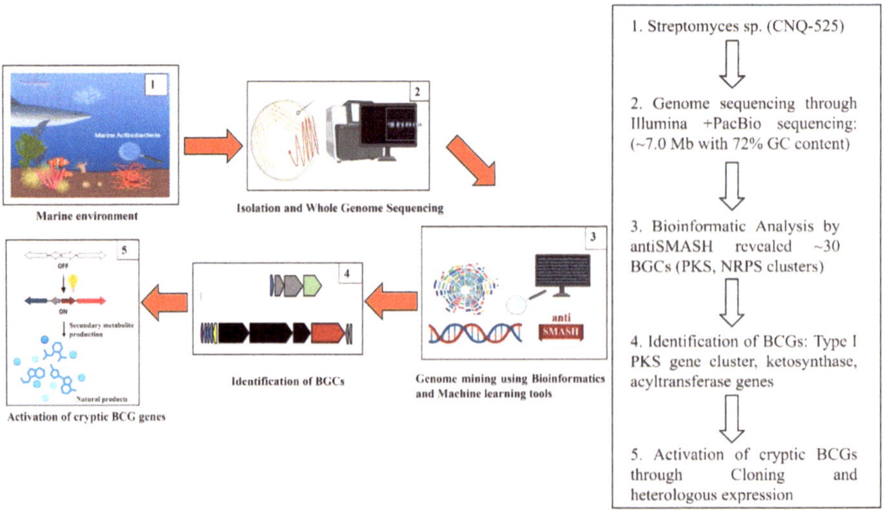

Fig. 2 Overall workflow of Genome mining

transferring the biosynthetic gene into genetically identified host strains under favorable conditions. Similarly, the silent genes can be activated by deletion of repressors, inserting inducible promoters, or replacing the native promoter of silent BGCs using genome editing tools like CRISPR–CAS9. Alternatively, through the One Strain Many Compounds (OSMAC) approach, we can trigger the activation of silent genes by modifying culture conditions, including pH, temperature, light, and oxygen supply, to induce secondary metabolite production. (5) Metabolite purification and characterization (Baltz, 2021; Mao et al., 2018; Saito & Arai, 2024). Therefore, the activation of cryptic gene clusters helps overcome the rediscovery of known compounds and facilitates the effective exploration of biosynthetic potential hidden in microbial genomes. The bioinformatic and machine learning tools (Table 2), used to detect biosynthetic genes with accuracy, depend on the high-quality metagenomic sequences for reliable outputs.

3.1 antiSMASH

Antibiotics & Secondary Metabolite Analysis Shell (antiSMASH) is a widely used bioinformatic tool designed for the analysis of a broad range of natural product biosynthetic gene clusters (BGCs) in archaeal, bacterial, and fungal genomes. It works based on the principle of profile hidden Markov model (pHMM) approach to discover novel secondary metabolites by detecting a broad range of BGCs, for instance NRPS, PKS type I and II, RiPPs. antiSMASH is noted for its feature of in-depth cluster-specific analyses, which facilitate the structural prediction and visualization of gene cluster architecture with detailed genomic annotations. The

Table 2 Common genome mining tools for BCG prediction and secondary metabolites identification

S. no	Genome mining tools	Web URL link	References
1	PRISM	http://magarveylab.ca/prism	Skinnider et al. (2017)
2	antiSMASH	http://antismash.secondarymetabolites.org/	Blin et al. (2019)
3	BAGEL	http://bagel4.molgenrug.nl/	van Heel et al. (2013)
4	ARTS	https://arts.ziemertlab.com	Mungan et al. (2020)
5	CLUSEAN	https://bitbucket.org/tilmweber/clusean	Weber et al. (2009)
6	2metDB	http://secmetdb.sourceforge.net/	Bachmann and Ravel (2009)
7	eSNaPD	http://esnapd2.rockefeller.edu/	Charlop-Powers et al. (2014)
8	PKMiner	http://pks.kaist.ac.kr/pkminer	Kim and Yi (2012)
9	ClusterFinder	https://github.com/petercim/ClusterFinder	Cimermancic et al. (2014)
10	SMURF	http://jcvi.org/smurf/index.php	Khaldi et al. (2010)
11	BiG SCAPE	https://git.wageningenur.nl/medema-group/BiG-SCAPE	Navarro-Muñoz et al. (2020)
12	BiG FAM	https://bigfam.bioinformatics.nl/	Kautsar et al. (2021)

tool often adds new BGC types, which catalyze the predication accuracy (Blin et al., 2023).

3.2 PRISM

PRediction Informatics for Secondary Metabolomes (PRISM) is an advanced bioinformatic tool developed for predicting chemical structures and biosynthetic gene clusters. It follows the principle of Hidden Markov Models (HMMs) for BCG and domain structure prediction and combinatorial rules. Compared to other genome mining tools, PRISM detects a wide array of BCG types, including novel families such as bisindoles, aminocoumarins, phosphonates, and various RiPP classes. It is a powerful complementary tool to antiSMASH, featuring an optimized backend infrastructure for high-throughput analysis of chemical structure prediction (Skinnider et al., 2017, 2020).

3.3 BAGEL

Bacteriocin Genome Mining Tool (BAGEL), designed to predict the BGCs that encode ribosomally synthesized and post-translationally modified peptides (RiPPs), such as bacteriocins, which are called antimicrobial peptides produced by bacteria. This tool specifically detects the alleged bacteriocin-encoding genes through motif and knowledge-based bacteriocin databases. This tool utilizes reference alignment methods, such as sequence similarity searches through BLAST, to identify homologous sequences in bacterial genomes. They specialize in detecting small and poorly conserved ORFs in DNA sequences for antimicrobial peptides (de Jong et al., 2006; Russell & Truman, 2020).

3.4 ClusterFinder

ClusterFinder is a genome mining tool that predicts the BGCs through a machine learning approach, specifically detecting NRPS and PKS genes, unlike the rule-based tools. It explores novel gene clusters by converting nucleotide sequences into Pfam protein domains, thereby determining whether these domains are part of the gene cluster. This works based on the hidden Markov model, and it will be used in parallel with other genome mining tools like antiSMASH, providing additional support for unearthing the potentially missed BGCs (Lee et al., 2020; Chavali & Rhee, 2018).

3.5 Comparative Genomics and Genome Sequencing Platforms

Comparative genomics is a valuable method for identifying BGCs across multiple genomes, which helps detect both unique and conserved biosynthetic pathways. Bioinformatics tools like BiG-SCAPE, CORASON, and antiSMASH utilize comparative genomics to prioritize novel clusters, group BGCs into gene cluster families, and trace their evolutionary relationships. Hence, this platform helps in recognizing gene duplication, modular rearrangements and horizontal gene transfer that contribute to biosynthetic diversity (Johnston et al., 2015). By comparing the genomes of related species, we can easily correlate genetic variations with chemical diversity, thereby guiding targeted discovery efforts. Genomes-to-Natural Products (GNP) platforms, by integrating genome data with LC-MS metabolomics, can systematize the prediction of polyketides and non-ribosomal peptides from biosynthetic assembly lines. Therefore, integrating comparative genome tools with metabolomics data enables the detection of significant secondary metabolites with precision (Bauman et al., 2021).

4 Limitations and Challenges in Current Genome Mining Approaches

Current genome mining approaches have several limitations that restrict their full potential. According to Wohlleben et al. (Lee et al., 2020), one of the disadvantages of traditional genome mining methods is that they are restricted to already known BGCs. The large BGCs will be fragmented into multiple contigs by sequencing assemblies, which will hinder the detection of complete pathways. Large-scale data handling requires bioinformatic expertise, and tools mainly focus on homologs, signature sequences, and known pathways, while leaving new or non-conventional pathways unaddressed. Sometimes these tools can generate putative BCGs, which will be difficult to prove their biosynthetic ability (Jensen et al., 2014; Teijaro et al., 2019). Predicting the exact chemical structure and bioactivity of a gene sequence remains difficult due to complex modifications and enzyme-substrate promiscuity.

5 Conclusion and Future Perspectives

The genome-mining technique, when combined with metabolomics and synthetic biology, is an effective strategy to unlock the biosynthetic potential of marine actinobacteria, which are a largely untapped source of novel natural compounds. Marine actinobacteria have unique adaptations to extreme environmental conditions, with potential sources of unexplored genomes. This multidisciplinary approach facilitates the exploration of novel secondary metabolites with significant potential for various biotechnological applications, thereby serving as an alternative source of antibacterial and anticancer agents. In the future, it would be beneficial to develop advanced Multiomics technologies and computational methods for biosynthetic gene detection. This collaborative initiative that combines genomics, metabolomics, ecology, and chemistry will be essential for fully utilizing the biosynthetic potential of marine actinobacteria. Ultimately, this will lead to the transformation of marine natural compounds into industrially valuable bioactive metabolites addressing urgent issues such as cancer therapy, antimicrobial resistance, and other medical challenges.

Acknowledgements The authors of this study gratefully acknowledge Vellore Institute of Technology, Vellore, for providing us with all the necessary financial and infrastructure support.

Conflict of Interest The authors declare that they have no conflict of interest.

References

Ahmad, S., Lohiya, S., Taksande, A., Meshram, R. J., Varma, A., & Vagha, K. (2024). A comprehensive review of innovative paradigms in microbial detection and antimicrobial resistance: Beyond traditional cultural methods. *Cureus., 16*(6), e61476.

Albarano, L., Esposito, R., Ruocco, N., & Costantini, M. (2020). Genome mining as new challenge in natural products discovery. In *Marine drugs*. MDPI AG.

Alvarez-Mico, X., Jensen, P. R., Fenical, W., & Hughes, C. C. (2013). Chlorizidine, a cytotoxic 5 *H*-Pyrrolo[2,1- *a*]isoindol-5-one-containing alkaloid from a marine *Streptomyces* sp. *Organic Letters, 15*, 988–991.

Bachmann, B. O., & Ravel, J. (2009). Chapter 8: Methods for in silico prediction of microbial polyketide and nonribosomal peptide biosynthetic pathways from DNA sequence data. *Methods in Enzymology, 458*, 181–217.

Baltz, R. H. (2021). Genome mining for drug discovery: Progress at the front end. *Journal of Industrial Microbiology & Biotechnology, 48*(9–10), kuab044.

Barka, E. A., Vatsa, P., Sanchez, L., Gaveau-Vaillant, N., Jacquard, C., Klenk, H.-P., et al. (2016). Taxonomy, physiology, and natural products of actinobacteria. *Microbiology and Molecular Biology Reviews, 80*, 1–43.

Bauman, K. D., Butler, K. S., Moore, B. S., & Chekan, J. R. (2021). Genome mining methods to discover bioactive natural products. *Natural Product Reports, 38*, 2100–2129.

Bhattacharjee, A., Sarma, S., Sen, T., Devi, M. V., Deka, B., & Singh, A. K. (2023). Genome mining to identify valuable secondary metabolites and their regulation in Actinobacteria from different niches. *The Archives of Microbiology, 205*, 127.

Biermann, F., Wenski, S. L., & Helfrich, E. J. N. (2022). Navigating and expanding the roadmap of natural product genome mining tools. *Beilstein Journal of Organic Chemistry, 18*, 1656–1671.

Blin, K., Kim, H. U., Medema, M. H., & Weber, T. (2019). Recent development of antiSMASH and other computational approaches to mine secondary metabolite biosynthetic gene clusters. *Briefings in Bioinformatics, 20*, 1103–1113.

Blin, K., Shaw, S., Augustijn, H. E., Reitz, Z. L., Biermann, F., Alanjary, M., et al. (2023). antiSMASH 7.0: New and improved predictions for detection, regulation, chemical structures and visualisation. *Nucleic Acids Research, 51*, W46–W50.

Bloudoff, K., & Schmeing, T. M. (2017). Structural and functional aspects of the nonribosomal peptide synthetase condensation domain superfamily: Discovery, dissection and diversity. *Biochimica et Biophysica Acta (BBA)—Proteins and Proteomics., 1865*, 1587–1604.

Charlop-Powers, Z., Owen, J. G., Reddy, B. V. B., Ternei, M. A., & Brady, S. F. (2014). Chemical-biogeographic survey of secondary metabolism in soil. *National Academy of Sciences of the United States of America, 111*, 3757–3762.

Chavali, A. K., & Rhee, S. Y. (2018). Bioinformatics tools for the identification of gene clusters that biosynthesize specialized metabolites. *Briefings in Bioinformatics, 19*, 1022–1034.

Chinnathambi, A., Salmen, S. H., Al-Garadi, M. A., Wainwright, M., & Ali Alharbi, S. (2023). Marine actinomycetes: An endless source of potentially therapeutic novel secondary metabolites and other bioactive compounds. *Journal of King Saud University, Science, 35*, 102931.

Cimermancic, P., Medema, M. H., Claesen, J., Kurita, K., Wieland Brown, L. C., Mavrommatis, K., et al. (2014). Insights into secondary metabolism from a global analysis of prokaryotic biosynthetic gene clusters. *Cell, 158*, 412–421.

de Jong, A., van Hijum, S. A. F. T., Bijlsma, J. J. E., Kok, J., & Kuipers, O. P. (2006). BAGEL: A web-based bacteriocin genome mining tool. *Nucleic Acids Research, 34*, W273–W279.

Fu, Y., Xu, Y., Ruijne, F., & Kuipers, O. P. (2023). Engineering lanthipeptides by introducing a large variety of RiPP modifications to obtain new-to-nature bioactive peptides. *FEMS Microbiology Reviews, 47*(3), fuad017.

Ghosh, S., Sarkar, T., Pati, S., Kari, Z. A., Edinur, H. A., & Chakraborty, R. (2022). Novel bioactive compounds from marine sources as a tool for functional food development. *Frontiers in Marine Science, 9*, 832957.

Gokulan, K., Khare, S., & Cerniglia, C. (2014). METABOLIC PATHWAYS I Production of secondary metabolites of bacteria. In *Encyclopedia of food microbiology* (pp. 561–569). Elsevier.

Han, S.-W., & Won, H.-S. (2024). Advancements in the application of ribosomally synthesized and post-translationally modified peptides (RiPPs). *Biomolecules, 14*, 479.

Huang, Z., Zhu, W., Bai, Y., Bai, X., & Zhang, H. (2024). Non-ribosomal peptide synthetase (NRPS)-encoding products and their biosynthetic logics in Fusarium. *Microbial Cell Factories, 23*, 93.

Hur, J. Y., Jeong, E., Kim, Y. C., & Lee, S. R. (2023). Strategies for natural product discovery by unlocking cryptic biosynthetic gene clusters in fungi. *Separations., 10*, 333.

Jensen, P. R., Chavarria, K. L., Fenical, W., Moore, B. S., & Ziemert, N. (2014). Challenges and triumphs to genomics-based natural product discovery. *Journal of Industrial Microbiology & Biotechnology, 41*, 203–209.

Johnston, C. W., Skinnider, M. A., Wyatt, M. A., Li, X., Ranieri, M. R. M., Yang, L., et al. (2015). An automated Genomes-to-Natural Products platform (GNP) for the discovery of modular natural products. *Nature Communications, 6*, 8421.

Julianti, E., Abrian, I. A., Wibowo, M. S., Azhari, M., Tsurayya, N., Izzati, F., et al. (2022). Secondary metabolites from marine-derived fungi and actinobacteria as potential sources of novel colorectal cancer drugs. *Marine Drugs, 20*, 67.

Kautsar, S. A., Blin, K., Shaw, S., Weber, T., & Medema, M. H. (2021). BiG-FAM: The biosynthetic gene cluster families database. *Nucleic Acids Research, 49*, D490–D497.

Khaldi, N., Seifuddin, F. T., Turner, G., Haft, D., Nierman, W. C., Wolfe, K. H., et al. (2010). SMURF: Genomic mapping of fungal secondary metabolite clusters. *Fungal Genetics and Biology., 47*, 736–741.

Khalifa, S. A. M., Elias, N., Farag, M. A., Chen, L., Saeed, A., Hegazy, M.-E. F., et al. (2019). Marine natural products: A source of novel anticancer drugs. *Marine Drugs, 17*, 491.

Kim, J., & Yi, G.-S. (2012). PKMiner: A database for exploring type II polyketide synthases. *BMC Microbiology, 12*, 169.

Kim, H., Hwang, J.-Y., Chung, B., Cho, E., Bae, S., Shin, J., et al. (2019). 2-Alkyl-4-hydroxyquinolines from a marine-derived streptomyces sp. inhibit hyphal growth induction in *Candida albicans*. *Marine Drugs, 17*, 133.

Lee, N., Hwang, S., Kim, J., Cho, S., Palsson, B., & Cho, B.-K. (2020). Mini review: Genome mining approaches for the identification of secondary metabolite biosynthetic gene clusters in Streptomyces. *Computational and Structural Biotechnology Journal, 18*, 1548–1556.

Liu, W., Li, E., Liu, L., Tian, F., Luo, X., Cai, Y., et al. (2023). Antifungal activity of compounds from Gordonia sp. WA8-44 isolated from the gut of *Periplaneta americana* and molecular docking studies. *Heliyon., 9*, e17777.

Mandragutti, T., & Sudhakar, G. (2023). Selective isolation and genomic characterization of biopolymer producer—A novel feature of halophile *Brachybacterium paraconglomeratum* MTCC 13074. *Journal, Genetic Engineering & Biotechnology, 21*, 24.

Manivasagan, P., Kang, K.-H., Sivakumar, K., Li-Chan, E. C. Y., Oh, H.-M., & Kim, S.-K. (2014). Marine actinobacteria: An important source of bioactive natural products. *Environmental Toxicology and Pharmacology, 38*, 172–188.

Mao, D., Okada, B. K., Wu, Y., Xu, F., & Seyedsayamdost, M. R. (2018). Recent advances in activating silent biosynthetic gene clusters in bacteria. *Current Opinion in Microbiology, 45*, 156–163.

Martín, J., Da, S., Sousa, T., Crespo, G., Palomo, S., González, I., Tormo, J., et al. (2013). Kocurin, the true structure of PM181104, an anti-methicillin-resistant *Staphylococcus aureus* (MRSA) thiazolyl peptide from the marine-derived bacterium *Kocuria palustris*. *Marine Drugs, 11*, 387–398.

Maurya, S. K., Niveshika, & Mishra, R. (2019). Importance of bioinformatics in genome mining of cyanobacteria for production of bioactive compounds. In *Cyanobacteria* (pp. 477–506). Elsevier.

Meenakshi, S., Hiremath, J., Meenakshi, M. H., & Shivaveerakumar, S. (2024). Actinomycetes: Isolation, cultivation and its active biomolecules. *Journal of Pure and Applied Microbiology, 18*, 118–143.

Morgan, R. N., Al, A. A., Alshahrani, M. Y., & Aboshanab, K. M. (2023). New insights on biological activities, chemical compositions, and classifications of marine actinomycetes antifouling agents. *Microorganisms., 11*, 2444.

Mothana, A. A., Al-Shamahy, H. A., Mothana, R. A., Khaled, J. M., Al-Rehaily, A. J., Al-Mahdi, A. Y., et al. (2022). Streptomyces sp. 1S1 isolated from Southern coast of the Red Sea as a renewable natural resource of several bioactive compounds. *Saudi Pharmaceutical Journal., 30*, 162–171.

Mukhtar, S., Zaheer, A., Aiysha, D., Malik, K. A., & Mehnaz, S. (2017). Actinomycetes: A source of industrially important enzymes. *Proteomics Bioinform., 10*, 316–319.

Mungan, M. D., Alanjary, M., Blin, K., Weber, T., Medema, M. H., & Ziemert, N. (2020). ARTS 2.0: Feature updates and expansion of the antibiotic resistant target seeker for comparative genome mining. *Nucleic Acids Research, 48*, W546–W552.

Navarro-Muñoz, J. C., Selem-Mojica, N., Mullowney, M. W., Kautsar, S. A., Tryon, J. H., Parkinson, E. I., et al. (2020). A computational framework to explore large-scale biosynthetic diversity. *Nature Chemical Biology, 16*, 60–68.

Ngamcharungchit, C., Chaimusik, N., Panbangred, W., Euanorasetr, J., & Intra, B. (2023). Bioactive metabolites from terrestrial and marine actinomycetes. In *Molecules*. Multidisciplinary Digital Publishing Institute (MDPI).

Paulus, C., Rebets, Y., Tokovenko, B., Nadmid, S., Terekhova, L. P., Myronovskyi, M., et al. (2017). New natural products identified by combined genomics-metabolomics profiling of marine Streptomyces sp. MP131-18. *Scientific Reports, 7*, 42382.

Pfeiffer, I. P.-M., Schröder, M.-P., & Mordhorst, S. (2024). Opportunities and challenges of RiPP-based therapeutics. *Natural Product Reports, 41*, 990–1019.

Prem Anand, K., Achuthananda Jayasri, M., & Suthindhiran, K. (2024). *Role and mechanism of action of extremophilic marine actinobacteria in plant growth promotion*. IntechOpen.

Robbins, T., Liu, Y.-C., Cane, D. E., & Khosla, C. (2016). Structure and mechanism of assembly line polyketide synthases. *Current Opinion in Structural Biology, 41*, 10–18.

Russell, A. H., & Truman, A. W. (2020). Genome mining strategies for ribosomally synthesised and post-translationally modified peptides. *Computational and Structural Biotechnology Journal, 18*, 1838–1851.

Sabatini, M., Comba, S., Altabe, S., Recio-Balsells, A. I., Labadie, G. R., Takano, E., et al. (2018). Biochemical characterization of the minimal domains of an iterative eukaryotic polyketide synthase. *The FEBS Journal, 285*, 4494–4511.

Saeed, A. F. U. H., Su, J., & Ouyang, S. (2021). Marine-derived drugs: Recent advances in cancer therapy and immune signaling. *Biomedicine & Pharmacotherapy, 134*, 111091.

Saito, S., & Arai, M. A. (2024). Methodology for awakening the potential secondary metabolic capacity in actinomycetes. *Beilstein Journal of Organic Chemistry, 20*, 753–766.

Salwan, R., & Sharma, V. (2020). Molecular and biotechnological aspects of secondary metabolites in actinobacteria. *Microbiological Research, 231*, 126374.

Sarkar, G., & Suthindhiran, K. (2022). Diversity and biotechnological potential of marine actinomycetes from India. *Indian Journal of Microbiology, 62*, 475–493.

Sarmiento-Tovar, A. A., Silva, L., Sánchez-Suárez, J., & Diaz, L. (2022). Streptomyces-derived bioactive pigments: Ecofriendly source of bioactive compounds. *Coatings, 12*, 1858.

Scherlach, K., & Hertweck, C. (2021). Mining and unearthing hidden biosynthetic potential. *Nature Communications, 12*, 3864.

Schorn, M. A., Alanjary, M. M., Aguinaldo, K., Korobeynikov, A., Podell, S., Patin, N., et al. (2016). Sequencing rare marine actinomycete genomes reveals high density of unique natural product biosynthetic gene clusters. *Microbiology (N Y)., 162*, 2075–2086.

Selim, M. S. M., Abdelhamid, S. A., & Mohamed, S. S. (2021). Secondary metabolites and biodiversity of actinomycetes. *Journal, Genetic Engineering & Biotechnology, 19*, 72.

Shi, S., Cui, L., Zhang, K., Zeng, Q., Li, Q., Ma, L., et al. (2022a). *Streptomyces marincola* sp. nov., a novel marine actinomycete, and its biosynthetic potential of bioactive natural products. *Frontiers in Microbiology, 13*, 860308.

Shi, T., Wang, Y.-F., Wang, H., & Wang, B. (2022b). Genus Nocardiopsis: A prolific producer of natural products. *Marine Drugs, 20*, 374.

Singh, T. A., Passari, A. K., Jajoo, A., Bhasin, S., Gupta, V. K., Hashem, A., et al. (2021). Tapping into actinobacterial genomes for natural product discovery. *Frontiers in Microbiology, 12*, 655620.

Skinnider, M. A., Merwin, N. J., Johnston, C. W., & Magarvey, N. A. (2017). PRISM 3: Expanded prediction of natural product chemical structures from microbial genomes. *Nucleic Acids Research, 45*, W49–W54.

Skinnider, M. A., Johnston, C. W., Gunabalasingam, M., Merwin, N. J., Kieliszek, A. M., MacLellan, R. J., et al. (2020). Comprehensive prediction of secondary metabolite structure and biological activity from microbial genome sequences. *Nature Communications, 11*, 6058.

Suresh, K. N., & Devi, C. S. (2025). Exploring the fibrinolytic potential of marine *Actinoalloteichus caeruleus* isolated from Bay of Bengal coast. *BMC Microbiology, 25*, 113.

Teijaro, C. N., Adhikari, A., & Shen, B. (2019). Challenges and opportunities for natural product discovery, production, and engineering in native producers versus heterologous hosts. *Journal of Industrial Microbiology & Biotechnology, 46*, 433–444.

Tizabi, D., & Hill, R. T. (2023). Micrococcus spp. as a promising source for drug discovery: A review. *Journal of Industrial Microbiology & Biotechnology, 50*(1), kuad017.

Udwary, D. W., Zeigler, L., Asolkar, R. N., Singan, V., Lapidus, A., Fenical, W., et al. (2007). Genome sequencing reveals complex secondary metabolome in the marine actinomycete *Salinispora tropica. National Academy of Sciences of the United States of America, 104*, 10376–10381.

van Heel, A. J., de Jong, A., Montalbán-López, M., Kok, J., & Kuipers, O. P. (2013). BAGEL3: Automated identification of genes encoding bacteriocins and (non-)bactericidal post-translationally modified peptides. *Nucleic Acids Research, 41*, W448–W453.

Wang, S., Li, X., Yang, W., & Huang, R. (2024). Exploring the secrets of marine microorganisms: Unveiling secondary metabolites through metagenomics. *Microbial Biotechnology, 17*, e14533.

Weber, T., Rausch, C., Lopez, P., Hoof, I., Gaykova, V., Huson, D. H., et al. (2009). CLUSEAN: A computer-based framework for the automated analysis of bacterial secondary metabolite biosynthetic gene clusters. *Journal of Biotechnology, 140*, 13–17.

Xu, B.-L., Wang, Y.-Y., & Dong, C.-M. (2023). Study on marine actinomycetes and analysis of their secondary metabolites. *Life Research., 6*, 18.

Yang, Z., He, J., Wei, X., Ju, J., & Ma, J. (2020). Exploration and genome mining of natural products from marine Streptomyces. In *The applied microbiology and biotechnology* (pp. 67–76). Springer.

Zhao, X. Q., Xu, X. N., & Chen, L. Y. (2016). Production of enzymes from marine actinobacteria. *Advances in Food and Nutrition Research, 78*, 137–151.
Zhou, H.-C., Shamala, L. F., Yi, X.-K., Yan, Z., & Wei, S. (2020). Analysis of terpene synthase family genes in *Camellia sinensis* with an emphasis on abiotic stress conditions. *Scientific Reports, 10*, 933.

Probiotics from Marine Actinobacteria and Applications in Aquaculture

Mohit Chauhan, Harsh Prajapati, Shivani V. Raval, Carmen Pire Galiana, Rosa Maria Martinez-Espinosa, and Vikram H. Raval

Abstract

Marine actinobacteria are a varied group of microorganisms that can survive in tough ocean environments. These environments have high salt levels, strong pressure, and changing amounts of nutrients. They are specially adapted to live in such challenging conditions. Their ability to adapt their metabolism allows them to make a wide variety of useful substances, making them an important asset for biotechnology, health, and environmental sustainability. These organisms act as hard survivors in nature and are becoming important for improving fish farming health, helping balance gut bacteria, and creating beneficial foods and health supplements. Marine actinobacteria are known for having many useful properties, like fighting bacteria, stopping fungi, reducing oxidation, and helping control inflammation. They also help in adjusting the immune system and how the body interacts with microbes. Their metabolites help create new products like synbiotics, therapeutic molecules, and biopharmaceuticals. Even though they have a lot of potential uses, there are still issues with checking how harmful they are, getting approval from regulators, and using them on a big scale. Future studies are focusing on using AI to shape the microbiome, creating custom probiotics tailored to individuals, and exploring new ways to find useful microbes to fully use their benefits. Marine actinobacteria are at the crossroads of marine ecology, biotechnology, and health research, providing big chances for creating sustainable and precise new bio-based solutions.

M. Chauhan · H. Prajapati · V. H. Raval (✉)
Department of Microbiology and Biotechnology, Gujarat University, Ahmedabad, India
e-mail: vhraval@gujaratuniversity.ac.in

S. V. Raval
Department of Biotechnology, M.G. Science Institute (Autonomous), Ahmedabad, India

C. P. Galiana · R. M. Martinez-Espinosa
Department of Biochemistry and Molecular Biology, Edaphology, and Agricultural Chemistry, Faculty of Sciences, University of Alicante, Alicante, Spain

K. Loganathan, Z. Li (eds.), *Marine Actinomycetes*,
https://doi.org/10.1007/978-981-95-9849-6_9

Graphical Abstract

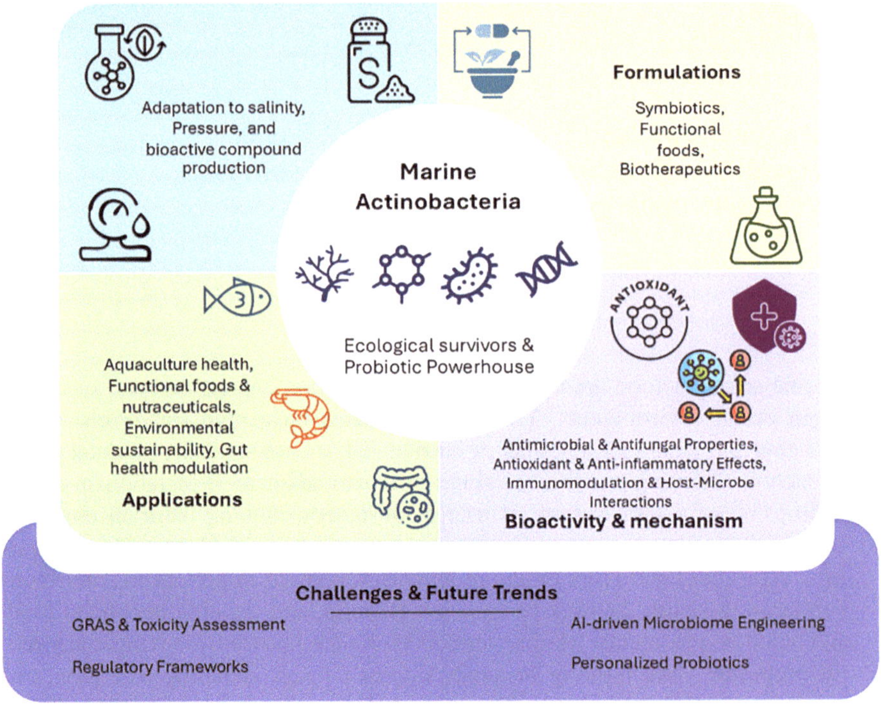

Keywords

Marine actinobacteria · Bioactive compounds · Probiotics · Functional foods · Immunomodulation · Microbiome engineering

1 Introduction

To meet the world's protein needs, one of the industries with the fastest growth is probiotic marine actinobacteria. Rise in aquatic farming has led to critical challenges, such as overuse of antibiotics and chemotherapeutic agents (Ahmed & Azra, 2022). This widespread application has led to the rise in multidrug-resistant (MDR) pathogens, immunodepression in aquatic animals, and environmental pollution. Along with this, there is an urgent demand for eco-friendly alternatives that support animal health, boost productivity, and preserve ecosystem integrity (James et al., 2023).

Probiotics have been used as a viable and sustainable alternative in aquaculture to suppress the negative effects of antibiotics. According to the World Health Organization (WHO) and Food and Agriculture Organization (FAO) of the United

Nations, Probiotics are live microorganisms that, when administered in adequate amounts, confer a health benefit on the host. Lactic acid bacteria (LAB) and *Bacillus* species have historically dominated the probiotic market (Martínez Cruz et al., 2012). The search for novel probiotic strains is crucial in combating developing diseases in aquaculture. As a result, a variety of different microorganisms, including yeasts, molds, bacteriophages, unicellular algae, and other non-conventional bacteria such as Actinobacteria, have been tested for probiotic activity (Ghosh et al., 2023).

Actinobacteria are gram-positive filamentous bacteria with high G + C contents in their genome. They have assorted, extraordinary appearances and, similar to fungi, were questionable in some situations. Actinobacteria typically live in soil and marine environments and are known to perform significant roles in biogeochemical cycles, bioremediation, and the synthesis of bacteriocins and other powerful bioactive substances like antibiotics and enzymes (Singh & Dubey, 2018). Actinobacteria should be investigated as a viable option for the aquaculture sector because of all these distinctive and varied characteristics.

They are also recognized as "Modern Actinobacteria" because of their multiple applications and can produce different secondary metabolites. *Streptomyces, Micromonospora*, and *Salinispora* are the most powerful actinobacterial genera due to their extensive secondary metabolite biosynthetic potential, ability to produce a wide range of antibiotics, enzymes, and other bioactive compounds, as well as their ecological adaptability and roles in antimicrobial and immunomodulatory activities, although *Streptomycetes* have received the most attention (James et al., 2023).

Streptomyces have unique physiological traits, such as their ability to secrete hydrolytic enzymes like amylase, protease, and lipase, which enhance feed digestion and nutrient absorption in aquaculture species after colonizing the host intestine (Tan et al., 2016). Strains from different environments exhibit a range of probiotic properties, including tolerance to gastrointestinal conditions, antimicrobial activity against fish pathogens, stimulation of immune responses, growth promotion, and improvement of water quality (Ghosh et al., 2023; Tan et al., 2016). These characteristics make Streptomyces a promising candidate for next-generation probiotics in aquaculture.

Despite their significant properties, Actinobacteria have not been extensively studied as probiotics in aquaculture. *Actinomycetes,* particularly *Streptomyces* and *Micromonospora,* are responsible for producing approximately 50% of microbial antibiotics (James et al., 2023). Actinobacteria are responsible for 45% of bioactive microbial compounds identified, out of 10,000 antibiotics. Actinobacteria are the source of most antibiotics used in aquaculture, specifically aminoglycosides, tetracyclines, and macrolides (James et al., 2023; Ranjani et al., 2016). Their distinct growth characteristics and complex metabolic traits make them ideal as bioaugmenters in aquaculture. Some of these compounds have anticancer properties.

Beyond their antimicrobial potential, Coral-associated actinomycetes can limit biofilm formation, including that of antibiotic-resistant *Staphylococcus aureus*, which can cause infection through medical devices (De Simeis & Serra, 2021; James et al., 2023). Actinomycetes may form biofilms and decompose complex polymers in the environment, making them valuable in composting and bioremediation efforts, particularly for hazardous pesticides (Jagannathan et al., 2021). Actinomycetes are being explored for their potential to boost agricultural output. Actinomycetes' diverse functional capacities may reflect their extensive repertoire of secondary metabolic pathways, a survival advantage that has made them important in the quest for novel bioactive chemicals (Jagannathan et al., 2021).

This chapter discusses the potential of marine Actinobacteria, particularly *Streptomyces* and *Micromonospora,* as next-generation probiotics in aquaculture. It explores their ecological functions, adaptation methods, and bioactive properties, with a focus on their uses to improve aquatic animal health, growth, and sustainability. The chapter goes on to explore their uses in human gut health, synergistic use with prebiotics, commercialization challenges, and novel probiotic development.

2 Adaptation Mechanisms of Actinobacteria in Marine Environments

2.1 Salt Tolerance

Marine actinobacteria are subjected to high and fluctuating salinity, which results in osmotic stress and dehydration. To thrive in such an environment, they accumulate osmolytes (Compatible solutes) such as glycine, betaine, ectoine, trehalose, and proline (Scholz et al., 2015). They are small organic molecules that stabilize protein and cellular functions without affecting metabolism. The biosynthesis of ectoine, a key osmo-protectant, involves core genes such as *ectA*, *ectB*, and *ectC*, but other genes like *ask*, *asd*, and *ectD* may also contribute to precursor synthesis and conversion to hydroxyectoine (Zhang et al., 2022).

Marine obligate genera such as *Salinispora* possess unique genetic adaptations that regulate osmolyte production, including dedicated gene clusters (Jensen, 2022). Actinobacteria also modify their membrane lipids and peptidoglycan cell walls to regulate ion permeability and maintain cellular integrity. In addition, polar growth and structural reorganization mechanisms in actinomycetes contribute to cellular adaptability under osmotic stress. Morphological flexibility also increases adaptability. *Arthrobacter* spp., for example, evolves from rod to mycelial morphologies, and some filamentous organisms can generate L-forms, which are wall-less cells capable of surviving in high osmotic conditions (Ramijan et al., 2018). Exopolysaccharide (EPS) synthesis and biofilm matrix formation conserve water and protect against desiccation and oxidative stress produced by high salt levels (Singh et al., 2021).

Actinobacteria generally use organic solutes, although some marine organisms may also use ion transport pathways to attain ionic equilibrium (Jagannathan et al., 2021). This strategy resembles the "salt-in" mechanism used by halophilic archaea and complements organic solute-based adaptations. Actinobacteria's salinity tolerance is polyphyletic, comprising biochemical, structural, and ecological strategies for survival in marine, hypersaline, and other salt-affected environments (DasSarma & DasSarma, 2015).

2.2 Pressure Adaptation

Marine planktonic Actinobacteria have undergone multiple evolutionary strategies for life in high-pressure deep-sea environments, although mechanisms of pressure tolerance are not described explicitly. Deep-sea genera have greater genomic GC content and longer intergenic regions, two adaptations that may help DNA stably provision under extreme pressure and chemical conditions (Roda-Garcia et al., 2023). They have metabolic pathways for the degradation of seemingly recalcitrant organic matter, and they also can metabolize nitrate and possibly nitrogen generated from the degradation of recalcitrant organic matter, which would aid their survival in nutrient-limited, oxygen-minimized waters and high-pressure environments.

In addition, they encode complex transcriptional regulatory proteins, like the unique WhiB paralogs, that could be involved in environmental sensing or the responses involving adaptation to multiple environmental and physiological stresses, such as high hydrostatic pressure (Suominen et al., 2021). Collectively, genomics and metabolic traits demonstrate a broad evolutionary response to deep oceans that are highly coupled with adverse environmental conditions, namely, high pressure. It is important to note that pressure-adaptive molecular mechanisms are not very well described at the time of writing this chapter.

2.3 Temperature Adaptation

Polyextremophilic actinobacteria, typically found in deserts and saline environments, often exhibit combined resistance to multiple stressors such as extreme temperatures, salinity, and desiccation—highlighting their versatile adaptive strategies (Mohammadipanah & Wink, 2016). As per the literature, to survive and grow in extreme environments, actinobacteria have good temperature adaptation mechanisms. Thermophilic actinobacteria live at 40–80 °C, relying on thermostable enzymes stabilized by ionic bonds and hydrophobic interactions, heat shock proteins such as GroEL and DnaK for protein refolding under thermal stress, and membranes enriched in saturated lipids for structural stability. Some examples are the genera *Amycolatopsis ruanii, Thermopolyspora,* and *Thermotunica* (Hui et al., 2021; Sen et al., 2022).

Conversely, psychrophilic actinobacteria, which survive at temperatures below 15 °C, produce cold-adapted enzymes with flexible structures suited for low kinetic energy environments, synthesize cold shock proteins for RNA folding and ribosomal function, and incorporate unsaturated fatty acids into their membranes to maintain fluidity. Furthermore, antifreeze proteins (AFPs) and exopolysaccharides (EPS) provide resistance against ice damage and UV radiation, as seen in Antarctic and deep-sea isolates like *Streptomyces* and *Arthrobacter* (Hui et al., 2021; Saralov, 2019).

Their genomic architecture also supports adaptation: psychrophiles often have larger genomes encoding a broader array of stress-response genes, while thermophiles tend to have compact, efficient genomes optimized for harsh conditions.

Actinobacteria thus demonstrate a broad and unique range of molecular strategies for thermal adaptation, enabling them to inhabit nearly every ecological niche on earth, and offering substantial promise for biotechnological applications (Hui et al., 2021; Salwan & Sharma, 2022; Sen et al., 2022).

2.4 pH and Oxidative Stress Tolerance

Marine actinobacteria can tolerate oxidative damage and pH variations in marine environments through a combination of physiological and molecular mechanisms. These include the use of proton pumps and membrane composition changes to regulate intracellular pH, as well as antioxidative defense systems to neutralize reactive oxygen species (ROS). Proton pumps and membrane composition changes are used to maintain their intracellular pH homeostasis, which controls ion flux and maintains cellular processes across a pH range (Behera & Das, 2023).

To combat oxidative stress brought on by reactive oxygen species (ROS), these bacteria produce antioxidative enzymes like catalase and superoxide dismutase (SOD). They also employ thiol-based redox systems, including the mycothiol and thioredoxin pathways, to preserve redox equilibrium and lessen oxidative damage (Jomova et al., 2023). To further increase their stress tolerance, marine actinobacteria produce siderophores and EPS that chelate heavy metals and form barriers that prevent ROS poisoning (Tang et al., 2024). Certain sigma factors (OxyR, σR-RsrA) control the coordinated expression of transcriptional regulators and stress response genes, allowing for a prompt reaction to environmental changes (Park et al., 2019). Omics research has provided insights into similar stress responses in polar marine bacteria, including metabolic alterations in lipid production and activation of antioxidant defense systems, indicating their resistance under pH and oxidative stress in severe conditions (Leprêtre et al., 2025). Together, these investigations highlight the vital physiological and molecular tools that marine actinobacteria use to endure and perform well in dynamic and demanding marine environments (Valliappan et al., 2014).

2.5 Nutrient Acquisition Strategies

Marine actinobacteria have evolved to survive in environments that are often nutrient-limited and characterized by extreme conditions. For example, they synthesize a range of hydrolytic enzymes, including cellulase, chitinase, protease, and lipase, to hydrolyze complex organic macromolecules into simpler forms that can be absorbed (Rodrigues et al., 2024; Siro et al., 2022). Another key adaptation for marine actinobacteria is the production of siderophores to scavenge iron, as it is a nutrient that is markedly limited in the marine environment. Siderophores bind to insoluble iron sources, which render the nutrients bioavailable. Siderophores can bind available iron from bound and unavailable forms in marine sediments through the solubilization of phosphates to allow for the continued provision of vital nutrients for cellular function. Marine actinobacteria utilize nutrient-rich host environments to form symbiotic relationships with marine animals such as coral and sponges to allow for biogeochemical cycles (Cui et al., 2022). These adaptations highlight the amazing metabolic capacity of marine actinobacteria to adapt and survive within the competitive marine ecosystem and further their associated genomic studies that describe biosynthetic gene clusters capable of producing cyclic structures characteristic of polyketide synthase (PKS) and non-ribosomal peptide synthetase (NRPS) (Helmi, 2025).

2.6 Symbiosis and Niche Adaptation

Marine actinobacteria have a diversity of symbiotic relationships and specialized niche adaptations that allow for their survival and functional roles within complex marine ecosystems. In Scleractinia corals, actinomycetes are putative hotspots for the production of specialized metabolites, which mediate interactions that can include protection against pathogens and nutrient recycling, thus emphasizing their symbiotic role in coral microbiomes (Siro et al., 2022). Analyses using genomics have shown that species-specific genomic islands that are full of secondary metabolite biosynthetic gene clusters are the basis of ecological differentiation and functional adaptation in *Salinispora* species that permit these actinobacteria to fill distinct niches and be symbionts or free living (Penn et al., 2009). To elaborate on niche specialization, uncultured UBA5794 actinobacteria have shown metabolic flexibility with genomic characteristics shaped from habitat types (sponge-associated and free-living) and exhibited divergent immune capabilities and energy acquisition strategies driving adaptations to their hosts or environmental conditions, which highlights the evolution from free-living ancestors to specialized symbionts (Huang et al., 2025). Moreover, *Micromonospora,* an endophytic marine actinobacterium, participates in symbiosis within nitrogen-fixing plant nodules and possesses genome characteristics related to the acquisition of nutrients, polysaccharide degradation, and secondary metabolites that contribute to its role as a plant growth promoter and as a niche-specific endophyte (Trujillo et al., 2015).

Together, these studies suggest that marine actinobacteria rely on a combination of mechanisms—including secondary metabolite production for ecological interactions (Siro et al., 2022), horizontal gene transfer and genomic island acquisition for functional diversification (Penn et al., 2009; Trujillo et al., 2015), and host- or habitat-specific physiological adaptations for metabolic flexibility and immune evasion to establish symbiotic relationships and occupy specialized ecological niches (Huang et al., 2025).

2.7 Genomic and Metabolic Plasticity

Actinobacteria found in marine environments, especially the genus *Salinispora,* show a significant genomic and metabolic plasticity that allows them to invade many ecological niches for a variety of bioactive compounds. Ziemert et al. (2014)

conducted a comparative genomic analysis of 75 strains of *Salinispora* and found substantial diversity in biosynthetic gene clusters (BGCs); they identified 124 unique pathways. Many of the pathways were either specific to the strain or were only found in a few strains, suggesting the relatively recent acquisition of metabolic potential via horizontal gene transfer (HGT) (Jagannathan et al., 2021). The BGCs were encoded by genomic islands associated with mobile genetic elements, suggesting a mechanism for the broad chemical diversity of *Salinispora* populations. Overall, there is evidence, for example, for *S. arenicola* with the biosynthesis of rifamycin and *S. tropica* with salinosporamide, that although lateral, there may also be vertical inheritance at the species level for some core pathway/biosynthetic potential (Letzel et al., 2017).

Secondary metabolites such as rifamycin, produced by *Salinispora arenicola*, and salinosporamide A from *S. tropica* have been shown to mediate competitive interactions and chemical defense in marine microbial communities, suggesting their specific ecological functions (Letzel et al., 2017).

Therefore, the evolutionary processes driving the metabolites' diversity in *Salinispora* are genetic strategies such as gene gain, gene loss, gene duplication, and divergence, coupled with a homologous recombination event at specific chromosomal locations, potentially at the levels of pathways (Ziemert et al., 2014). While the evolutionary processes may vary at the levels of pathways or genetic traits, the capacity for plasticity remains consistent for *Salinispora* to adapt and function in multiple marine environments, resulting in metabolic investment contributing to the ecological diversity and resilience of *Salinispora* populations (Rotter et al., 2021).

In a wider context, Kumar and colleagues discuss the metabolic versatility of marine actinobacteria, which contain a plethora of bioactive compounds such as polyketides, lipopeptides, and alkaloids, which have antibacterial, antifungal, antiviral, and anticancer activities. The authors note that genome-based approaches (like metagenomics and bioinformatics) have greatly aided in the discovery of new BGCs, because these biosynthetic pathways often encode for specialized secondary metabolites that help microbes survive extreme conditions in the marine environment. The authors go on to cover the challenges of accessing the metabolic potential of these organisms, including their lack of culturing and the difficulty in synthesizing very complex natural products, while also proposing synthetic biology and an interdisciplinary approach for the future (Kumar et al., 2024).

Comparative genomic studies confirm that *Salinispora* species exhibit conserved, yet adaptable biosynthetic gene clusters shaped by horizontal gene transfer and genome rearrangement, supporting both ecological adaptation and biosynthetic innovation (Ziemert et al., 2014; Letzel et al., 2017).

2.8 Secondary Metabolite Gene Clusters Underpinning Probiotic Traits

Marine actinobacteria have a remarkable genomic potential for secondary metabolite biosynthesis, which is to a great extent due to their extensive collection of biosynthetic gene clusters (BGCs). Some of these are polyketide synthases (PKS), non-ribosomal peptide synthetases (NRPS), ribosome-synthesized and post-translationally modified peptides (RiPPs), and siderophore clusters. In contrast to actinobacteria found on land, their saltwater counterparts have developed specialized structural forms of the BGCs to adapt to high salinity, high pressure, and limiting nutrient marine conditions (Manivasagan et al., 2014). These adaptations give them an edge in competition by enabling them not only to produce metabolites for protecting themselves but also for colonizing host organisms like fish and crustaceans successfully, an essential aspect of probiotics. Genome mining research conducted with antiSMASH and other platforms has identified a treasure trove of "cryptic" or "silent" BGCs in marine *Streptomyces, Nocardiopsis, and Salinispora,* most of which are anticipated to contain new antimicrobials, signaling molecules, and colonization factors (Prazdnova et al., 2025; Stincone & Brandelli, 2020).

In addition to antimicrobial function, secondary metabolite gene clusters have direct involvement in probiotic processes like stress resistance and immune modulation of hosts. For example, siderophore clusters augment competitive exclusion through iron deprivation of pathogenic bacteria, and bacteriocin clusters facilitate direct inhibition of aquaculture pathogens such as *Vibrio* spp. Likewise, osmoprotectant gene clusters (e.g., genes involved in ectoine biosynthesis) facilitate probiotic survival during osmotic stress in fish gut environments. Recent comparative genomics of marine actinobacterial probiotics from shrimp ponds showed enrichment in BGCs related to antioxidant metabolites, indicating an evolutionary connection between environmental adaptation and host-beneficial characteristics (Pickard et al., 2017).

Current omics-based studies further emphasize the importance of these gene clusters in functional probiotic expression. Transcriptomic studies have demonstrated that specific BGCs are induced during host colonization, suggesting their involvement in probiotic–host interaction and not just environmental persistence. Furthermore, synthetic biology methodologies are increasingly being employed to induce silent clusters, facilitating the discovery of new metabolites with double antimicrobial and immunomodulatory activities (Subramani & Aalbersberg, 2012). Therefore, genomic diversity among marine actinobacteria creates a fundamental platform for their probiotic bioactivities, making them next-generation candidates for sustainable aquaculture and human health interventions (Table 1).

3 Bioactivity and Mechanism of Marine Actinobacterial Probiotics

A wide variety of microbial communities that are specifically suited to high pressure, salinity, and low nutrient availability can be found in marine environments. The creation of structurally unique and pharmacologically promising compounds that are uncommon in terrestrial organisms is frequently the outcome of these adaptations. Novel antibiotics, anticancer chemicals, and immunomodulatory drugs have been demonstrated to be produced by marine actinobacteria, specifically

Table 1 Stress response strategies of marine actinobacteria: mechanisms, ecological roles, and taxonomic diversity

Type of stress	Mechanism of adaptation	Organisms name	Ecological benefits	References
High salinity	Synthesis of compatible solutes (e.g., ectoine, proline)	*Salinispora tropica, Streptomyces albidoflavus*	Maintains turgor and enzyme function in saline conditions	Giri and Varma (2019) and Ventosa et al. (1998)
	Na^+/H^+ antiporters for ionic balance	*Streptomyces rochei*	Prevention of toxic ion accumulation, regulation of cytoplasmic pH	Gohel et al. (2018)
	Production of EPS/biofilm	*Actinopolyspora halophila*	Protects cells from desiccation and salt shock	Kar and Singhal (2015)
Temperature extremes	Heal/cold shock proteins (e.g., DnaK, GroEL)	*Thermopolyspora flexuosa, Streptomyces thermoviolaceus*	Under stress, protein folding	Shivlata and Satyanarayana (2015)
	Change in membrane lipid composition	*Streptomyces sp. strain CNQ-509*	Maintain fluidity of the membrane	Wong et al. (2017)
	Cold-active enzymes in psychrotrophs	*Rhodococcus sp.* from polar seawater	Catalysis at low temperature	De Santi et al. (2014) and Furhan (2020)
Oxidative stress	Enzymatic antioxidants (SOD, catalase)	*Nocardiopsis dassonvillei*	Detoxification of ROS	Siddharth and Rai (2019)
	Non-Enzymatic antioxidants (carotenoids, melanin)	*Streptomyces griseus*	Protection from ROS, UV, and heavy metals	Tan et al. (2017)
UV radiation	Melanin and carotenoid production	*Actinomadura madurae, Streptomyces bikiniensis*	Absorbs UV, prevents DNA damage	Subramani and Aalbersberg (2012)
	DNA repair systems (photolyases, excision repair)	Marine *Streptomyces* spp.	Genome stability under solar radiation	Nicholson et al. (2000) and Ramírez-Gamboa et al. (2022)
Desiccation	Sporulation (endospores, exospores)	*Micromonospora spp., Streptomyces coelicolor*	Long-term survival in dry sediments or biofilm niches	Lemfack et al. (2020) and Maldonado and Quintana (2015)
	Retention of water through the EPS matrix	*Nocardia spp.*	Microenvironmental hydration	Donlan, (2002) and Hodar (2024)

(continued)

Table 1 (continued)

Type of stress	Mechanism of adaptation	Organisms name	Ecological benefits	References
Nutrient limitation	Extracellular enzyme secretion	*Streptomyces* spp., *Micromonospora* spp.	Nutrient scavenging from complex organic substances	Barka et al. (2016)
	Siderophore production	*Streptomyces tritolerans*	Enhance iron uptake, overcome pathogens	Wang and Li (2025)
	Stringent response and global regulators	*Streptomyces lividans*	Survival and differentiation in starvation	Lee et al. (2020) and Sivapragasam and Grove (2019)
High pressure	Synthesis of unsaturated lipids	*Streptomyces* spp. (deep-sea isolate)	Maintains membrane fluidity under hydrostatic pressure	Sandoval-Calderón et al. (2017)
	Baro-stable enzymes	Deep-sea *Rhodococcus* spp.	Enables metabolic activity under pressure	Poli et al. (2017)
	Biofilm formation for communal	Deep-sea *Streptomyces* spp	Structural and metabolic protection	Lo Giudice and Rizzo (2022)
Heavy metal stress	Efflux pumps	*Streptomyces mirabilis*, *Actinobacteria sp.*	Detoxification by active metal transport	Behera and Das (2023) and Gahlawat and Choudhury (2019)
	Chelators (Siderophores, Metallothioneins)	*Nocardiopsis alba*	Immobilizes toxic metals	Presentato et al. (2020)

Streptomyces and *Salinispora* (Manivasagan et al., 2014). The ecological roles that marine-derived strains play in symbiosis with marine hosts like sponges, corals, and algae—where they frequently function as natural probiotics, suppressing pathogenic colonization while boosting host immunity—are what set them apart from their terrestrial relatives, in addition to the structural novelty of their metabolites.

3.1 Antimicrobial Properties of Marine Actinobacterial Metabolites

One of the strongest bioactivities of marine probiotic actinobacteria is their ability to produce antimicrobial metabolites. These actinobacteria have developed to compete in multispecies microbial communities of the sea, within which competition for nutrients and space forces the production of effective antimicrobial compounds.

Probiotic strains of *Streptomyces, Micromonospora*, and *Nocardiopsis* have been shown to inhibit fish and shrimp pathogens, including *Vibrio harveyi, V. parahaemolyticus, Aeromonas hydrophila*, and *Edwardsiella tarda*. Antimicrobial action is usually mediated through bacteriocins, polyketides, lipopeptides, and phenolic compounds, which interfere with cell wall integrity, quorum sensing, or biofilm formation. All these activities are vital in aquaculture, where bacterial infection usually causes huge economic losses (Jagannathan et al., 2021).

Along with broad-spectrum antimicrobial action, marine actinobacterial metabolites tend to be specific toward aquaculture pathogens without influencing beneficial microbiota. Such selectivity increases their probiotic activity, enabling them to colonize the intestines of the host and inhibit opportunistic pathogens. For instance, *Streptomyces rubrolavendulae* extracted from marine sediments has been found to produce antimicrobial peptides that selectively target *Vibrio spp.* Without inhibiting host-associated lactic acid bacteria (Vladkova et al., 2022). This attribute not only enhances survival and growth levels of aquatic animals but also minimizes the use of synthetic antibiotics in keeping with sustainable aquaculture.

One highly desirable property is their anti-biofilm activity. *Vibrio* species, which are infamous for biofilm formation on shrimp and fish surfaces, become extremely resistant to traditional antibiotics. Metabolites of marine actinobacteria, like diketopiperazines and prodiginines, have been shown to inhibit biofilm formation and interfere with quorum-sensing networks of these pathogens. Through cell–cell interference, actinobacterial probiotics suppress pathogen virulence instead of applying lethal pressure, which assists in preventing the risk of antimicrobial resistance (Claverías et al., 2015). These observations highlight the promise of marine actinobacteria as probiotic agents with intrinsic pathogen-control activity.

3.2 Novel Antibiotics, Antifungals, and Their Mode of Action

Marine actinobacteria are one of the most structurally novel antibiotic-rich sources, numerous of which exhibit distinctive modes of action from terrestrial-derived molecules. Around 40% of all microbial secondary metabolites known to date are from actinobacteria, with a notable proportion coming from marine-derived strains (Claverías et al., 2015). New polyketide, aminoglycoside, and glycopeptide classes have been described from *Salinispora* and *Streptomyces* species with activities against multidrug-resistant pathogens. Notably, these metabolites tend to act via mechanisms different from traditional antibiotics, e.g., disturbing membrane potential, inhibiting new ribosomal targets, or repressing efflux pump systems (Subramani & Aalbersberg, 2012).

Antifungal ability of marine actinobacteria also contributes to their probiotic activity in aquaculture. Saprolegnia and Fusarium spp. are pathogenic fungi that infect fish hatcheries and pond ecosystems, causing serious losses. Marine-derived *Streptomyces* species produce polyene macrolides and azole-like metabolites that inhibit fungal growth by interfering with ergosterol biosynthesis and destabilizing fungal cell membranes. In addition, actinobacteria also generate enzymes like chitinases and glucanases that can lyse fungal cell walls, boosting their antifungal activity (Stincone & Brandelli, 2020).

The novelty of these antifungals and antibiotics not only offers a substitute for chemical therapy in aquaculture but also inhibits the development of resistance. Local *in situ* production of metabolites through probiotic use of antibiotic-producing actinobacteria provides continuous secretion at the host–pathogen boundary, which sustains pathogen suppression without the need for high doses externally applied. Targeted *in situ* production thus presents a sustainable, resistance-preventing alternative to pathogen control in marine and aquaculture environments.

3.3 Antioxidant and Anti-inflammatory Activities

Oxidative stress is a prevalent problem in aquaculture, caused by environmental fluctuations, intensive aquaculture rearing, and pathogen infection. Marine actinobacteria yield a variety of antioxidant secondary metabolites such as carotenoids, phenolic acids, flavonoid-like metabolites, and distinctive marine alkaloids. These metabolites have the potential to scavenge reactive oxygen species (ROS), hence reducing host tissue damage caused by oxidative stress. For instance, marine *Streptomyces* spp. extracts have shown high free-radical scavenging ability, with considerable inhibition of lipid peroxidation in stressed fish models. This attribute improves host survival and growth performance and is a major probiotic characteristic (Vladkova et al., 2022).

Anti-inflammatory activities are intimately associated with antioxidant mechanisms. In aquaculture species, protracted inflammation due to infection or environmental stressors compromises immunity and growth. Actinobacterial probiotics have also been demonstrated to suppress pro-inflammatory cytokines like TNF-α, IL-1β, and IL-6 while at the same time promoting anti-inflammatory mediators like IL-10. These control mechanisms maintain the integrity of intestinal barriers and diminish tissue damage. Some metabolites operate by inhibiting NF-κB signaling, a dominant pathway in inflammatory management (Manivasagan et al., 2014). Cumulatively, antioxidant and anti-inflammatory activities emphasize the two-fold role of marine actinobacteria: they not only inhibit pathogens, but also stimulate host resistance to environmental and physiological stress. This two-pronged mechanism sets them apart from traditional antibiotics and makes them holistic biotherapeutics in aquaculture systems.

3.4 Role in Oxidative Stress Reduction and Inflammation Control

Streptomyces specialis 208DD-067, a marine isolate, produces various streptoglyceride family (A-H) chemicals with proven anti-inflammatory properties (Moreira de Gouveia et al., 2024). To diminish the inflammatory activities generated by particulate matter, streptinone, a novel indanone derivative from *Streptomyces massiliensis* of marine origin, repressed the gene expression of pro-inflammatory cytokines (TNF-α, IL-6, and IL-1β). The substances also suppressed pro-inflammatory pathways, including the TLR-mediated NF-κB pathway, by reducing the phosphorylation and nuclear translocation of IκB-α, p65, and p50 (Undabarrena et al., 2016). ManA, ManB, Asu, and Col—compounds produced from manumycin were able to inhibit the expression of pro-inflammatory genes, such as IL-1β and TNF-α, thereby reducing the production of IL-1β and TNF-α, respectively, and strengthening the anti-inflammatory impact (Moreira de Gouveia et al., 2024).

Strains of *Streptomyces gramineus*, which are linked to the lichen *Leptogium trichophorum*, generate actinofuranones D–I, which have anti-inflammatory properties (Moreira de Gouveia et al., 2024). These metabolites could suppress the release of TNF-α and IL-6, and their anti-inflammatory actions were linked to the hydroxyl substituent's placement on unsaturated alkyl chains, which may modify the physicochemical characteristics of medications and their biological activities (Undabarrena et al., 2016).

3.5 Immunomodulatory Effects and Host–Microbe Interactions

The probiotic action of marine actinobacteria also involves direct modulation of host immunity. Experimental research involving fish models has shown that dietary supplementation with actinobacterial probiotics improves both innate and adaptive immunity. For example, dietary supplementation with *Streptomyces chartreusis* stimulated increases in lysozyme activity, complement protein content, and respiratory burst activity of phagocytes. These increases directly translate to enhanced resistance against prevalent aquaculture pathogens (Ghosh et al., 2023). On the cellular level, marine actinobacteria engage with gut epithelial cells, where they induce mucin production and enhance intestinal barrier function. This is a physical and biochemical barrier to pathogen colonization. Moreover, actinobacterial probiotics modulate cytokine expression, promoting a balanced Th1/Th2 immune profile in the host. This facilitates optimal pathogen clearance without uncontrolled inflammation that can damage host tissues (Engevik et al., 2019). In addition to immune stimulation, host–microbe interactions driven by actinobacteria participate in competitive exclusion of pathogens. Through adhesion to intestinal surfaces, these probiotics take up ecological niches and release localized antimicrobials, thus blocking pathogen colonization. This multifaceted immunomodulation thus highlights their probiotic function not just as microbial competitors but also as collaborators in modulating host immunity and homeostasis.

3.6 Interaction with Immune Cells, Cytokine Regulation, Gut Immunity

The cells that act during this phase of the immune response are natural killer lymphocytes, neutrophils, and macrophages. The latter are one of the most important classes of antigen-presenting cells (APCs) that play roles in innate immunity and regulate adaptive immunity. Cytokines secreted by activated macrophages cause inflammatory responses and are essential for host defenses against invading pathogens (Li et al., 2018). The tumor cells recruit and activate other cells at the site of infection. In response to this stimulus, macrophages increase the production of reactive nitrogen intermediates such as nitric oxide (NO), reactive oxygen species (ROS), and pro-inflammatory cytokines such as tumor necrosis factor alpha (TNF) and interleukin 6 (IL-6). Cytokines are low molecular weight, soluble proteins released by a variety of cells, including natural killer cells, lymphocytes, and macrophages. They mediate and convey every reaction observed during the immune response. Among the groups they fall under are colony-stimulating factors (CSFs), transforming growth factors (TGFs), interleukins (ILs), lymphokines, mono-kines, interferons (IFNs), and tumor necrosis factor (TNF-α) (Moreira de Gouveia et al., 2024).

The development of Immunotherapeutics based on natural chemicals, such as those derived from Streptomyces spp., primarily targets cytokines, which are regulators of the immune response (Undabarrena et al., 2016). By inhibiting the activation of microglial cells in the pro-inflammatory M1 phenotype, which release pro-inflammatory cytokines like pro-inflammatory interleukins (IL) and TNF-α, as well as nitric oxide (NO) and ROS, *Streptomyces* spp. produce anhydroesfoliamycin. If left unchecked and prolonged, this pro-inflammatory activity can result in neurodegenerative disorders, including Parkinson's and Alzheimer's (Moreira de Gouveia et al., 2024). The novel exopolysaccharide (EPS) ebosine, which was discovered from Streptomyces sp. 139, showed a comparable impact. Both transcriptional and post-transcriptional levels of inflammatory activity were found to be mitigated by the metabolite, which also reduced the mRNA expression levels of genes encoding pro-inflammatory cytokines IL-23, IL-22, IL-17α, IL-6, IL-1β, and TNF-α (Undabarrena et al., 2016).

3.7 Environmental Bioremediation and Indirect Probiotic Benefits

Apart from the direct impacts on host health, marine actinobacteria also help promote aquaculture sustainability by remediating the environment. These microorganisms have enzymatic mechanisms that can degrade organic wastes, mitigate ammonia build-up, and degrade toxic aquaculture pond metabolites. By enhancing water quality, they indirectly minimize stress on aquatic animals and reduce the possibility of opportunistic infections. For example, *Nocardiopsis* spp. Strains isolated from shrimp farms are found to break down complex polysaccharides and hydrocarbons, improving the balance of pond ecosystems (Binda et al., 2018). Bioremediation is also extended to pathogen management in the environment. Through modified nutrient cycling and organic load reduction, marine actinobacteria establish an environment that is not suitable for the growth of pathogens. This ecosystem service enhances their probiotic function since they not only promote the health of hosts but also enhance the rearing environment. Marine actinobacteria, therefore, act as "dual probiotics," conferring both host-associated and ecosystem benefits, an added advantage over traditional probiotic strains being widely applied in aquaculture.

3.8 Emerging Applications and Future Directions

The probiotic potential of marine actinobacteria is growing at a furious pace, fueled by genomics, systems biology, and synthetic biology advancements. One promising direction is the formulation of synbiotics that pair actinobacterial probiotics with prebiotics like marine polysaccharides that facilitate their colonization and metabolic activity. Another hot area is the application of actinobacteria-derived metabolites as antibiotics and chemical antioxidant substitutes in aquaculture feeds (Bredholdt et al., 2007).

Prospects for the future also include the identification of antiviral and antiparasitic agents. Whereas antibacterial activity has been the main interest, preliminary findings suggest that marine actinobacteria inhibit viral replication and parasite attachment in aquatic hosts. Omics-driven discovery platforms, such as metagenomics and metabolomics, are anticipated to reveal novel classes of bioactives possessing multi-targeting activity (Subramani & Aalbersberg, 2012). In the future, the incorporation of probiotic actinobacteria into precision aquaculture—where microbiome modulation is informed by real-time monitoring and AI-driven management—has the potential to transform disease control strategies. These novel applications position marine actinobacteria as a linchpin in sustainable aquaculture and biotherapeutics.

4 Potential Applications in Human Gut Health

Compared to the quantity of human cells found in our bodies. The human body is densely colonized by different types of bacteria on almost every surface, including the skin, genitourinary system, mouth cavity, respiratory tract, ear, and gastrointestinal tract. With a vast microbial ecology, the gastrointestinal tract (GIT) is by far the most colonized organ; it is believed that the colon alone contains more than 70% of all the microorganisms in the human body (Vyas & Ranganathan, 2012).

Numerous species, including bacteria and unicellular eukaryotes, live inside the human body. Numerous bacterial phyla, such as *Firmicutes, Actinobacteria, and Bacteroidetes*, predominate in the human gut. It is estimated that there are up to 10 14 bacterial cells in the human microbiota, which is ten times more than there are human cells in our bodies. (Vyas & Ranganathan, 2012). The term "microbiota," "microflora," or "normal flora" refers to this enormous population of microorganisms that live alongside the host (Dekaboruah et al., 2020). A microbiota is a group of microorganisms that live in a particular habitat, such as bacteria, viruses, archaea, and some unicellular eukaryotes (D'Argenio & Salvatore, 2015).

Over 90% of the gut microbiome's relative abundance is made up of Firmicutes and Bacteroidetes, and their interaction is essential for preserving gut homeostasis. Proteobacteria and Actinobacteria make up the remaining 10% (Binda et al., 2018). Different bio-compartmentalization with unique and stable microbial communities is a characteristic of the various gastrointestinal areas. As opposed to the colon, where bacterial colonization is encouraged by the low redox potential and the slow transit, the stomach and small intestine have high peristaltic activity, bile and pancreatic secretion, and an acidic environment that prevents stable bacterial colonization (Arboleya et al., 2016).

Human physiology, immune system development, digestion, and detoxification processes all depend on these communities. Indeed, several of these gut-dwelling microbes encode proteins that play vital roles in the host's health, like the enzymes needed for the hydrolysis of indigestible food ingredients and the production of 49 vitamins (D'Argenio & Salvatore, 2015). Notably, the bacterial microbiota in the human digestive system is linked to both health and illness (Li et al., 2018).

4.1 Modulation of Gut Microbiota and Gut–Brain Axis Implication

Actinobacteria represent a small yet influential phylum within the human gut microbiota. Dominated by *Bifidobacterium* during infancy—largely promoted by human milk oligosaccharides (HMOs), they facilitate immune maturation and pathogen defense. Beyond bifidobacteria, genera like *Collinsella*, *Corynebacterium*, and *Adlercreutzia* participate in metabolizing dietary polyphenols, bile acids, and plant-derived carbohydrates, contributing to the metabolic network and cross-feeding interactions with other microbial phyla (Arboleya et al., 2016). As individuals progress from infancy through adulthood, Actinobacteria's relative abundance stabilizes at around 8–10% of the microbiota but can decline with ageing. Reduced levels have been associated with frailty, metabolic disorders, and gut inflammation—highlighting their status as health indicators. In contrast, gut dysbiosis in conditions such as obesity and inflammatory bowel disease often sees diminished bifidobacterial populations.

Emerging research explores marine actinobacteria (e.g., *Streptomyces*), which, while not classical gut residents, produce potent bioactive metabolites—antimicrobials, antioxidants, and immunomodulators—that mimic probiotic benefits. Early models suggest they could complement traditional commensals and assist in restoring gut balance under dysbiosis, introducing novel avenues for next-generation probiotics (O'Callaghan & van Sinderen, 2016).

The gut barrier comprises a physical layer (mucus, epithelial tight junctions) and functional defenses (peristalsis, antimicrobial peptides, and GALT). Disruption of these defenses can lead to increased permeability ("leaky gut") and systemic inflammation, triggering chronic immune activation (Chelakkot et al., 2018). Actinobacteria, particularly Bifidobacterium, reinforce this barrier via short-chain fatty acid (SCFA) production—especially acetate—which strengthens tight junction proteins and enhances mucus secretion. Acetate production protects against pathogens like E. coli and Shigella, and boosts mucin expression (Fukuda et al., 2011). Additionally, exopolysaccharides (EPS) from certain bifidobacteria form protective biofilms and stimulate host signaling pathways that further preserve barrier integrity (Binda et al., 2018). Marine actinobacteria introduce unique EPS and bioactives that can similarly enhance barrier health under stress. Their compounds may increase mucus layer viscosity, resist pathogen adhesion, and mitigate oxidative stress in epithelial cells—potentially offering next-generation probiotic support for maintaining barrier integrity (Manivasagan et al., 2014).

Gut Actinobacteria are pivotal in processing otherwise indigestible dietary components. Bifidobacterium species express glycosyl hydrolases that break down complex carbohydrates like inulin, FOS, GOS, and starches into SCFAs, with acetate influencing systemic metabolism and gut signaling. Additionally, bifidobacteria transform linoleic acid into conjugated linoleic acids (CLAs), which confer anti-obesity, anti-inflammatory, and anticarcinogenic benefits (Arboleya et al., 2016). They also contribute to vitamin biosynthesis (e.g., folate, biotin), highlighting multi-pronged metabolic support to the host. Marine actinobacteria expand this metabolic toolbox by producing enzymes capable of breaking down marine-specific polysaccharides (e.g., alginates, carrageenans), generating bioactive oligosaccharides with antioxidant and prebiotic properties. Incorporating these strains into probiotic formulations could enhance digestive resilience through salt-tolerant and novel metabolic functions (Manivasagan et al., 2014).

Actinobacteria—especially *B. longum subsp. infantis*—play essential roles in immune modulation via gut-epithelial signaling, IgA induction, and dendritic cell interactions. They engage TLRs to balance pro- and anti-inflammatory responses while bolstering T-regulatory cell populations and suppressing cytokines like MCP-1 and IL-6, effectively dampening systemic inflammation. Furthermore, bifidobacteria-derived metabolites, especially acetate, reach extra-intestinal sites, influencing immune function in tissues such as the liver and adipose, revealing their systemic immunomodulatory reach (Arboleya et al., 2016). Notably, marine actinobacteria produce immunomodulatory secondary metabolites (e.g., actinomycin derivatives) that suppress inflammatory cytokines, offering potential for precise probiotic immunotherapies targeting gut inflammation and broader immune dysregulation (Norouzi et al., 2018).

Microbial communication via the gut–brain axis occurs through neural, endothelial, and immune-mediated channels. Dysbiosis—such as loss of bifidobacteria—can alter blood–brain barrier permeability, neurotransmitter balance, and inflammatory signaling, laying the path for mood disorders and cognitive decline (Loh et al., 2024). Clinically, *B. breve* CCFM1025 has improved depression symptoms via modulation of tryptophan metabolism and gut microbiota. Another human study found that *B. longum* 1714 enhanced sleep quality, vitality, and social function after four weeks, further supporting its role as a psychobiotic strain. Emerging research also highlights that probiotics may attenuate neuroinflammation and support cognition via FMT and gut-brain interventions (Solanki et al., 2023). Mechanistically, SCFAs, neurotransmitter modulation (serotonin, GABA), vagus nerve activation, and immune regulation mediate these effects (Hwang & Oh, 2025). Though evidence is nascent, marine actinobacteria-derived neuroactive compounds—antioxidant, anti-inflammatory, cytoprotective—may hold promise in shaping the gut-brain axis, particularly under stress or neuroinflammatory conditions. Their novel structures offer compelling leads for future psychobiotic development (Fakhri et al., 2021).

5 Role in Aquaculture as Next-Gen Probiotics and in Functional Foods and Nutraceuticals

Marine actinobacteria are being recognized as next-generation probiotics and valuable contributors to the functional foods and nutraceutical industry. Their particular adaptation to marine conditions provides them with a large reservoir of bioactive chemicals and metabolic capabilities, allowing them to deliver health advantages to both aquaculture and humans. As the industry evolves toward sustainable, bio-based solutions, marine actinobacteria emerge as attractive agents for boosting growth, improving immunity, and preventing disease in a variety of host environments (Fig. 1).

5.1 Disease Prevention and Growth Promotion in Fish and Shrimp

Marine actinobacteria have proven to be beneficial in aquaculture species. This is due to their numerous physiological and biochemical activities aiding in disease control and growth enhancement (Das et al., 2008). Marine actinobacteria are

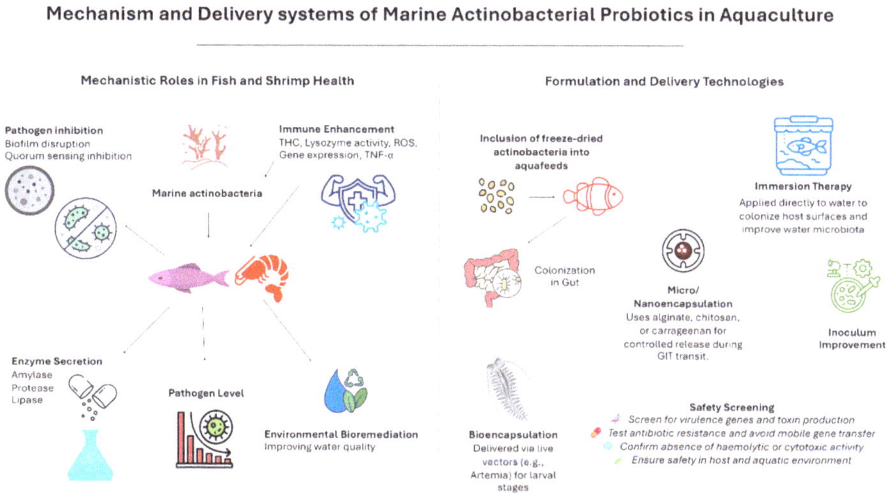

Fig. 1 Mechanism and delivery systems of marine actinobacterial probiotics in aquaculture

naturally occurring biocontrol agents by synthesizing bioactive metabolites that are antibacterial, immunomodulatory, and growth-enhancing, while maintaining the aquatic environment—these environmental aspects help to achieve optimal rearing conditions for the aquaculture species. They synthesize important bioactive metabolites, including streptomycin, polyketides, and siderophores that have important antibacterial, antifungal, and antiviral capabilities (Manivasagan et al., 2014). These bioactive metabolites may be used to suppress infections caused by important pathogens that can lead to diseases associated with growth reductions, including *Vibrio harveyi*, *V. parahaemolyticus*, *Aeromonas hydrophila*, and White Spot Syndrome Virus (WSSV). Some *Streptomyces* species produce bacteriocins as well as enzymes like chitinases and proteases that degrade the physical structures of pathogens, particularly targeting bacterial biofilms, which can increase biosecurity when using actinobacteria in aquaculture ponds (Kah Sem et al., 2023). Many of these metabolites also disrupt the quorum-sensing ability of harmful bacteria, reducing colonization and pathogenicity.

Immunomodulation is another important mechanism by which marine actinobacteria enhance disease prevention in aquatic animals. Actinobacteria's probiotic effect significantly improves innate immunity in shrimp and fish by acting on recognized immune systems. Actinobacteria supplementation improves health parameters in shrimp, including total hemocyte counts (THC), lysozyme activity, and respiratory burst activity (Abo-Al-Ela et al., 2025; James et al., 2023). It also stimulates reactive oxygen species (ROS) production and upregulates immune genes like prophenoloxidase and tumor necrosis factor-alpha (TNF-α), leading to increased resistance to pathogens and lower mortality rates. Actinobacteria disrupted the regulation of cytokine levels in the organism, allowing the host's pro-inflammatory and anti-inflammatory responses to be better balanced (Liu et al., 2010). Marine actinobacteria had a competitive effect in the gastrointestinal tracts of shrimp and fish by colonizing attachment sites and consuming available nutrients, preventing pathogens from properly establishing in areas where these opportunistic pathogens might otherwise colonize. Actinobacteria provide a competitive exclusion for a more balanced microbiota, which is correlated with a decreased risk of opportunistic infection (Ghosh et al., 2023; Goh et al., 2023). Actinobacteria also contribute significantly to disease suppression in aquaculture systems in combination with their known ability to produce antimicrobial compounds, and in some cases are present in minimal numbers, which can also predictably reduce microbial burden. Actinobacteria effects become more apparent during the vulnerable larval and juvenile stages (Khanjani et al., 2022).

Besides their disease-mitigating effects, marine actinobacteria also support fish and shrimp growth and development by regulating nutrient digestion and bioavailability (James et al., 2023). They produce important, extracellular enzymes to digest complex feed ingredients, which improves digestibility. The enzymes chewed by marine actinobacteria are extracellular amylase, protease, and lipase. In turn, the digestion of carbohydrates, protein, and fat by these enzymes improves Feed conversion ratio (FCR) and the Specific growth ratio (SGR). For example, introducing the actinobacteria, *Streptomyces,* to shrimp meal enhances protein digestibility, giving amino acids needed for rapid growth (Wang et al., 2024). Marine actinobacteria alter the gut microbiota, promoting beneficial microbial communities that aid digestion, energy production, and immunity. This intestinal stability causes energy to be shifted away from battling an active pathogen and toward boosting growth and reducing metabolic stress (Tolas et al., 2025). Marine actinobacteria contribute to a healthier aquatic environment via bioremediation of toxic nitrogenous wastes such as ammonia and nitrites. Microbial bioremediation assists with improving water quality, physiological stress levels in cultured animals, and survival and growth (Khanjani et al., 2022). The mentioned appeal of the anti-pathogenic, immunostimulatory, digestive, and environmental activities of marine actinobacteria would benefit production as part of aquaculture health management.

5.2 Formulation and Delivery of Live Biotherapeutics

To successfully develop marine actinobacterial probiotics for use in both aquaculture and human health, it is essential to focus on formulation strategies that ensure stability, viability, efficacy, and specificity across diverse biological and environmental conditions (Wang et al., 2024). Whether used in aquaculture ponds or incorporated into functional foods and nutraceuticals, marine actinobacteria face challenges such as extreme temperatures, pH variation, digestive enzyme exposure, and mechanical stress during processing, storage, and delivery.

Innovative technologies, including freeze-drying, microencapsulation, and nanoparticle-based systems, have emerged as critical tools to protect probiotic viability and ensure targeted delivery to host organisms. These technologies are applied in both aquaculture and human probiotic development. For example, freeze-drying allows long-term preservation of marine *Streptomyces* strains for both incorporation into shrimp feed and encapsulated capsules or powder-based supplements for human use (Venkatesh et al., 2024).

In aquaculture, delivery strategies include the incorporation of probiotics into feed pellets, which enhances nutrient bioavailability, gut colonization, and pathogen resistance in fish and shrimp. Immersion therapies, where probiotics are suspended in water and introduced into rearing tanks, also support skin and gut microbiota development, while simultaneously improving water quality through bioremediation (Rhee et al., 2018).

In human health, similar principles apply but through different delivery formats. Encapsulation technologies using alginate or chitosan biopolymers protect probiotics from gastric acids and bile salts during transit through the gastrointestinal tract. This enables the controlled release of viable cells or bioactive metabolites at the intestinal mucosa, where health benefits are realized (Wang et al., 2024). Additionally, hydrogels derived from marine polysaccharides (e.g., carrageenan) have been explored for extended release and intestinal adhesion, showing promise in the development of marine-derived probiotic supplements and functional foods (Junaid et al., 2024). Emerging technologies such as nanoparticle delivery systems further refine precision targeting, allowing for controlled delivery of actinobacterial metabolites or postbiotics to specific tissue sites—a promising frontier for both personalized nutrition and therapeutic interventions. Safety considerations remain a key factor across both sectors. Marine actinobacterial strains must be rigorously assessed for pathogenicity, antibiotic resistance, and regulatory compliance, whether they are intended for use in aquafeeds or human nutraceuticals. Recent studies suggest that many marine actinobacteria meet Generally Recognized as Safe (GRAS) criteria and offer a safe, effective alternative to synthetic additives or conventional probiotics (Fachri et al., 2024). In both domains, the flexibility in formulation and delivery methods, combined with the storage stability of marine actinobacteria, supports their potential as a reliable class of next-generation probiotics. Their broad application spectrum—from aquaculture systems to human functional foods—strengthens their translational value and relevance in modern biotechnology.

6 Synergistic Effect with Prebiotics

Aquaculture is anticipated to become the primary source of fish worldwide in the upcoming ten years. However, as aquatic production has increased, infectious diseases have increased significantly, and the effectiveness of antibiotics used to treat these diseases has also increased. It is anticipated to become the primary source of fish worldwide in the upcoming ten years. However, as aquatic production has increased, infectious diseases have increased significantly, and the effectiveness of antibiotics used to treat these diseases has also increased. In addition to bacteria, yeasts, and other fungi, actinomycetes are thought to play a major role in the decomposition and recycling of organic materials (Barka et al., 2016). Today, a variety of choices are accessible, such as postbiotics, prebiotics, and probiotics (Ghosh et al., 2023).

Aquaculture is an industry worth billions of dollars that is growing quickly. However, illness is a major worry in this sector. Aquatic animals are closely and continuously impacted by the changes and makeup of their surroundings. Their infections, which can grow to densities high enough to induce illness or weaken the immune system of the animals, are also supported by the water environment. Furthermore, poor seed conditions or overstocking greatly exacerbate the breakdown of the "host–pathogen–environment" balance, which in turn leads to disease outbreaks (Das et al., 2008).

Actinomycetes are Gram-positive prokaryotes that are abundant in nature and have a great capacity for secondary metabolite coding. They also contain significant levels of guanine and cytosine (G + C, which make up about 70%). And nearly half of all known bioactive compounds come from Actinomycetes (Li et al., 2021).

It is known that actinomycetes produce a wide variety of hydrolytic enzymes. Therefore, it is impossible to overlook actinobacteria's ecological function as probiotics in aquaculture. Actinobacteria, like other probiotic microbes, have been shown to have a variety of positive effects on aquatic creature growth, immunity, and water quality (Ghosh et al., 2023). The six main classes that make up the phylum Actinobacteria are *Acidimicrobiia, Coriobacteria*, *Nitriliruptoria, Rubrobacteria*, and *Thermoleophilia*. The largest group of the six classes is called Actinobacteria, which contains 15 orders, including *Actinomycetales, Actinopolysporales, Bifidobacteriales, Catenulisporales, Corynebacteriales, Glycomycetales, Jiangellales, Kineosporales, Micrococcales, Micromonosporales, Propionibacteriales, Pseudonocardiales, Streptomycetales, Streptosporangiales*, etc. Gram-positive bacteria belonging to the class Actinobacteria have distinctive morphological traits, including complex fragmenting hyphal, rod, and coccus morphologies. Their unusual, varied appearance and striking similarity to mushrooms were dubious in a number of circumstances (Kim et al., 2021).

Prebiotics, probiotics, and synbiotics have drawn more attention in light of this because of their potential to help manage disorders linked to the microbiome. According to one definition, probiotics are "live microorganisms which when administered in adequate amounts confer a health benefit on the host" (Cerdó et al., 2019). Lactic acid-producing bacteria from the genera *Lactobacillus* and *Bifidobacterium* make up the majority of them at the moment. Synbiotics are blends of probiotics and prebiotics, while prebiotics are dietary ingredients that promote the growth of good bacteria. A number of pathways are believed to be responsible for the health-promoting effects of probiotics, including intestinal colonization, immunological homeostasis promotion, intestinal mucosal barrier enforcement, and improvement of the microbiota profile in cases of dysbiosis (Jačan et al., 2020).

More specifically, little is known about how synbiotics, on the one hand, alter antibiotics' capacity to induce dysbiosis, and whether probiotics or synbiotics, on the other hand, affect any antibiotic's short- or long-term effects (Markowiak & Ślizewska, 2017). Insufficient knowledge regarding how probiotics and antibiotics interact to influence the gut microbiota's structure. In order to prevent or treat antibiotic-induced adverse effects, like antibiotic-associated diarrhea, it is challenging to fully utilize the therapeutic potential of probiotics and synbiotics (Petrariu et al., 2023). Actinobacteria are shown to have significant roles in biogeochemical cycles, bioremediation, and the synthesis of bacteriocins and other powerful bioactive substances, including antibiotics and new enzymes. They typically live in soil and marine environments. Because of all these distinctive and varied characteristics, Actinobacteria should be investigated as a viable option for the aquaculture sector. Based on their capacity to generate various secondary metabolites, they were given the new moniker "Modern Actinobacteria" for their versatility (James et al., 2023).

One of the industries in agriculture that is expanding the fastest to meet the global population's demand for protein is aquaculture. High stocking density is causing an increase in disease outbreaks in aquaculture, which are then followed by the buildup of organic and inorganic wastes and a variety of microbial diseases. Nevertheless, the widespread use of chemotherapeutics and antibiotics causes them to accumulate in aquatic environments, increasing the prevalence of multiple antibiotic resistance (Kim et al., 2021).

Eventually, as efforts to find new metabolites from other microbial sources continue, the researchers discover the presence of Actinobacteria in a variety of settings, including deserts, mangroves, and marine environments. This cleared the path for the isolation of novel Actinobacteria, particularly those belonging to taxa other than *Streptomyces*, which are referred to as "rare Actinobacteria" and have varying bioactive potentials (James et al., 2023). *Streptomyces* accounted for over 60% of newly discovered antibiotics as of the 1970s (Kim et al., 2021). Probiotics, prebiotics, and synbiotics have all been utilized and researched in a variety of illness areas. Numerous investigations have revealed a connection between a modified gut microbiota and a number of illnesses that are especially common in the twenty-first century (Markowiak & Ślizewska, 2017).

6.1 Designing Effective Synbiotic Combinations for Enhanced Efficacy

"Synbiotics" are commonly defined as the contemporaneous administration of probiotics and prebiotics as a single therapeutic component. It was Gibson and Roberfroid who originally used this word (Gibson & Roberfroid, 2008). In synbiotic therapies, prebiotics are typically given to help sustain probiotic viability, enhance gastrointestinal tract colonization, and increase probiotic efficacy. The probiotic bacteria's survival rate is thought to be increased by the prebiotic component in the gut, which acts as an in-situ energy source to support them in the gut environment (Goh et al., 2022).

Additionally, prebiotics encourage the formation of additional good bacteria that support the health of the host. Furthermore, a synbiotic treatment's prebiotic component may also serve as a protective agent that reduces probiotic inactivation, especially during drying. These low molecular weight carbohydrates stabilize cell protein structures by forming hydrogen bonds with water, which are eliminated during desiccation (Hai, 2015).

Many synbiotics have been demonstrated to improve the overall health condition of shrimps more effectively than probiotics or prebiotics alone since their introduction to the shrimp culture. Applying synbiotics to treated shrimps resulted in a notable increase in immunological function, survival, and growth rate. Synbiotics (*P. piscicida* and MOS) produced the highest shrimp survival rates (95%); prebiotics (87%), probiotics (84%), and the control group (79%) (Goh et al., 2022).

In addition to its direct impact on the gut microbiota, synbiotics supplementation has a significant effect on shrimp growth promotion. Shrimps with a synbiotic intervention had the highest SGR (31%). This was significantly higher than that of the probiotics (28.8%), prebiotics (26.9%), and the control group (27.6%). The other study demonstrated that the synbiotic combination of *Bacillus sp.* and sweet potato extract prebiotics decreased the FCR and contributed to total weight gain, which is also in line with this outcome (Hai, 2015).

7 Challenges and Regulatory Perspectives in Marine Probiotic Development and Commercialization

The main challenges related to the development of probiotics in general, and marine probiotics particularly, are mainly connected to (Lei et al., 2025; Garg et al., 2024):

(i) The nature of the marine probiotic strains: Identifying and characterizing marine probiotic strains can be complex due to the unique environmental conditions in marine ecosystems. Strain-specific data on safety, efficacy, and benefits are required for regulatory approval and commercialization.
(ii) Lack of standardization in probiotic formulations: including dosage, delivery method, and concentration. This can complicate regulatory approval processes and lead to inconsistencies in product efficacy.
(iii) Scientific evidence confirming probiotic benefits in human/animal health. The regulatory agencies require robust clinical and preclinical data to support health claims. Many marine probiotics are still in the early stages of research, and a lack of well-controlled trials makes it difficult to meet the stringent standards for claims in global markets.

Considering all the previous factors, regulation and regulatory agencies arise as a key factor for the development and commercialization of marine probiotics. These regulations differ across regions due to varying health and safety standards, scientific evidence requirements, and market demands. This significant variation makes it difficult to develop and commercialize globally all those formula-based on marine probiotics. To make visible the diversity of regulations, a summary of them in different areas of the world is offered below, focusing on major regulatory agencies like the U.S. Food and Drug Administration (FDA), European Food Safety Authority (EFSA), and others in key markets.

7.1 U.S. Regulatory Framework

In the United States, marine probiotics are primarily regulated by the FDA and fall under the following categories (https://www.fda.gov/):

- Food Additives: If marine probiotics are intended to be added to food products, they must be evaluated and approved as food additives under the Federal Food, Drug, and Cosmetic Act (FDCA). This requires safety data demonstrating that the probiotic strains do not pose harm when consumed in the quantities specified.
- Dietary Supplements: Many marine probiotics are marketed as dietary supplements, and these products must comply with the Dietary Supplement Health and Education Act (DSHEA). The manufacturer is responsible for ensuring the safety and efficacy of the probiotic, though the FDA can intervene if the product is found to be unsafe or misbranded.

- Generally Recognized as Safe (GRAS): For probiotics intended for food applications, manufacturers may petition the FDA to grant GRAS status. This requires scientific evidence showing that the probiotic is safe when used in food products.
- Health Claims: Any health claims made about the probiotic, such as improving gut health or immune function, must be substantiated by scientific evidence. The FDA evaluates these claims and may require manufacturers to submit pre-market approval for health claims. Additionally, structure-function claims (e.g., "supports digestive health") do not require FDA approval but must not be misleading.

7.2 European Union Regulatory Landscape

In the EU, marine probiotics are primarily regulated by the European Food Safety Authority (EFSA), and the European Commission enforces regulations for food supplements and novel foods.

- Novel Foods Regulation: If the marine probiotic strain is not traditionally consumed in the EU, it may be classified as a novel food and require authorization before it can be marketed. EFSA evaluates the safety of novel foods, and a positive opinion from EFSA is required before approval (EFSA, 2021, 2023).
- Health Claims: EFSA has stringent requirements for health claims. The EU Regulation (EC) No 1924/2006 governs health claims related to foods. Only claims that are substantiated by scientific evidence and authorized by the European Commission can be used. As of 2023, the EFSA has been cautious in approving health claims for probiotics due to insufficient evidence (EFSA, 2021, 2023).
- General Food Law: Probiotics in food products must adhere to the General Food Law (Regulation (EC) No 178/2002), which includes general safety requirements.

7.3 Asia-Pacific Regulatory Perspectives

In the Asia-Pacific region, countries like Japan, South Korea, and Australia have distinct regulations for probiotics, influenced by local food safety standards and probiotic research.

- Japan: The Food Sanitation Law governs the safety of probiotics, and Functional Foods (foods that claim to offer health benefits) must pass the Food for Specified Health Uses (FOSHU) approval process. Probiotics marketed in Japan require approval by the Ministry of Health, Labour, and Welfare (MHLW). Japan has specific categories for probiotics, and approval includes clinical trial data supporting efficacy claims (Amagase, 2008).
- Australia and New Zealand: Probiotics are regulated by the Food Standards Australia New Zealand (FSANZ), and health claims for probiotics are assessed under the Food Standards Code. FSANZ also mandates safety assessments for

novel probiotic strains. Probiotics marketed as dietary supplements must be listed on the Australian Register of Therapeutic Goods (ARTG) if therapeutic claims are made (https://www.foodstandards.gov.au/).

7.4 International Guidelines and Harmonization Efforts

There is growing interest in harmonizing probiotic regulations globally to ensure safety and efficacy while facilitating trade. Some initiatives and guidelines include:

- Codex Alimentarius: The Codex Alimentarius Commission, managed by the Food and Agriculture Organization (FAO) and World Health Organization (WHO), is working toward providing international standards for the safety, quality, and labeling of probiotics in food products. However, Codex does not currently have specific guidelines for probiotics, but has published general guidelines for health foods (https://www.fao.org/fao-who-codexalimentarius/en/).
- WHO/FAO Probiotic Guidelines: The WHO/FAO Expert Consultation on probiotics offers guidelines on the safe use of probiotics in food, suggesting that probiotics should be from well-characterized strains with proven safety records. It emphasizes the need for a risk-based approach to regulation.

8 Safety Assessment and Regulatory Considerations

As mentioned in the case of marine probiotic development and commercialization, the assessment of safety concerns and their relative regulations also vary based on the region analyzed, which results in difficulties in the development of probiotic formulations. The following section highlights the most relevant tests and guidelines for safety assessment, as well as the key regulatory frameworks for safety assessment.

8.1 GRAS Status, Toxicity Tests, Guidelines for Commercialization

There is no blanket GRAS status for all marine probiotics. Each strain or product must either rely on a history of safe use in food (or food supplements) or be evaluated via scientific procedures (including toxicology) to support a GRAS notification. Some non-probiotic microbial-derived or fermentation products (including marine-source ones) have obtained GRAS status for specific end uses (e.g., feed, food). But usage is limited to the specific strain, process, dosage/specification (Burdock & Carabin, 2004). Table 2 summarizes definitions about safety assessment and regulatory considerations for marine probiotics focused on GRAS in the U.S. (https://www.fda.gov/food/food-ingredients-packaging/

Table 2 Key regulatory frameworks and safety paradigms

Regulatory/safety regime	Definition	Relevance to marine probiotics/new strains
GRAS (Generally Recognized as Safe) U.S. FDA	A status under U.S. law (FDandC Act, sections 201(s) and 409) for substances (including microbial-derived) added to food. If a substance is GRAS, it is exempted from being treated as a food additive, provided there is scientific consensus among qualified experts that its use is safe under intended conditions.	A marine probiotic strain intended for food must be evaluated under GRAS standards (through either historical safe use or scientific procedures). There are lists of microorganisms/substances that have been affirmed or notified as GRAS. For marine-derived organisms, if there is limited prior use, more data will likely be required.
Qualified Presumption of Safety (QPS) EFSA (EU)	A "pre-assessment" framework for microorganisms used in food, feed, enzymes, etc. If a species/genus is on the QPS list, safety assessments for new strains of that species can be simplified or "fast-tracked," though strain-level assessments are still required.	Some marine probiotic strains are included in the QPS list. The safety concerns (virulence, antibiotic resistance, toxigenic potential) remain key qualifiers. Full evaluation is needed for strains that are not in QPS list.
Live Biotherapeutic Product (LBP) / Medicinal Product Regulation	For live microorganisms intended for disease prevention or treatment rather than general supplementation. In the EU and the USA, products making therapeutic claims or intended for vulnerable populations must follow medicinal product/biologic product regulatory pathways.	If a marine probiotic is intended for a health / medical claim, or for clinical use, this pathway applies; more stringent safety, clinical trial, manufacturing, and quality standards are required.

generally-recognized-safe-gras) and QPS in the EU (https://www.efsa.europa.eu/en/applications/qps-assessment).

Regarding toxicity/safety test types, when evaluating a marine probiotic (or any novel live microorganism) for food/feed or health-adjacent uses, the safety package should cover multiple lines of evidence (Table 3).

Marine-sourced microorganisms in general and marine probiotics particularly bring extra safety challenges (on top of "ordinary probiotics") due to: (i) potential for secondary metabolites/marine toxins (some marine archaea/bacteria/algae produce toxins). Any novel marine strain must be screened for known toxins, and its biosynthetic potential assessed; (ii) unusual environmental resistances, potential for horizontal gene transfer in marine environments; (iii) differences in physiology: salt tolerance, spore formation, etc. These can affect stability, viability, and impact safety (e.g., survival in gut, potential to colonize unintended sites), and (iv) risk of allergenicity might be higher for marine-derived proteins in some cases.

To bring a marine probiotic to market safely and compliantly, developers should generally follow a phased, evidence-based path, thus guaranteeing best practices for commercialization. The best practices include the following key steps (Table 4):

Table 3 Summary of toxicity/safety test types

Safety/toxicity-aspect	Standards
Strain identification and characterization	Must identify genus/species (and subspecies if relevant) using reliable molecular and phenotypic tools; also deposit strain in recognized culture collections.
Assessment of virulence	For species known to include virulent members, tests must show that the candidate strain lacks these properties. Hemolysis assays, toxin gene screening, cytotoxicity tests.
Antibiotic / antimicrobial resistance (AMR)	Check for the presence of acquired resistance genes, ensure they are not transferable, ensure susceptibility to clinically relevant antibiotics; rule out production of antimicrobials that could have adverse effects.
Metabolic or biochemical activities that might pose a risk	For example, bile salt deconjugation, D-lactate production, production of biogenic amines, etc. If the strain has undesirable metabolic capabilities, this needs to be quantified, and risk assessed.
In vitro toxicity/ cytotoxicity	Assays using cell lines to test for cytotoxic effects; sometimes other in vitro tests (e.g., for genotoxicity, mutagenicity, endotoxin levels), depending on risk.
In vivo/animal studies	As relevant, using models to test for infectivity, pathogenicity, repeated-dose toxicity, etc. Also, sometimes observations of behavior, digestion, and immune responses. Animal challenge, feed trials.
Allergenicity	If the microorganism (or its products) could provoke allergic responses, protein components, etc., should be assessed (in silico, in vitro, or in vivo).
Stability, viability, dosage, shelf life	The product must maintain viability (if live) until the end of shelf life; dosage (CFU, etc.) must be clearly defined; manufacturing process must ensure purity (contaminant microorganisms, endotoxins, etc.)

Finally, it is important to clarify that there are current gaps and unresolved issues related to the safety of marine probiotics. Thus, there is still no unified international standard for what constitutes sufficient safety for marine probiotics, especially for less-well-studied marine taxa. For many marine probiotic candidates, data on longer-term usage, environmental impacts (for aquaculture), and interspecies transfer of genes (AMR, virulence) are sparse. Finally, regulatory bodies are cautious about approving health claims; therefore, the threshold of proof is high (especially in the EU).

9 Future Trends and Translational Potential

The field of marine probiotics is undergoing rapid innovation, and future trends show promising directions for translational applications across human health, aquaculture, and environmental sustainability. Emerging trends and the translational potential of marine probiotics, with a focus on personalization, microbiome

Table 4 Best practices and guidelines for commercialization of safe marine probiotics

Step	Description
1. Preliminary characterization	Taxonomy/phylogeny, strain-level identification (molecular methods). Genome sequencing (to detect virulence genes, AMR, toxin biosynthesis genes). *In-vitro* characterization: metabolic profile, potential harmful metabolites, hemolysis, cytotoxicity.
2. Safety toxicology	*In-vitro* safety assays (cytotoxicity, genotoxicity) under relevant conditions. Animal (or model) studies if needed: dose-ranging, repeated exposure, infectivity, etc. For aquaculture or feed uses: trials in target species to assess safety (growth, health, immune responses).
3. Efficacy trials/health benefit claims	Human clinical trials claim human health benefits (or veterinary trials for animal probiotics). Clear demonstration of benefit at a specific dosage, in a specific form.
4. Manufacturing and quality control	Adherence to GMP or equivalent (for dietary supplements/food/feed depending on jurisdiction). Ability to ensure purity: absence of contaminants, unwanted organisms, endotoxins. Stability testing: viability across shelf life, effect of storage, and processing.
5. Regulatory dossier preparation	As examples. For U.S.: GRAS notice or affirmation; for products intended as therapeutics, IND/BLA or equivalent. For EU: QPS evaluation (if applicable), novel food/feed or feed additive authorization, etc.
6. Labeling, claims, post-market monitoring	Claims must be supported by evidence; in EU, many probiotic health claims have been rejected due to insufficient proof. Labeling of viable count, strain designation, storage conditions, potential allergens, contraindications, if known. Monitor for adverse events in consumers or animals; maintain traceability of strains.

engineering, and AI-driven discovery, will be the next key areas to focus efforts on shortly in this area.

9.1 Personalized Marine Probiotics

This concept implies tailoring probiotic marine-derived strains (including engineered strains) to individual human microbiome profiles, genetics, lifestyle, and health conditions to improve human wellness. This personalized approach to probiotic use is relatively new and seeks to overcome the limitations of widespread probiotic use (for example, the human gut microbiome is highly individualized; "one-size-fits-all" probiotics often yield inconsistent benefits). On the other hand, marine microorganisms may offer unique metabolic or immunomodulatory functions not found in terrestrial strains, providing novel therapeutic angles (e.g., neuroprotective, anti-inflammatory, antioxidant) (Manan, 2025). Some of the applications of personalized marine probiotics are: (i) precision nutrition: Marine probiotics matched to individuals with dysbiosis, obesity, IBD, etc.; (ii) skin microbiome modulation: Marine-derived bacteria adapted to high salinity may benefit skin barrier function and (iii) bioactive delivery platforms: Marine probiotic strains engineered to deliver peptides, polyunsaturated fatty acids (PUFAs), or other therapeutic molecules (Eze et al., 2023; Manan, 2025). In the context of environmental protection, some interesting applications of marine probiotics concern the bioremediation of oil spills, heavy-metal contamination, or plastic degradation.

However, this personalized use of marine probiotics is not exempt from limitations and challenges that will need to be studied in the coming years: regulatory hurdles in using engineered or novel strains (engineered strains are subject to stricter evaluation (e.g., GMO regulations and environmental release of engineered marine strains must be risk-assessed); need for robust bioinformatics platforms to match probiotic strains to individual microbiome signatures, safety and allergenicity of marine-derived strains in human use.

9.2 AI-Driven Probiotic Discovery

The marine environment holds vast microbial biodiversity, and over 99% is uncultured. Besides, traditional screening is time-consuming, costly, and limited in scope. For these reasons, artificial intelligence (AI) and machine learning (ML) are powerful tools that can accelerate the discovery, screening, and optimization of marine probiotic strains. By utilizing AI, the prediction of probiotic functionality from genomic/metagenomic data, the identification of beneficial metabolites, and safety prediction (toxicity, virulence, and AMR potential) can be accurately addressed (Han et al., 2025; McCoubrey et al., 2022). Thanks to AI-Driven Tools, studies aiming probiotics discovery can integrate genome mining (tools like antiSMASH, DeepBGC, or BiG-SCAPE predict biosynthetic gene clusters; e.g., for antibiotics, antioxidants, polysaccharides), metabolite prediction (deep learning models trained on LC–MS/MS data to find novel postbiotic compounds), host-microbe interaction models (predict how a marine strain may modulate host immunity, metabolism, or gut barrier) and personalized probiotic design (AI models integrating microbiome data + clinical metadata to match marine strains to individuals) (Abdul Manan 2025; Min et al., 2024). Consequently, startups and biotech companies are investing in AI-powered probiotic R&D platforms, particularly in the functional foods and skin health sectors.

As a future outlook, it is important to highlight that a new era of "next-gen biotics" is currently starting to develop post-biotics, para-probiotics, and meta-biotics from marine sources with specific molecular targets, encompassing the expansion of marine microbiome databases and integrating omics disciplines (metagenomics, metabolomics, proteomics) and machine learning to enable high-throughput discovery and strain design (Cuevas-González et al., 2020; Collado et al., 2021).

Acknowledgement VHR gratefully acknowledges funding support from UGC-BSR [F.30-521/2020(BSR)]. The research fellowships and associateships awarded to MC, HP, and SVR by DBT [DBTHRDPMU/JRF/BET-23/I/2023-24/266], New Delhi, CSIR JRF [24D/03/00666], and GSBTM [GSBTM/JD(R&D)/662/2022-23/00292647], respectively, are righteously acknowledged. We also acknowledge infrastructural support from DST-FIST at the Department of Microbiology & Biotechnology, Gujarat University. Authors CPG, RMM-E gratefully acknowledge the funds and support received under VIGROB309 from the University of Alicante, Spain. These fundings and contributions were vital for research work.

References

Abdul Manan, M. (2025). Progress in probiotic science: Prospects of functional probiotic-based foods and beverages. In *International Journal of Food Science* (Vol. 2025, Issue 1, p. 5567567). John Wiley and Sons Ltd. https://doi.org/10.1155/ijfo/5567567

Abo-Al-Ela, H. G., Mahdi, S., Angthong, P., & Rungrassamee, W. (2025). Probiotic modulation of key immune macromolecules in shrimp. *Microbial Pathogenesis, 203*, 107463. https://doi.org/10.1016/j.micpath.2025.107463

Ahmed, N., & Azra, M. N. (2022). Aquaculture production and value chains in the COVID-19 pandemic. *Current Environmental Health Reports, 9*(3), 423–435. https://doi.org/10.1007/s40572-022-00364-6

Amagase, H. (2008). Current marketplace for probiotics: A Japanese perspective. *Clinical Infectious Diseases, 46*(2), S73–S75. https://doi.org/10.1086/523338

Arboleya, S., Watkins, C., Stanton, C., & Ross, R. P. (2016). Gut bifidobacteria populations in human health and aging. In *Frontiers in microbiology* (Vol. 7, Issue AUG). Frontiers Media S.A. https://doi.org/10.3389/fmicb.2016.01204

Barka, E. A., Vatsa, P., Sanchez, L., Gaveau-Vaillant, N., Jacquard, C., Klenk, H.-P., Clément, C., Ouhdouch, Y., & van Wezel, G. P. (2016). Taxonomy, physiology, and natural products of actinobacteria. *Microbiology and Molecular Biology Reviews, 80*(1), 1–43. https://doi.org/10.1128/MMBR.00019-15

Behera, S., & Das, S. (2023). Potential and prospects of Actinobacteria in the bioremediation of environmental pollutants: Cellular mechanisms and genetic regulations. *Microbiological Research, 273*, 127399. https://doi.org/10.1016/j.micres.2023.127399

Binda, C., Lopetuso, L. R., Rizzatti, G., Gibiino, G., Cennamo, V., & Gasbarrini, A. (2018). Actinobacteria: A relevant minority for the maintenance of gut homeostasis. In *Digestive and liver disease* (Vol. 50, Issue 5, pp. 421–428). Elsevier B.V. https://doi.org/10.1016/j.dld.2018.02.012

Bredholdt, H., Galatenko, O. A., Engelhardt, K., Fjaervik, E., Terekhova, L. P., & Zotchev, S. B. (2007). Rare actinomycete bacteria from the shallow water sediments of the Trondheim fjord, Norway: Isolation, diversity and biological activity. *Environmental Microbiology, 9*(11), 2756–2764. https://doi.org/10.1111/j.1462-2920.2007.01387.x

Burdock, G. A., & Carabin, I. G. (2004). Generally recognized as safe (GRAS): History and description. *Toxicology Letters, 150*(1), 3–18. https://doi.org/10.1016/j.toxlet.2003.07.004

Cerdó, T., García-Santos, J. A., Bermúdez, M. G., & Campoy, C. (2019). The role of probiotics and prebiotics in the prevention and treatment of obesity. In *Nutrients* (Vol. 11, Issue 3). MDPI AG. https://doi.org/10.3390/nu11030635

Chelakkot, C., Ghim, J., & Ryu, S. H. (2018). Mechanisms regulating intestinal barrier integrity and its pathological implications. In *Experimental and molecular medicine* (Vol. 50, Issue 8). Nature Publishing Group. https://doi.org/10.1038/s12276-018-0126-x

Claverías, F. P., Undabarrena, A., González, M., Seeger, M., & Cámara, B. (2015). Culturable diversity and antimicrobial activity of Actinobacteria from marine sediments in Valparaíso bay, Chile. *Frontiers in Microbiology, 6*(JUL). https://doi.org/10.3389/fmicb.2015.00737

Collado, M. C., Seppo, S., & Vinderola, G. (2021). Postbiotics: Defining the impact of inactivated microbes and their metabolites on promotion of health. In O. Koren & S. Rautava (Eds.), *The human microbiome in early life* (pp. 257–268., ISBN 9780128180976). Academic. https://doi.org/10.1016/B978-0-12-818097-6.00011-0

Cuevas-González, P. F., Liceaga, A. M., & Aguilar-Toalá, J. E. (2020). Postbiotics and paraprobiotics: From concepts to applications. *Food Research International, 136*, 109502. https://doi.org/10.1016/j.foodres.2020.109502

Cui, K., Xu, T., Chen, J., Yang, H., Liu, X., Zhuo, R., Peng, Y., Tang, W., Wang, R., Chen, L., Zhang, X., Zhang, Z., He, Z., Wang, X., Liu, C., Chen, Y., & Zhu, Y. (2022). Siderophores, a potential phosphate solubilizer from the endophyte Streptomyces sp. CoT10, improved phosphorus mobilization for host plant growth and rhizosphere modulation. *Journal of Cleaner Production, 367*, 133110. https://doi.org/10.1016/j.jclepro.2022.133110

D'Argenio, V., & Salvatore, F. (2015). The role of the gut microbiome in the healthy adult status. *Clinica Chimica Acta, 451*, 97–102. https://doi.org/10.1016/j.cca.2015.01.003

Das, S., Ward, L. R., & Burke, C. (2008). Prospects of using marine actinobacteria as probiotics in aquaculture. In *Applied microbiology and biotechnology* (Vol. 81, Issue 3, pp. 419–429). Springer. https://doi.org/10.1007/s00253-008-1731-8

DasSarma, S., & DasSarma, P. (2015). Halophiles and their enzymes: Negativity put to good use. *Current Opinion in Microbiology, 25*, 120–126. https://doi.org/10.1016/j.mib.2015.05.009

De Santi, C., Tedesco, P., Ambrosino, L., Altermark, B., Willassen, N.-P., & de Pascale, D. (2014). A new alkaliphilic cold-active esterase from the psychrophilic marine bacterium Rhodococcus sp.: Functional and structural studies and biotechnological potential. *Applied Biochemistry and Biotechnology, 172*(6), 3054–3068. https://doi.org/10.1007/s12010-013-0713-1

De Simeis, D., & Serra, S. (2021). Actinomycetes: A never-ending source of bioactive compounds—An overview on antibiotics production. *Antibiotics, 10*(5), 483. https://doi.org/10.3390/antibiotics10050483

Dekaboruah, E., Suryavanshi, M. V., Chettri, D., & Verma, A. K. (2020). Human microbiome: An academic update on human body site-specific surveillance and its possible role. *Archives of Microbiology, 202*(8), 2147–2167. https://doi.org/10.1007/s00203-020-01931-x

Donlan, R. M. (2002). Biofilms: Microbial life on surfaces. *Emerging Infectious Diseases, 8*(9), 881–890. https://doi.org/10.3201/eid0809.020063

Engevik, M. A., Luk, B., Chang-Graham, A. L., Hall, A., Herrmann, B., Ruan, W., Endres, B. T., Shi, Z., Garey, K. W., Hyser, J. M., & Versalovic, J. (2019). Bifidobacterium dentium fortifies the intestinal mucus layer via autophagy and calcium signaling pathways. *MBio, 10*(3). https://doi.org/10.1128/mBio.01087-19

European Food Safety Authority (EFSA). (2021). *Guidance on the submission of health claims*. EFSA Journal.

Eze, O. C., Berebon, D. P., Emencheta, S. C., Evurani, S. A., Okorie, C. N., Balcão, V. M., & Vila, M. M. D. C. (2023). Therapeutic potential of marine probiotics: A survey on the anticancer and antibacterial effects of *Pseudoalteromonas* spp. *Pharmaceuticals, 16*(8), 1091. https://doi.org/10.3390/ph16081091

Fachri, M., Amoah, K., Huang, Y., Cai, J., Alfatat, A., Ndandala, C. B., Shija, V. M., Jin, X., Bissih, F., & Chen, H. (2024). Probiotics and paraprobiotics in aquaculture: A sustainable strategy for enhancing fish growth, health and disease prevention—A review. *Frontiers in Marine Science, 11*. https://doi.org/10.3389/fmars.2024.1499228

Fakhri, S., Yarmohammadi, A., Yarmohammadi, M., Farzaei, M. H., & Echeverria, J. (2021). Marine natural products: Promising candidates in the modulation of gut-brain axis towards neuroprotection. In *Marine drugs* (Vol. 19, Issue 3). MDPI. https://doi.org/10.3390/MD19030165

Fukuda, S., Toh, H., Hase, K., Oshima, K., Nakanishi, Y., Yoshimura, K., Tobe, T., Clarke, J. M., Topping, D. L., Suzuki, T., Taylor, T. D., Itoh, K., Kikuchi, J., Morita, H., Hattori, M., & Ohno, H. (2011). Bifidobacteria can protect from enteropathogenic infection through the production of acetate. *Nature, 469*(7331), 543–549. https://doi.org/10.1038/nature09646

Furhan, J. (2020). Adaptation, production, and biotechnological potential of cold-adapted proteases from psychrophiles and psychrotrophs: Recent overview. *Journal of Genetic Engineering and Biotechnology, 18*(1), 36. https://doi.org/10.1186/s43141-020-00053-7

Gahlawat, G., & Choudhury, A. R. (2019). A review on the biosynthesis of metal and metal salt nanoparticles by microbes. *RSC Advances, 9*(23), 12944–12967. https://doi.org/10.1039/C8RA10483B

Garg, V., Velumani, D., Lin, Y. C., & Haye, A. (2024). A comprehensive review of probiotic claims regulations: Updates from Asia-Pacific, United States, and Europe. *PharmaNutrition, 30*, 100423. https://doi.org/10.1016/j.phanu.2024.100423

Ghosh, K., Harikrishnan, R., Mukhopadhyay, A., & Ringø, E. (2023). Fungi and actinobacteria: Alternative probiotics for sustainable aquaculture. In *Fishes* (Vol. 8, Issue 12). Multidisciplinary Digital Publishing Institute (MDPI). https://doi.org/10.3390/fishes8120575

Gibson, G. R., & Roberfroid, M. (Eds.). (2008). *Handbook of prebiotics*. CRC Press.

Giri, B., & Varma, A. (2019). *Microorganisms in saline environments: Strategies and functions* (Soil biology volume 56 (Issue May)). https://doi.org/10.1007/978-3-030-18975-4

Goh, J. X. H., Tan, L. T. H., Law, J. W. F., Ser, H. L., Khaw, K. Y., Letchumanan, V., Lee, L. H., & Goh, B. H. (2022). Harnessing the potentialities of probiotics, prebiotics, synbiotics, paraprobiotics, and postbiotics for shrimp farming. In *Reviews in aquaculture* (Vol. 14, Issue 3, pp. 1478–1557). Wiley. https://doi.org/10.1111/raq.12659

Goh, J. X. H., Tan, L. T.-H., Law, J. W.-F., Khaw, K.-Y., Zengin, G., Chan, K. G., Letchumanan, V., Lee, L.-H., & Goh, B. H. (2023). Probiotics: Comprehensive exploration of the growth promotion mechanisms in shrimps. *Progress In Microbes & Molecular Biology, 6*(1). https://doi.org/10.36877/pmmb.a0000324

Gohel, S. D., Sharma, A. K., Dangar, K. G., Thakrar, F. J., & Singh, S. P. (2018). Biology and applications of halophilic and haloalkaliphlic actinobacteria. In *Extremophiles: From biology to biotechnology* (p. 437). CRC Press.

Hai, N. V. (2015). The use of probiotics in aquaculture. In *Journal of Applied Microbiology* (Vol. 119, Issue 4, pp. 917–935). https://doi.org/10.1111/jam.12886

Han, X., Liu, Q., Li, Y., Zhang, M., Liu, K., Kwok, L. Y., Zhang, H., & Zhang, W. (2025). Synergizing artificial intelligence and probiotics: A comprehensive review of emerging applications in health promotion and industrial innovation. *Trends in Food Science & Technology, 159*, 104938. https://doi.org/10.1016/j.tifs.2025.104938

Helmi, N. R. (2025). Exploring the diversity and antimicrobial potential of actinomycetes isolated from different environments in Saudi Arabia: A systematic review. *Frontiers in Microbiology, 16*(March), 1–28. https://doi.org/10.3389/fmicb.2025.1568899

Hodar, J. H. (2024). *Anti-adhesive and anti-biofilm potential of extracellular biosurfactants produced by mycolate actinobacteria of the suborder Corynebacterineae against biofilm-forming human pathogens*. https://doi.org/10.17869/enu.2024.3789827

Huang, J., Zheng, X., Yu, T., Ali, M., Wiese, J., Hu, S., Huang, L., & Huang, Y. (2025). Diverse lifestyles and adaptive evolution of uncultured UBA5794 actinobacteria, a sister order of "*Candidatus actinomarinales*". *Environmental Microbiomes, 20*(1), 39. https://doi.org/10.1186/s40793-025-00701-w

Hui, M. L.-Y., Tan, L. T.-H., Letchumanan, V., He, Y.-W., Fang, C.-M., Chan, K.-G., Law, J. W.-F., & Lee, L.-H. (2021). The extremophilic actinobacteria: From microbes to medicine. *Antibiotics, 10*(6), 682. https://doi.org/10.3390/antibiotics10060682

Hwang, Y. K., & Oh, J. S. (2025). Interaction of the Vagus nerve and serotonin in the gut–brain axis. In *International Journal of Molecular Sciences* (Vol. 26, Issue 3). Multidisciplinary Digital Publishing Institute (MDPI). https://doi.org/10.3390/ijms26031160

Jačan, A., Kashofer, K., Zenz, G., Fröhlich, E. E., Reichmann, F., Hassan, A. M., & Holzer, P. (2020). Synergistic and antagonistic interactions between antibiotics and synbiotics in modifying the murine fecal microbiome. *European Journal of Nutrition, 59*(5), 1831–1844. https://doi.org/10.1007/s00394-019-02035-z

Jagannathan, S. V., Manemann, E. M., Rowe, S. E., Callender, M. C., & Soto, W. (2021). Marine actinomycetes, new sources of biotechnological products. In *Marine drugs* (Vol. 19, Issue 7). MDPI. https://doi.org/10.3390/md19070365

James, G., Prasannan Geetha, P., Thavarool Puthiyedathu, S., & Vattringal Jayadradhan, R. K. (2023). Applications of Actinobacteria in aquaculture: Prospects and challenges. In *3 Biotech* (Vol. 13, Issue 2). Springer. https://doi.org/10.1007/s13205-023-03465-7

Jensen, P. R. (2022). Microbe profile: *Salinispora tropica*: Natural products and the evolution of a unique marine bacterium. *Microbiology, 168*(4). https://doi.org/10.1099/mic.0.001163

Jomova, K., Raptova, R., Alomar, S. Y., Alwasel, S. H., Nepovimova, E., Kuca, K., & Valko, M. (2023). Reactive oxygen species, toxicity, oxidative stress, and antioxidants: Chronic diseases and aging. *Archives of Toxicology, 97*(10), 2499–2574. https://doi.org/10.1007/s00204-023-03562-9

Junaid, P. M., Dar, A. H., Dash, K. K., Rohilla, S., Islam, R., Shams, R., Pandey, V. K., Srivastava, S., Panesar, P. S., & Zaidi, S. (2024). Polysaccharide-based hydrogels for microencapsulation of bioactive compounds: A review. *Journal of Agriculture and Food Research, 15*, 101038. https://doi.org/10.1016/j.jafr.2024.101038

Kah Sem, N. A. D., Abd Gani, S., Chong, C. M., Natrah, I., & Shamsi, S. (2023). Management and mitigation of vibriosis in aquaculture: Nanoparticles as promising alternatives. In *International Journal of Molecular Sciences* (Vol. 24, Issue 16, pp. 12542). Multidisciplinary Digital Publishing Institute (MDPI). https://doi.org/10.3390/ijms241612542

Kar, J. R., & Singhal, R. S. (2015). Investigations on the ideal mode of cell disruption in extremely halophilic *Actinopolyspora halophila* (MTCC 263) for efficient release of glycine betaine and trehalose. *Biotechnology Reports, 5*, 89–97. https://doi.org/10.1016/j.btre.2014.12.005

Khanjani, M. H., Mohammadi, A., & Emerenciano, M. G. C. (2022). Microorganisms in biofloc aquaculture system. *Aquaculture Reports, 26*, 101300. https://doi.org/10.1016/j.aqrep.2022.101300

Kim, G. W., Lee, D. H., Jeon, Y. H., Yoo, J., Kim, S. Y., Lee, S. W., Cho, H. Y., & Kwon, S. H. (2021). Glutamine synthetase as a therapeutic target for cancer treatment. In *International Journal of Molecular Sciences* (Vol. 22, Issue 4, pp. 1–17). MDPI AG. https://doi.org/10.3390/ijms22041701

Koutsoumanis, K., Allende, A., Álvarez-Ordóñez, A., Bolton, D., Bover-Cid, S., Chemaly, M., de Cesare, A., Hilbert, F., Lindqvist, R., Nauta, M., Peixe, L., Ru, G., Simmons, M., Skandamis, P., Suffredini, E., Cocconcelli, P. S., Fernández Escámez, P. S., Maradona, M. P., Querol, A., … Herman, L. (2023). Update of the list of qualified presumption of safety (QPS) recommended microorganisms intentionally added to food or feed as notified to EFSA. *EFSA Journal, 21*(1). https://doi.org/10.2903/j.efsa.2023.7747

Kumar, R., Kumar, S., Gupta, A. K., Singh, S., Kumar, G., Narsinghani, T., Babu, M. R., & Sindhu, R. K. (2024). Recent progress in bioactive discovery from marine microorganisms and future directions. *Biochemical and Cellular Archives*. https://doi.org/10.51470/BCA.2024.24.1-S.3295

Lee, Y., Lee, N., Hwang, S., Kim, K., Kim, W., Kim, J., Cho, S., Palsson, B. O., & Cho, B.-K. (2020). System-level understanding of gene expression and regulation for engineering secondary metabolite production in Streptomyces. *Journal of Industrial Microbiology and Biotechnology, 47*(9–10), 739–752. https://doi.org/10.1007/s10295-020-02298-0

Lei, G., Khan, A., Budryn, G., & Grzelczyk, J. (2025). Probiotic products from laboratory to commercialization. *Trends in Food Science & Technology, 155*, 104807. https://doi.org/10.1016/j.tifs.2024.104807

Lemfack, M.-C., Bahl, H., Piechulla, B., & Magnus, N. (2020). The domain of bacteria and their volatile metabolic potential. In *Bacterial volatile compounds as mediators of airborne interactions* (pp. 1–38). Springer Singapore. https://doi.org/10.1007/978-981-15-7293-7_1

Leprêtre, M., Hamar, J., Urias, M. B., & Kültz, D. (2025). Comparative proteomics of salinity stress responses in fish and aquatic invertebrates. *Proteomics*. https://doi.org/10.1002/pmic.202400255

Letzel, A., Li, J., Amos, G. C. A., Millán-Aguiñaga, N., Ginigini, J., Abdelmohsen, U. R., Gaudêncio, S. P., Ziemert, N., Moore, B. S., & Jensen, P. R. (2017). Genomic insights into specialized metabolism in the marine actinomycete *Salinispora*. *Environmental Microbiology, 19*(9), 3660–3673. https://doi.org/10.1111/1462-2920.13867

Li, J., Li, Y., Zhou, Y., Wang, C., Wu, B., & Wan, J. (2018). Actinomyces and alimentary tract diseases: A review of its biological functions and pathology. In *BioMed research international* (Vol. 2018). Hindawi Limited. https://doi.org/10.1155/2018/3820215

Li, J., Si, H., Du, H., Guo, H., Dai, H., Xu, S., & Wan, J. (2021). Comparison of gut microbiota structure and Actinobacteria abundances in healthy young adults and elderly subjects: A pilot study. *BMC Microbiology, 21*(1). https://doi.org/10.1186/s12866-020-02068-z

Liu, J., Morgan, M., Hutchison, K., & Calhoun, V. D. (2010). A study of the influence of sex on genome wide methylation. *PLoS ONE, 5*(4). https://doi.org/10.1371/journal.pone.0010028

Lo Giudice, A., & Rizzo, C. (2022). Bacteria associated with benthic invertebrates from extreme marine environments: Promising but underexplored sources of biotechnologically relevant molecules. *Marine Drugs, 20*(10), 617. https://doi.org/10.3390/md20100617

Loh, J. S., Mak, W. Q., Tan, L. K. S., Ng, C. X., Chan, H. H., Yeow, S. H., Foo, J. B., Ong, Y. S., How, C. W., & Khaw, K. Y. (2024). Microbiota–gut–brain axis and its therapeutic applications in neurodegenerative diseases. In *Signal transduction and targeted therapy* (Vol. 9, Issue 1). Springer. https://doi.org/10.1038/s41392-024-01743-1

Maldonado, L. A., & Quintana, E. T. (2015). Unexpected properties of micromonosporae from marine origin. *Advances in Microbiology, 05*(06), 452–456. https://doi.org/10.4236/aim.2015.56046

Manan, M. A. (2025). The role of probiotics in personalized therapeutics: Advances in gut microbe-driven interventions. *The Microbe, 8*, 100497. https://doi.org/10.1016/j.microb.2025.100497

Manivasagan, P., Kang, K. H., Sivakumar, K., Li-Chan, E. C. Y., Oh, H. M., & Kim, S. K. (2014). Marine actinobacteria: An important source of bioactive natural products. In *Environmental toxicology and pharmacology* (Vol. 38, Issue 1, pp. 172–188). Elsevier. https://doi.org/10.1016/j.etap.2014.05.014

Markowiak, P., & Ślizewska, K. (2017). Effects of probiotics, prebiotics, and synbiotics on human health. In *Nutrients* (Vol. 9, Issue 9). MDPI AG. https://doi.org/10.3390/nu9091021

Martínez Cruz, P., Ibáñez, A. L., Monroy Hermosillo, O. A., & Ramírez Saad, H. C. (2012). Use of probiotics in aquaculture. *International Scholarly Research Notices, 2012*(1), 916845.

McCoubrey, L. E., Seegobin, N., Elbadawi, M., Hu, Y., Orlu, M., Gaisford, S., & Basit, A. W. (2022). Active Machine learning for formulation of precision probiotics. *International Journal of Pharmaceutics, 616*, 121568. https://doi.org/10.1016/j.ijpharm.2022.121568. Epub 2022 Feb 9. PMID: 35150845.

Min, U., Jin, Y. J., Jang, Y. J., Lim, J., & Kim, B. Y. (2024). Personalized probiotic strategy considering bowel habits: Impacts on gut microbiota composition and alleviation of gastrointestinal symptoms via Consti-Biome and Sensi-Biome. *Frontiers in Nutrition, 11*, 1302093. https://doi.org/10.3389/fnut.2024.1302093

Mohammadipanah, F., & Wink, J. (2016). Actinobacteria from arid and desert habitats: Diversity and biological activity. *Frontiers in Microbiology, 6*. https://doi.org/10.3389/fmicb.2015.01541

Moreira de Gouveia, M. I., Bernalier-Donadille, A., & Jubelin, G. (2024). *Enterobacteriaceae* in the human gut: Dynamics and ecological roles in health and disease. In *Biology* (Vol. 13, Issue 3, p. 142). Multidisciplinary Digital Publishing Institute (MDPI). https://doi.org/10.3390/biology13030142

Nicholson, W. L., Munakata, N., Horneck, G., Melosh, H. J., & Setlow, P. (2000). Resistance of Bacillus endospores to extreme terrestrial and extraterrestrial environments. *Microbiology and Molecular Biology Reviews, 64*(3), 548–572. https://doi.org/10.1128/MMBR.64.3.548-572.2000

Norouzi, H., Danesh, A., Mohseni, M., & Khorasgani, M. R. (2018). Marine actinomycetes with probiotic potential and bioactivity against multidrug-resistant bacteria. *The International Journal of Molecular and Cellular Medicine*. https://doi.org/10.22088/IJMCM.BUMS.7.1.44

O'Callaghan, A., & van Sinderen, D. (2016). Bifidobacteria and their role as members of the human gut microbiota. In *Frontiers in microbiology* (Vol. 7, Issue JUN). Frontiers Media S.A. https://doi.org/10.3389/fmicb.2016.00925

Park, J., Lee, J., & Roe, J. (2019). SigR, a hub of multilayered regulation of redox and antibiotic stress responses. *Molecular Microbiology, 112*(2), 420–431. https://doi.org/10.1111/mmi.14341

Penn, K., Jenkins, C., Nett, M., Udwary, D. W., Gontang, E. A., McGlinchey, R. P., Foster, B., Lapidus, A., Podell, S., Allen, E. E., Moore, B. S., & Jensen, P. R. (2009). Genomic islands link secondary metabolism to functional adaptation in marine Actinobacteria. *The ISME Journal, 3*(10), 1193–1203. https://doi.org/10.1038/ismej.2009.58

Petrariu, O. A., Barbu, I. C., Niculescu, A. G., Constantin, M., Grigore, G. A., Cristian, R. E., Mihaescu, G., & Vrancianu, C. O. (2023). Role of probiotics in managing various human diseases, from oral pathology to cancer and gastrointestinal diseases. In *Frontiers in microbiology* (Vol. 14). Frontiers Media SA. https://doi.org/10.3389/fmicb.2023.1296447

Pickard, J. M., Zeng, M. Y., Caruso, R., & Núñez, G. (2017). Gut microbiota: Role in pathogen colonization, immune responses, and inflammatory disease. *Immunological Reviews, 279*(1), 70–89. https://doi.org/10.1111/imr.12567

Poli, A., Finore, I., Romano, I., Gioiello, A., Lama, L., & Nicolaus, B. (2017). Microbial diversity in extreme marine habitats and their biomolecules. *Microorganisms, 5*(2), 25. https://doi.org/10.3390/microorganisms5020025

Prazdnova, E. V., Mazanko, M. S., Shevchenko, V. N., Skripnichenko, R. V., Kulikov, M. P., Golovko, L. S., Grigoriev, V. A., Maltseva, T. A., Kulikova, D. B., & Rudoy, D. V. (2025). Genomic characterization of four novel probiotic strains with enzymatic activity and their effects on carp (*Cyprinus carpio*). *Animals: An Open Access Journal from MDPI, 15*(13), 1998. https://doi.org/10.3390/ani15131998

Presentato, A., Piacenza, E., Turner, R. J., Zannoni, D., & Cappelletti, M. (2020). Processing of metals and metalloids by actinobacteria: Cell resistance mechanisms and synthesis of metal(loid)-based nanostructures. *Microorganisms, 8*(12), 2027. https://doi.org/10.3390/microorganisms8122027

Ramijan, K., Ultee, E., Willemse, J., Zhang, Z., Wondergem, J. A. J., van der Meij, A., Heinrich, D., Briegel, A., van Wezel, G. P., & Claessen, D. (2018). Stress-induced formation of cell wall-deficient cells in filamentous actinomycetes. *Nature Communications, 9*(1), 5164. https://doi.org/10.1038/s41467-018-07560-9

Ramírez-Gamboa, D., Díaz-Zamorano, A. L., Meléndez-Sánchez, E. R., Reyes-Pardo, H., Villaseñor-Zepeda, K. R., López-Arellanes, M. E., Sosa-Hernández, J. E., Coronado-Apodaca, K. G., Gámez-Méndez, A., Afewerki, S., Iqbal, H. M. N., Parra-Saldivar, R., & Martínez-Ruiz, M. (2022). Photolyase production and current applications: A review. *Molecules, 27*(18), 5998. https://doi.org/10.3390/molecules27185998

Ranjani, A., Dhanasekaran, D., & Gopinath, P. M. (2016). An introduction to actinobacteria. In *Actinobacteria—Basics and biotechnological applications*. InTechOpen. https://doi.org/10.5772/62329

Rhee, C., Kim, H., Aalfin Emmanuel, S., Kim, H. G., Won, S., Bae, J., Bai, S. C., & Koh, S. C. (2018). Microbial community analysis of an eco-friendly recirculating aquaculture system for olive flounder (*Paralichthys olivaceus*) using complex microbial probiotics. *Korean Journal of Microbiology, 54*(4), 369–378. https://doi.org/10.7845/kjm.2018.8085

Roda-Garcia, J. J., Haro-Moreno, J. M., & López-Pérez, M. (2023). Evolutionary pathways for deep-sea adaptation in marine planktonic Actinobacteriota. *Frontiers in Microbiology, 14*. https://doi.org/10.3389/fmicb.2023.1159270

Rodrigues, R., de Souza, A. Q. L., Feitoza, M. D. O., Alves, T. C. L., Barbosa, A. N., da Silva de Souza Santiago, S. R. S., & de Souza, A. D. L. (2024). Biotechnological potential of actinomycetes in the 21st century: A brief review. In *Antonie van Leeuwenhoek, International Journal of General and Molecular Microbiology* (Vol. 117, Issue 1). https://doi.org/10.1007/s10482-024-01964-y

Rotter, A., Barbier, M., Bertoni, F., Bones, A. M., Cancela, M. L., Carlsson, J., Carvalho, M. F., Cegłowska, M., Chirivella-Martorell, J., Conk Dalay, M., & Cueto, M. (2021). The essentials of marine biotechnology. *Frontiers in Marine Science, 8*. https://doi.org/10.3389/fmars.2021.629629

Salwan, R., & Sharma, V. (2022). Genomics of prokaryotic extremophiles to unfold the mystery of survival in extreme environments. *Microbiological Research, 264*, 127156. https://doi.org/10.1016/j.micres.2022.127156

Sandoval-Calderón, M., Guan, Z., & Sohlenkamp, C. (2017). Knowns and unknowns of membrane lipid synthesis in streptomycetes. *Biochimie, 141*, 21–29. https://doi.org/10.1016/j.biochi.2017.05.008

Saralov, A. I. (2019). Adaptivity of archaeal and bacterial extremophiles. *Microbiology, 88*(4), 379–401. https://doi.org/10.1134/S0026261719040106

Scholz, A., Stahl, J., de Berardinis, V., Müller, V., & Averhoff, B. (2015). Osmotic stress response in *Acinetobacter baylyi*: Identification of a glycine-betaine biosynthesis pathway and regulation of osmoadaptive choline uptake and glycine betaine synthesis through a choline-responsive BetI repressor. *Environmental Microbiology*. https://doi.org/10.1111/1462-2920.13148

Sen, G., Ghosh, S., Sarkar, I., & Sen, A. (2022). Perspectives on extremophilic actinobacteria—A review. *NBU Journal of Plant Sciences, 14*(1), 1–7. https://doi.org/10.55734/nbujps.2022.v14i01.001

Shivlata, L., & Satyanarayana, T. (2015). Thermophilic and alkaliphilic Actinobacteria: Biology and potential applications. *Frontiers in Microbiology, 6*. https://doi.org/10.3389/fmicb.2015.01014

Siddharth, S., & Rai, V. R. (2019). Isolation and characterization of bioactive compounds with antibacterial, antioxidant, and enzyme inhibitory activities from marine-derived rare actinobacteria, Nocardiopsis sp. SCA21. *Microbial Pathogenesis, 137*, 103775. https://doi.org/10.1016/j.micpath.2019.103775

Singh, R., & Dubey, A. K. (2018). Diversity and applications of endophytic actinobacteria of plants in special and other ecological niches. *Frontiers in Microbiology, 9*, 1767.

Singh, S., Datta, S., Narayanan, K. B., & Rajnish, K. N. (2021). Bacterial exo-polysaccharides in biofilms: Role in antimicrobial resistance and treatments. *Journal of Genetic Engineering and Biotechnology, 19*(1), 140. https://doi.org/10.1186/s43141-021-00242-y

Siro, G., Pipite, A., Christi, K., Srinivasan, S., & Subramani, R. (2022). Marine actinomycetes associated with stony corals: A potential hotspot for specialized metabolites. *Microorganisms, 10*(7), 1–32. https://doi.org/10.3390/microorganisms10071349

Sivapragasam, S., & Grove, A. (2019). The link between purine metabolism and production of antibiotics in streptomyces. *Antibiotics, 8*(2), 76. https://doi.org/10.3390/antibiotics8020076

Solanki, R., Karande, A., & Ranganathan, P. (2023). Emerging role of gut microbiota dysbiosis in neuroinflammation and neurodegeneration. In *Frontiers in neurology* (Vol. 14). Frontiers Media S.A. https://doi.org/10.3389/fneur.2023.1149618

Stincone, P., & Brandelli, A. (2020). Marine bacteria as a source of antimicrobial compounds. In *Critical reviews in biotechnology* (Vol. 40, Issue 3, pp. 306–319). Taylor and Francis Ltd. https://doi.org/10.1080/07388551.2019.1710457

Subramani, R., & Aalbersberg, W. (2012). Marine actinomycetes: An ongoing source of novel bioactive metabolites. *Microbiological Research, 167*(10), 571–580. https://doi.org/10.1016/j.micres.2012.06.005

Suominen, S., Dombrowski, N., Sinninghe Damsté, J. S., & Villanueva, L. (2021). A diverse uncultivated microbial community is responsible for organic matter degradation in the Black Sea sulphidic zone. *Environmental Microbiology, 23*(6), 2709–2728. https://doi.org/10.1111/1462-2920.14902

Tan, L. T.-H., Chan, K.-G., Lee, L.-H., & Goh, B.-H. (2016). Streptomyces bacteria as potential probiotics in aquaculture. *Frontiers in Microbiology, 7*. https://doi.org/10.3389/fmicb.2016.00079

Tan, L. T. H., Chan, K. G., Khan, T. M., Bukhari, S. I., Saokaew, S., Duangjai, A., Pusparajah, P., Lee, L. H., & Goh, B. H. (2017). Streptomyces sp. MUM212 as a source of antioxidants with radical scavenging and metal chelating properties. *Frontiers in Pharmacology, 8*(MAY), 1–18. https://doi.org/10.3389/fphar.2017.00276

Tang, H., Xiang, G., Xiao, W., Yang, Z., & Zhao, B. (2024). Microbial mediated remediation of heavy metals toxicity: Mechanisms and future prospects. *Frontiers in Plant Science, 15*. https://doi.org/10.3389/fpls.2024.1420408

Tolas, I., Zhou, Z., Zhang, Z., Teame, T., Olsen, R. E., Ringø, E., & Rønnestad, I. (2025). A fishy gut feeling—Current knowledge on gut microbiota in teleosts. *Frontiers in Marine Science, 11*. https://doi.org/10.3389/fmars.2024.1495373

Trujillo, M. E., Riesco, R., Benito, P., & Carro, L. (2015). Endophytic actinobacteria and the interaction of micromonospora and nitrogen fixing plants. *Frontiers in Microbiology, 6*. https://doi.org/10.3389/fmicb.2015.01341

Undabarrena, A., Beltrametti, F., Claverías, F. P., González, M., Moore, E. R. B., Seeger, M., & Cámara, B. (2016). Exploring the diversity and antimicrobial potential of marine actinobacteria from the Comau Fjord in Northern Patagonia, Chile. *Frontiers in Microbiology, 7*(JUL). https://doi.org/10.3389/fmicb.2016.01135

Valliappan, K., Sun, W., & Li, Z. (2014). Marine actinobacteria associated with marine organisms and their potentials in producing pharmaceutical natural products. *Applied Microbiology and Biotechnology, 98*(17), 7365–7377. https://doi.org/10.1007/s00253-014-5954-6

Venkatesh, G. P., Kuruvalli, G., Syed, K., & Reddy, V. D. (2024). An updated review on probiotic production and applications. *Gastroenterology Insights, 15*(1), 221–236. https://doi.org/10.3390/gastroent15010016

Ventosa, A., Nieto, J. J., & Oren, A. (1998). Biology of moderately halophilic aerobic bacteria. *Microbiology and Molecular Biology Reviews, 62*(2), 504–544. https://doi.org/10.1128/MMBR.62.2.504-544.1998

Vladkova, T., Georgieva, N., Staneva, A., & Gospodinova, D. (2022). Recent progress in antioxidant active substances from marine biota. In *Antioxidants* (Vol. 11, Issue 3). MDPI. https://doi.org/10.3390/antiox11030439

Vyas, U., & Ranganathan, N. (2012). Probiotics, prebiotics, and synbiotics: Gut and beyond. In *Gastroenterology research and practice*. https://doi.org/10.1155/2012/872716

Wang, M., & Li, H. (2025). Structure, function, and biosynthesis of siderophores produced by Streptomyces species. *Journal of Agricultural and Food Chemistry, 73*(8), 4425–4439. https://doi.org/10.1021/acs.jafc.4c08231

Wang, Q., He, Y., & Tan, Q. (Eds.). (2024). *Metabolic research in aquatic animal nutrition, physiology and disease*. MDPI. https://doi.org/10.3390/books978-3-0365-9986-1

Wong, J., Rios-Solis, L., & Keasling, J. D. (2017). Microbial production of isoprenoids. In *Consequences of microbial interactions with hydrocarbons, oils, and lipids: Production of fuels and chemicals* (pp. 359–382). Springer. https://doi.org/10.1007/978-3-319-50436-0_219

Zhang, H., Liang, Z., Zhao, M., Ma, Y., Luo, Z., Li, S., & Xu, H. (2022). Metabolic engineering of *Escherichia coli* for ectoine production with a fermentation strategy of supplementing the amino donor. *Frontiers in Bioengineering and Biotechnology, 10*. https://doi.org/10.3389/fbioe.2022.824859

Ziemert, N., Lechner, A., Wietz, M., Millán-Aguiñaga, N., Chavarria, K. L., & Jensen, P. R. (2014). Diversity and evolution of secondary metabolism in the marine actinomycete genus *Salinispora*. *Proceedings of the National Academy of Sciences, 111*(12). https://doi.org/10.1073/pnas.1324161111

Bioremediation of Marine Actinobacteria

Pratheep Thangaraj, Peng Zeng,
and Manikandan Rajendran

Abstract

Recently, marine actinobacteria have been getting significant attention for bioremediation of environmental contamination due to their exclusive metabolic abilities and tolerance and resistance to the environmental contaminants' toxicity. Marine actinobacteria are an eco-friendly, promising, and sustainable approach to bioremediation of environmental contamination such as heavy metals, dyes, pesticides, and hydrocarbons. These actinobacteria synthesize bioactive enzymes and metabolites for transforming the toxic molecules into non-toxic form and decomposing pesticides, hydrocarbons, and polyethylene. This book chapter aims to provide detailed information on the bioremediation of heavy metals, dyes, pesticides, and hydrocarbons by marine actinobacteria. For example, various studies have confirmed that actinobacteria species, including *Rhodococcus* spp. and *Streptomyces* spp., can remediate heavy metals from soil and water. Micromonospora, Rhodococcus, Cellulosimicrobium, and Gordonia are promising species for the bioremediation of hydrocarbon contamination. This book chapter also deals with recent research based on heavy metals, textile dyes, pesticides, and hydrocarbons by Marine actinobacteria. These incorporated findings emphasize that the potential of marine actinobacteria is cost-effective, eco-friendly bioremediation of environmental pollutants, whereas further research needs to standardize technologies and validation in pilot scale to field application. This chapter provides an overview of recent research on the application of marine actinobacteria in the remediation of heavy metals, textile dyes, pesticides, and hydrocarbons. The compiled findings highlight their potential as eco-friendly bioremediation agents while also underscoring the need for further

P. Thangaraj · M. Rajendran (✉)
Department of Biotechnology, Rathinam College of Arts and Science,
Coimbatore, Tamil Nadu, India
e-mail: manikandan.bio@rathinam.in

P. Zeng
Colleges of Ecology and Environment, Central South University of Forestry and Technology,
Changsha, China

K. Loganathan, Z. Li (eds.), *Marine Actinomycetes*,
https://doi.org/10.1007/978-981-95-9849-6_10

research to standardize technologies and validate their efficacy from pilot-scale studies to large-scale field applications.

Keywords

Marine actinobacteria · Bioremediation · Environmental contamination · Heavy metals · Pesticide · Hydrocarbons

1 Introduction

Marine environments are progressively contaminated by hydrocarbons, heavy metals, plastics, and radioactive substances caused by industrialization, shipping, oil spills, and land-based waste discharge. These pollutants threaten marine ecosystems and human health, demanding the advancement of efficient, sustainable, and eco-friendly remediation strategies. From the group of biological agents, marine actinobacteria are Gram-positive bacteria and have high G + C content and considerable interest due to their versatility in metabolism, resistance to harsh environmental conditions, and ability to degrade different contaminants (Fig. 1). Marine actinobacteria surpass other microbes in bioremediation activities owing to their ability to function under adverse levels of salinity, pH, and temperature. Compared to terrestrial types, marine actinobacteria synthesize distinctive bioactive metabolites and supernatural enzymes for the decomposition of complex hydrocarbons, heavy metals, and synthetic polymers, including polyethylene and polystyrene (Oliveira et al., 2022; Lv et al., 2024). Hydrocarbon pollution is a very intractable contamination in marine ecosystems. Marine actinobacteria belonging to genera such as Rhodococcus, Gordonia, Micromonospora, and Cellulosimicrobium exhibit high hydrocarbon-degrading potential, and *Rhodococcus ruber* effectively degrades fluorene (Fernandes et al., 2025) and transforms plastics. *Streptomyces gougerotti* and *Micromonospora matsumotoense* degrade low-density polyethylene and convert carbon sources from plastics into biodegradable polyhydroxyalkanoates, representing a promising model for circular bioeconomy applications (Oliveira et al., 2022). The potential of actinobacteria extends to heavy metal and radionuclide remediation. *Actinomadura* spp. have been biosorbing uranium around nuclear power plant sites by exploiting extracellular polymeric substances (EPSs) and diverse adsorption mechanisms (Sivaperumal et al., 2022). In aquaculture and brackish systems, marine actinobacterial species act as probiotics and bioremediators. By improving water quality, degrading wastes, and preventing pathogenic biofilm development through quorum quenching, they contribute toward sustainable aquaculture management (James et al., 2023; Preena et al., 2019). Furthermore, their diversity and function in the realms of marine, cultivable genera such as Streptomyces and Micromonospora produce biosurfactants and lipases, which degrade and emulsify the hydrocarbon (Blandón et al., 2022). Besides, marine actinobacteria are responsible for plastic degradation by producing enzymes such as esterase, cutinase, and laccase. The marine actinobacterial strain *Rhodococcus pyridinivorans* P23 expresses an esterase capable of hydrolyzing polyethylene terephthalate, shining a

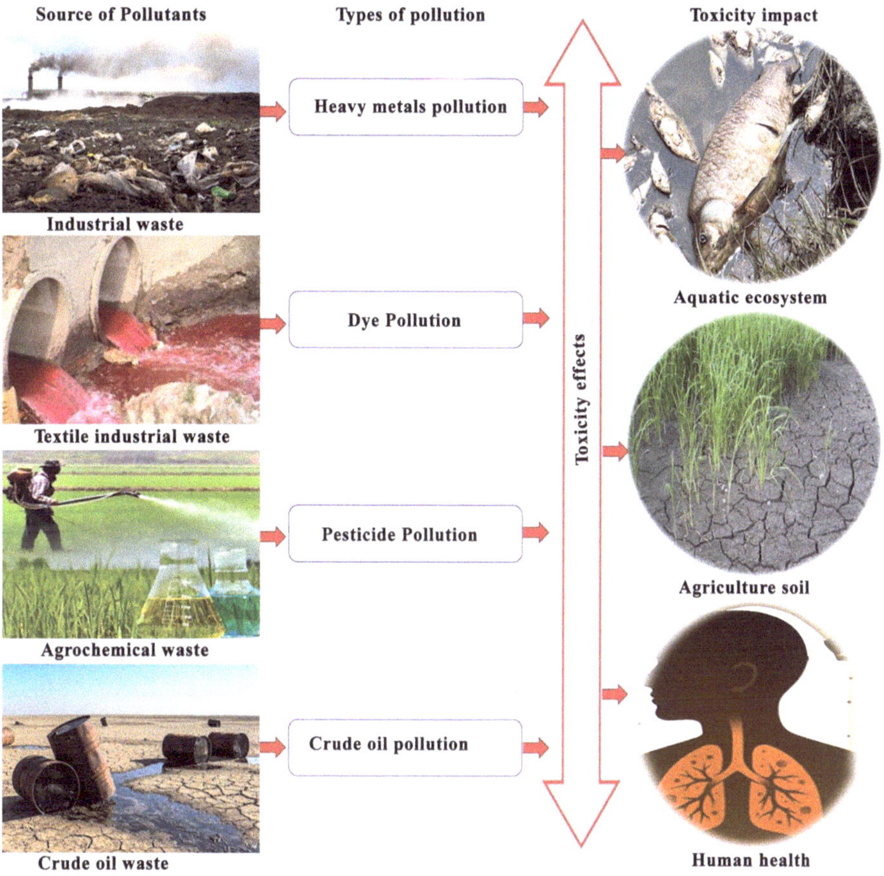

Fig. 1 Pollutant sources, type of pollution, and the effect of pollution on the aquatic ecosystem, agricultural soil, and human health

light on molecular pathways involved with plastic degradation in marine settings (Guo et al., 2023). Marine actinobacteria conceivably figure prominently within integrated bioremediation techniques, considering recent developments in microbial biotechnologies. Either by welding these microbes with physicochemical methods or by employing them with other functional organisms, international efforts could go a long way to enhance pollutant removal in marine environments. Undoubtedly, their broad enzymatic and biosynthetic repertoire renders them profound collaborators in the fight against twenty-first century pollution problems. In this chapter, exploring and assessing the potential of marine actinobacteria in the remediation of environmental pollution, especially degradation of heavy metals, dyes, pesticides, and crude oil, emphasizes the diversity, metabolic capabilities, and applications of marine-derived actinobacterial strains for purposes of bio-restoration of contaminated marine and coastal environments through a green and sustainable approach.

2 Role of Marine Actinobacteria in Environmental Remediation

Actinobacteria show exceptional metabolic adaptability, allowing them to break down different organic and inorganic contaminants. Their endurance to harsh environmental conditions and an incredible capacity to biosynthesize secondary metabolites with detrimental activities further exalt them to be principal agents of the time for sustaining remediation measures.

2.1 Mechanisms of Bioremediation by Actinobacteria

Actinobacteria have a wide range of catabolic enzymes, including oxygenases, peroxidases, laccases, and hydrolases, to break down complex pollutants. Among these enzymatic systems, the transformation of Persistent Organic Pollutants (POPs) such as polycyclic aromatic hydrocarbons (PAHs), pesticides, and synthetic dyes into less toxic or nontoxic substances is widely documented (Behera & Das, 2023; Makarani & Kaushal, 2025). Streptomyces express azo reductase and laccase, breaking azo bonds in dyes such as acid orange and Congo red (Bautista-Pinzón et al., 2024). *Rhodococcus* spp. undergo dioxygenase-mediated ring cleavage processes to effectively degrade PAH (Ma et al., 2023). Heavy metals, such as chromium, arsenic, and lead, are immobilized through the biosorption mechanism by which metals are surface-bound by functional groups of the bacterial cell wall and bioaccumulate intracellularly. The secretion of EPSs further enhances this immobilization capacity by providing a matrix for pollutant entrapment (Makarani & Kaushal, 2025). Siderophores act by capturing iron, acting as a metal scavenging detoxifying agent, aiding in nutrient uptake and heavy metal detoxification. Toxic metals bound to contaminated sites become immobilized, and bioavailability is reduced; hence, toxicity is reduced. EPSs contribute to the adsorption of pollutants to some extent and biofilm formation; the latter protects bacterial cells from environmental stress, thereby supporting enhanced remediation (Behera & Das, 2023).

2.2 Target Pollutants and Degradation Capacity

Actinobacteria can degrade complex organic compounds such as petroleum hydrocarbons, pharmaceuticals, pesticides, and industrial dyes. Volcanic site isolated actinobacterial strains showed the highest tolerance and degradation potential against non-steroidal anti-inflammatory drugs such as diclofenac and ibuprofen (Soria-Camargo et al., 2025). Streptomyces species in the bioaugmented biobed have performed 70% atrazine removal from agricultural soils (Saez et al., 2022). Actinobacteria have been shown to remediate environments contaminated with heavy metals. The actinobacteria from co-contaminated Argentinian soils revealed tolerance against Cr(VI) and phenanthrene, which makes them ideal candidates for multi-pollutant clean-up (Antezana et al., 2023). Moreover, species of Nocardioides have shown high efficiency in the degradation of hard-to-remove pollutants such as nitrogen heterocycles and polymeric substances (Ma et al., 2023). It has further been revealed that actinobacteria could transform emerging contaminants such as aflatoxin B1 via novel enzymes like dye-decolorizing peroxidases found in Paracoccus and Actinobacteria-rich consortia (Hu et al., 2024). This widens the scope of actinobacterial applications in modern pollution challenges.

2.3 Biotechnological Enhancements

The enhancement of pollutant degradation rates and broadening of activity spectra have been reported in the combination of various actinobacteria strains forming defined consortia. Such synergistic interactions enable the cells to degrade complex pollutants sequentially or simultaneously (Antezana et al., 2023). Actinobacteria are immobilized in biocarriers, biofilms, and biomixtures to maintain and enhance their activity. Bioaugmentation of biobeds with *Streptomyces* spp. successfully used to treat pesticide-contaminated runoff *in situ* (Saez et al., 2022). Recent advances in omics technology, such as metagenomics, metatranscriptomics, and proteomics, allowed a deeper understanding of the metabolism and genetics used by actinobacteria in bioremediation (Makarani & Kaushal, 2025; Hsiao et al., 2021). Especially the discovery of functional genes, such as aedA and aedB in Rhodococcus, has allowed for the development of biomarkers of estrogen degradation in estuarine sediments (Hsiao et al., 2021). Now, genetic engineering is also being developed to improve these pathways for large industrial use.

3 Bioremediation of Heavy Metal Contamination by Marine Actinobacteria

Heavy metal contamination in marine environments is a growing environmental problem due to industrial discharges, mining activities, and agricultural runoff. These metals, including cadmium, lead, chromium, and arsenic, persist and cannot be biodegraded; hence, all of them have some level of toxicity to aquatic ecosystems. Bioremediation through Streptomyces due to their extreme resistance and metal transformation capabilities has gained attention as an alternative eco-friendly solution (Hamdan et al., 2021).

3.1 Biosorption Potential of Marine Actinobacteria

Streptomyces rochei can adsorb heavy metals such as Cr^{6+}, Cd^{2+}, and Pb^{2+} from industrial effluents (Hamdan et al., 2021). Under optimal conditions, biosorption efficiency was reported at approximately 85%, thereby confirming the species' potential for detoxification of these metals and subsequent water quality improvement for agricultural reuse. Rare and metabolically diverse bacterial taxa, including actinobacteria found in marine sediments, and some groups show adaptation to extreme, toxic environments. In a metagenomic and culture-based investigation of heavily contaminated marine sediments in the Gulf of Naples, Dell'Anno et al. (2021) found that actinobacteria were among the leading groups reacting to metal (Cu, Cr, Cd, and As) stress. They isolated rare culturable strains resistant to high concentrations of mixed contaminants, bringing to light the great possibility of these organisms being used for bioaugmentation practices in targeted marine sediment remediation. Duran et al. (2015) studied actinobacterial assemblages across coastal marine sediments at differing levels of pollution in the West Istrian Sea. The research showed that distinct actinobacterial taxa associated depending upon the pollutants, *Rhodococcus* spp. with PAHs, and *Streptomyces* spp. with clusters of metals (Hg, Cd, Cu, Pb, and Zn). Such ecological specificity has the implication of the existence of functional microbial guilds able to adapt to heavy metal pressure and even grow there, which implies an aiding role for native actinobacteria in ecological restoration.

3.2 Consortium-Based Bioremediation

Bioremediation can be further compounded by microbial consortia. According to Aparicio et al. (2019), a marine actinobacterial consortium composed of *Streptomyces* spp. and *Amycolatopsis tucumanensis* had successfully removed Cr(VI) and lindane from soils co-contaminated by these chemicals. The success of the bioremediation was supported by ecotoxicological bioassays employing plant and earthworm species, which confirmed restoration of soil health. Latterly, augmentation with actinobacteria, phytoremediation with *Brassica napus*, and vermiremediation with *Eisenia fetida* were synergistically applied for the clean-up of Cr(VI) and lindane-contaminated soils (Aparicio et al., 2021). Hence, integrated approaches suggest that marine actinobacteria may be used as suitable central agents in a multipronged remediation scheme. In rhizosphere studies, *Streptomyces* spp. were reported to be dominant under high Cd exposure in the Cd-hyperaccumulating plant *Sedum alfredii* (Hou et al., 2018). The presence of these actinobacteria significantly altered the bacterial community structure and enhanced Cd tolerance, suggesting possible cross-environment applications of marine actinobacteria for plant-assisted remediation.

4 Bioremediation of Dye Contamination by Marine Actinobacteria

Dye contamination is a major environmental problem, mostly released from industrial activities such as textile manufacturing, food processing, and other dye-utilizing industries. Dye pollution affects both aquatic ecosystems and human health. Textile industries are the main sources of dye contamination due to the release of huge volumes of untreated wastewater into the environment. In addition, dye producers, tanneries, and food industries are also major sources of dye pollution in the environment (Kapanga et al., 2024). Dye leaching in soil is affecting the soil fertility, reducing physicochemical properties and microbial population. For example, Peng et al. (2024) reported that the dye contamination significantly reduced the soil health, fungal diversity, and affected the soil's nutrient cycle. In aquatic ecosystems, dye contamination affects the organisms' growth and reproduction by inducing oxidative stress. Humans are exposed to dye hazards through the food chain and polluted water, leading to high health risks such as cancer and skin disease (Dutta et al., 2024).

Therefore, marine actinobacteria are a promising agent for the biodegradation of dye pollution. Actinobacteria have a unique ability to degrade and detoxify different synthetic dyes, which are prevalent contaminants due to industrial activities. The biodegradation process includes the breakdown of dye complex molecules into less toxic substances, thereby decreasing their environmental and ecosystem impact. For example, *Nocardiopsis* sp. effectively decolorizes and biodegrades 85.6% of azo dyes such as Reactive Orange-16 through enzymatic pathways within 24 h (Chittal et al., 2019). The biodegradation process can be increased through biostimulation, which standardizes the environmental conditions, such as pH, weather, and salinity, to favor microbial activities (Rathore et al., 2021). Different marine actinobacteria, including *Halobacillus trueperi* and *Marinobacter excellens*, have showed 85.20% and 72.90% Mordant Orange 1 and Acid Orange decolorization, respectively (Thorat & Shertate, 2017). Moreover, *Marinobacter* sp. decreased 80.78% of Reactive Blue 171 (Shertate & Thorat, 2015). Actinobacteria can produce enzymes, such as Flavin Mononucleotide (FMN) and Flavin Adenine Dinucleotide (FAD)-dependent peroxidase and azo-reductase, which effectively break down azo linkages and degrade anthraquinone dyes under high pH, temperature, and dye concentration (Moopantakath & Kumavath, 2018). Similarly, *Marinobacter* sp. was reported to remove 80.78% of Reactive Blue 171 (Shertate & Thorat, 2015). Actinobacteria can produce enzymes like FMN- and FAD-dependent peroxidase and azo-reductase, which effectively break down azo linkages and degrade anthraquinone dyes under high pH, temperature, and dye concentration (Moopantakath & Kumavath, 2018). Marine actinobacteria, including genera such as Nocardiopsis, Rhodococcus, and Streptomyces, possess exclusive enzymatic systems for breaking down synthetic dyes. In addition to decolorization, the bioremediation process mediated by marine actinobacteria also results in a substantial reduction in chemical oxygen demand, indicating a concurrent decrease in the overall environmental pollution load. Therefore, utilizing marine actinobacteria for dye bioremediation offers an economical, eco-friendly, and promising approach to addressing dye contamination. Moreover, integrating biodegradation with complementary treatment methods such as nanoparticles, biomaterials, and biochar could further enhance overall efficiency. Continued research and development are essential to optimize these processes and fully harness the potential of marine actinobacteria in dye remediation (Table 1).

Table 1 Marine actinobacteria involved in the bioremediation of dye contamination

Actinobacteria species	Name of the dye	Decolorization/ removal (%)	References
Halobacillus trueperi and *Marinobacter excellens*	Mordant orange 1 and acid orange	85.20 and 72.90	Thorat and Shertate (2017)
Nocardiopsis sp. (VITVAMB 1)	Reactive orange-16 (azo dye)	85.6	Chittal et al. (2019)
Streptomyces bacillaris	Malachite green (MG), methyl violet (MV), crystal violet (CV), cotton blue (CB)	MG 94.7, MV 91.8, CV 86.6, CB 68.4	Adenan et al. (2022)
Nocardiopsis and *Streptomyces alba*	MG, MV, CV, CB (triphenylmethane dyes)	MG 97.0, MV 95.1, CV 95.8, CB 83.8	Li et al. (2014)
Streptomyces	MG, MV, CV, CB	MG 95, MV 92, CV 87, CB 68	Adenan et al. (2021)
Dietzia maris	Remazol brilliant blue R	*86%*	Vieira et al. (2021)
Marinobacter sp.	Reactive blue 171	80.78%	Shertate and Thorat (2015)

5 Bioremediation of Pesticide Contamination by Marine Actinobacteria

Pesticide pollution is a major concern for the environment and human health. While it is beneficial for agricultural yield, it often accumulates in non-targeted soil and waterbodies, leading to water and soil pollution. The translocation of pesticides in the aquatic ecosystem, agricultural soil, and humans through the food chain, it poses severe health risks. For example, AbuQamar et al. (2024) showed that the suspended particles are accumulating in the aquatic system and contaminating and degrading the aquatic ecosystem. Yao et al. (2019) confirmed the presence of pesticide particles in fish tissues. Soil microbial population is also affected by pesticide contamination compared to uncontaminated soil (Uneze et al., 2024). However, the bioaccumulation of pesticides in aquatic ecosystems and soil can enhance hazardous compounds, such as nitrates and ammonium, threatening the environmental ecosystem and human health. Moreover, it is considered carcinogenic, teratogenic, and genotoxic and causes chronic diseases (Singh et al., 2018). Therefore, the effective management of pesticide pollution includes complete monitoring and regulatory measures to protect the environment and human health. Strategies such as bioremediation, proper procedures for pesticide application, and the use of alternative non-toxic materials for pest control can help to decrease the pollution levels in soil and waterbodies (AbuQamar et al., 2024). Marine actinobacteria play a vital role in pesticide bioremediation through the breakdown of toxic pesticides into non-toxic compounds, thereby decreasing environmental contamination. For example, Despotovic et al. (2021) showed that the marine actinobacteria secrete phosphotriesterase enzyme that breaks down organophosphate pesticides and is utilized as a phosphorus source. Spreptomyces have the ability to degrade the carbaryl pesticide 36.20% within 30 days of treatment. Microbial consortia of actinobacteria increase 60–80% w of pesticide residue degradation in contaminated soil (Ordu-Pac & Bosede, 2024). On the other hand, genetically modified Actinobacteria effectively degraded pesticide pollutants (Sehrawat et al., 2021). Although marine actinobacteria are a promising agent for bioremediation of pesticide pollution, challenges such as environmental variability and microbial interaction are majorly addressed to optimize the application. Future research needs to focus on ecological understanding, microbial taxa, and developing advanced bioremediation methods that can be applied in different environments. Marine actinobacteria mediated remediation approach is a very suitable management for pesticide contamination in soil and water bodies.

6 Bioremediation of Crude Oil Contamination by Marine Actinobacteria

Crude oil pollution is an important environmental issue with far-reaching effects on aquatic ecosystems, agricultural soil, and humans. The pollution arises from different sources, including industrial waste, oil exploration activity, and accidental spills, leading to modifications in the chemical, physical, and biological properties of exaggerated environments. The crude oil pollution is diverse, affecting the aquatic ecosystem, soil fertility, microbial diversity, and human health. Crude oil pollution creates reproductive problems in animals in the aquatic ecosystem. For example, crude oil contamination induces toxicity effects on testes histopathology, reduces the production of sperm, and alters the sex hormone content, finally affecting fertility in male goldfish (Mahlouji et al., 2024). Crude oil pollution can alter the soil properties and enhance the liquid limits of soils, decreasing their dry density (Oborie & Opukumo, 2024). Moreover, crude oil pollution significantly affects the microbial diversity and communities in soil (Godwin & Inu, 2024). Furthermore, crude oil pollution poses a serious problem to human health due to the occurrence of toxic combinations such as PAHs. The crude oil pollution can lead to soil toxicity and affect plant growth and development (Buzmakov et al., 2019). In soil, some microorganisms utilize hydrocarbons as a nutrient source for growth and development, leading to the degradation of hydrocarbons (Antonioli et al., 2023). Hence, different degradation and remediation approaches are employed to address crude oil pollution, including biodegradation and bioremediation. These techniques use microorganisms to degrade or remove crude oil. Additionally, phytoremediation uses plants to absorb and remove crude oil contamination (Ogidi & Akpan, 2024). On the other hand, some other methods, such as electrokinetic remediation and thermal desorption, are cost-effective and environmentally effective techniques.

However, biological degradation/remediation techniques are offering a promising solution for crude oil contamination. Particularly, actinobacteria have the ability to degrade crude oil (hydrocarbons). For example, Rhodococcus and Gordonia showed a significant role in the breakdown of different crude oil components such

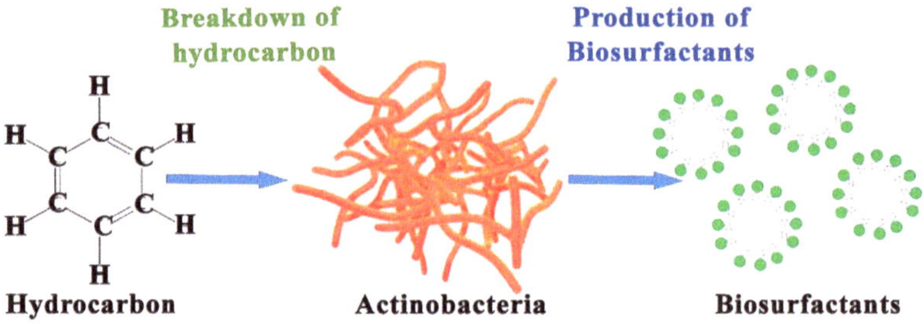

Fig. 2 Biodegradation pathway of hydrocarbons by Actinobacteria

as hydrocarbons and alkanes. Actinobacteria can utilize crude oil as a carbon source and produce biosurfactants, which increase their crude oil degradation and remediation efficiency. For instance, *Gordonia rubripertincta* and *Rhodococcus erythropolis* utilize the hydrocarbon as a carbon source, produce biosurfactants, and enhance the degradation ability (Pidgorskyi & Nogina, 2016). Actinobacteria exhibit changes in hydrophobicity in the cell surface and increase their interaction with hydrophobic oil molecules. Actinobacteria can produce some enzymes for the degradation of hydrocarbons, such as iso-alkanes and n-alkanes. These are commonly present in crude oil (Pidgorskyi & Nogina, 2016). *Actinomyces israelii* and *Actinomyces viscosus* showed enhanced growth in oil-contaminated media, which indicated that actinobacteria utilize hydrocarbons (Olajuyigbe & Ehiosun, 2016). Actinobacteria are a potential agent for the biodegradation of hydrocarbons at optimized environmental conditions. Further research needs to understand the metabolic and genetic pathways of actinobacteria to increase their application in biodegradation and bioremediation strategies, making them a healthier solution for hydrocarbon contamination (Fig. 2).

7 Conclusion

Marine actinobacteria are highly versatile and effective agents for the bioremediation of toxic heavy metals, textile industrial dyes, pesticides, and hydrocarbons, due to their ability to tolerate and resist a wide range of environmental contaminants through various mechanisms. Although they have their promising potential, further research is needed to develop innovative technologies for contaminant removal, particularly addressing challenges connected to pollutant complexity, biosafety, and the role of transgenic plants and genetically modified microorganisms in bioremediation of heavy metals, textile dyes, pesticides, and hydrocarbons. Future research requires focusing on employing multidisciplinary investigational approaches to generate datasets that increase analytical precision.

Acknowledgments The authors, Pratheep Thangaraj and Manikandan Rajendran, gratefully acknowledge the support received from the Department of Biotechnology (DBT), Government of India, under the DBT Star College Scheme for providing the necessary infrastructure and resources to carry out this research in the laboratory of the Department of Biotechnology at Rathinam College of Arts and Science.

References

AbuQamar, S. F., El-Saadony, M. T., AlKafaas, S. S., Elsalahaty, M. I., Elkafas, S. S., Mathew, B., Aljasmi, A. N., Alhammadi, H. S., Salem, H. M., Abd El-Mageed, T. A., Zaghloul, R., Mosa, W. F. A., Ahmed, A. E., Elrys, A. S., Saad, A. M., Al-Saeed, F. A., & El-Tarabily, K. A. (2024). Ecological impacts and management strategies of pesticide pollution on aquatic life and human beings. *Marine Pollution Bulletin, 206*, 116613. https://doi.org/10.1016/j.marpolbul.2024.116613

Adenan, N. H., Lim, Y. Y., & Ting, A. S. Y. (2021). Identification and optimization of triphenylmethane dyes removal by Streptomyces sp. from forest soil. *Sustainable Environment Research, 31*(1), 8.

Adenan, N. H., Lim, Y. Y., & Ting, A. S. Y. (2022). Removal of triphenylmethane dyes by *Streptomyces bacillaris*: A study on decolorization, enzymatic reactions and toxicity of treated dye solutions. *Journal of Environmental Management, 318*, 115520.

Antezana, P. E., Colin, V. L., Bourguignon, N., Benimeli, C. S., & Fuentes, M. S. (2023). Applied of actinobacteria consortia-based bioremediation to restore co-contaminated systems. *Research in Microbiology, 174*(4), 104028. https://doi.org/10.1016/j.resmic.2023.104028

Antonioli, R., Poloni, J. de F, Escalona, M. A. R., & Dorn, M. (2023). Functional response of microbial communities in lab-controlled oil-contaminated marine sediment. *Molecular Omics*. https://doi.org/10.1039/d3mo00007a

Aparicio, J. D., Garcia-Velasco, N., Urionabarrenetxea, E., Soto, M., Álvarez, A., & Polti, M. A. (2019). Evaluation of the effectiveness of a bioremediation process in experimental soils polluted with chromium and lindane. *Ecotoxicology and Environmental Safety, 181*, 255–263. https://doi.org/10.1016/j.ecoenv.2019.06.019

Aparicio, J. D., Lacalle, R. G., Artetxe, U., Urionabarrenetxea, E., Becerril, J. M., Polti, M. A., Garbisu, C., & Soto, M. (2021). Successful remediation of soils with mixed contamination of chromium and lindane: Integration of biological and physico-chemical strategies. *Environmental Research, 194*, 110666. https://doi.org/10.1016/j.envres.2020.110666

Bautista-Pinzón, F., Fonseca-Ordoñez, J., Falla-Obando, M., Gonzales-Tuta, J., & Diaz-Barrera, L. (2024). *Streptomyces* as a novel biotool for azo pigments remediation in contaminated scenarios. *Frontiers in Bioscience (Elite Edition), 16*(3), 29. https://doi.org/10.31083/j.fbe1603029

Behera, S., & Das, S. (2023). Potential and prospects of actinobacteria in the bioremediation of environmental pollutants: Cellular mechanisms and genetic regulations. *Microbiological Research, 273*, 127399. https://doi.org/10.1016/j.micres.2023.127399

Blandón, L. M., Marín, M. A., Quintero, M., Jutinico-Shubach, L. M., Montoya-Giraldo, M., Santos-Acevedo, M., & Gómez-León, J. (2022). Diversity of cultivable bacteria from deep-sea sediments of the Colombian Caribbean and their potential in bioremediation. *Antonie Van Leeuwenhoek, 115*(3), 421–431. https://doi.org/10.1007/s10482-021-01706-4

Buzmakov, S. A., Egorova, D. O., & Gatina, E. (2019). Effects of crude oil contamination on soils of the Ural region. *Journal of Soils and Sediments, 19*(1), 38–48. https://doi.org/10.1007/S11368-018-2025-0

Chittal, V., Gracias, M., Anu, A., Saha, P., & Rao, K. B. (2019). Biodecolorization and biodegradation of azo dye reactive orange-16 by marine Nocardiopsis sp. *Iranian Journal of Biotechnology, 17*(3), e1551. https://doi.org/10.29252/IJB.1551

Dell'Anno, F., Rastelli, E., Tangherlini, M., Corinaldesi, C., Sansone, C., Brunet, C., Balzano, S., Ianora, A., Musco, L., Montereali, M. R., & Dell'Anno, A. (2021). Highly contaminated marine sediments can host rare bacterial taxa potentially useful for bioremediation. *Frontiers in Microbiology, 12*, 584850. https://doi.org/10.3389/fmicb.2021.584850

Despotovic, D., Aharon, E., Trofimyuk, O., Dubovetskyi, A., Cherukuri, K. P., Ashani, Y., Leader, H., Castelli, A., Fumagalli, L., Savidor, A., Levin, Y., Longo, L. M., Segev, E., & Tawfik, D. S. (2021). Widespread utilization of diverse organophosphate pollutants by marine bacteria. *bioRxiv*. https://doi.org/10.1101/2021.10.18.464763

Duran, R., Bielen, A., Paradžik, T., Gassie, C., Pustijanac, E., Cagnon, C., Hamer, B., & Vujaklija, D. (2015). Exploring actinobacteria assemblages in coastal marine sediments under contrasted human influences in the West Istria Sea, Croatia. *Environmental Science and Pollution Research, 22*(20), 15215–15229. https://doi.org/10.1007/s11356-015-4240-1

Dutta, S., Adhikary, S., Bhattacharya, S., Roy, D., Chatterjee, S., Chakraborty, A., Banerjee, D., Ganguly, A., Nanda, S., & Rajak, P. (2024). Contamination of textile dyes in aquatic environment: Adverse impacts on aquatic ecosystem and human health, and its management using bioremediation. *Journal of Environmental Management, 353*, 120103.

Fernandes, C. F., da Silva Iúdice, T. N., Bezerra, N. V., & Pontes, A. N. (2025). Biodegradation of oil-derived hydrocarbons by marine actinobacteria: A systematic review. *Environmental Pollution, 367*, 125509. https://doi.org/10.1016/j.envpol.2024.125509

Godwin, U. A., & Inu, N. U. (2024). Effect of crude oil contamination on microbial community structure and urease activity in Coastal Plain Sands of Uyo, Akwa Ibom State, Nigeria. *African Journal of Environment and Natural Science Research*. https://doi.org/10.52589/ajensr-28grzz4k

Guo, W., Duan, J., Shi, Z., Yu, X., & Shao, Z. (2023). Biodegradation of PET by the membrane-anchored PET esterase from the marine bacterium *Rhodococcus pyridinivorans* P23. *Communications Biology, 6*(1), 1090. https://doi.org/10.1038/s42003-023-05470-1

Hamdan, A. M., Abd-El-Mageed, H., & Ghanem, N. (2021). Biological treatment of hazardous heavy metals by *Streptomyces rochei* ANH for sustainable water management in agriculture. *Scientific Reports, 11*, 9314. https://doi.org/10.1038/s41598-021-88843-y

Hou, D., Lin, Z., Wang, R., Ge, J., Wei, S., Xie, R., Wang, H., Wang, K., Hu, Y., Yang, X., Lu, L., & Tian, S. (2018). Cadmium exposure–*Sedum alfredii* planting interactions shape the bacterial community in the hyperaccumulator plant rhizosphere. *Applied and Environmental Microbiology, 84*(12), e02797–e02717. https://doi.org/10.1128/AEM.02797-17

Hsiao, T. H., Chen, Y. L., Meng, M., Chuang, M. R., Horinouchi, M., Hayashi, T., Wang, P. H., & Chiang, Y. R. (2021). Mechanistic and phylogenetic insights into actinobacteria-mediated oestrogen biodegradation in urban estuarine sediments. *Microbial Biotechnology, 14*(3), 1212–1227. https://doi.org/10.1111/1751-7915.13798

Hu, S., Xu, C., Lu, P., Wu, M., Chen, A., Zhang, M., Xie, Y., & Han, G. (2024). Widespread distribution of the DyP-carrying bacteria involved in the aflatoxin B1 biotransformation in Proteobacteria and Actinobacteria. *Journal of Hazardous Materials, 478*, 135493. https://doi.org/10.1016/j.jhazmat.2024.135493

James, G., Prasannan Geetha, P., Thavarool Puthiyedathu, S., & Vattringal Jayadradhan, R. K. (2023). Applications of actinobacteria in aquaculture: Prospects and challenges. *3 Biotech, 13*(2), 42. https://doi.org/10.1007/s13205-023-03465-7

Kapanga, P. M., Nyakairu, G. W. A., Nkanga, C. I., Lusamba, S. N., Tshimanga, R. M., & Shehu, Z. (2024). A review of dye effluents polluting African surface water: Sources, impacts, physicochemical properties, and treatment methods. *Discover Water, 4*(1), 85.

Li, G., Peng, L., Ding, Z., Liu, Y., Gu, Z., Zhang, L., & Shi, G. (2014). Decolorization and biodegradation of triphenylmethane dyes by a novel *Rhodococcus qingshengii* JB301 isolated from sawdust. *Annals of Microbiology, 64*(4), 1575–1586.

Lv, S., Li, Y., Zhao, S., & Shao, Z. (2024). Biodegradation of typical plastics: From microbial diversity to metabolic mechanisms. *International Journal of Molecular Sciences, 25*(1), 593. https://doi.org/10.3390/ijms25010593

Ma, J., Zhuang, Y., Wang, Y., Zhu, N., Wang, T., Xiao, H., & Chen, J. (2023). Update on new trend and progress of the mechanism of polycyclic aromatic hydrocarbon biodegradation by *Rhodococcus*, based on the new understanding of relevant theories: A review.

Environmental Science and Pollution Research, 30(41), 93345–93362. https://doi.org/10.1007/s11356-023-28894-y
Mahlouji, M., Alavi, S. M. H., Ghasemi, J. B., Jalili, A. H., Mozanzadeh, M. T., Zhang, S., Shazada, N. E., Butts, I. A. E., Hoseinifar, S. H., & Linhart, O. (2024). Crude oil–induced reproductive disorders in male goldfish: Testicular histopathology, sex steroid hormones, and sperm swimming kinematics. *Journal of Applied Toxicology*. https://doi.org/10.1002/jat.4745/
Makarani, N., & Kaushal, R. S. (2025). Advances in actinobacteria-based bioremediation: Mechanistic insights, genetic regulation, and emerging technologies. *Biodegradation, 36*(2), 24. https://doi.org/10.1007/s10532-025-10118-4
Moopantakath, J., & Kumavath, R. (2018). *Bio-augmentation of actinobacteria and their role in dye decolorization* (pp. 297–304). Elsevier. https://doi.org/10.1016/B978-0-444-63994-3.00020-5
Oborie, E., & Opukumo, A. W. (2024). Effect of crude oil contamination on compaction and consistency of some clay soils from parts of Yenagoa, Bayelsa State, Nigeria. *Journal of Civil Construction and Environmental Engineering*. https://doi.org/10.11648/j.jccee.20240901.13
Ogidi, O. I., & Akpan, U. M. (2024). Microbial and phytoremediation of crude oil-contaminated soil. In *Eco-restoration of polluted environment* (pp. 110–128). https://doi.org/10.1201/9781003423393-6
Olajuyigbe, F. M., & Ehiosun, K. I. (2016). Assessment of crude oil degradation efficiency of newly isolated actinobacteria reveals untapped bioremediation potentials. *Bioremediation Journal, 20*(2), 133–143. https://doi.org/10.1080/10889868.2015.1113926
Oliveira, J., Almeida, P. L., Sobral, R. G., Lourenço, N. D., & Gaudêncio, S. P. (2022). Marine-derived actinomycetes: Biodegradation of plastics and formation of PHA bioplastics—A circular bioeconomy approach. *Marine Drugs, 20*(12), 760. https://doi.org/10.3390/md20120760
Ordu-Pac, O. P., & Bosede, A. A. (2024). *Bioremediation potential of indigenous bacteria in pesticide-contaminated soils: A sustainable solution for agriculture*. Preprint. https://doi.org/10.2196/preprints.70299
Peng, Q., Zheng, H., Xu, H., Cheng, S., Yu, C., Wu, J., Meng, K., & Xie, G. (2024). Response of soil fungi to textile dye contamination. *Environmental Pollution, 359*, 124577.
Pidgorskyi, V. S., & Nogina, T. M. (2016). Biodegradation of petroleum hydrocarbons by actinobacteria and acinetobacteria strains producing biosurfactant. *Mikrobiolohichnyĭ Zhurnal, 78*(6), 92–103. https://doi.org/10.15407/MICROBIOLJ78.06.092
Preena, P. G., Achuthan, C., Kumar, V. J. R., Boobal, R., Deepa, G. D., Puthumana, J., Poulose, S., Surekhamol, I. S., & Singh, I. S. B. (2019). Community composition of marine and brackish water ammonia-oxidizing consortia developed for aquaculture application. *Water Science and Technology, 79*(5), 1017–1028. https://doi.org/10.2166/wst.2019.029
Rathore, D. S., Sheikh, M., & Singh, S. P. (2021). *Marine actinobacteria: New horizons in bioremediation* (pp. 425–449). Springer. https://doi.org/10.1007/978-981-15-4439-2_20
Saez, J. M., González, S. K., Ocante, T. A. L., Bigliardo, A. L., Briceño, G. E., & Benimeli, C. S. (2022). Actinobacteria bioaugmentation and substrate evaluation for biobeds useful for the treatment of atrazine residues in agricultural fields. *Journal of Environmental Management, 320*, 115870. https://doi.org/10.1016/j.jenvman.2022.115870
Sehrawat, A., Phour, M., Kumar, R., & Sindhu, S. S. (2021). *Bioremediation of pesticides: An eco-friendly approach for environment sustainability* (pp. 23–84). Springer. https://doi.org/10.1007/978-981-15-7447-4_2
Shertate, R. S., & Thorat, P. R. (2015). Biodegradation and detoxification of a textile azo dye reactive blue 171 by Marinobacter sp. NB-8. *Indian Journal of Geo-Marine Sciences, 44*(10), 1604–1607. http://nopr.niscair.res.in/bitstream/123456789/34987/1/IJMS%2044%2810%29%201604-1607.pdf
Singh, N. S., Sharma, R., Parween, T., & Patanjali, P. K. (2018). *Pesticide contamination and human health risk factor* (pp. 49–68). Springer. https://doi.org/10.1007/978-3-319-64501-8_3
Sivaperumal, P., Kamala, K., Subramanian, K., Aruni, W., Shanmugam, R., & Rajaram, R. (2022). Baseline assessment of marine actinobacterial diversity around the nuclear power plant sites,

India and its application to uranium remediation. *Environmental Research, 212*, 113135. https://doi.org/10.1016/j.envres.2022.113135

Soria-Camargo, C., Can-Ubando, L. C., Manzanares-Leal, G. L., Sánchez-Reyes, A., Dávila-Ramos, S., Batista-García, R. A., & Ramírez-Durán, N. (2025). Tolerance to NSAIDs in Actinobacteria from a Mexican volcano crater: Genomics and bioremediation potential. *Journal of Basic Microbiology, 65*(4), e2400772. https://doi.org/10.1002/jobm.202400772

Thorat, P. R., & Shertate, R. S. (2017). Decolorization and biodegradation of acid orange 63 by *Marinobacter excellens* AAO-10 in marine environmental conditions. *International Journal of Pharma and Bio Sciences, 2*(1), 7–16.

Uneze, D. P., Kugbenu, G. J., & Obire, O. (2024). Influence of pesticides contamination on microbial population of selected farmlands. *British Journal of Environmental Sciences, 12*(5), 14–56. https://doi.org/10.37745/bjes.2013/vol12n51456

Vieira, G. A. L., Cabral, L., Otero, I. V. R., Ferro, M., de Faria, A. U., de Oliveira, V. M., Bacci, M., & Sette, L. D. (2021). Marine associated microbial consortium applied to RBBR textile dye detoxification and decolorization: Combined approach and metatranscriptomic analysis. *Chemosphere, 267*, 129190.

Yao, K. S., Coulibaly, S., Atse, B. C., Trokourey, A., & Tidou, A. S. (2019). Evaluation of the risk of contamination by pesticides. *African Journal of Applied/General Agriculture, 8*(1), 73–81. https://doi.org/10.4314/AGA.V8I1

Marine Actinobacteria-Derived Enzymes: Sustainable Production From Waste Materials and Industrial Applications

Vaishali R. Majithiya, Nishtha R. Vaghela, Sangeeta D. Gohel, and Satya P. Singh

Abstract

Oceanic microorganisms, particularly marine Actinobacteria such as *Nocardiopsis*, *Streptomyces*, *Micromonospora*, and *Salinospora*, are underexplored for their enzymes. Studies have established that they possess a range of enzymes with thermostability, haloalkalitolerance, and pressure resistance, making them highly suitable for industrial applications. These microorganisms produce diverse enzymes, including proteases, cellulases, amylases, and lipases, and they can function under extreme conditions. Furthermore, using agro-industrial waste, such as rice bran, sugarcane bagasse, and fish processing residues, as substrate for growth and enzyme production projects a sustainable and cost-effective solution. Rich in organic compounds, these materials are ideal substrates for microbial growth, supporting waste valorization. The optimization of fermentation conditions of temperature, pH, and substrate concentration in solid-state fermentation and submerged fermentation is essential to maximize enzyme production. Statistical approaches like response surface methodology have improved yield and efficiency, reducing enzyme production costs and promoting environmental sustainability. Enzyme production by marine actinobacteria using agro-industrial waste offers a potential green pathway of an eco-friendly and economically viable solution for industrial processes, aiding ecological conservation and waste management efforts.

V. R. Majithiya
Department of Biosciences, Saurashtra University, Rajkot, Gujarat, India

Faculty of Science, Atmiya University, Rajkot, Gujarat, India

N. R. Vaghela · S. D. Gohel (✉) · S. P. Singh (✉)
Department of Biosciences, Saurashtra University, Rajkot, Gujarat, India
e-mail: sdgohel@sauuni.ac.in

K. Loganathan, Z. Li (eds.), *Marine Actinomycetes*,
https://doi.org/10.1007/978-981-95-9849-6_11

Keywords

Marine actinomycetes · Green biotechnology · Extremozymes · Agro-industrial waste

1 Introduction

Actinomycetes are Gram-positive, high G + C-containing, sporulating bacteria, exhibiting a morphology ranging from simple rods and cocci to complex, branched filamentous structures (mycelia) that produce various spores. The morphological diversity and morphogenesis are key aspects of their characteristics. Their ecological habitats include soil, fresh water, black water, lakes, compost, sewage, and the marine environment. They are not only present in ambient seawater but also associated with marine hosts, including invertebrates, vertebrates, seaweeds, algae, and mangroves (Vaghela & Gohel, 2023; Azman et al., 2015). Actinobacteria play a vital role in nutrient cycling, breaking down organic matter and fixing nitrogen in their environments. They possess potent defense mechanisms by producing diverse bioactive compounds, acting as natural antibiotics and protective agents in their habitats and associated hosts. In plant tissues, they can survive as endophytes, contributing to nutrient assimilation and growth. Actinobacteria are present on the exterior or interior of insects and protect the host and its nutritional resources against pathogens. They can survive in extreme environments like the deep ocean and drive nutrient cycling under harsh conditions (Singh & Reddy, 2014). Actinomycetes have adapted to marine environments due to unique genetic and metabolic traits. Haloalkaliphilic, thermophilic, and pressure-resistant actinomycetes are a rich source of enzymes of superior properties (Atif et al., 2024; Gohel et al., 2023; Kumar et al., 2022; Chauhan et al., 2021; Rathore et al., 2021; Abdukhakimova et al., 2021; Shukla & Singh, 2016; Thakrar et al., 2016; Gohel & Singh, 2012a, b).

These adaptations not only allow them to survive but also to outcompete other microbes in harsh marine conditions. Their ecological importance lies in maintaining the microbial balance and promoting symbiotic relationships with marine invertebrates, thereby influencing the health and productivity of marine ecosystems (Majithiya & Gohel, 2022; Alouadi et al., 2022). Actinobacteria are a potential source of natural antibiotics and various anticancer, antihelminthic, antifungal, and immunosuppressive drugs. The notable anti-infective compounds include streptomycin, tetracyclines, chloramphenicol, neomycin, erythromycin, vancomycin, kanamycin, rifamycin, gentamycin, and daptomycin (Mahajan & Balachandran, 2012). The extracellular enzymes, such as amylase, protease, gelatinase, cellulase, chitinase, pectinase, lipase, and asparaginase, are produced by actinomycetes under the extremely competitive environment (Rathore et al., 2021; Gohel et al., 2015; Gohel & Singh, 2013). The enzymes from marine actinomycetes function and maintain stability under extreme conditions, such as alkaline pH, salinity, organic solvents, and high temperatures (Silveira et al., 2023; Yaradoddi & Kontro, 2022; Vijayakumar, 2021; Salwan & Sharma, 2020; Thakrar et al., 2018; Dangar et al., 2018; Sharma & Singh, 2016; Gohel & Singh, 2015; Sharma et al., 2012).

2 Enzymatic Potential of Marine Actinobacteria

Marine Actinobacteria, particularly the genera *Streptomyces*, *Micromonospora*, *Salinispora*, *Nocardiopsis*, and *Actinokineospora*, are prolific producers of a diverse array of industrially significant enzymes (Majithiya et al., 2025; Castro-Falcón et al., 2025; Rathore et al., 2022; Chen et al., 2021). These organisms can survive in extreme oceanic environments and produce enzymes with unique properties of halotolerance, thermostability, and activity under pH extremes. Such extremozymes are highly desirable in various industrial processes such as detergent formulation, development of enzyme-based molecular techniques, food processing, and so on (Salas-Bruggink et al., 2024; Bhatt et al., 2024). Proteases are among the most extensively studied enzymes, widely applicable in detergents, leather processing, and the food industry (Vaghela & Gohel, 2025b; Ashraf et al., 2023; Matter et al., 2023; Thakrar et al., 2023; Mahnashi et al., 2022; Dwivedi et al., 2021; Chauhan et al., 2021; Thakrar & Singh, 2019; Rathore et al., 2019; Gohel & Singh, 2018a, b).

For example, *Streptomyces radiopugnans* VITSD8, associated with the marine brown tube sponge *Agelas conifera,* produces a novel fibrinolytic protease with improved fibrinolysis activity and stability compared to streptokinase (Dhamodharan, 2019). A protease from *Nocardiopsis dassonvillei* strain VCS-4 associated with seaweed *Caulerpa scalpeliformis* was highly stable at 80 °C and effective for the removal of blood stain (Majithiya et al., 2025; Majithiya & Gohel, 2025). Keratinases, a specific class of proteases, degrade various insoluble keratin substrates. Notably, *Streptomyces* sp. G11C from the Chilean marine actinobacterial culture collection exhibited the highest keratinolytic activity, achieving complete feather degradation in liquid culture (Valencia et al., 2021). Furthermore, a keratinase-producing *Nocardia* sp., isolated from the Thoothukudi shoreline in Tamil Nadu, India, possessed significant antimicrobial activity against *Streptococcus mutans*, *Salmonella* typhi, and *Klebsiella pneumoniae* (Subash et al., 2024). Amylases are another significant enzyme widely applied in the detergent, pharmaceutical, and food industries. Marine *Streptomyces rochei* KR108310 isolated from the Gulf of Suez produced a thermo-halotolerant α-amylase which maintained stability at 70 °C for 40 min in 1 M NaCl. Moreover, the enzyme was highly stable in Cu^{+2}, Zn^{+2}, and Fe^{+2} (Al-Agamy et al., 2021). Beyond proteases and amylases, marine actinobacteria are also known to produce chitinases, cellulases, and xylanases, having potential applications in agriculture, bioconversion, and the paper and textile industries. In this context, the genomic analysis of *Glycomyces harbinensis* JCM7347, *Cellulomonas cellulans* YCWD3, *Curtobacterium luteum* JCM1480, *Promicromonospora sukumoe* JCM6845, *Actinoplanes brasiliensis* JCM3196, *Actinokineospora riparia* JCM747, and *Kibdelosporangium aridum* subsp. *aridum* JCM7912 revealed the presence of family 19 chitinase genes (Kawase et al., 2004). A moderately thermostable chitinase of *Streptomyces mutabilis* IMA8 hydrolyzed chitin extracted from *Charybdis smithii* (Rajendran et al., 2024). *Micromonospora aurantiaca*-derived chitinase was stable in 5 mM Mg^{2+} and 10 mM dithiothreitol and displayed optimal activity at pH 5.0 and 55 °C (Guo et al., 2024). Similarly, a xylanase derived from marine *Streptomyces viridochromogenes* exhibited the highest activity at 70 °C, retaining 80% of the activity at 60 °C for 1 h. Additionally, this enzyme was stable in alkaline pH, demonstrating its potential in the pulp and paper industry (Liu et al., 2013).

Cellulases degrade plant-derived hemicellulose and cellulose under saline conditions, potentially contributing to biofuel and paper production. For example, a lignocellulose-degrading *Streptomyces griseoaurantiacus*, isolated from coastal marine areas of Andhra Pradesh, India, possesses highly thermostable endoglucanase and β-glucosidase (Kumar, 2015). These enzymes exhibit catalytic efficiency under marine-mimicking conditions, capable of functioning under adverse industrial conditions.

Recently, focus has been on lesser-known yet industrially relevant enzymes (Ashaolu et al., 2025). Esterase encoding a 325-amino-acid protein (SAestA) belonging to the family IV of lipolytic enzymes was produced by the obligate marine *Salinispora arenicola* CNP193. It exhibited optimal activity at 30 °C and pH 9.0 (Fang et al., 2015). A lipase from marine *Streptomyces* sp. strain W007 was thermostable, retaining 80% of its initial activity after 1 h at 60 °C. Furthermore, it was also stable in Cu^{2+}, Ca^{2+}, Ni^{2+}, and Mg^{2+} ions (Yuan et al., 2016). Laccases from *Streptomyces* sp. are copper-containing oxidases employed in dye decolorization and environmental bioremediation (Kaur et al., 2022). A tyrosinase produced from a marine *Streptomyces espinosus* strain LK4 was suitable for removing phenolic compounds from aqueous solution (Roy et al., 2014). Phytase in livestock and fish feeds enhances the nutritive values, particularly in monogastric livestock and fish (Lamid et al., 2018). The phytases from *Streptomyces luteogriseus* are used in animal feed to increase phosphorus bioavailability and minimize eutrophication (Aly et al., 2015). Prenyltransferase from marine *Streptomyces* sp. CNQ-509 synthesizes bioactive compounds, such as prenylated flavonoids and alkaloids of pharmacological significance as antimicrobial, anticancer, and antioxidant agents (Zhao et al., 2016). Similarly, another enzyme, nitrile hydratase, is derived from *Rhodococcus erythropolis* from deep oceanic waters. This organism grows at 4 °C, with 0–4% NaCl, under 40–60 MPa pressure (Yadav et al., 2025; Brandao et al., 2003). Furthermore, amidases from the marine actinomycete, *Salinispora arenicola* CNS-205, display the potential for environmental remediation with substrate specificity to short-chain aliphatic amide substrates (acetamide, isobutyramide, and propenamide) (Yanling et al., 2018). Furthermore, haloperoxidases from *Marinactinospora thermotolerans* have been shown to possess potent antimicrobial and antifouling properties (Yang et al., 2025). L-asparaginase, a therapeutic enzyme with significant medical, industrial, and research relevance, is particularly known for its anticancer properties. L-asparaginase from *Pseudonocardia endophytica* VUK-10 from Nizampatnam mangrove sediments has a high activity with L-asparagine (Kiranmayi et al., 2014). Thus, a vast enzymatic arsenal establishes the potential of marine actinobacteria for future green biotechnological innovations.

3 Waste Materials as Substrates for Enzyme Production

The booming global enzyme market requires cost-effective alternatives to synthetic media for microorganisms to produce enzymes. In the wake of rapid urbanization, the reuse of waste materials generated from agriculture, food processing, and animal husbandry has gained significant attention. This aligns with the objectives of the Sustainable Development Goals (SDGs), particularly SDG 6 (Clean Water and Sanitation), SDG 9 (Industry, Innovation and Infrastructure), and SDG 12 (Responsible Consumption and Production). Waste from crop harvests, animal manure, kitchen waste, and by-products from food industries, including fruit and vegetable processing, oil extraction, and juice production, is effective for producing industrially significant enzymes. Generating these waste materials not only adversely affects production costs but also raises waste management and environmental issues. Agricultural residues, such as coffee peel, sugarcane bagasse, wheat bran, corn cobs, and rice husk, are produced in large quantities and serve as useful substrates for microbial growth. These residues can be profitably used, particularly in agriculture-dependent countries, by converting them into valuable products. The sources of energy and moisture present in these wastes are ideal for microbial growth. For instance, rice bran contains many nutrients such as carbohydrates (34–62%), lipids (20%), protein (11–15%), crude fiber (7–11%), and ash (7–10%) (Das et al., 2025). Sugarcane bagasse, a by-product of the sugarcane industry, contains cellulose (45%), hemicellulose (32%), and lignin (17%) with little ash. Rice bran and sugarcane bagasse have been used to produce amylase (Silva et al., 2025). Similarly, mango seeds, which contain a high concentration of oleic and stearic acids, are also a vital source of carbohydrates (58–80%), protein (6–13%), and lipids (6–16%) (Torres-León et al., 2016). An amylase was produced from the kernel of the mango using *Fusarium* species (Kumar et al., 2013). The marine waste, including the gills and fins of fish and crustacean shells, is protein-rich and can produce proteases and other enzymes (Ifuku et al., 2025; Liaqat & Eltem, 2018). *Stenotrophomonas maltophilia* produces chitinase with 6677.07 U/mg specific activity, using shrimp residues (Majengbasan et al., 2025). Recently, a protease-producing *Nocardiosis dosenvali* strain VCS-4 could grow and produce protease using agro-industrial waste (Majithiya & Gohel, 2025). The finely chopped straws of wheat, millet, sorghum, groundnut shells, groundnut cake, soybean meal, and gram flour were used to produce protease. Groundnut cakes emerged as the most effective substrate, yielding high levels of protease at 408 U/mL, followed by soybean meal with 106 U/mL. The enzyme production using agricultural and marine-based wastes offers a cost-effective and sustainable alternative to conventional methods. Table 1 represents enzyme production using waste substrates.

Table 1 Marine actinobacteria-derived enzymes and corresponding agro-industrial waste substrates used for enzyme production

Actinobacteria	Enzyme	Enzymatic condition/properties	Waste material used	References
Nocardiopsis dassonvillei strain VCS-4	Protease	Pmax of 408.42 U/mL, generation time of 9.74 h, specific growth rate of 0.361 g/mL/h, and Qp of 68.07 U/mL/h	Groundnut cake	Majithiya and Gohel (2025)
Streptomyces sp. AM9-01	Lipase	Lipase activity: 94.6 ± 1.6 U/mL	Residual babassu mesocarp	Junior et al. (2025)
Nocardiopsis synnemataformans NBRM9	Cellulase	Cellulase activity: 13.20 U/mL, optimal activity at 50 °C, pH 9.0, and 10% salinity, V_{max} of 20.19 ± 1.88 U/mL, and a K_m of 0.25 ± 0.07 mM	Bean straw	El-Sayed et al. (2024)
Nocardiopsis sp.	α-Amylase	pH 8.0, 28 °C, and 5% salt concentration	Potato peel	Dobariya et al. (2023)
Stenotrophomonas maltophilia	Chitinase	Chitinase activity: 5.01 U/mL, the stability of the enzyme was 95% at 45 °C for 168 h, optimum pH: 6.5	Shrimp shell waste	Subramanian et al. (2022)
Streptomyces speibonae TKU048 using	Protease	Protease activity: 38.36 ± 0.28 U/mL, pH:5.8, temperature: 70 °C	Tuna head powder proteinaceous wastes from fishery	Tran et al. (2022)
Streptomyces sp.	Pectinase, polygalacturonase, xylanase, cellulase	Pectinase: 4.24 ± 0.05 U/mL, polygalacturonase (PGase): 2.88 ± 0.05 U/mL, xylanase: 9.43 ± 0.16 U/mL, and cellulase: 1.62 ± 0.09 U/mL	Agro-waste mixture of orange peel, pomegranate peel, and pumpkin pulp and seeds	Shrestha et al. (2022)
Streptomyces thermocarboxydus TKU045	Xylanase	Temperature:0–70 °C and pH 5–8	Wheat bran	Tran et al. (2021)
Streptomyces sp. KP314280	Amylase	Amylase activity: 12.19 U/mL	Orange fruit peel waste	Ousaadi et al. (2021)
Streptomyces rochei subsp. *chromatogenes* NEAE-K	L-asparaginase	Enzyme activity: 175.96 U/gds	Wheat bran and soybean meal	El-Naggar and El-Shweihy (2020)

Streptomyces violascens OC125–8	Lipase	Optimum lipase activity at 40 °C and pH 8.0, V_{max} and K_m values were 0.61 μmol/min/mg and 0.259 mM, respectively	Highest activity toward olive oil (100%), sunflower oil (90%), and burned sunflower oil (55%), while corn oil (44%) and burned olive oil (15%)	Boran et al. (2019)
Promicromonospora sp. VP111	Cellulase (CMCase, FPase, and β-glucosidase)	Enzymatic activity of CMCase: 15.9 ± 1.3 U/g, FPase: 24.6 ± 0.5 U/g, β-glucosidase: 6.9 ± 0.0 U/g using wheat straw. Similarly enzymatic activity of CMCase: 24.2 ± 3.5 U/g, FPase: 28.3 ± 4.0 U/g, β-glucosidase: 10.5 ± 0.0 U/g using sugarcane bagasse	Wheat straw, sugarcane bagasse	Thomas et al. (2016)
Streptomyces sp. MSC702	Xylanase	Sugar cane bagasse: 9.32 IU/mL, wheat bran: 29.91 IU/mL, rice bran: 30.15 IU/mL, corn cob: 6.23 IU/mL, wheat straw: 28.40 IU/mL	Sugar cane bagasse, wheat bran, rice bran, corn cob, wheat straw	Singh et al. (2012)
Streptomyces sp. MSC702	Amylase	Sugar cane bagasse: 201.14 IU/mL, wheat bran: 261.48 IU/mL, rice bran: 373.89 IU/mL, corn cob: 368.87 IU/mL, wheat straw: 234.54 IU/mL	Sugar cane bagasse, wheat bran, rice bran, corn cob, wheat straw	Singh et al. (2012)
Streptomyces sp. *MDS*	Cellulase	Cellulase activity of 90.15 U/mL using CMC; 41.78 U/mL using rice straw; 6.42 U/mL using avicel; 1.85 U/mL using filter paper; 16.40 U/mL using wood straw)	Rice straw, avicel, wood straw	Saratale et al. (2012)
Nocardiopsis sp. KNU	Cellulase	Cellulase activity of 18.4 U/mL using CMC; 249 U/mL using xylan; 51.4 U/mL using rice straw; 2.47 U/mL using avicel	Rice straw and avicel	Saratale and Oh (2011)
Streptomyces viridosporus T7A	Xylanase	Xylanase production: 423.9 U/g	10% soybean meal, 20% Napier grass, and 70% sugarcane bagasse	Alberton et al. (2009)
Streptomyces sp. Ab106	Cellulase-free xylanase	Temperature: 50 °C and pH: 7.0	Sugarcane bagasse	Techapun et al. (2002)

4 Fermentation and Bioprocess Optimization

Fermentation and bioprocess optimization are essential for optimum enzyme production from marine microorganisms, especially when utilizing sustainable waste materials as substrates. Solid-state fermentation (SSF) and submerged fermentation (SmF) are extensively used as primary approaches. SSF, which employs low-moisture agro-industrial or seafood processing wastes, often results in higher yields of enzymes by better mimicking the natural, solid marine environment. For instance, *Streptomyces speibonae* strain TKU048 produced protease using fishery waste (Tran et al., 2022), whereas *Streptomyces* sp. MAB18 used Chicken feather and peptone as a substrate under SSF conditions (Manivasagan et al., 2013). Both SmF and SSF require optimization of environmental conditions for enzyme production. The variables influencing the process are pH, dissolved oxygen, temperature, agitation/static condition, humidity, incubation time, micro- and macronutrients, and proper selection of substrates, whether refined chemicals or crude waste materials (Ferreira et al., 2007). Traditionally, these parameters have been optimized using the one-variable-at-a-time method, which involves altering a single factor while keeping others constant. Although straightforward and simple, this approach requires many experimental sets and is time-consuming (Rathore & Singh, 2021). More effective statistical methods, such as response surface methodology (RSM), have recently been widely adopted. Designs such as the Box-Behnken design (BBD), central composite design (CCD), and the Taguchi method are commonly used, often with specialized software. Such approaches significantly reduce the number of experimental sets to identify and pinpoint the most effective combination of the variables to maximize enzyme production (Vaghela & Gohel, 2023; Dobariya et al., 2023; Shukla & Singh, 2023; Thakrar et al., 2023; Kikani et al., 2023; Szpisják-Gulyás et al., 2023; Dwivedi et al., 2021; Chauhan et al., 2021). The production of chitinase by a mangrove-associated *Streptomyces olivaceus* (MSU3) under SmF was optimized by the statistical design of the BBD (Sanjivkumar et al., 2020). The optimized conditions included NaCl concentration (3.05% w/v), pH 6.5, temperature 47 °C, and an inoculum size of 2.75%, with an incubation period of 90 h. The model demonstrated a high coefficient of determination ($R^2 = 0.9680$) and was statistically highly significant ($P < 0.0001$) (Sanjivkumar et al., 2020). Similarly, xylanase production was optimized using the BBD by a mangrove-associated actinobacterium, *Streptomyces variabilis* (MAB3). Initially, xylanase production by the selected strain was carried out through SmF using birchwood xylan as the substrate. Subsequent statistical optimization via BBD resulted in a 5.30-fold increase in xylanase production compared to the conventional one-factor-at-a-time approach. The optimized conditions involved a basal medium containing birchwood xylan (2.0% w/v) at pH 8.2, a temperature of 46.5 °C, an inoculum size of 2%, and an incubation period of 68 h. Analysis of variance revealed a high coefficient of determination ($R^2 = 0.9490$) for the respective responses, with a highly significant model ($P < 0.0001$) (Sanjivkumar et al., 2018). Enzymes from marine actinobacteria are sensitive to osmotic pressure. Low-cost organic nitrogen sources such as shrimp, shell powder, and gelatin have been found to enhance the production of chitinases and proteases in marine-derived *Nocardiopsis* and *Streptomyces*, respectively (Wang et al., 2023; Majithiya & Gohel, 2025). Co-production of amylase and protease by *Streptomyces lopnurensis* KaM5 using a two-phase statistical approach was investigated (Rathore & Singh, 2021). The optimized medium comprised starch (2.0%), yeast extract (1.5%), gelatin (0.5%), peptone (0.5%), and NaCl (6.5%), at pH 9. Furthermore, the production of an α-amylase in *Bacillus licheniformis* TSI-14 was optimized using RSM and CCD (Shukla & Singh, 2015). The highest α-amylase production was obtained at 50 °C and pH 7.0, in 2 days of growth.

When scaling up from shake flasks to larger bioreactors, maintaining parameters such as oxygen transfer rates and foam management becomes essential to preserve cell viability and enzyme activity. Furthermore, advanced strategies like fed-batch fermentation have successfully boosted marine enzyme yields, offering a promising approach for sustainable industrial applications (Zou et al., 2024).

5 Applications of Marine Actinobacterial Enzymes

Enzymes are in the center of metabolism, affecting various processes of life. Recently, enzymes have gained significant attention in diverse industrial processes due to their specific and rapid action, substrate-to-product bioconversion efficiency, and ability to conserve raw materials, energy, and chemicals compared to conventional manufacturing processes. Enzymes can provide environmentally friendly (green) alternatives to hazardous chemicals, thus supporting the shift toward sustainability in industrial practices (Victorino da Silva Amatto et al., 2022). The enzymes from marine actinobacteria, in particular, have high potential in industrial applications due to their unique properties of salt tolerance, thermostability, and substrate versatility (Gohel et al., 2023; Salwan & Sharma, 2018; Singh et al., 2013). Such enzymes include proteases, amylases, cellulases, lipases, keratinase, asparaginase, and chitinases, which are used in a wide range of industries, such as food, detergent, pharmaceuticals, and bioenergy production (Vaghela & Gohel, 2023; Gohel et al., 2023; Shukla & Singh, 2023; Thakrar et al., 2023; de Souza et al., 2022; Majithiya & Gohel, 2022; Ramesh et al., 2020). Additionally, actinomycetes from extreme environments produce enzymes with novel properties of high thermal and salt stability, variable pH range, substrate specificity, and resistance to organic solvents. These enzymatic features are crucial for high production and meeting industrial requirements for commercialization and societal benefit (Salwan & Sharma, 2018). Xylanases, particularly endo-xylanases and β-xylosidases, are of high industrial interest as they hydrolyze xylan, a major component of hemicellulose (Bajpai, 2022). Marine actinobacterial enzymes are helpful in the food industry to process and improve the texture of food products, such as cheese, meat, and fish, by breaking down proteins and starches (Farooq et al., 2021). The marine actinobacterial enzymes, especially L-asparaginase, which has anti-carcinogenic properties, have pharmaceutical significance. L-asparaginase inhibits cancer cell growth by cleaving L-asparagine into ammonia and aspartic acid. While fungi mainly produce L-asparaginase, some actinobacterial species have also been reported to produce

Table 2 Enzyme-producing companies and their enzyme products

Company name	Enzymes produced	Website
Novozymes A/S (Denmark)	>700 kinds of enzymes including amylases, proteases, lipases, cellulases, xylanases, lactases, phytases	www.novozymes.com
DuPont (USA)	Cellulases, xylanases, amylases, proteases, phytases, lipases	www.dupont.com
DSM (Netherlands)	Lipases, amylases, proteases, lactases, cellulases	ww.dsm-firmenich.com
Amano Enzyme Inc. (Japan)	Lipases, proteases, amylases, glucosidases, esterases	www.amano-enzyme.com
EnzymeWorks Inc. (USA/China)	Lipases, oxidases, dehydrogenases, hydrolases	www.enzymeworking.com
BASF (Germany)	Phytases (Natuphos® E), xylanases and glucanases (Natugrain® TS), mannanases (Natupulse® TS)	www.nutrition.basf.com

this enzyme (Majithiya & Gohel, 2022; Chand et al., 2020). The hydrolytic enzymes such as amylases, proteases, and lipases are used in detergents, especially in high-salinity conditions (Vaghela & Gohel, 2025a). Furthermore, marine Actinobacterial enzymes are also vital in bioremediation as they degrade marine pollutants, including oils, plastics, and heavy metals, thus playing a crucial role in environmental cleanup. In bioenergy, these enzymes are vital for converting lignocellulosic biomass into biofuels by breaking down complex polysaccharides (Lewin et al., 2016). Due to their ability to function under extreme conditions such as high salinity, temperature, and pH, marine actinobacterial enzymes provide sustainable and cost-effective solutions for industrial processes. Table 2 includes the commercial production of enzymes by different companies.

6 Conclusion

Marine actinomycetes are a distinctive reservoir of enzymes. Their adaptation to harsh oceanic environments facilitates the production of novel extremozymes exhibiting characteristics such as thermostability and haloalkalitolerance. This chapter demonstrates their enormous enzymatic potential in numerous industrial applications. Moreover, this chapter highlights sustainable enzyme production, utilizing agro-industrial waste as substrates, supporting the global SDGs. Increased enzyme yields and commercial viability depend on optimized fermentation processes facilitated by the statistical designs, such as RSM. This green catalytic approach provides environmentally friendly substitutes for traditional chemical processes in various industries, including food processing, pharmaceutical, and bioremediation. Overall, marine actinomycetes have significant potential in the green biotechnological arena.

Acknowledgments VRM and NRV acknowledge the ScHeme Of Developing High quality research (SHODH) Fellowships from the Government of Gujarat. SDG gratefully acknowledges funding received from the University Grants Commission - Basic Scientific Research (UGC-BSR) Start up Grant, the Early Career Research Grant from Department of Science and Technology - Science and Engineering Research Board (DST-SERB), Government of India, and the Seed Money project from Internal Quality Assurance Cell (IQAC), Saurashtra University. SPS acknowledges the Department of Biotechnology (DBT) Multi-Institutional Project, Ministry of Earth Sciences (MoES) Networking Project, UGC-BSR Faculty Fellowship, and International Travel Fellowships from DST-SERB, Council of Scientific and Industrial Research (CSIR), DBT, and UGC. Facilities created under the University Grants Commission - Centre for Advanced Study (UGC-CAS) and Department of Science and Technology - Fund for Improvement of S & T Infrastructure (DST-FIST) are duly acknowledged. We are grateful to Saurashtra University for the necessary infrastructure and support.

References

Abdukhakimova, D., Markhametova, Z., Shamkeeva, S., Zhulamanova, A., Trenozhnikova, L., Berezin, V., & Azizan, A. (2021). Characterizationharacterization of extremophilic actinomycetes strains as sources of antimicrobial agents. In C. Barreiro & J. L. Barredo (Eds.), *Antimicrobial therapies: Methods and protocols* (pp. 59–75). Springer US.

Al-Agamy, M. H., Alhuzani, M. R., Kelany, M. S., & Hamed, M. M. (2021). Production and partial characterization of α-amylase enzyme from marine actinomycetes. *BioMed Research International, 2021*(1), 5289848.

Alberton, L. R., Vandenberghe, L. P. D. S., Assmann, R., Fendrich, R. C., Rodriguéz-León, J., & Soccol, C. R. (2009). Xylanase production by *Streptomyces viridosporus* T7A in submerged and solid-state fermentation using agro-industrial residues. *Brazilian Archives of Biology and Technology, 52*, 171–180.

Alouadi, Y., Riahi, A., Ismail, A., Cherif, W., Mraouna, R., & Bour, M. E. (2022). Antimicrobial activity and enzymatic production of aquatic invertebrate-associated bacteria. In *Euro-Mediterranean conference for environmental integration* (pp. 285–290). Springer.

Aly, M. M., Sanaa Tork, S. T., Al-Garni, S. M., & Kabli, S. A. (2015). Production and characterization of phytase from *Streptomyces luteogriseus* R10 isolated from decaying wood samples. *International Journal of Agriculture and Biology, 17*(3), 515–522.

Ashaolu, T. J., Malik, T., Soni, R., Prieto, M. A., & Jafari, S. M. (2025). Extremophilic microorganisms as a source of emerging enzymes for the food industry: A review. *Food Science & Nutrition, 13*(1), e4540.

Ashraf, M., Hussain, N., Baqar, Z., Kumar, A., Ferreira, L. F. R., & Iqbal, H. M. (2023). Bioprospecting microbial proteases in various industries/sectors. In H. M. Iqbal (Ed.), *Microbial biomolecules* (pp. 301–324). Academic.

Atif, F., Maqsood, N., Ali, W., Ali, W., & Irfan, M. (2024). Extremophiles and their enzymatic diversity and biotechnological potential. *Systems Microbiology and Biomanufacturing, 4*(3), 833–849.

Azman, A. S., Othman, I., Velu, S. S., Chan, K. G., & Lee, L. H. (2015). Mangrove rare actinobacteria: Taxonomy, natural compound, and discovery of bioactivity. *Frontiers in Microbiology, 6*, 856.

Bajpai, P. (2022). Chapter 5—Sources, production, and classification of xylanases. In *Microbial xylanolytic enzymes. Progress in biochemistry and biotechnology* (pp. 69–97). https://doi.org/10.1016/B978-0-323-99636-5.00003-8

Bhatt, H. B., Sani, R. K., Amoozegar, M. A., & Singh, S. P. (2024). Editorial: Extremozymes: Characteristics, structure, protein engineering and applications. *Frontiers in Microbiology, 15*, 1423463. https://doi.org/10.3389/fmicb.2024.1423463

Boran, R., Ugur, A., Sarac, N., & Ceylan, O. (2019). Characterisation of *Streptomyces violascens* OC125-8 lipase for oily wastewater treatment. *3 Biotech, 9*(1), 5.

Brandao, P. F., Clapp, J. P., & Bull, A. T. (2003). Diversity of nitrile hydratase and amidase enzyme genes in *Rhodococcus erythropolis* recovered from geographically distinct habitats. *Applied and Environmental Microbiology, 69*(10), 5754–5766.

Castro-Falcón, G., Guillén-Matus, D. G., Silva, E. B. D., Guo, W., Ross, A., Sá Magalhães Serafim, M., et al. (2025). Structure elucidation, biosynthetic gene cluster distribution, and biological activities of ketomemicin analogs in Salinispora. *Marine Drugs, 23*(3), 126.

Chand, S., Mahajan, R. V., Prasad, J. P., Sahoo, D. K., Mihooliya, K. N., Dhar, M. S., et al. (2020). A comprehensive review on microbial l-asparaginase: Bioprocessing, characterization, and industrial applications. *Biotechnology and Applied Biochemistry, 67*(4), 619–647. https://doi.org/10.1002/bab.1888

Chauhan, J. V., Mathukiya, R., Singh, S. P., & Gohel, S. D. (2021). Two steps purification, biochemical characterization, thermodynamics and structure elucidation of thermostable alkaline serine protease from *Nocardiopsis alba* strain OM-5. *International Journal of Biological Macromolecules, 169*, 39–50. https://doi.org/10.1016/j.ijbiomac.2020.12.061

Chen, J., Xu, L., Zhou, Y., & Han, B. (2021). Natural products from actinomycetes associated with marine organisms. *Marine Drugs, 19*(11), 629.

Dangar, K. G., Kalasava, A. B., Dave, A. V., & Singh, S. P. (2018). Molecular diversity of *Nocardiopsis alba* sp. isolated from the coastal region of Gujarat, India. *Journal of Cell & Tissue Research, 18*(3), 6559–6570.

Das, P. P., Gul, M. Z., Weber, A. M., Srivastava, R. K., Marathi, B., Ryan, E. P., & Ghazi, I. A. (2025). Rice bran extraction and stabilization methods for nutrient and phytochemical biofortification, nutraceutical development, and dietary supplementation. *Nutrition Reviews, 83*(4), 692–712.

de Souza, H. F., Borges, L. A., Gonçalves, V. D. D. P., Dos Santos, J. V., Bessa, M. S., Carosia, M. F., et al. (2022). Recent advances in the application of xylanases in the food industry and production by actinobacteria: A review. *Food Research International, 162*, 112103.

Dhamodharan, D. (2019). Novel fibrinolytic protease producing *Streptomyces radiopugnans* VITSD8 from marine sponges. *Marine Drugs, 17*(3), 164.

Dobariya, A., Mankad, G. P., Ramavat, H., & Singh, S. P. (2023). Efficacy of the fruit and vegetable peels as substrates for the growth and production of α-amylases in marine actinobacteria. *Applied Biochemistry and Biotechnology, 195*(12), 7603–7623.

Dwivedi, P., Sharma, A. K., & Singh, S. P. (2021). Biochemical properties and repression studies of an alkaline serine protease from a haloalkaliphilic actinomycete, *Nocarpdiopsis dassonvillei* subsp. *albirubida* OK-14. *Biocatalysis and Agricultural Biotechnology, 35*, 1–7. https://doi.org/10.1016/j.bcab.2021.102059

El-Naggar, N. E. A., & El-Shweihy, N. M. (2020). Bioprocess development for L-asparaginase production by *Streptomyces rochei*, purification and in-vitro efficacy against various human carcinoma cell lines. *Scientific Reports, 10*(1), 7942.

El-Sayed, M. H., Elsayed, D. A., & Gomaa, A. E. R. F. (2024). *Nocardiopsis synnemataformans* NBRM9, an extremophilic actinomycete producing extremozyme cellulase, using lignocellulosic agro-wastes and its biotechnological applications. *AIMS Microbiology, 10*(1), 187.

Fang, Y., Wang, S., Liu, S., & Jiao, Y. (2015). Discovery a novel organic solvent tolerant esterase from *Salinispora arenicola* CNP193 through genome mining. *International Journal of Biological Macromolecules, 80*, 334–340.

Farooq, M. A., Ali, S., Hassan, A., Tahir, H. M., Mumtaz, S., & Mumtaz, S. (2021). Biosynthesis and industrial applications of α-amylase: A review. *Archives of Microbiology, 203*, 1281–1292.

Ferreira, S. L. C., Bruns, R. E., da Silva, E. G. P., Dos Santos, W. N. L., Quintella, C. M., David, J. M., et al. (2007). Statistical designs and response surface techniques for the optimization of chromatographic systems. *Journal of Chromatography A, 1158*(1–2), 2–14.

Gohel, S. D., & Singh, S. P. (2012a). Cloning and expression of alkaline protease genes from two salt-tolerant alkaliphilic actinomycetes in *E. coli*. *International Journal of Biological Macromolecules, 50*(3), 664–671. https://doi.org/10.1016/j.ijbiomac.2012.01.039

Gohel, S. D., & Singh, S. P. (2012b). Single step purification, characteristics and thermodynamic analysis of a highly thermostable alkaline protease from a salt-tolerant alkaliphilic actinomycete, *Nocardiopsis alba* OK-5. *Journal of Chromatography B, 889–890*, 61–68. https://doi.org/10.1016/j.jchromb.2012.01.031

Gohel, S. D., & Singh, S. P. (2013). Characteristics and thermodynamics of a thermostable protease from a salt-tolerant alkaliphilic actinomycete. *International Journal of Biological Macromolecules, 56*, 20–27.

Gohel, S., & Singh, S. P. (2015). Thermodynamics of a Ca2+ dependent highly thermostable alkaline protease from a haloalkliphilic actinomycete. *International Journal of Biological Macromolecules, 74*, 421–429. https://doi.org/10.1016/j.ijbiomac.2014.08.008

Gohel, S. D., & Singh, S. P. (2018a). Thermodynamics of a Ca2+ dependent, highly thermostable and detergent compatible purified alkaline serine protease from *Nocardiopsis xinjiangensis* strain OM-6. *International Journal of Biological Macromolecules, 113*, 565–574.

Gohel, S. D., & Singh, S. P. (2018b). Molecular phylogeny and diversity of the salt-tolerant alkaliphilic actinobacteria inhabiting Coastal Gujarat, India. *Geomicrobiology Journal, 35*(9), 775–789. https://doi.org/10.1080/01490451.2018.1471107

Gohel, S. D., Sharma, A. K., Dangar, K. G., Thakrar, F. J., & Singh, S. P. (2015). Antimicrobial and biocatalytic potential of haloalkaliphilic actinobacteria. In *Halophiles: Biodiversity and sustainable exploitation* (pp. 29–55). Springer.

Gohel, S. D., Majithiya, V. R., & Singh, S. P. (2023). Genetic and physiological diversity of marine actinobacteria from the Okha Coast, Gujarat, India. *Geomicrobiology Journal, 40*(6), 590–604.

Guo, H. Z., Wang, D., Yang, H. T., Wu, Y. L., Li, Y. C., Xia, G. H., & Zhang, X. Y. (2024). Heterologous expression and characterization of a pH-stable chitinase from *Micromonospora aurantiaca* with a potential application in chitin degradation. *Marine Drugs, 22*(6), 287.

Ifuku, S., Kaminaka, H., & Shams, M. I. (2025). Nanochitin from crab shells: Production, chemical modification, composite materials, and physiological functions. *Macromolecular Rapid Communications, 46*(9), 2400765.

Junior, A. A., dos Anjos, T. N., Gutarra, M. L., Nascimento, R. P. D., & Itabaiana, I., Jr. (2025). Production of lipase from Streptomyces spp. AM9-01 by solid-state fermentation of residual babassu mesocarp and immobilization in Accurel® MP1000. *Reactions, 6*(1), 11.

Kaur, R., Salwan, R., & Sharma, V. (2022). Structural properties, genomic distribution of laccases from Streptomyces and their potential applications. *Process Biochemistry, 118*, 133–144.

Kawase, T., Saito, A., Sato, T., Kanai, R., Fujii, T., Nikaidou, N., et al. (2004). Distribution and phylogenetic analysis of family 19 chitinases in actinobacteria. *Applied and Environmental Microbiology, 70*(2), 1135–1144.

Kikani, B. A., Patel, R. K., Thumar, J. T., Bhatt, H. B., Rathore, D., Koladiya, G., & Singh, S. P. (2023). Solvent tolerant enzymes in extremophiles: Adaptations and applications. *International Journal of Biological Macromolecules*. Advance online publication. https://doi.org/10.1016/j.ijbiomac.2023.124051

Kiranmayi, M. U., Poda, S., & Vijayalakshmi, M. (2014). Production and optimization of L-asparaginase by an actinobacterium isolated from Nizampatnam mangrove ecosystem. *Journal of Environmental Biology, 35*(5), 799.

Kumar, A. K. (2015). UV mutagenesis treatment for improved production of endoglucanase and β-glucosidase from newly isolated thermotolerant actinomycetes, *Streptomyces griseoaurantiacus*. *Bioresources and Bioprocessing, 2*, 1–10.

Kumar, D., Yadav, K. K., Muthukumar, M., & Garg, N. (2013). Production and characterization of [alpha]-amylase from mango kernel by *Fusarium solani* NAIMCC-F-02956 using submerged fermentation. *Journal of Environmental Biology, 34*(6), 1053.

Kumar, A., Singh, A. K., Bilal, M., & Chandra, R. (2022). Extremophilic ligninolytic enzymes: Versatile biocatalytic tools with impressive biotechnological potential. *Catalysis Letters, 152*(8), 2302–2326.

Lamid, M., Al-Arif, A., Asmarani, O., & Warsito, S. H. (2018, April). Characterization of phytase enzymes as feed additive for poultry and feed. In *IOP conference series: Earth and environmental science* (Vol. 137, No. 1, p. 012009). IOP Publishing.

Lewin, G. R., Carlos, C., Chevrette, M. G., Horn, H. A., McDonald, B. R., Stankey, R. J., et al. (2016). Evolution and ecology of actinobacteria and their bioenergy applications. *Annual Review of Microbiology, 70*(1), 235–254.

Liaqat, F., & Eltem, R. (2018). Chitooligosaccharides and their biological activities: A comprehensive review. *Carbohydrate Polymers, 184*, 243–259.

Liu, Z., Zhao, X., & Bai, F. (2013). Production of xylanase by an alkaline-tolerant marine-derived *Streptomyces viridochromogenes* strain and improvement by ribosome engineering. *Applied Microbiology and Biotechnology, 97*, 4361–4368.

Mahajan, G. B., & Balachandran, L. (2012). Antibacterial agents from actinomycetes—A review. *Frontiers in Bioscience-Elite, 4*(1), 240–253.

Mahnashi, M. H., Muddapur, U. M., Turakani, B., Shaikh, I. A., Al Awadh, A. A., Alshahrani, M. M., & Shakeel Iqubal, S. M. (2022). A review on versatile eco-friendly applications of microbial proteases in biomedical and industrial applications. *Science of Advanced Materials, 14*(4), 622–637.

Majengbasan, O. S., Unuofin, J. O., Daramola, M. O., Iwarere, S. A., Semenya, K., & Odeniyi, O. A. (2025). Chitinous waste depolymerization and biocontrol potential of *Stenotrophomonas maltophilia* 3E chitinase against mycophytopathogens and *Anopheles gambiae* instar 3. *Bioresource Technology Reports, 30*, 102095.

Majithiya, V. R., & Gohel, S. D. (2022). Actinobacteria associated with marine invertebrates: Diversity and biological significance. In *Actinobacteria-diversity, applications and medical aspects*. IntechOpen.

Majithiya, V. R., & Gohel, S. D. (2025). Agro-industrial waste utilization, medium optimization, and immobilization of economically feasible halo-alkaline protease produced by *Nocardiopsis dassonvillei* strain VCS-4. *Applied Biochemistry and Biotechnology, 197*(1), 545–569.

Majithiya, V. R., Ghoghari, A. M., & Gohel, S. D. (2025). Purification, characterization, structural elucidation, and industrial applications of thermostable alkaline protease produced by seaweed-associated *Nocardiopsis dassonvillei* strain VCs-4. *International Journal of Biological Macromolecules, 305*, 141147.

Manivasagan, P., Venkatesan, J., Sivakumar, K., & Kim, S. K. (2013). Production, characterization and antioxidant potential of protease from *Streptomyces* sp. MAB18 using poultry wastes. *BioMed Research International, 2013*, 496586.

Matter, I. R., Al-Omari, A. W., & Mohammed, N. (2023). Industrial applications of microbial protease: A review. *Academic Science Journal, 1*(3), 141–160.

Ousaadi, M. I., Merouane, F., Berkani, M., Almomani, F., Vasseghian, Y., & Kitouni, M. (2021). Valorization and optimization of agro-industrial orange waste for the production of enzyme by halophilic Streptomyces sp. *Environmental Research, 201*, 111494.

Rajendran, K., Krishnamoorthy, M., Karuppiah, K., Ethiraj, K., & Sekar, S. (2024). Chitinase from *Streptomyces mutabilis* as an effective eco-friendly biocontrol agent. *Applied Biochemistry and Biotechnology, 196*(1), 18–31.

Ramesh, A., Harani Devi, P., Chattopadhyay, S., & Kavitha, M. (2020). Commercial applications of microbial enzymes. In *Microbial enzymes: Roles and applications in industries* (pp. 137–184). Springer Singapore.

Rathore, D. S., & Singh, S. P. (2021). Kinetics of growth and co-production of amylase and protease in novel marine actinomycete, *Streptomyces lopnurensis* KaM5. *Folia Microbiologica, 66*(3), 303–316.

Rathore, D. S., Sheikh, M. A., Gohel, S. D., & Singh, S. P. (2019). Isolation strategies, abundance and characteristics of the marine actinomycetes of Kachhighadi, Gujarat, India. *Journal of the Marine Biological Association of India, 61*(1), 21–27. https://doi.org/10.6024/jmbai.2019.61.1.2028-11

Rathore, D. R., Sheikh, M., Gohel, G. D., & Singh, S. P. (2021). Genetic and phenotypic heterogeneity of the *Nocardiopsis alba* strains of sea water. *Current Microbiology, 78*, 1377–1387. https://doi.org/10.1007/s00284-021-02420-0

Rathore, D. S., Sharma, A. K., Dobariya, A., Ramavat, H., & Singh, S. P. (2022). Cultivation and diversity of marine actinomycetes: Molecular approaches and bioinformatics tools. In *Actinobacteria: Microbiology to synthetic biology* (pp. 215–240). Springer Nature Singapore.
Roy, S., Das, I., Munjal, M., Karthik, L., Kumar, G., Kumar, S., & Rao, K. V. B. (2014). Isolation and characterization of tyrosinase produced by marine actinobacteria and its application in the removal of phenol from aqueous environment. *Frontiers in Biology, 9*, 306–316.
Salas-Bruggink, D. I., Sánchez-San Martín, J., Leiva, G., & Blamey, J. M. (2024). Extremozymes: Challenges and opportunities on the road to novel enzymes production. *Process Biochemistry, 143*, 323–336.
Salwan, R., & Sharma, V. (2018). The role of actinobacteria in the production of industrial enzymes. In *New and future developments in microbial biotechnology and bioengineering* (pp. 165–177). Elsevier.
Salwan, R., & Sharma, V. (2020). Molecular and biotechnological aspects of secondary metabolites in actinobacteria. *Microbiological Research, 231*, 126374.
Sanjivkumar, M., Silambarasan, T., Balagurunathan, R., & Immanuel, G. (2018). Biosynthesis, molecular modeling and statistical optimization of xylanase from a mangrove associated actinobacterium *Streptomyces variabilis* (MAB3) using Box-Behnken design with its bioconversion efficacy. *International Journal of Biological Macromolecules, 118*, 195–208.
Sanjivkumar, M., Vijayalakshmi, K., Silambarasan, T., Sholkamy, E. N., & Immanuel, G. (2020). Biosynthesis, statistical optimization and molecular modeling of chitinase from crab shell wastes by a mangrove associated actinobacterium Streptomyces olivaceus (MSU3) using Box-Behnken design and its antifungal effects. *Bioresource Technology Reports, 11*, 100493.
Saratale, G. D., & Oh, S. E. (2011). Production of thermotolerant and alkalotolerant cellulolytic enzymes by isolated Nocardiopsis sp. KNU. *Biodegradation, 22*, 905–919.
Saratale, G. D., Saratale, R. G., & Oh, S. E. (2012). Production and characterization of multiple cellulolytic enzymes by isolated Streptomyces sp. MDS. *Biomass and Bioenergy, 47*, 302–315.
Sharma, A. K., & Singh, S. P. (2016). Effect of amino acids on the repression of alkaline protease synthesis in haloalkaliphilic *Nocardiopsis dassonvillei*. *Biotechnology Reports, 12*, 40–51. https://doi.org/10.1016/j.btre.2016.10.004
Sharma, A. K., Gohel, S., & Singh, S. P. (2012). Actinobase: A database on molecular diversity, phylogeny and biocatalytic potential of salt tolerant alkaliphilic actinomycetes. *Bioinformation, 8*(11), 535–538. https://doi.org/10.6026/97320630008535
Shrestha, S., Chio, C., Khatiwada, J. R., Kognou, A. L. M., & Qin, W. (2022). Formulation of the agro-waste mixture for multi-enzyme (pectinase, xylanase, and cellulase) production by mixture design method exploiting Streptomyces sp. *Bioresource Technology Reports, 19*, 101142.
Shukla, R., & Singh, S. P. (2015). Production optimization, purification and characterization of alpha-amylase from thermophilic *Bacillus licheniformis* TSI-14. *Starch/Stärke, 67*(1–2), 1–11. https://doi.org/10.1002/star.201500046
Shukla, R., & Singh, S. P. (2016). Structural and catalytic properties of immobilized alpha-amylase from *Laceyella sacchari* TSI-2. *International Journal of Biological Macromolecules, 85*, 208–216. https://doi.org/10.1016/j.ijbiomac.2015.12.079
Shukla, R., & Singh, S. P. (2023). Purification and characterization of an amylase from a newly isolated *Geobacillus Thermoleovorans* TTIO4 from hot spring of Tuwa-Timba. *Starch*. Advance online publication. https://doi.org/10.1002/star.202200254
Silva, A. S., Neves, K. M., Freitas, R. F., das Chagas, T. P., Salay, L. C., da Silva, E. G. P., et al. (2025). Valorization of cocoa and peach-palm wastes for the production of amylases by *Pleurotus pulmonarius* CCB19 and its application as an additive in commercial detergents. *Applied Biochemistry and Biotechnology*, 1–24. https://doi.org/10.1007/s12010-024-05147-3
Silveira, M. A. V., Batista dos Santos, S. M., Okamoto, D. N., de Melo, I. S., Juliano, M. A., Ribeiro Chagas, J., & Vasconcellos, S. P. (2023). Atlantic Forest's and Caatinga's semiarid soils and their potential as a source for halothermotolerant actinomycetes and proteolytic enzymes. *Environmental Technology, 44*(11), 1566–1578.
Singh, R. P., & Reddy, C. R. K. (2014). Seaweed–microbial interactions: Key functions of seaweed-associated bacteria. *FEMS Microbiology Ecology, 88*(2), 213–230.

Singh, R., Kapoor, V., & Kumar, V. (2012). Utilization of agro-industrial wastes for the simultaneous production of amylase and xylanase by thermophilic actinomycetes. *Brazilian Journal of Microbiology, 43*, 1545–1552.

Singh, S. P., Thumar, J. T., Gohel, S. D., Kikani, B., Shukla, R., Sharma, A., & Dangar, K. (2013). Actinomycetes from marine habitats and their enzymatic potential. In *Marine enzymes for biocatalysis* (pp. 191–214). Woodhead Publishing.

Subash, H. A., Santhosh, K., Kannan, K., & Pitchiah, S. (2024). Extraction of keratin degrading enzyme from marine Actinobacteria of Nocardia sp and their antibacterial potential against oral pathogens. *Oral Oncology Reports, 9*, 100184.

Subramanian, K., Balaraman, D., Panangal, M., Nageswara Rao, T., Perumal, E., Kumarappan, A., et al. (2022). Bioconversion of chitin waste through *Stenotrophomonas maltophilia* for production of chitin derivatives as a Seabass enrichment diet. *Scientific Reports, 12*(1), 4792.

Szpisják-Gulyás, N., Al-Tayawi, A. N., Horváth, Z. H., László, Z., Kertész, S., & Hodúr, C. (2023). Methods for experimental design, central composite design and the Box–Behnken design, to optimise operational parameters: A review. *Acta Alimentaria, 52*(4), 521–537.

Techapun, C., Charoenrat, T., Watanabe, M., Sasaki, K., & Poosaran, N. (2002). Optimization of thermostable and alkaline-tolerant cellulase-free xylanase production from agricultural waste by thermotolerant Streptomyces sp. Ab106, using the central composite experimental design. *Biochemical Engineering Journal, 12*(2), 99–105.

Thakrar, F. J., & Singh, S. P. (2019). Catalytic, thermodynamic and structural properties of an immobilized and highly thermostable alkaline protease from a haloalkaliphilic actinobacteria, *Nocardiopsis alba* Tata-5. *Bioresource Technology, 278*, 150–158. https://doi.org/10.1016/j.biortech.2019.01.058

Thakrar, F. J., Gohel, S. D., & Singh, S. P. (2016). Repression of the synthesis of alkaline proteases of salt-tolerant alkaliphilic actinomycetes. *Asian Journal of Biochemical and Pharmaceutical Research, 6*(4). https://doi.org/10.1007/s12257-009-0087-y

Thakrar, F. J., Kikani, B. A., Sharma, A. K., & Singh, S. P. (2018). Stability of alkaline proteases from haloalkaliphilic actinobacteria probed by circular dichroism spectroscopy. *Applied Biochemistry and Microbiology, 54*(6), 591–602. https://doi.org/10.1134/S0003683818100022

Thakrar, F., Koladiya, G., & Singh, S. P. (2023). Heterologous expression and structural elucidation of a highly thermostable alkaline serine protease from haloalkaliphilic actinobacterium, *Nocardiopsis sp.* Mit-7. *Applied Biochemistry and Biotechnology*. Advance online publication. https://doi.org/10.1007/s12010-023-04472-3

Thomas, L., Ram, H., Kumar, A., & Singh, V. P. (2016). Production, optimization, and characterization of organic solvent tolerant cellulases from a lignocellulosic waste-degrading actinobacterium, Promicromonospora sp. VP111. *Applied Biochemistry and Biotechnology, 179*, 863–879.

Torres-León, C., Rojas, R., Contreras-Esquivel, J. C., Serna-Cock, L., Belmares-Cerda, R. E., & Aguilar, C. N. (2016). Mango seed: Functional and nutritional properties. *Trends in Food Science & Technology, 55*, 109–117.

Tran, T. N., Doan, C. T., & Wang, S. L. (2021). Conversion of wheat bran to xylanases and dye adsorbent by *Streptomyces thermocarboxydus*. *Polymers, 13*(2), 287.

Tran, T. N., Doan, C. T., Nguyen, V. B., Nguyen, A. D., & Wang, S. L. (2022). Conversion of fishery waste to proteases by *Streptomyces speibonae* and their application in antioxidant preparation. *Fishes, 7*(3), 140.

Vaghela, N., & Gohel, S. (2023). Medicinal plant-associated rhizobacteria enhance the production of pharmaceutically important bioactive compounds under abiotic stress conditions. *Journal of Basic Microbiology, 63*(3–4), 308–325.

Vaghela, N. R., & Gohel, S. D. (2025a). Liquid formulation of halo-alkali-thermo-tolerant rhizobacteria for enhanced growth of mung bean crops under abiotic stresses. *Scientific Reports, 15*(1), 21330.

Vaghela, N. R., & Gohel, S. D. (2025b). The rhizospheric alkali-halotolerant Bacillus sp. KhSb-159 enhanced growth parameters of the mung bean crop. *Current Microbiology, 82*(3), 114.

Valencia, R., González, V., Undabarrena, A., Zamora-Leiva, L., Ugalde, J. A., & Cámara, B. (2021). An integrative bioinformatic analysis for keratinase detection in marine-derived Streptomyces. *Marine Drugs, 19*(6), 286.

Victorino da Silva Amatto, I., Gonsales da Rosa-Garzon, N., Antonio de Oliveira Simoes, F., Santiago, F., Pereira da Silva Leite, N., Raspante Martins, J., & Cabral, H. (2022). Enzyme engineering and its industrial applications. *Biotechnology and Applied Biochemistry, 69*(2), 389–409.

Vijayakumar, N. (2021). Review on halophilic microbes and their applications. *Bulletin of Environmental Pharmacology and Life Sciences, 10*(12), 23–36.

Wang, J. L., Chen, Y. C., Deng, J. J., Mo, Z. Q., Zhang, M. S., Yang, Z. D., et al. (2023). Synergic chitin degradation by Streptomyces sp. SCUT-3 chitinases and their applications in chitinous waste recycling and pathogenic fungi biocontrol. *International Journal of Biological Macromolecules, 225*, 987–996.

Yadav, K., Rani, V., Anand, G., Yadava, U., & Yadav, D. (2025). Bioinformatic-driven research in microbial enzymes: An overview. *Microbial Enzymes: Production, Purification and Industrial Applications, 2*, 739–759.

Yang, M., Wang, N., Zhang, R., Khan, S., Hou, B., & Sand, W. (2025). Biomimetic haloperoxidases for antifouling on the surface of marine materials: A review. *Materials Advances, 6*, 479–489.

Yanling, M., Xinfeng, Z., & Rong, Z. (2018). A novel amidase signature family amidase from the marine actinomycete *Salinispora arenicola* CNS-205. *bioRxiv*, 342527.

Yaradoddi, J. S., & Kontro, M. H. (2022). Actinobacteria: Basic adaptation to harsh environments. In S. K. Singh & S. P. Singh (Eds.), *Actinobacteria: Ecology, diversity, classification and extensive applications* (pp. 69–88). Springer Nature Singapore.

Yuan, D., Lan, D., Xin, R., Yang, B., & Wang, Y. (2016). Screening and characterization of a thermostable lipase from marine Streptomyces sp. strain W007. *Biotechnology and Applied Biochemistry, 63*(1), 41–50.

Zhao, X. Q., Xu, X. N., & Chen, L. Y. (2016). Production of enzymes from marine actinobacteria. *Advances in Food and Nutrition Research, 78*, 137–151.

Zou, L., Qi, Z., Cheng, H., Yu, B., Li, Y. Y., & Liu, J. (2024). Advanced anaerobic digestion of household food waste pretreated by in situ-produced mixed enzymes via solid-state fermentation: Feasibility and application perspectives. *Environmental Research, 252*(Pt 4), 119137.

Recombinant Enzymes From Marine Actinomycetes: Genome Mining, Cloning, Expression Systems, and Biotechnological Significance

Hasti Ramavat, Hitarth B. Bhatt, Ramesh Kothari, and Satya P. Singh

Abstract

Marine actinomycetes represent a unique and underexplored reservoir of novel enzymes with valuable properties. However, their cultivation and low enzyme yield often hinder commercial applications. This book chapter reviews strategies for overcoming these limitations by producing recombinant enzymes. Modern genome mining has led to the discovery of promising biocatalytic genes directly from microbial genomic data. Subsequently, we discuss cloning strategies, including restriction-based, ligation-independent, and Gibson assembly methods, emphasizing the critical role of codon optimization for heterologous expression. A comparative analysis of various expression systems—including *Escherichia coli, Streptomyces, Bacillus*, and *Pichia pastoris*—is presented to highlight their respective advantages and suitability for different enzyme types. Finally, the properties and diverse biotechnological applications of recombinant enzymes from marine actinomycetes, from pharmaceuticals to bioremediation, are described.

Keywords

Marine actinomycetes · Halophiles-recombinant enzymes · Genome mining · Gene cloning · Expression systems

H. Ramavat · R. Kothari · S. P. Singh (✉)
Department of Biosciences, Saurashtra University, Rajkot, Gujarat, India

H. B. Bhatt
Department of Biological Sciences, P. D. Patel Institute of Applied Sciences, Charotar University of Science and Technology (CHARUSAT), Changa, Gujarat, India

K. Loganathan, Z. Li (eds.), *Marine Actinomycetes*,
https://doi.org/10.1007/978-981-95-9849-6_12

1 Introduction

Marine ecosystems comprising diverse habitats harbor a wide variety of microorganisms that are exposed to extreme and dynamic conditions of high salinity, variable temperatures, high hydrostatic pressure, and limited nutrient availability (Nichols et al., 2005; Singh et al., 2013; Raval et al., 2018; Poli et al., 2017; Fulke et al., 2025). Marine actinomycetes play a crucial role in nutrient cycling, exhibiting exceptional biosynthetic potential (Girão et al., 2022). They are prolific producers of secondary metabolites and industrially relevant enzymes, suitable in biotechnology, pharmaceuticals, and industrial biocatalysis (Fiedler et al., 2005; Singh et al., 2013; Farda et al., 2022). Unlike their terrestrial counterparts, marine actinomycetes thrive in unique ecological niches such as sediments, sea sponges, algae surfaces, and deep-sea environments. Consequently, they serve as an underexplored but promising reservoir for drug discovery, biocatalysis, and biotechnological innovations (Rathore et al., 2020; Gohel et al., 2015; Jensen et al., 2005).

The marine environment exhibits distinctive selective pressures that drive the evolution of novel biosynthetic pathways and unique enzymatic properties (Rotter et al., 2021; Kennedy et al., 2011; Rathore et al., 2022). The ecological versatility leads to high capacity to synthesize structurally diverse natural products and robust enzymes with unique properties (Zhang & Kim, 2010; Subramani & Sipkema, 2019; Olano et al., 2009; Jagannathan et al., 2021). The antibiotics, anticancer agents, immunosuppressants, antivirals, and anti-inflammatory molecules from marine actinomycetes are chemically distinct from those derived from terrestrial counterparts.

Besides secondary metabolites, marine actinomycetes are a prolific source of industrial enzymes capable of functioning under conditions where the majority of conventional enzymes fail (Gonzalez et al., 2020; Jagannathan et al., 2021; Gohel & Singh, 2018; Sharma et al., 2020; Kikani & Singh, 2011). They produce salt-tolerant, thermostable, or cold-active lipases, proteases, amylases, cellulases, and chitinases. Such enzymes are highly desirable in many industries that involve high salinity, variable pH, or extreme temperatures (Mehnaz et al., 2017; Sharma et al., 2020; Gohel et al., 2023).

Despite their immense potential, the applications of marine actinomycetes are limited due to difficulties in cultivation, slow growth rates, and low enzyme yields under laboratory conditions (de Souza Rodrigues et al., 2024). These limitations, therefore, necessitate alternative approaches. Recombinant enzyme technology, a powerful tool, can overcome most of these bottlenecks. Recombinant DNA technology related to the cloning and heterologous expression of enzyme-encoding genes from marine actinomycetes into fast-growing, genetically stable hosts such as *Escherichia coli*, *Bacillus subtilis*, or engineered *Streptomyces* strains (Xu & Wright, 2019). This approach facilitates large-scale enzyme production, bypassing the cultivation of the slow-growing native marine organisms. Moreover, metagenomic techniques allow access to novel enzyme genes from unculturable actinomycetes (Mahapatra et al., 2020; Raiyani & Singh, 2020; Purohit et al., 2022). Advances in cultivation strategies, genome mining, and recombinant DNA technologies are expanding our ability to harness their biosynthetic and enzymatic potential (Bhatt et al., 2024).

2 Marine Actinomycetes: A Reservoir of Novel Biocatalysts

Extremophilic microorganisms from marine environments produce extremozymes, which remain active and stable under harsh conditions of extreme temperature, pH, pressure, and salinity (Suriya et al., 2016; Di Donato et al., 2018). These biocatalysts are superior to mesophilic counterparts, making them valuable for industrial applications where conventional enzymes would be denatured or inactivated (Elleuche et al., 2014). Marine actinomycetes have evolved sophisticated enzyme systems and metabolic machinery adapted to complex marine environments (Singh et al., 2013). However, the vast marine microbiome remains largely unexplored for biotechnologically valuable species and novel biocatalysts (Di Donato et al., 2018). These extremozymes have significant potential for sustainable processes, including biofuel production, bioremediation, and chemical, food, and pharmaceutical applications (Suriya et al., 2016; Di Donato et al., 2018).

Halophilic and halotolerant microorganisms produce enzymes with exceptional stability and activity at high salt concentrations, making them valuable for biotechnological applications (DasSarma & DasSarma, 2015; Singh et al., 2010; Dobariya et al., 2023; Dodiya et al., 2008). These enzymes are characterized by many acidic over basic amino acid residues, creating a negatively charged surface, enhancing solubility and functionality in low water activity (Mevarech et al., 2000; DasSarma & DasSarma, 2015). Tightly bound water dipoles neutralize high negative surface charge, while salt stabilization occurs through low-affinity binding to specific sites on the protein surface (Mevarech et al., 2000). Salt-stable proteases from halophilic bacteria are commercially significant and account for approximately 65% of the global enzyme market (Mokashe et al., 2018; Margesin & Schinner, 2001).

Marine actinomycetes are a significant source of industrially important enzymes with diverse application potential across multiple sectors (Qeshmi et al., 2020; Mehnaz et al., 2017; Sharma et al., 2020; Gohel et al., 2023). The thermostable nature of these enzymes adds significant advantages for industrial processes, including reduced risk of microbial contamination, lower substrate viscosity, improved transfer rates, and enhanced substrate solubility (Bruins et al., 2001; Kikani & Singh, 2011; Sharma et al., 2020).

Deep-sea environments, representing high pressure, harbor specialized microorganisms called piezophiles (Zhang et al., 2015). These organisms produce unique enzymes with high pressure tolerance (Kato et al., 2007a, b; Jin et al., 2019; Abe & Horikoshi, 2001; Zhang et al., 2015).

3 Genome Mining

Genome mining has revolutionized natural product discovery by exploiting the vast genomic sequence data available in public databases to identify novel bioactive compounds. This computational approach analyzes biosynthetic gene clusters (BGCs) that remain "silent" or "cryptic" in traditional cultivation methods, representing an untapped reservoir of biochemical diversity (Li et al., 2018; Zerikly & Challis, 2009). The strategy includes multiple complementary approaches of bioinformatics predictions, metabolomic comparisons, and genetic manipulations to target specific chemical features or biological properties (Li et al., 2018; Bauman et al., 2021). Genome mining directly associates natural product structures with synthetic pathways, facilitating biosynthesis studies and combinatorial approaches (Li et al., 2018). These approaches are valuable for marine natural products, where traditional bioassay-guided isolation strategies are limited despite the ocean's immense biodiversity and novel chemistry (Chu et al., 2020). AntiSMASH, a leading automated pipeline, employs sequence similarity-based detection of core biosynthetic enzymes and pattern recognition to identify BGCs (Blin et al., 2019). Other tools, such as PRISM and MultiGeneBlast, leverage comparative genomics and machine learning techniques to predict BGCs and infer chemical structures of encoded natural products (Thokchom et al., 2024). The genome mining workflow involves systematic stages, including data selection, functional annotation, phylogenetic analysis, and comparative analysis (Nuhamunada et al., 2024). The key challenges are determining cluster boundaries, assessing novelty through database comparisons, and classifying clusters into families (Adamek et al., 2016).

The computational approaches predict the nature of secondary metabolites before laboratory work. Genome mining for secondary metabolite discovery is based on identifying conserved domains within polyketide synthases (PKSs) and non-ribosomal peptide synthetases (NRPSs), which are multi-modular megasynthases responsible for synthesizing diverse bioactive compounds (Anand & Mohanty, 2011). The enzymatic assembly lines contain highly conserved catalytic domains that serve as biosynthetic signatures, enabling computational prediction of novel natural products from genome sequences (Anand & Mohanty, 2011; Nikolouli & Mossialos, 2012). Advances in bioinformatics tools and genome sequencing have accelerated the discovery of novel bioactive compounds through genome-mining approaches (Nikolouli & Mossialos, 2012). Genomic studies have revealed unexpected enzymatic diversity and expanded the known distribution of PKS/NRPS systems across all three domains of life (Wang et al., 2015). Sophisticated computational tools like SBSPKSv2 now incorporate structure-based sequence analysis and chemical space searching capabilities, utilizing profile-based searches and HMM profiles for enhanced domain identification and biosynthetic pathway analysis (Khater et al., 2017).

Marine *Streptomyces* are a significant reservoir of bioactive compounds. The genome mining is a powerful strategy to unlock their hidden biosynthetic potential (Yang et al., 2020; Chu et al., 2020). Whole-genome sequencing reflects that the capacity for secondary metabolite synthesis in marine Streptomyces has been substantially underestimated. Many BGCs remain silent under standard laboratory conditions (Yang et al., 2020; Jeong et al., 2025). Genome mining integrates bioinformatics to discover natural products based on gene cluster sequences and biosynthetic pathways, enhancing new compound discovery from marine Streptomyces (Yang et al., 2020; Chu et al., 2020; Bauman et al., 2021).

Sequencing of 21 marine actinomycete strains revealed that rare genera, including *Nocardiopsis* and *Saccharomonospora*, possess high densities of unique BGCs with substantial novelty compared to existing databases (Schorn et al., 2016). Genome mining analysis of three *Nocardiopsis* genomes identified 36 BGCs, with many showing little similarity to previously described clusters (Ouchene et al., 2024). These rare actinomycetes have much lower isolation frequencies than *Streptomyces* yet represent untapped resources for antibiotic discovery (Tiwari & Gupta, 2012). They are widely distributed across volcanic areas, caves, and marine settings, projecting them as promising sources for unique enzymes (Aly et al., 2019).

Genome mining revealed vast untapped potential for secondary metabolites in microorganisms. However, it faces challenges with cryptic or silent BGCs that remain unexpressed under standard laboratory conditions (Baral et al., 2018; Brakhage & Schroeckh, 2011; Ren et al., 2017; Lee et al., 2020). The silent BGCs cannot be isolated for functional validation without activation (Baral et al., 2018). Various strategies, such as gene knockouts, promoter exchange, and transcription factor overexpression, have been attempted to activate the dormant pathways (Brakhage & Schroeckh, 2011). Additional methods encompass pleiotropic, BGC-specific, and targeted genome-wide approaches, as well as epigenetic strategies and co-cultivation techniques that simulate natural habitats (Baral et al., 2018; Brakhage & Schroeckh, 2011). Both native host and heterologous expression systems are employed for BGC activation (Ren et al., 2017). Despite computational advances in BGC identification, linking these clusters to their encoded products remains challenging for novel bioactive compound discovery (Lee et al., 2020).

Genome mining appears to be a transformative approach for discovering microbial natural products and enzymes, enhancing efficiency compared to traditional experimental methods (Bachmann et al., 2014; Wang et al., 2025; Zhu et al., 2025). Integrating artificial intelligence and machine learning has improved the pace and precision of BGC identification and prediction (Zhu et al., 2025; Zhang et al., 2025). Synthetic biology approaches are crucial to bridge the gap between genomic potential and phenotypic reality by engineering cryptic metabolite BGCs and editing host genomes to activate silent clusters (Li, 2023). Despite notable progress, current approaches face limitations of data scarcity, model generalizability, and the challenge of linking BGCs to their products (Li, 2023; Zhang et al., 2025). Nevertheless, the synergistic combination of computational tools and experimental strategies continues to accelerate natural product discovery.

4 Recombinant Enzymes and Cloning Strategies

The production of recombinant enzymes relies on molecular cloning strategies to efficiently express the target genes in suitable host systems. Advancements in molecular biology have led to the evolution of cloning techniques, expression vectors, and host systems tailored to maximize enzyme yield, solubility, and catalytic efficiency (Celie et al., 2016). This section discusses cloning strategies, including restriction-based, ligation-independent, and Gibson assembly approaches. Furthermore, codon optimization for heterologous expression and an overview of widely used expression systems such as *Escherichia coli*, *Streptomyces*, *Bacillus*, and *Pichia pastoris* are presented (Bertero et al., 2017).

5 Cloning Techniques

5.1 Restriction Enzyme-Based Cloning

Restriction enzyme-based cloning is a classical yet widely used technique in recombinant DNA technology. It involves digesting the target gene and plasmid vector with specific restriction endonucleases and ligation using T4 DNA ligase (Green & Sambrook, 2021).

The classical cloning workflow involves the following steps:

1. *Restriction Digestion*
   Restriction enzymes are chosen based on their presence in multiple cloning sites of the vector and absence within the internal coding region of the gene. Use of two different enzymes allows directional cloning, ensuring correct insert orientation.
   The vector and insert DNA are digested using the selected enzymes under optimal buffer and temperature conditions. Depending on enzyme compatibility, sequential or simultaneous digestion may be performed.
2. *Purification of Digested Fragments and Ligation*
   Digested products are purified by gel extraction to remove enzymes and undesirable fragments. T4 DNA ligase catalyzes the formation of phosphodiester bonds between the 5′ phosphate and 3′ hydroxyl termini of vector and insert. Typical insert-to-vector molar ratios range from *3:1 to 5:1*.
3. *Transformation and Screening*
   Recombinant plasmids are inserted into competent host cells, followed by screening via colony polymerase chain reaction (PCR), restriction digestion analysis, or sequencing to confirm correct insertion.

5.2 Ligation-Independent Cloning

Ligation-independent cloning (LIC), an advanced molecular cloning approach, enables the seamless insertion of target genes into expression vectors without DNA ligase or restriction endonucleases. LIC exploits complementary single-stranded overhangs generated through controlled exonuclease activity or PCR-based primer design, allowing precise and efficient assembly of recombinant DNA constructs (Aslanidis & De Jong, 1990; Camilo & Polikarpov, 2014).

LIC generates long complementary single-stranded overhangs on the vector and insert DNA. The overhangs are created through controlled exonuclease activity or primer design during PCR. These fragments anneal specifically and efficiently through base pairing. Upon transformation into competent host cells, the host's DNA repair machinery seals the nicks, resulting in a stable recombinant plasmid. Unlike restriction enzyme-based cloning, LIC, being sequence-independent, eliminates the need for restriction sites and DNA ligation and allows scar-free assembly of genes for expression (Bertero et al., 2017).

The general workflow for LIC includes the following steps:

1. *Primer Designing and Amplification of Insert:*
   Primers are designed to amplify the target gene with terminal sequences homologous to the linearized vector. Typically, 15–20 bp overlaps ensure efficient annealing. The gene of interest is amplified by a high-fidelity DNA polymerase, followed by purification of the PCR product to remove excess primers, nucleotides, and enzymes.
2. *Vector Preparation and Generation of Overhangs:*
   The expression vector is linearized by restriction digestion at a unique site or by PCR amplification, followed by purification to remove uncut plasmid. The insert and vector are ligated by T4 DNA polymerase in the absence of dNTPs, allowing its $3' \rightarrow 5'$ exonuclease activity to create complementary overhangs. The reaction is terminated by adding a single dNTP to halt further digestion.
3. *Annealing:*
   Equimolar amounts of the treated vector and insert are mixed under conditions that promote hybridization of the complementary overhangs, resulting in stable vector–insert complexes.
4. *Transformation, Screening, Verification:*
   The annealed DNA is introduced into competent host cells (e.g., *E. coli*). Host DNA repair mechanisms ligate the nicks and establish the recombinant plasmid. Transformants are screened on antibiotic-containing plates and confirmed through colony PCR, restriction digestion, or sequencing.

6 Gibson Assembly

Gibson Assembly is an advanced and highly versatile cloning method that allows the seamless and simultaneous joining of multiple DNA fragments in a single isothermal reaction. Unlike restriction enzyme-based or LIC methods, Gibson Assembly enables the construction of scarless DNA molecules without the need for restriction sites, ligases, or specific host repair systems, making it particularly useful for assembling complex enzyme expression constructs and synthetic pathways (Avilan, 2023).

Gibson Assembly relies *on* three enzymatic activities in a single reaction mixture at a constant temperature (typically 50 °C). 5′ Exonuclease creates single-stranded 3′ overhangs by chewing back the 5′ ends of DNA fragments, exposing complementary regions of homology (usually 15–30 bp). DNA Polymerase extends from the annealed overhangs, filling gaps between the fragments to generate a continuous double-stranded molecule. The third enzyme, DNA Ligase, seals the nicks in the sugar-phosphate backbone, resulting in a covalently closed, scarless recombinant DNA construct. As all reactions co-occur under isothermal conditions, Gibson Assembly enables the precise and seamless assembly of multiple fragments in a single tube.

7 Gibson Assembly

The standard workflow of Gibson Assembly involves the following steps:

1. *Design of Overlapping Fragments:*
   Vector and insert(s) are designed with 15–30 bp overlapping sequences complementary to adjacent fragments. Overlaps are introduced via PCR primers or synthetic gene synthesis.
2. *Preparation of DNA Fragments and Assembly Reaction:*
   DNA fragments are generated by high-fidelity PCR or obtained synthetically. The vector backbone is linearized by restriction digestion or PCR. All fragments are combined in the Gibson Assembly Master Mix, which contains 5′ exonuclease, high-fidelity DNA polymerase, and ligase. Incubation at 50 °C for 15–60 min facilitates exonuclease digestion, annealing of overlapping regions, polymerase-mediated gap filling, and ligation.
3. *Transformation and Screening:*
   The assembled construct is directly transformed into competent cells (e.g., *E. coli*), followed by selection on appropriate antibiotic media. Positive clones are screened by PCR or confirmed through sequencing.

8 Codon Optimization for Heterologous Expression

Codon optimization, a critical step in molecular cloning and recombinant protein production, involves modifying the nucleotide sequence of a gene to improve its expression in a heterologous host without altering the encoded amino acid sequence (Gustafsson et al., 2004). The degeneracy of the genetic code allows multiple codons to encode the same amino acid; however, organisms exhibit a strong bias for specific codons, reflecting differences in their tRNA abundance and translational machinery. When expressing a gene in a non-native host, mismatches between the gene's codon usage and the host's preferred codons can lead to inefficient translation, reduced protein yield, and impaired functional folding. Codon optimization addresses these limitations by adapting gene sequences to the host-specific codon preferences to improve transcriptional efficiency, translational accuracy, and overall recombinant enzyme production (Mirzahoseini et al., 2011).

8.1 Rationale for Codon Optimization

Heterologous expression involves transferring gene/s from one organism into a host such as *Escherichia coli*, *Bacillus subtilis*, *Streptomyces*, or *Pichia pastoris* (Berini, 2014). The codon usage of the source organism may differ significantly from that of the chosen host, leading to the following impacts.

Ribosome stalling: Low-abundance tRNAs for rare codons slow down translation elongation.
Premature termination: Rare codon clusters increase ribosome pausing and misincorporation of amino acids.
Misfolding and aggregation: Inefficient translation kinetics impair co-translational folding, leading to inclusion body formation.
Low mRNA stability: Non-optimal codons can create mRNA secondary structures that impede ribosome binding or initiate degradation.

Codon optimization aims to resolve these issues while maintaining the recombinant protein's amino acid sequence and enzymatic function.

9 Key Parameters in Codon Optimization

Effective codon optimization considers multiple molecular and genetic factors beyond simple codon usage substitution (Gomez-Escribano & Bibb, 2014):

(a) *Codon Usage Bias*
- Different organisms exhibit unique codon preferences that correlate with their tRNA pools. For example, *E. coli* favors codons such as AAA (lysine) and GGT (glycine), whereas eukaryotic genes often use AAG (lysine) or GGC (glycine). Optimization involves replacing rare codons with synonymous high-frequency codons specific to the target host.

(b) *GC Content*
- Optimal GC content ensures gene stability and efficient transcription. For bacterial hosts, GC content between 30% and 70% is generally preferred. Extreme GC levels can hinder transcription and increase mRNA secondary structures, reducing translation efficiency.

(c) *mRNA Secondary Structures*
- Stable hairpins or stem-loop structures near the *ribosome binding site* or start codon impede translation initiation. Optimization includes minimizing stable secondary structures in the 5′ untranslated region and near the initiation codon.

(d) *Rare Codon Clustering*
- Consecutive rare codons can deplete low-abundance tRNAs and stall translation. Distributed codon optimization avoids long clusters of rare codons that disrupt translational kinetics.

(e) *Codon Harmonization*

- Instead of uniformly replacing all rare codons, codon harmonization strategically retains specific slow codons to mimic natural translation rates. This preserves co-translational folding kinetics, especially for enzymes with complex tertiary structures or multiple disulfide bonds.

(f) *Incorporation of Translational Enhancers*

- Addition of Kozak sequences (eukaryotic) or Shine-Dalgarno sequences (prokaryotic) to improve ribosome recruitment.

Codon optimization is now facilitated by computational tools and algorithms (e.g., GenScript, OptimumGene™, and IDT Codon Optimization Tool) that predict mRNA stability and folding, ensuring high-level expression of soluble, functional enzymes.

9.1 Expression Systems—*E. coli, Streptomyces, Bacillus, and Pichia pastrosis*

The production of recombinant proteins has emerged as a prominent approach in today's biotechnology, enabling large-scale synthesis of enzymes, therapeutic proteins, and industrial biocatalysts. An appropriate expression system is crucial for protein yield, folding, secretion, and post-translational modifications (Sriyapai et al., 2011). Among the most widely used microbial hosts, *Escherichia coli*, *Streptomyces*, *Bacillus*, and the methylotrophic yeast *Pichia pastoris* represent versatile systems for fundamental studies and industrial applications (Baghban et al., 2018; Vojnovic et al., 2024).

Escherichia coli remains the most common prokaryotic host for recombinant protein production due to its rapid growth, inexpensive cultivation, and well-established molecular biology tools (Baneyx & Mujacic, 2004; Raval et al., 2015; Bhatt & Singh, 2020). Expression in *E. coli* is supported by a wide range of vectors and promoters, such as T7, lac, and arabinose systems, which provide tight and inducible regulation. Despite its advantages, *E. coli* has certain limitations. Proteins requiring complex folding, disulfide bond formation, or glycosylation may aggregate into insoluble inclusion bodies, necessitating refolding strategies or fusion tags to enhance solubility (Baneyx & Mujacic, 2004; Berkmen, 2012). Additionally, endotoxin contamination is a significant concern for pharmaceutical applications. Nonetheless, *E. coli* is highly suited for producing enzymes of bacterial origin, particularly when post-translational modifications are not essential.

In *E. coli*, the system's speed, cost-effectiveness, and mature genetic tools have been used successfully for soluble production and biochemical characterization of marine actinobacterial enzymes. Notable examples include the heterologous expression of a codon-optimized L-glutamate oxidase (Sl-LGOX) from the marine actinomycete *Streptomyces lydicamycinicus* in *E. coli* BL21(DE3), which yielded soluble, Ni-NTA-purifiable enzyme displaying high specificity for L-glutamate and broad stability across pH and temperature ranges (Oikawa & Yamanaka, 2025). Another instance is the thermostable xylanase XynS14 from *Actinomadura* sp. S14, where a mature Open Reading Frame (ORF) was expressed in *E. coli* to produce an active enzyme for comparative enzymology alongside thc alternative yeast expression (Sriyapai et al., 2011). Furthermore, alkaline proteases from salt-tolerant alkaliphilic *Nocardiopsis* sp. have been cloned and expressed in *E. coli* (Gohel & Singh, 2012; Rohamare et al., 2015; Thakrar et al., 2023).

In contrast to *E. coli*, *Streptomyces* species are equipped with an efficient secretory apparatus that enables the extracellular release of heterologous proteins, streamlining downstream purification processes. A striking demonstration of this compatibility is the heterologous expression of LipIAF5.2, an alkaline and thermostable lipase from marine metagenomic DNA, in *Streptomyces lividans*. Under the control of the *S. coelicolor* groEL2 heat-shock promoter, the enzyme was efficiently secreted and purified (Meilleur et al., 2009). Similarly, phospholipase D genes from marine-derived Streptomyces species such as *S. racemochromogenes* and *S. antibioticus* have been expressed in *S. lividans* using shuttle vectors and optimized promoters, achieving robust extracellular enzyme production (Nakazawa et al., 2011).

Members of the genus *Bacillus*, particularly *Bacillus subtilis*, are another vital platform for recombinant protein production (Zhang et al., 2020). These Gram-positive bacteria are non-pathogenic, well characterized, and naturally secrete proteins into the extracellular milieu. The absence of endotoxins makes *Bacillus* especially suitable for food-grade enzyme production. Additionally, their ability to secrete large quantities of heterologous proteins reduces the need for cell lysis during downstream processing. *B. subtilis* and *B. licheniformis* have been widely employed in the enzyme industry, producing proteases, lipases, amylases, and cellulases. However, proteolytic degradation of heterologous proteins can be problematic in *Bacillus*. However, engineered protease-deficient strains have significantly addressed this issue.

The Generally Recognized as Safe (GRAS) status and powerful secretion machinery of *Bacillus* spp. make them attractive hosts. Several enzymes of marine actinobacteria have been transferred into *Bacillus* for scale-oriented secretion. For instance, a Glycine - Aspartate - Serine - Leucine (GDSL) -family lipase (OUC-Sb-lip12) from *Streptomyces bacillaris* was expressed in *Bacillus subtilis* and applied for enzymatic hydrolysis of astaxanthin esters, demonstrating functional secretion and utility in product processing (Gao et al., 2021). The *Bacillus* platform, therefore, offers a pragmatic route to extracellular production of marine actinobacterial hydrolases.

The methylotrophic yeast *Pichia pastoris* is among the most successful eukaryotic expression systems, particularly for enzymes requiring post-translational modifications. *P. pastoris* combines the advantages of bacterial and higher eukaryotic systems. It grows to high cell densities on inexpensive substrates, utilizes strong and tightly regulated promoters such as AOX1, and secretes recombinant proteins into the culture medium (Juturu & Wu, 2018). *P. pastoris* glycosylation pathways have been engineered to produce human-like glycoforms, expanding its application in biopharmaceuticals. Compared to *Saccharomyces cerevisiae*, *Pichia* often produces proteins with fewer hypermannosylation, making it suitable for industrial enzymes and therapeutic proteins.

For enzymes requiring eukaryotic secretion, disulfide bonds, or particular folding environments, *Pichia pastoris* appears superior for marine actinobacterial enzymes. A thermostable esterase HydS14 from *Actinomadura* sp. S14 was cloned into *P. pastoris* under AOX1 control (Sriyapai et al., 2015). Likewise, an endo-1,4-β-xylanase, NhX1 from *Nocardiopsis halotolerans*, was cloned into pPIC9m and produced in *P. pastoris* in a 10 L fermenter (Lisov et al., 2024). The comparative case of XynS14 (Actinomadura S14) expressed in *E. coli* and *P. pastoris* further exemplifies host choice. The yeast yielded a higher activity, while *E. coli* enabled rapid cytosolic expression for initial screening (Sriyapai et al., 2011).

Taken together, these marine-actinobacteria-derived cases establish a clear pattern for host selection. When post-translational complexity is low, *E. coli* appears in the first line for rapid soluble expression and initial biochemical study. *Streptomyces* provides a native-compatible background for the expression of actinobacterial enzyme clusters. *Bacillus* offers a GRAS route to secretory production for scaling up specific hydrolases, and *Pichia pastoris* enables high-yield secretory expression and favorable folding for thermostable marine enzymes.

10 Recombinant Enzymes From Marine Actinomycetes: Properties and Applications

Marine actinomycetes possess unique genetic diversity and metabolic pathways, making them a promising source for novel recombinant enzymes. Adapting to extreme marine conditions gives these enzymes superior functionality, including tolerance to high salt, a wide range of pH and temperatures, and specific substrate activity (Gohel & Singh, 2018). Heterologous expression, a genetic engineering technique, allows for the efficient, large-scale production of these enzymes in host organisms like *E. coli*, making their use in industries more feasible and cost-effective (Oikawa & Yamanaka, 2025).

Table 1 enlists recombinant enzymes from marine actinomycetes, detailing their properties and applications (Table 1).

Thus, genetic engineering can produce these robust enzymes in higher yields for various industrial and biotechnological applications.

While marine environments harbor vast microbial diversity that represents an enormous reservoir of novel enzymes with biotechnological potential, there are significant challenges in developing the marine-derived biocatalysts. Limitations in culturing marine microorganisms under laboratory conditions restrict enzyme discovery (Kennedy et al., 2011; Debashish et al., 2005). Further, heterologous expression of marine enzymes faces challenges of rare codon usage, secondary mRNA structures, protein misfolding, and aggregation (Salas-Bruggink et al., 2024). Despite these hurdles, genomic and metagenomic approaches offer solutions for marine biodiscovery (Kennedy et al., 2011). Developments in expression systems and the emergence of artificial intelligence can potentially overcome challenges (Salas-Bruggink et al., 2024).

Table 1 Recombinant enzymes from marine actinomycetes

Sr. no.	Recombinant enzyme	Source organism	Property of recombinant enzyme	Applications	References
1.	L-glutamte oxidase	*Streptomyces lydicamycinicus* NBRC 110027	Temperature and pH stability, substrate specific	Biocatalysis and various industrial process	Oikawa and Yamanaka (2025)
2.	Dmxorosin BGC	*Streptomyces thermolilacinus* SPC6	Halotolerant, cryptic in nature	Production of dmxorosin, exploiting cryptic genes, pharmaceutical industry	Dong et al. (2025)
3.	α-Amylase	*Nocardiopsis aegyptia* HDN19-252	Halotolerant, stable at alkaline pH and low temperature	High-temperature sterilization or cooking processes, detergent industry for green detergent	Liu et al. (2025)
4.	Alkaline serine protease	*Nocardiopsis* sp. Mit-7 (NCIM-5746)	Haloalkaliphilic, thermostable	Detergent formulations, bioremediation, food processing, textile industry, pharmaceutical industry	Thakrar et al. (2023)
5.	β-glucosidase	Streptomyces sp. SCSIO 04578	Halotolerant, glucose-tolerant, mesophilic	Biomass conversion, food industry	Mai et al. (2013)
6.	Chitinase	*Micromonospora aurantiaca*	pH stability, broad substrate specificity	Chitin waste bioconversion, bioactive oligosaccharide production, waste water treatment	Guo et al. (2024)
7.	Alkaline serine proteases	Strain OM-6: *Brachystreptospora xinjiangensis* Strain OK-5: *Nocardiopsis alba*	Haloalkaliphilic, thermostable	Detergent industry, food and textile industry	Gohel and Singh (2012)

11 Conclusion

Marine actinomycetes offer a vast and largely untapped reservoir of novel enzymes. The unique properties of these biocatalysts make them highly suitable for industrial applications. Beyond traditional cultivation methods, genome mining and recombinant DNA technology have emerged as powerful tools for discovering, isolating, and producing enzymes and secondary metabolites. This integrated approach overcomes the limitations of culturing these microbes and unlocks their full genetic potential. Furthermore, combining bioinformatics and genetic engineering can solidify marine actinomycetes' position as a critical resource for sustainable and efficient biocatalysts. Codon optimization and other developments in expression systems would enable us to harness the unique traits of enzymes for diverse applications. The continued focus on robust enzymes underscores the vital role of marine biodiversity in accelerating sustainable industrial processes.

Acknowledgments HR acknowledges the ScHeme of Developing High Quality Research Fellowship by the Government of Gujarat. HB acknowledges the Senior Research Fellowship awarded by Council of Scientific & Industrial Research (CSIR) under Direct scheme and Meritorious Fellowship awarded by University Grants Comission (UGC) at the Department of Biosciences, Saurashtra University. SPS acknowledges the Multi-Institutional Project by Department of Biotechnology (DBT), Ministry of Earth Sciences (MoES) Networking Project, University Grants Comission (UGC) - Basic Science Research (BSR) Faculty Fellowship, and International Travel Fellowships from Department of Science & Technology (DST) - Science & Engineering Research Board (SERB), CSIR, DBT, and UGC. Furthermore, SPS and RKK acknowledge the facilities and support under the University Grants Comission (UGC) -Center for Advanced study (CAS), Department of Science & Technology (DST) - Fund for Fund for improvement of Science & Technology (FIST), and Saurashtra University.

References

Abe, F., & Horikoshi, K. (2001). The biotechnological potential of piezophiles. *Trends in Biotechnology, 19*(3), 102–108.

Adamek, M., Spohn, M., Stegmann, E., & Ziemert, N. (2016). Mining bacterial genomes for secondary metabolite gene clusters. In *Antibiotics: Methods and protocols* (pp. 23–47). Springer.

Aly, M. M., Bahamdain, L. A., & Aba, S. A. (2019). Unexplored extreme habitats as sources of novel and rare actinomycetes with enzyme and antimicrobial activities. *IOSR Journal of Pharmacy and Biological Sciences, 14*, 45–54.

Anand, S., & Mohanty, D. (2011). Computational methods for identification of novel secondary metabolite biosynthetic pathways by genome analysis. In *Handbook of research on computational and systems biology: Interdisciplinary applications* (pp. 380–405). IGI Global.

Aslanidis, C., & De Jong, P. J. (1990). Ligation-independent cloning of PCR products (LIC-PCR). *Nucleic Acids Research, 18*(20), 6069–6074.

Avilan, L. (2023). Assembling multiple fragments: The Gibson Assembly. In *DNA manipulation and analysis* (pp. 45–53). Springer US.

Bachmann, B. O., Van Lanen, S. G., & Baltz, R. H. (2014). Microbial genome mining for accelerated natural products discovery: Is a renaissance in the making. *Journal of Industrial Microbiology and Biotechnology, 41*(2), 175–184.

Baghban, R., Farajnia, S., Ghasemi, Y., Mortazavi, M., Zarghami, N., & Samadi, N. (2018). New developments in *Pichia pastoris* expression system, review and update. *Current Pharmaceutical Biotechnology, 19*(6), 451–467.
Baneyx, F., & Mujacic, M. (2004). Recombinant protein folding and misfolding in *Escherichia coli*. *Nature Biotechnology, 22*(11), 1399–1408.
Baral, B., Akhgari, A., & Metsä-Ketelä, M. (2018). Activation of microbial secondary metabolic pathways: Avenues and challenges. *Synthetic and Systems Biotechnology, 3*(3), 163–178.
Bauman, K. D., Butler, K. S., Moore, B. S., & Chekan, J. R. (2021). Genome mining methods to discover bioactive natural products. *Natural Product Reports, 38*(11), 2100–2129.
Berini, F. (2014). *Heterologous expression, characterization and applications of metagenome-and genome-sourced enzymes*. University of Insubria.
Berkmen, M. (2012). Production of disulfide-bonded proteins in *Escherichia coli*. *Protein Expression and Purification, 82*(1), 240–251.
Bertero, A., Brown, S., & Vallier, L. (2017). Methods of cloning. In *Basic science methods for clinical researchers* (pp. 19–39). Academic.
Bhatt, H. B., & Singh, S. P. (2020). Cloning, expression, and structural elucidation of a biotechnologically potential alkaline serine protease from a newly isolated haloalkaliphilic *Bacillus lehensis* JO-26. *Frontiers in Microbiology, 11*, 941.
Bhatt, H. B., Sani, R. K., Amoozegar, M. A., & Singh, S. P. (2024). Extremozymes: Characteristics, structure, protein engineering and applications. *Frontiers in Microbiology, 15*, 1423463.
Blin, K., Kim, H. U., Medema, M. H., & Weber, T. (2019). Recent development of antiSMASH and other computational approaches to mine secondary metabolite biosynthetic gene clusters. *Briefings in Bioinformatics, 20*(4), 1103–1113.
Brakhage, A. A., & Schroeckh, V. (2011). Fungal secondary metabolites–strategies to activate silent gene clusters. *Fungal Genetics and Biology, 48*(1), 15–22.
Bruins, M. E., Janssen, A. E., & Boom, R. M. (2001). Thermozymes and their applications: A review of recent literature and patents. *Applied Biochemistry and Biotechnology, 90*(2), 155–186.
Camilo, C. M., & Polikarpov, I. (2014). High-throughput cloning, expression and purification of glycoside hydrolases using Ligation-Independent Cloning (LIC). *Protein Expression and Purification, 99*, 35–42.
Celie, P. H., Parret, A. H., & Perrakis, A. (2016). Recombinant cloning strategies for protein expression. *Current Opinion in Structural Biology, 38*, 145–154.
Chu, L., Huang, J., Muhammad, M., Deng, Z., & Gao, J. (2020). Genome mining as a biotechnological tool for the discovery of novel marine natural products. *Critical Reviews in Biotechnology, 40*(5), 571–589.
DasSarma, S., & DasSarma, P. (2015). Halophiles and their enzymes: Negativity put to good use. *Current Opinion in Microbiology, 25*, 120–126.
de Souza Rodrigues, R., de Souza, A. Q. L., Feitoza, M. D. O., Alves, T. C. L., Barbosa, A. N., da Silva Santiago, S. R. S., & de Souza, A. D. L. (2024). Biotechnological potential of actinomycetes in the 21st century: A brief review. *Antonie Van Leeuwenhoek, 117*(1), 82.
Debashish, G., Malay, S., Barindra, S., & Joydeep, M. (2005). Marine enzymes. *Marine Biotechnology I, 96*, 189–218.
Di Donato, P., Buono, A., Poli, A., Finore, I., Abbamondi, G. R., Nicolaus, B., & Lama, L. (2018). Exploring marine environments for the identification of extremophiles and their enzymes for sustainable and green bioprocesses. *Sustainability, 11*(1), 149.
Dobariya, A., Mankad, G. P., Ramavat, H., & Singh, S. P. (2023). Efficacy of the fruit and vegetable peels as substrates for the growth and production of α-amylases in marine actinobacteria. *Applied Biochemistry and Biotechnology, 195*(12), 7603–7623.
Dodia, M. S., Rawal, C. M., Bhimani, H. G., Joshi, R. H., Khare, S. K., & Singh, S. P. (2008). Purification and stability characteristics of an alkaline serine protease from a newly isolated Haloalkaliphilic bacterium sp. AH-6. *Journal of Industrial Microbiology and Biotechnology, 35*(2), 121–131.

Dong, M., Feng, H., Zhang, W., & Ding, W. (2025). Direct cloning and heterologous expression of the dmxorosin biosynthetic gene cluster from *Streptomyces thermolilacinus* SPC6, a halotolerant actinomycete isolated from the desert in China. *International Journal of Molecular Sciences, 26*(4), 1492.

Elleuche, S., Schröder, C., Sahm, K., & Antranikian, G. (2014). Extremozymes—Biocatalysts with unique properties from extremophilic microorganisms. *Current Opinion in Biotechnology, 29*, 116–123.

Farda, B., Djebaili, R., Vaccarelli, I., Del Gallo, M., & Pellegrini, M. (2022). Actinomycetes from caves: An overview of their diversity, biotechnological properties, and insights for their use in soil environments. *Microorganisms, 10*(2), 453.

Fiedler, H. P., Bruntner, C., Bull, A. T., Ward, A. C., Goodfellow, M., Potterat, O., et al. (2005). Marine actinomycetes as a source of novel secondary metabolites. *Antonie Van Leeuwenhoek, 87*(1), 37–42.

Fulke, A. B., Sharma, N., & Nadekar, J. (2025). Darkness to discovery: A comprehensive mini-review on culturable and non-culturable microbial diversity from deep sea. *Microbial Ecology, 88*(1), 77.

Gao, K., Wang, X., Jiang, H., Sun, J., & Mao, X. (2021). Identification of a GDSL lipase from *Streptomyces bacillaris* and its application in the preparation of free astaxanthin. *Journal of Biotechnology, 325*, 280–287.

Girão, M., Ribeiro, I., & Carvalho, M. D. F. (2022). Actinobacteria from marine environments: A unique source of natural products. In *Natural products from actinomycetes: Diversity, ecology and drug discovery* (pp. 1–45). Springer Singapore.

Gohel, S. D., & Singh, S. P. (2012). Cloning and expression of alkaline protease genes from two salt-tolerant alkaliphilic actinomycetes in *E. coli*. *International Journal of Biological Macromolecules, 50*(3), 664–671.

Gohel, S. D., & Singh, S. P. (2018). Thermodynamics of a Ca2+ dependent, highly thermostable and detergent compatible purified alkaline serine protease from *Nocardiopsis xinjiangensis* strain OM-6. *International Journal of Biological Macromolecules, 113*, 565–574.

Gohel, S. D., Sharma, A. K., Dangar, K. G., Thakrar, F. J., & Singh, S. P. (2015). Antimicrobial and biocatalytic potential of haloalkaliphilic actinobacteria. In *Halophiles: Biodiversity and sustainable exploitation* (pp. 29–55). Springer.

Gohel, S. D., Majithiya, V. R., & Singh, S. P. (2023). Genetic and physiological diversity of marine actinobacteria from the Okha Coast, Gujarat, India. *Geomicrobiology Journal, 40*(6), 590–604.

Gomez-Escribano, J. P., & Bibb, M. J. (2014). Heterologous expression of natural product biosynthetic gene clusters in *Streptomyces coelicolor*: From genome mining to manipulation of biosynthetic pathways. *Journal of Industrial Microbiology and Biotechnology, 41*(2), 425–431.

González, V., Vargas-Straube, M. J., Beys-da-Silva, W. O., Santi, L., Valencia, P., Beltrametti, F., & Cámara, B. (2020). Enzyme bioprospection of marine-derived actinobacteria from the Chilean coast and new insight in the mechanism of keratin degradation in Streptomyces sp. G11C. *Marine Drugs, 18*(11), 537.

Green, M. R., & Sambrook, J. (2021). Cloning polymerase chain reaction (PCR) products: Making T vectors. *Cold Spring Harbor Protocols, 2021*(6), pdb-prot101295.

Guo, H. Z., Wang, D., Yang, H. T., Wu, Y. L., Li, Y. C., Xia, G. H., & Zhang, X. Y. (2024). Heterologous expression and characterization of a pH-stable chitinase from *Micromonospora aurantiaca* with a potential application in chitin degradation. *Marine Drugs, 22*(6), 287.

Gustafsson, C., Govindarajan, S., & Minshull, J. (2004). Codon bias and heterologous protein expression. *Trends in Biotechnology, 22*(7), 346–353.

Jagannathan, S. V., Manemann, E. M., Rowe, S. E., Callender, M. C., & Soto, W. (2021). Marine actinomycetes, new sources of biotechnological products. *Marine Drugs, 19*(7), 365.

Jensen, P. R., Mincer, T. J., Williams, P. G., & Fenical, W. (2005). Marine actinomycete diversity and natural product discovery. *Antonie Van Leeuwenhoek, 87*(1), 43–48.

Jeong, H., Choe, Y., Nam, J., & Ban, Y. H. (2025). A guide to genome mining and genetic manipulation of biosynthetic gene clusters in Streptomyces. *Journal of Microbiology, 63*(4), e2409026.

Jin, M., Gai, Y., Guo, X., Hou, Y., & Zeng, R. (2019). Properties and applications of extremozymes from deep-sea extremophilic microorganisms: A mini review. *Marine Drugs, 17*(12), 656.

Juturu, V., & Wu, J. C. (2018). Heterologous protein expression in *Pichia pastoris*: Latest research progress and applications. *Chembiochem, 19*(1), 7–21.

Kato, C., Sato, T., Abe, F., Ohmae, E., Tamegai, H., & Nakasone, K. (2007a). Discoveries of deep-sea piezophiles, and their pressure adapted enzymes. *High Pressure Bioscience and Biotechnology, 1*(1), 114–121.

Kato, C., Sato, T., Ishii, A., Kawano, H., Abe, F., Ohmae, E., & Nakasone, K. (2007b). Proteins, Under high-pressure environments-discoveries of deep-sea piezophiles, and their pressure adapted enzymes. In *Proceedings of international symposium on extremophiles and their applications international symposium on extremophiles and their applications 2005* (pp. 148–155). Extremobiosphere Research Center, JAMSTEC.

Kennedy, J., O'leary, N. D., Kiran, G. S., Morrissey, J. P., O'Gara, F., Selvin, J., & Dobson, A. D. W. (2011). Functional metagenomic strategies for the discovery of novel enzymes and biosurfactants with biotechnological applications from marine ecosystems. *Journal of Applied Microbiology, 111*(4), 787–799.

Khater, S., Gupta, M., Agrawal, P., Sain, N., Prava, J., Gupta, P., et al. (2017). SBSPKSv2: Structure-based sequence analysis of polyketide synthases and non-ribosomal peptide synthetases. *Nucleic Acids Research, 45*(W1), W72–W79.

Kikani, B. A., & Singh, S. P. (2011). Single step purification and characterization of a thermostable and calcium independent α-amylase from *Bacillus amyloliquifaciens* TSWK1-1 isolated from Tulsi Shyam hot spring reservoir, Gujarat (India). *International Journal of Biological Macromolecules, 48*(4), 676–681.

Lee, N., Hwang, S., Kim, J., Cho, S., Palsson, B., & Cho, B. K. (2020). Mini review: Genome mining approaches for the identification of secondary metabolite biosynthetic gene clusters in Streptomyces. *Computational and Structural Biotechnology Journal, 18*, 1548–1556.

Li, L. (2023). Next-generation synthetic biology approaches for the accelerated discovery of microbial natural products. *Engineering Microbiology, 3*(1), 100060.

Li, Z., Zhu, D., & Shen, Y. (2018). Discovery of novel bioactive natural products driven by genome mining. *Drug Discoveries & Therapeutics, 12*(6), 318–328.

Lisov, A. V., Belova, O. V., Belov, A. A., Lisova, Z. A., Nagel, A. S., Shadrin, A. M., et al. (2024). Expression in *Pichia pastoris* of thermostable endo-1, 4-β-xylanase from the actinobacterium *Nocardiopsis halotolerans*: Properties and use for saccharification of xylan-containing products. *International Journal of Molecular Sciences, 25*(16), 9121.

Liu, F., Zheng, X., Liao, W., Ren, X., Ma, C., Zhang, G., et al. (2025). Heterologous expression and biochemical characterization of a new α-amylase from *Nocardiopsis aegyptia* HDN19-252 of Antarctic animal origin. *Marine Drugs, 23*(4), 159.

Mahapatra, G. P., Raman, S., Nayak, S., Gouda, S., Das, G., & Patra, J. K. (2020). Metagenomics approaches in discovery and development of new bioactive compounds from marine actinomycetes. *Current Microbiology, 77*(4), 645–656.

Mai, Z., Yang, J., Tian, X., Li, J., & Zhang, S. (2013). Gene cloning and characterization of a novel salt-tolerant and glucose-enhanced β-glucosidase from a marine Streptomycete. *Applied Biochemistry and Biotechnology, 169*(5), 1512–1522.

Margesin, R., & Schinner, F. (2001). Potential of halotolerant and halophilic microorganisms for biotechnology. *Extremophiles, 5*(2), 73–83.

Mehnaz, D. S., Aiysha, D., Zaheer, A., & Mukhtar, S. (2017). Actinomycetes: A source of industrially important enzymes. *Journal of Proteomics & Bioinformatics, 10*(12), 316–319.

Meilleur, C., Hupé, J. F., Juteau, P., & Shareck, F. (2009). Isolation and characterization of a new alkali-thermostable lipase cloned from a metagenomic library. *Journal of Industrial Microbiology and Biotechnology, 36*(6), 853–861.

Mevarech, M., Frolow, F., & Gloss, L. M. (2000). Halophilic enzymes: Proteins with a grain of salt. *Biophysical Chemistry, 86*(2–3), 155–164.

Mirzahoseini, H., Mafakheri, S., Soltan Mohammadi, N., Enayati, S., & Mortazavidehkordi, N. (2011). Heterologous proteins production in *Escherichia coli*: An investigation on the effect of codon usage and expression host optimization. *Cell Journal (Yakhteh), 12*(4), 453–458.

Mokashe, N., Chaudhari, B., & Patil, U. (2018). Operative utility of salt-stable proteases of halophilic and halotolerant bacteria in the biotechnology sector. *International Journal of Biological Macromolecules, 117*, 493–522.

Nakazawa, Y., Sagane, Y., Sakurai, S. I., Uchino, M., Sato, H., Toeda, K., & Takano, K. (2011). Large-scale production of phospholipase D from *Streptomyces racemochromogenes* and its application to soybean lecithin modification. *Applied Biochemistry and Biotechnology, 165*(7), 1494–1506.

Nichols, C. M., Guezennec, J., & Bowman, J. P. (2005). Bacterial exopolysaccharides from extreme marine environments with special consideration of the southern ocean, sea ice, and deep-sea hydrothermal vents: A review. *Marine Biotechnology, 7*(4), 253–271.

Nikolouli, K., & Mossialos, D. (2012). Bioactive compounds synthesized by non-ribosomal peptide synthetases and type-I polyketide synthases discovered through genome-mining and metagenomics. *Biotechnology Letters, 34*(8), 1393–1403.

Nuhamunada, M., Mohite, O. S., Phaneuf, P. V., Palsson, B. O., & Weber, T. (2024). BGCFlow: systematic pangenome workflow for the analysis of biosynthetic gene clusters across large genomic datasets. *Nucleic Acids Research, 52*(10), 5478–5495.

Oikawa, T., & Yamanaka, K. (2025). Heterologous expression and enzymological characterization of L-glutamate oxidase from the marine actinomycete *Streptomyces lydicamycinicus* NBRC 110027. *Bioscience, Biotechnology, and Biochemistry, 89*(3), 371–381.

Olano, C., Mendez, C., & Salas, J. A. (2009). Antitumor compounds from marine actinomycetes. *Marine Drugs, 7*(2), 210–248.

Ouchene, R., Zaatout, N., & Suzuki, M. T. (2024). An overview on Nocardiopsis species originating from North African biotopes as a promising source of bioactive compounds and in silico genome mining analysis of three sequenced genomes. *Journal of Basic Microbiology, 64*(9), e2400046.

Poli, A., Finore, I., Romano, I., Gioiello, A., Lama, L., & Nicolaus, B. (2017). Microbial diversity in extreme marine habitats and their biomolecules. *Microorganisms, 5*(2), 25.

Purohit, M. K., Rathore, D. S., Koladiya, G., Pandey, S., & Singh, S. P. (2022). Comparative analysis of the catalysis and stability of the native, recombinant and metagenomic alkaline proteases in organic solvents. *Environmental Science and Pollution Research, 29*(53), 80968–80982.

Qeshmi, F. I., Homaei, A., Fernandes, P., Hemmati, R., Dijkstra, B. W., & Khajeh, K. (2020). Xylanases from marine microorganisms: A brief overview on scope, sources, features and potential applications. *Biochimica et Biophysica Acta (BBA)-Proteins and Proteomics, 1868*(2), 140312.

Raiyani, N. M., & Singh, S. P. (2020). Taxonomic and functional profiling of the microbial communities of Arabian Sea: A metagenomics approach. *Genomics, 112*(6), 4361–4369.

Rathore, D. S., Sheikh, M., & Singh, S. P. (2020). Marine actinobacteria: New horizons in bioremediation. In *Recent developments in microbial technologies* (pp. 425–449). Springer Nature Singapore.

Rathore, D. S., Sharma, A. K., Dobariya, A., Ramavat, H., & Singh, S. P. (2022). Cultivation and diversity of marine actinomycetes: Molecular approaches and bioinformatics tools. In *Actinobacteria: Microbiology to synthetic biology* (pp. 215–240). Springer Nature Singapore.

Raval, V. H., Rawal, C. M., Pandey, S., Bhatt, H. B., Dahima, B. R., & Singh, S. P. (2015). Cloning, heterologous expression and structural characterization of an alkaline serine protease from sea water haloalkaliphilic bacterium. *Annals of Microbiology, 65*(1), 371–381.

Raval, V. H., Bhatt, H. B., & Singh, S. P. (2018). Adaptation strategies in halophilic bacteria. In *Extremophiles* (pp. 137–164). CRC Press.

Ren, H., Wang, B., & Zhao, H. (2017). Breaking the silence: New strategies for discovering novel natural products. *Current Opinion in Biotechnology, 48*, 21–27.

Rohamare, S., Gaikwad, S., Jones, D., Bhavnani, V., Pal, J., Sharma, R., & Chatterjee, P. (2015). Cloning, expression and in silico studies of a serine protease from a marine actinomycete (Nocardiopsis sp. NCIM 5124). *Process Biochemistry, 50*(3), 378–387.

Rotter, A., Barbier, M., Bertoni, F., Bones, A. M., Cancela, M. L., Carlsson, J., et al. (2021). The essentials of marine biotechnology. *Frontiers in Marine Science, 8*, 629629.

Salas-Bruggink, D. I., Sánchez-San Martín, J., Leiva, G., & Blamey, J. M. (2024). Extremozymes: Challenges and opportunities on the road to novel enzymes production. *Process Biochemistry, 143*, 323–336.

Schorn, M. A., Alanjary, M. M., Aguinaldo, K., Korobeynikov, A., Podell, S., Patin, N., & Moore, B. S. (2016). Sequencing rare marine actinomycete genomes reveals high density of unique natural product biosynthetic gene clusters. *Microbiology, 162*(12), 2075–2086.

Sharma, A. K., Kikani, B. A., & Singh, S. P. (2020). Biochemical, thermodynamic and structural characteristics of a biotechnologically compatible alkaline protease from a haloalkaliphilic, *Nocardiopsis dassonvillei* OK-18. *International Journal of Biological Macromolecules, 153*, 680–696.

Singh, S. P., Purohit, M. K., Raval, V. H., Pandey, S., Akbari, V. G., & Rawal, C. M. (2010). Capturing the potential of Haloalkaliphilic bacteria from the saline habitats through culture dependent and metagenomic approaches. In A. Méndez-Vilas (Ed.), *Current research, technology and education topics in applied microbiology and microbial biotechnology* (pp. 81–87). FORMATEX.

Singh, S. P., Thumar, J. T., Gohel, S. D., Kikani, B., Shukla, R., Sharma, A., & Dangar, K. (2013). Actinomycetes from marine habitats and their enzymatic potential. In *Marine enzymes for biocatalysis* (pp. 191–214). Woodhead Publishing.

Sriyapai, T., Somyoonsap, P., Matsui, K., Kawai, F., & Chansiri, K. (2011). Cloning of a thermostable xylanase from Actinomadura sp. S14 and its expression in *Escherichia coli* and *Pichia pastoris*. *Journal of Bioscience and Bioengineering, 111*(5), 528–536.

Sriyapai, P., Kawai, F., Siripoke, S., Chansiri, K., & Sriyapai, T. (2015). Cloning, expression and characterization of a thermostable esterase HydS14 from Actinomadura sp. strain S14 in *Pichia pastoris*. *International Journal of Molecular Sciences, 16*(6), 13579–13594.

Subramani, R., & Sipkema, D. (2019). Marine rare actinomycetes: A promising source of structurally diverse and unique novel natural products. *Marine Drugs, 17*(5), 249.

Suriya, J., Bharathiraja, S., Krishnan, M., Manivasagan, P., & Kim, S. K. (2016). Extremozymes from marine actinobacteria. *Advances in Food and Nutrition Research, 79*, 43–66.

Thakrar, F. J., Koladiya, G. A., & Singh, S. P. (2023). Heterologous expression and structural elucidation of a highly thermostable alkaline serine protease from haloalkaliphilic Actinobacterium, Nocardiopsis sp. Mit-7. *Applied Biochemistry and Biotechnology, 195*(12), 7583–7602.

Thokchom, S., Mukherjee, S., Ningthoujam, D. S., & Banerjee, S. (2024). Identification of biosynthetic gene clusters using bioinformatic tools. *Flora and Fauna, 30*(1), 163–166.

Tiwari, K., & Gupta, R. K. (2012). Rare actinomycetes: A potential storehouse for novel antibiotics. *Critical Reviews in Biotechnology, 32*(2), 108–132.

Vojnovic, S., Aleksic, I., Ilic-Tomic, T., Stevanovic, M., & Nikodinovic-Runic, J. (2024). Bacillus and Streptomyces spp. as hosts for production of industrially relevant enzymes. *Applied Microbiology and Biotechnology, 108*(1), 185.

Wang, H., Sivonen, K., & Fewer, D. P. (2015). Genomic insights into the distribution, genetic diversity and evolution of polyketide synthases and nonribosomal peptide synthetases. *Current Opinion in Genetics & Development, 35*, 79–85.

Wang, M., Chen, L., Zhang, Z., & Wang, Q. (2025). Recent advances in genome mining and synthetic biology for discovery and biosynthesis of natural products. *Critical Reviews in Biotechnology, 45*(1), 236–256.

Xu, M., & Wright, G. D. (2019). Heterologous expression-facilitated natural products' discovery in actinomycetes. *Journal of Industrial Microbiology and Biotechnology, 46*(3–4), 415–431.

Yang, Z., He, J., Wei, X., Ju, J., & Ma, J. (2020). Exploration and genome mining of natural products from marine Streptomyces. *Applied Microbiology and Biotechnology, 104*(1), 67–76.

Zerikly, M., & Challis, G. L. (2009). Strategies for the discovery of new natural products by genome mining. *Chembiochem, 10*(4), 625–633.
Zhang, C., & Kim, S. K. (2010). Research and application of marine microbial enzymes: Status and prospects. *Marine Drugs, 8*(6), 1920–1934.
Zhang, Y., Li, X., Bartlett, D. H., & Xiao, X. (2015). Current developments in marine microbiology: High-pressure biotechnology and the genetic engineering of piezophiles. *Current Opinion in Biotechnology, 33*, 157–164.
Zhang, K., Su, L., & Wu, J. (2020). Recent advances in recombinant protein production by *Bacillus subtilis*. *Annual Review of Food Science and Technology, 11*(1), 295–318.
Zhang, Y., Zong, J., Liu, Y., Zhou, K., Shi, H., Yin, W. B., ... & Chen, Y. (2025). Progress on targeted discovery of microbial natural products based on the predictions of both structure and activity. *Natural Product Reports*, *42*(10), 1651–1663.
Zhu, S., Xu, H., Liu, Y., Hong, Y., Yang, H., Zhou, C., & Tao, L. (2025). Computational advances in biosynthetic gene cluster discovery and prediction. *Biotechnology Advances, 79*, 108532.

Extremozymes From Marine Actinobacteria: Current Status and Applications

Riya S. Jasani, Shruti P. Buddhdev, Jenisha K. Ranpariya, Rushit Shukla, Hitarth B. Bhatt, and Satya P. Singh

Abstract

Marine ecosystems are vast, physicochemically dynamic environments characterized by extremes of salinity, temperature, pH, pressure, and nutrient availability. Such heterogeneity has driven the evolution of metabolically versatile and resilient microorganisms, including marine actinobacteria—a prolific yet underexplored group. These bacteria produce extremozymes, enzymes known for their stability, efficiency, and robustness under harsh conditions, such as proteases, amylases, cellulases, lipases, esterases, and xylanases. This chapter critically reviews their ecological distribution, adaptive mechanisms, and enzyme characteristics and highlights their potential applications in food, pharmaceuticals, textiles, bioenergy, detergents, and emerging biotechnologies.

Keywords

Halophilic enzymes · Extremozymes · Enzymatic adaptation · Biotechnological applications · Marine actinomycetes

R. S. Jasani · S. P. Buddhdev · J. K. Ranpariya
Department of Life Sciences, Atmiya University, Rajkot, Gujarat, India

R. Shukla (✉)
Department of Microbiology, Christ College, Rajkot, Gujarat, India

H. B. Bhatt (✉)
Department of Biological Sciences, P. D. Patel Institute of Applied Sciences, Charotar University of Science and Technology, (CHARUSAT), Changa, Gujarat, India

S. P. Singh (✉)
Department of Biosciences, Saurashtra University, Rajkot, Gujarat, India

K. Loganathan, Z. Li (eds.), *Marine Actinomycetes*,
https://doi.org/10.1007/978-981-95-9849-6_13

1 Introduction

Enzymes are important to modern industrial biotechnology because of their catalytic efficiency, eco-friendly nature, and substrate specificity. Unlike chemical catalysts, enzymes enable faster reactions under mild conditions, significantly reducing energy consumption and minimizing hazardous by-products (Kumar & Singh, 2024). These characteristics make them ideal for green chemistry and sustainable industrial processes.

The global enzyme market, which was valued at US$5.6 billion in 2019, is expected to expand at an average yearly rate of 6.4% between 2020 and 2027, driven by rising environmental concerns, stricter regulations on chemical processing, and increasing demand for cleaner and cost-effective technologies. Nowadays, a wide range of industries, including food and feed, textiles, detergents, leather, pharmaceuticals, and bioenergy, use enzymes as active ingredients, additives, or processing aids (Adiguzel et al., 2019; Kikani et al., 2023).

Enzymes have been derived from animals, plants, and microbes to meet evolving needs (Hassanein et al., 2011). Among these sources, microbial enzymes have gained particular prominence because of their faster growth rates, ease of genetic manipulation, scalability, and ability to function in various process conditions (Suriya et al., 2016). However, many industrial substrates, including lignocellulosic biomass, lipid-rich effluents, and polymeric waste, are recalcitrant under the mild aqueous conditions favored by conventional enzymes (Suriya et al., 2016; Salam et al., 2025). This challenge has sparked growing interest in extremozymes, enzymes derived from extremophilic microorganisms that flourish in extreme environments.

Extremophiles such as halophiles, thermophiles, acidophiles, alkaliphiles, psychrophiles, and barophiles inhabit hostile niches on Earth, including hypersaline lakes, deep-sea hydrothermal vents, acidic hot springs, polar ice caps, and desert soils (Bhatt et al., 2018; Sysoev et al., 2021). These organisms have evolved unique biochemical mechanisms to survive high salinity, extreme temperatures, pressure, pH, and radiation (Bhatt & Singh, 2022). Therefore, the extremozymes they produce are naturally stable and catalytically active under conditions that would denature most enzymes (Karan et al., 2020; Raval et al., 2022; Laye et al., 2022; Pasqualetti et al., 2022).

Among extremophiles, marine microorganisms, particularly actinobacteria, are promising sources of industrially viable enzymes. Over 70% of Earth's surface is covered by the marine biosphere, a reservoir of unique microbial diversity shaped by extreme environmental gradients (Suriya et al., 2016). Marine actinobacteria have drawn attention due to their genetic richness, metabolic versatility, and ability to produce a broad spectrum of industrially essential enzymes (Singh et al., 2016; Makarani & Kaushal, 2025).

Structural characteristics such as reinforced hydrophobic cores, salt bridges, and unique surface amino acid motifs enable these enzymes to function and maintain stability in extreme conditions (Raval et al., 2014; Chaturvedi & Khurana, 2019; Bhatt & Singh, 2020; Aguiar et al., 2022; Pizarro et al., 2022; de Souza et al., 2022; Kikani et al., 2023; Majithiya et al., 2025). Robust nature is crucial in biofuel production, pharmaceutical synthesis, detergent formulation, waste treatment, and leather processing. The advancements in metagenomics, high-throughput screening, and protein engineering have accelerated the discovery, functional characterization, and commercial development of robust biocatalysts (de Souza et al., 2022; Hassanein et al., 2011). This chapter highlights the ability of marine actinobacteria to thrive in extreme environments and to produce extremozymes suitable for a wide range of industrial applications and sustainable biotechnology.

2 Ecology of Marine Actinobacteria

The marine environment is one of Earth's largest and most diverse ecosystems, hosting a wide range of microorganisms. Marine environments reflect geophysical, nutritional, and chemical differences, influencing the dynamics of microorganisms. Its habitats are numerous, ranging from the ocean surface to the seafloor and depths exceeding 10,000 m (Rampelotto, 2013). However, certain marine habitats, like remote or deep-sea zones, are difficult to explore for microbial diversity. Actinobacteria are highly abundant among the dominant bacterial groups in marine environments (Chen et al., 2016; Zamora-Quintero et al., 2022; Prins et al., 2024).

Advancements in molecular methods have shown the existence of truly marine actinobacteria. Culture-independent analyses of Arctic and Atlantic deep-sea sediments revealed more than 20 actinobacterial genera. Simultaneously, 44 strains, mostly from genera such as *Microbacterium, Brachybacterium,* and *Brevibacterium,* were cultured using traditional methods. Many of these isolates have shown antibacterial, anticancer, and anti-inflammatory activities (Ribeiro et al., 2023). Actinobacteria have also been known to reside in marine life such as oysters and sponges, which make compounds such as piericidin A_1 and ilamycins with strong antimicrobial and anticancer properties, respectively (Cera et al., 2022). Moreover, genus *Salinispora* has been identified as an obligate marine actinobacterium, requiring seawater for growth (Román-Ponce et al., 2020). Similarly, the research on the diversity and biocatalytic potential of marine actinobacteria has been conducted in the coastal areas and saline habitats of Gujarat, India (Singh et al., 2013; Thumar & Singh, 2009; Rathore & Singh, 2021; Dobariya et al., 2023; Gohel et al., 2023; Rathore et al., 2021; Sharma et al., 2020).

Streptomyces strains from the Mediterranean Sea have shown the production of secondary metabolites, with the potential to biosynthesize silver nanoparticles, which are effective against drug-resistant bacteria and even cancer cells (Khalil et al., 2022). Furthermore, eight new species were identified, belonging to the genera *Rhodococcus, Salinispora, Kocuria, Kineococcus, Arthrobacter,* and *Janibacter* from the Andaman Islands (Meena et al., 2021). Some strains have demonstrated the ability to degrade persistent organic pollutants and heavy metals, thereby contributing to the detoxification of marine environments. *Streptomyces* sp. breaks down pesticides and stores heavy metals (lead and cadmium) (Jagannathan et al., 2021). Knowledge of the ecology and adaptations of marine actinobacteria is crucial for explaining potential biotechnological applications.

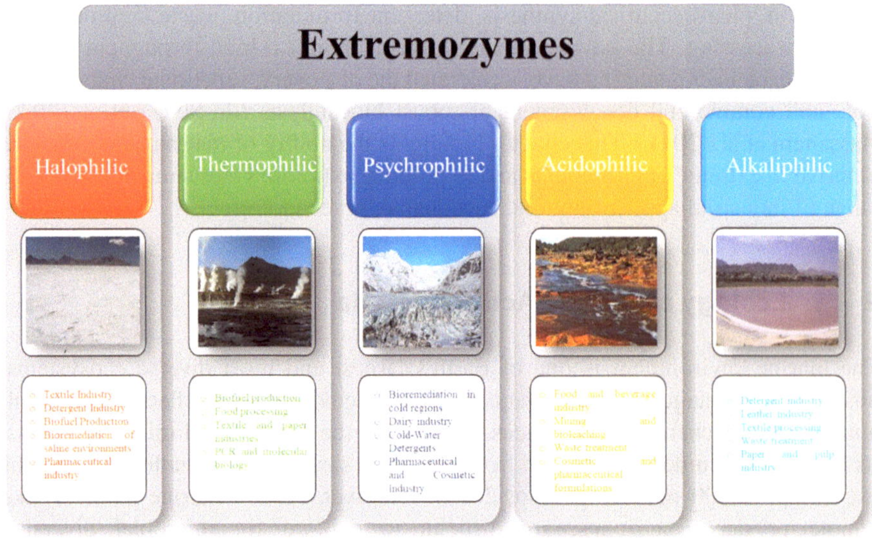

Fig. 1 Types of extremozymes and its applications

3 Extremozymes

Microbial enzymes are distinguished by their high productivity, consistent performance, and cost-effectiveness, making them well-suited for a wide range of applications. In extreme environments with high temperatures, salinity, or pH, traditional chemical methods have been the only available option. Nevertheless, the enzymes and other metabolites of extremophiles can play vital roles (Angelakis et al., 2024; Elleuche et al., 2014). These organisms thrive in harsh environments, including hot springs, deep oceans, glaciers, alkaline waters, salt flats, radioactive zones, and deserts (Rekadwad et al., 2023). Extremozymes can further be classified as thermozymes, halozymes, alkalizymes, acidozymes, psychrozymes, barozymes, and solvent-stable enzymes (Fig. 1). The extremozymes have potential roles in pharmaceuticals, biofuels, food processing, detergents, cosmetics, bioremediation, and so on (Samanta et al., 2022).

4 Extremozymes From Marine Actinobacteria

Exploring marine ecosystems has revealed a vast reservoir of bioactive compounds and enzymes with considerable industrial and pharmaceutical potential. Marine actinobacteria are prolific producers of biotechnologically significant enzymes among the diverse microorganisms inhabiting these environments. Such enzymes are adapted to function under extreme conditions. Table 1 summarizes representative extremozymes derived from marine actinobacteria, along with their optimal physicochemical conditions and potential industrial applications.

Table 1 Properties and applications of different extremozymes reported from Actinobacteria

Enzyme	Organisms	Source	Salt optima	pH optima	Temp. optima	Other properties	Application	References
Amylase	*Laceyella sacchari* TSI-2	TulsiShyam hot spring	–	7	70 °C	Solvent stable	Detergent industry	Shukla and Singh (2016)
Protease	*Nocardiopsis dassonvillei*	Kelambakkam salt pan	–	7	36 °C	Stable with surfactants, detergents, bleaching compounds	Detergents, leather processing	Majithiya et al. (2025)
	Streptomyces cellulosae AU-10	Salt-pan soil	–	9–11	40 °C	Metalloenzyme; surfactant and oxidizer-tolerant; olive oil hydrolysis	Detergent additive	Boran (2018)
	Actinopolyspora sp. VITSDK2	Marine saltern, India	–	10.0	60 °C	Stable pH 4–12; activated by Mn^{2+}, Mg^{2+}, Ca^{2+}	Industrial proteolysis, waste protein removal	Suthindhiran et al. (2014)
Lipases	*Streptomyces* sp. W007	Marine sediment	5% NaCl	6.0–9.0	40 °C	Thermostable at 60 °C, solvent- and metal-ion-tolerant	Biodiesel, detergents, biocatalysts	Yuan et al. (2016)
	Haliclona simulans		5 M	7	40 °C	Halotolerant, extremely salt-stable	Oil waste degradation, detergent, high-salt biotech	Selvin et al. (2012)
	Streptomyces exfoliatus DSMZ 41693	Marine sediment	–	8.5	60 °C	Thermostable, solvent & detergent stable	Polyester degradation, synthetic esters	Rodríguez-Alonso et al. (2023)

(continued)

Table 1 (continued)

Enzyme	Organisms	Source	Salt optima	pH optima	Temp. optima	Other properties	Application	References
Esterase	*Streptomyces lividans* TK64	Soil marine isolate	–	–	40 °C	Highly thermostable; metal-ion resilient	Fine chemical synthesis, chiral resolution	Wang et al. (2016)
	Bacillus sp.	Deep-sea sediment	–	–	40 °C	>99% enantiomeric excess, organic co-solvent-tolerant	Chiral lactate production	Huang et al. (2016)
Cellulases	*Streptomyces* sp.	Sediment sample, Mangrove coastal areas, Maharashtra-Goa border	–	5	55 °C	–	Paper, vegetable, seed coat, and rice straw degradation, and clarification of fruit juice	Kulkarni and Maurya (2017)
	Actinoalloteichus sp. MHA15	Sediment sample, Havelock Island, Andaman	–	6.5	37 °C	Dissolved oxygen-4.5 ml/L at the 3rd hour	Microbial fuel cells/ Bioelectricity production	Gobalakrishnan and Bhuvaneswari (2019)
	Nocardiopsis dassonvillei PSY13	Water samples, Southern Ocean	–	7.5	10 °C	–	–	Sivasankar et al. (2022)
Xylanase	*Streptomyces olivaceus (MSU3)*	sediment sample of mangrove environment	1.5%	8	40	Metal dependent (Fe2+)	Production of reducing sugar and bioethanol	Sanjivkumar et al. (2018a, b)

Other enzymes	*Streptomyces* (Laccase)	–	–	4–5	50 °C	Thermostable; less metal inhibition	Industrial detoxification, Pulp bleaching	Fernandes et al. (2013)
	Streptomyces albidoflavus (DyP peroxidase)	–	–	Alkaline	50–60 °C	No mediator required; H_2O_2-stable	Textile effluent degradation	Musengi et al. (2020)
	Streptomyces coelicolor (DyP-type peroxidase)	–	Ionic-stress-tolerant	3.5	30 °C	Ultra-high catalytic efficiency oxidizes lignin model substrates	Lignin remodeling, biorefinery feedstock valorization	Pupart et al. (2023)
	Saccharomonospora viridis DSM 43017(DyP-type peroxidase)	CGMCC	Alkalotolerant; salt compatible	**7**	70 °C	Retains >63% activity at 50–80 °C; broad pH (4–9) stability	Kraft pulp biobleaching, dye degradation	Yu et al. (2014)

4.1 Amylase

Amylase, discovered in the nineteenth century, is a groundbreaking enzyme that has undergone extensive scientific investigation. It is widely distributed in nature and is found in humans, plants, animals, and microbes.

In general, amylase catalyzes the hydrolytic breakdown of starch and related compounds, such as dextrin, amylose, and amylopectin, converting them into mono-, di-, and oligosaccharides. Amylases, based on the mode of action, have been classified into three major classes: alpha (endoamylases acting on α-1-4 linkage), beta (exoamylases acting on α-1-4 linkage), and gamma (gluco) (acting on α-1-4 and α-1-6 linkage). α-Amylases are widely studied among all three classes. The α-amylases consist of around 450–500 amino acids with three domain architectures—A, B, and C. Domain A, with (α/β) 8-barrel configuration, acts as the catalytic center, while Domain B is a loop inserted between β-strand and α-helix acting as a calcium-binding loop for structural stability, and Domain C forms a C-terminal Greek-key β-barrel associated with substrate interaction and enzyme stability. Marine actinobacterial α-amylases share this conserved fold but also possess adaptations for salt tolerance, cold activity, or alkaliphilicity, allowing the enzymes to function in extreme marine conditions (Singh et al., 2013).

Besides their critical role in physiological processes, amylases are widely used in various industrial processes (Tiwari et al., 2015). It is among the most widely produced enzymes, accounting for over a quarter of global enzyme production (Shukla & Singh, 2016). The significant applications of amylases are in detergents, food, fermentation, and the pharmaceutical industry. With increased demand, the need for enzymes with industrially suitable characteristics has arisen. For instance, detergent industries require amylase that can catalyze optimally at high salt and a wide pH range, whereas the alcohol-producing industry requires thermally stable enzymes (Sundarram & Murthy, 2014; Shukla & Singh, 2016, 2023; Bhatt et al., 2020). Gene manipulation and protein engineering approaches have helped to develop unique catalytic and stability traits in enzymes (Ashaolu et al., 2025; Chatterjee et al., 2023).

Several enzymes with robust characteristics have been isolated and studied. However, only a few have been commercialized. Among the most commercially valuable enzymes, amylases have been studied in marine microbes. Improving enzyme production in marine actinobacteria has been investigated. Several marine *Streptomyces* spp. were studied using statistical tools to enhance amylase production (Suthindhiran et al., 2014). Furthermore, fruit and vegetable peels proved suitable substrates for amylase production in many marine *Nocardiopsis* strains isolated from Gujarat, India (Dobariya et al., 2023). Optimal amylase production was achieved with a 2% inoculum, at pH 8, 28 °C, and a 5% salt concentration (Dobariya et al., 2023). Amylases from marine actinobacteria can function at high salt concentrations and extreme pH values without requiring calcium. Thus, they are most suitable for many industrial applications.

4.2 Protease

Proteases account for more than 60% of the global industrial enzyme market, highlighting their extensive use across a variety of industries (Muffler et al., 2015; Naveed et al., 2021). Proteases are used in waste treatment, bioremediation, wood processing, meat tenderization, and food production because they degrade proteins. Due to their effectiveness, versatility, and wide substrate specificity, they are also significant in the leather, pharmaceutical, and detergent sectors (Solanki et al., 2021; Sujitha & Shanthi, 2023).

Bacteria are the primary microbial source of proteases due to their rapid growth, metabolic versatility, and ease of genetic manipulation. These benefits contribute to high enzyme yields and cost-effective large-scale production, often outperforming fungal and animal systems (Raval et al., 2022; Bhatt & Singh, 2020; Solanki et al., 2021; Singh et al., 2013). In particular, bacterial alkaline proteases are often used in detergent formulations due to their strong activity against proteinaceous stains under alkaline and high-temperature conditions (Gohel & Singh, 2018; Salwan & Sharma, 2019; Sharma et al., 2020; Rathore & Singh, 2021). Extremophilic bacteria, including halophiles, haloalkaliphiles, alkaliphiles, and thermophiles, offer advantages by producing catalytically efficient proteases and being structurally stable under harsh industrial environments. Marine actinobacteria, especially genera such as *Nocardiopsis*, *Streptomyces,* and *Micromonospora,* are known to produce extremozymes with remarkable alkaline and halotolerant properties (Das et al., 2022; Hazari et al., 2022; Sharma et al., 2021). These robust enzymes are suited to the leather, pharmaceutical, detergent, and environmental sectors.

An alkaline protease from the marine *Nocardiopsis dassonvillei* strain VCs-4 exhibited an optimal activity of 362.97 U/mL under haloalkaliphilic conditions. The protease was stable at 80 °C and in the presence of surfactants, detergents, and bleaching agents, effectively removing blood stains and hydrolyzing gelatin from used X-ray films, demonstrating strong industrial potential (Majithiya et al., 2025). Similar to this, *Streptomyces* sp. XE-1 produced an alkaline protease tolerant to oxidizing agents, detergents, and organic solvents with stability across a broad pH range (Xu et al., 2015). Another marine strain, *Streptomyces* sp. MAB18 secreted an extracellular protease using poultry waste, active at pH 8–10 and 50–60 °C, stable up to pH 12 and 12% NaCl. Interestingly, it also exhibited antioxidant activity, indicating its dual use in animal feed as a protein supplement and antioxidant (Manivasagan et al., 2013).

4.3 Lipases

Lipases (EC 3.1.1.3) are broad hydrolytic enzymes that catalyze the breakdown of ester bonds in triglycerides, releasing glycerol and free fatty acids. Beyond hydrolysis, they also facilitate esterification and transesterification reactions in non-aqueous media, expanding their utility in various industrial processes (Casas-Godoy et al., 2018; Pizarro et al., 2022). Microbial lipases, particularly those from marine actinobacteria, have also attracted considerable attention due to their ability to function in extreme environments, including low temperatures and elevated pressure. These characteristics make them valuable in cosmetics, food processing, detergents, paper manufacturing, and organic synthesis (Rotter et al., 2021; Mehta et al., 2021).

Streptomyces species are the most thoroughly researched lipase-producing actinobacteria (Panyachanakul et al., 2020; Nithya et al., 2018). However, a number of other genera have also been found to be promising lipase producers, including *Micromonospora, Micrococcus, Kocuria, Corynebacterium, Streptosporangium, Actinomadura, Nocardiopsis, Nocardia, Elytrosporangium,* and *Actinopolyspora* (Hamedi et al., 2019; Mansour et al., 2015). The thermostable lipase MAS1 derived from marine *Streptomyces* sp. W007 exhibited maximum enzymatic activity at 40 °C and retained more than 80% of its activity after 1 h of incubation at 60 °C. Additionally, the enzyme was resistant to several metal ions and organic solvents, remained stable across a wide pH range (6.0–9.0), and effectively hydrolyzed triacylglycerols and diacylglycerols (Yuan et al., 2016).

Similarly, an alkaline lipase from *Streptomyces cellulosae* AU-10 was stable between pH 9.0 and 11.0 and 30 and 50 °C, and it was most active at pH 9.0 and 40 °C. It demonstrated metalloenzyme activity, hydrolyzed olive and sunflower oils effectively, and either retained or increased activity in the presence of oxidizers, surfactants, and commercial detergents such as Pril and Ariel (Boran, 2018).

4.4 Esterase

Esterases, classified within the lipolytic hydrolase family, are enzymes capable of catalyzing the hydrolysis and synthesis of ester bonds. They are particularly valued for their broad substrate specificity, independence from cofactors, and exceptional stability in organic solvents. These robust traits make them highly attractive biocatalysts for diverse industrial applications, including biodiesel and biopolymer synthesis, detergents, food processing, textiles, paper manufacturing, and environmental bioremediation (Barzkar et al., 2021; Dong et al., 2017; Parte et al., 2017).

A halotolerant esterase from *Rhodococcus* sp. LKE-028 exhibits thermoalkaliphilic and salt-tolerant characteristics. With a specific activity of 861 U/mg and a molecular weight of approximately 38 kDa, this enzyme demonstrates promising properties for use under extreme industrial conditions (Kumar et al., 2012). Similarly, Zhang et al. (2015) reported a novel alkaline esterase from South China Sea sediment that catalyzes at high pH but is susceptible to organic solvents. A cold-adapted esterase from *Pseudonocardia antitumoralis* SCSIO 01299 possesses exceptional enantioselectivity (ee > 99%) in surfactants. Combined with its cold-adaptation, stereo-selectivity, and surfactant compatibility, the enzyme exhibits surfactant-assisted kinetic resolution, particularly for the synthesis of chiral intermediates in the pharmaceutical and fine chemical industries (Cao et al., 2016).

4.5 Cellulases

Cellulases are enzymes that break down cellulose into smaller sugars likc glucose. Cellulases are exoglucanase, endoglucanase, and glucosidase. Endoglucanases break cellulose chains at internal points, creating more open ends, while exoglucanases act on ends to produce short sugar chains such as cellodextrins and cellobiose. Furthermore, glucosidases convert short chains into simple glucose (An et al., 2015). These enzymes are significant across many industries, including textile manufacturing, biofuel production, and bioremediation (John et al., 2022). The demand for industrial enzymes is gradually increasing. However, due to the high cost of production, large-scale production and application of enzymes are limited. *Cellulomonas uda* NCIM 2353 and *Paenibacillus elgii*, normal soil flora, produce cellulases for lignocellulose degradation. Their enzymes often lack stability under harsh conditions of high salinity, extreme pH, or elevated temperatures (Doan et al., 2024; Swathy et al., 2020). Alternatively, marine actinobacteria can also be a promising source of robust cellulases (Trivedi et al., 2016; John et al., 2022).

Cellulase-producing marine actinobacteria from various habitats have been studied in recent years. Marine sediments have been reported to contain various actinomycete genera with cellulolytic activity, such as *Actinoalloteichus* sp. MHA15 (Gobalakrishnan & Bhuvaneswari, 2019), *Streptomyces albidoflavus* (Fatokun et al., 2016), *Nocardiopsis dassonvillei* PSY13 (Sivasankar et al., 2022), and *Streptomyces* sp. NIOTNAVKKMA02 (John et al., 2022). They play an important role in biofuel production, converting lignocellulosic biomass into fermentable sugars for bioethanol. A cellulase from *Bacillus* sp. SR22, isolated from coral reefs, exhibited high salt tolerance, suggesting its potential for bioethanol production (Birolli et al., 2019). In the textile industry, cellulases from *Streptomyces* sp. are used to soften fabrics and remove excess dyes, thereby serving as an eco-friendly alternative to chemical treatments (Jayasekara et al., 2019). Additionally, these enzymes are used in the paper and pulp, food, and detergent industries (Ramírez et al., 2016; Singh et al., 2016).

4.6 Xylanase

Xylanase, an important hydrolytic enzyme, was first reported in 1955 and was initially termed as pentosanases (EC 3.2.1.8), which was later designated as endo-1,4-β-xylanase. Xylanase is a hydrolytic enzyme belonging to the family of glycosidases. It hydrolyzes xylan by cleaving a β-1,4 xylosidic linkage between xylan monomers, thus liberating xylose sugar. Xylan, a vital component of hemicellulose, is the second most abundant renewable biomass present in nature (Beg et al., 2001). Xylan is a branched heteropolysaccharide with a β-β-1,4-linked D-xylopyranosyl backbone that is sometimes substituted with other side chains (Collins et al., 2005). Being a complex polysaccharide, xylan requires several hydrolytic enzymes for its complete biodegradation. Xylanase plays a vital role in converting the complex polymer xylan into the simple sugar xylose (Pal & Khanum, 2010).

Bacteria and fungi widely produce xylanase. *Streptomyces* spp., members of the actinobacteria, have also been reported to produce xylanase. A few species of Streptomyces, when cultivated on xylan (sole carbon source), secreted different types of xylanases (Li et al., 2000). *Streptomyces viridosporus* T7A produces xylanase and cellulase from the lignocellulose waste (Ramachandra et al., 1987). Many Streptomyces strains produce xylanases. Some other xylanase-producing *Streptomyces* species are *Streptomyces halste* (Kaneko et al., 2000), *Streptomyces roseiscleroticus* (Georis et al., 2000), *Streptomyces lividans* (Ragauskas et al., 1994), *Streptomyces cyancus* SN32 (Ninawe et al., 2006), *Streptomyces olivaceus* (MSU3) (Sanjivkumar et al., 2017), and *Streptomyces variabilis* (MAB3) (Sanjivkumar et al., 2018a, b).

Ammonium sulfate precipitation is usually the first step in xylanase purification, followed by different chromatographic techniques. Chapla et al. (2012) purified xylanase from Paenibacillus sp. using gel permeation chromatography. A xylanase from *Streptomyces olivaceus* (MSU3) was purified up to 4.27-fold with 15.57% yield by ammonium sulfate precipitation, Diethylaminoethyl (DEAE)-anion exchange chromatography, and gel permeation chromatography on Sephadex G-75 (Sanjivkumar et al., 2017). A xylanase from *Jonesia denitiricans* BN-13 was purified up to 15-fold with a yield of 7.5% using ion-exchange chromatography with Q-sepharose (Lv et al., 2008). The investigated xylanases have varying molecular weights, for instance, 22 kDa for *Jonesia denitiricans* BN-13 (Lv et al., 2008), 29 kDa for *Cellulosimicrobium cellulans* (CKMX1) (Walia et al., 2014), and 36 kDa for *Aspergillus niger* (Coral et al., 2002).

Xylanases are commercially valuable enzymes having applications in diverse fields. In addition to the food and feed industries, xylanase plays a crucial role in the paper and pulp industries, as it improves product quality relative to chemical bleaching and thus reduces environmental damage (Beg et al., 2001; Viikari, 1994; Bhat, 2000). Xylanase also contributes significantly to several industrial processes, like the separation of starch from gluten and the retting of flax and other materials. Xylanase is also used to treat waste generated by agriculture and the food industry (Prade, 1995; Mielenz, 2001; Saha, 2003). Thus, xylanases have proven to be valuable enzymes and may be used in many other processes in the future.

4.7 Other Enzymes

In addition to the hydrolases discussed above, extremophilic actinobacteria produce a diverse range of non-hydrolase enzymes. Such as those catalyzing oxidation–reduction or the transfer of specific functional groups. Some oxidoreductases include cholesterol oxidase, peroxidase, catechol 1,2-dioxygenase, tyrosinase, and laccase. A cholesterol oxidase from a thermophilic strain, *Saccharopolyspora hirsuta* VKM Ac-666^T, metabolizes steroids (Lobastova et al., 2021). Cholesterol oxidases are widely used to monitor serum and food cholesterol levels to assess atherosclerosis, lipid-related disorders, and cardiac diseases (El-Naggar &

El-Shweihy, 2020). Another oxidoreductase, laccase, uses anilines, polyphenols, and phenols as substrates and is employed in various sectors, including bioremediation, cork production, teeth whitening, and textile coloring (Espina et al., 2021). A psychrotolerant *Arthrobacter* sp. isolated from Antarctica synthesizes catechol 1,2-dioxygenase, with potential applications in the bioremediation of phenolic compounds in cold environments (Lee et al., 2022). Furthermore, a tyrosinase from the marine actinobacterium *Streptomyces espinosus* strain LK4 efficiently removes phenolic compounds from water (Roy et al., 2014). Table 1 summarizes representative extremozymes derived from marine actinobacteria, highlighting their optimal physicochemical conditions and potential industrial uses.

5 Adaptations of Extremozymes Under Harsh Conditions

Marine actinobacteria isolated from diverse and extreme oceanic habitats such as deep-sea sediments, mangrove ecosystems, hydrothermal vents, marine caves, salt marshes, and deserts are prolific producers of extremozymes (Subramani & Aalbersberg, 2013). These microorganisms are evolutionarily adapted to extremes in pH, salinity, temperature, and nutrient availability (Yaradoddi & Kontro, 2021). Halophilic extremozymes have evolved distinct structural and biochemical features that enable catalysis and stability. For instance, such enzymes have more glutamate and aspartate residues on their surfaces to combat dehydrating environments, thereby favoring water-solubilization of proteins (Bhatt & Singh, 2020; Oren, 2013). Moreover, the enzymes exhibit weak hydrophobic interactions that enhance their flexibility (Brininger et al., 2018).

Actinobacterial extremozymes are lipases, amylases, proteases, chitinases, cellulases, and xylanases produced by marine actinobacteria that can tolerate multiple extremities due to evolutionary adaptation. *Streptomyces albidofalvus* isolated from a hot spring of Telangana produces an α-amylase at 55 °C and pH 9 (Cherifa et al., 2023). Similarly, authors have reported that haloalkaliphilic actinobacteria *Nocardiopsis* sp. Mit-7 can optimally produce recombinant protease at 65 °C and is active at pH 7–13 (Thakrar et al., 2023). These adaptive characteristics of actinobacteria toward extreme environments reflect their significance as potential reservoirs of industrially relevant extremozymes.

6 Applications of Extremozymes From Marine Actinobacteria

Extremozymes produced by marine actinobacteria exhibit exceptional stability and catalytic efficiency, making them highly desirable for a range of processes. The robustness of these enzymes reduces dependence on controlled processing environments. It enhances the efficiency and sustainability of various applications, ranging from pharmaceuticals and agriculture to bioenergy and environmental remediation.

As demand for eco-friendly, high-performance biocatalysts continues to rise, extremozymes represent a promising frontier for developing innovative solutions to modern industrial challenges.

6.1 Food Industry

Enzymes, owing to their high reaction rates and specificity, have become a favored choice in the food processing industry. Several microbial products, such as enzymes, pigments, and other secondary metabolites, have played a crucial role in improving the quality and taste of various food products. Enzymes have diverse applications, ranging from fruit and vegetable juice clarification to the brewing and baking industries (van Oort, 2010; Simpson et al., 2012).

Enzymes used for commercial purposes can be obtained either by extraction from living tissues or by microbial fermentation (Dodge, 2010). According to one report, in 2014, approximately 260 enzymes were made available in the European Union, of which 91% were obtained through microbial fermentation (Fraatz et al., 2014).

However, the commercial production of the enzymes is constrained by stringent process conditions. Many of these problems can be efficiently addressed by promoting the enzymes from extremophiles. Several processes in the food industry have been improved using extremophilic enzymes. For instance, haloalkaliphilic xylanase is now widely used in fruit and vegetable processing to improve maceration and juice clarification, thereby reducing juice viscosity (Bhat, 2000; Galante et al., 1998; Wong & Saddler, 1993). Furthermore, in the bakery industry, xylanase, amylase, and cellulase play crucial roles in improving dough elasticity, facilitating handling, and enhancing bread quality (Chauhan et al., 2023; Dutron et al., 2004; Bhat, 2000; Maat, 1992). Xylanase reduces the non-starchy content in the animal feeds and thus improves the intestinal viscosity, resulting in improved utilization of nutrition (Bhat, 2000; Mathlouthi et al., 2002, 2003). Xylanase, upon hydrolyzing xylan, produces xylooligosaccharides, a few of which exhibit prebiotic effects and may be useful in developing functional foods.

In addition to xylanases, enzymes such as pectinases, carboxymethylcellulases, and amylases are used as clarifiers (Motta et al., 2013) and improve filtration efficiency during brewing processes (Wang et al., 2016). Some extremophilic enzymes, such as halophilic amylase, have been applied in the starch liquefaction process for the preparation of sugar syrups. Furthermore, some amylolytic enzymes have been reported to be used to reduce the viscosity of sugar syrups and fruit juices during starch saccharification (Pandey et al., 2000; Pandey & Singh, 2012; Singh et al., 2013). Amylase is also crucial as an anti-staling agent of breads (Bock, 2024).

Cellulases, in addition to clarifying juice, convert cellulose into fermentable sugars suitable for human consumption. Additionally, cellulase can be used to produce bioethanol and valuable organic compounds by degrading certain agricultural residues (Hardiman et al., 2010). Cellulase is widely obtained from Streptomyces and other cellulolytic bacteria (Joo & Chang, 2005; Bui, 2014). Other applications of enzymes in fields such as textile processing, paper and pulp production, additives in animal feed, and antioxidants in food have been reported earlier (Jones et al., 2004; Gohel et al., 2015).

Furthermore, halophilic hydrolytic enzymes can function under adverse conditions required for industrial processes. Extremophilic enzymes can revolutionize industrial processes by serving as alternatives to conventional methods.

6.2 Pharmaceutical Industry

Marine actinobacteria are prolific producers of enzymes relevant to pharmaceutical applications, including ribonucleases, cholesterol oxidase, lipase, laccase, and esterases. These enzymes create compounds in their purest form, which is important in synthesizing medicines (Zhao et al., 2016; Chandra et al., 2020; González et al., 2020; Barzkar et al., 2021). A recent review highlights that actinobacteria, particularly Streptomyces spp., serve as precise biocatalysts. They handle regio-, chemo-, and enantio-selective reactions perfectly for creating chiral pharmaceutical intermediates (Salama et al., 2024). *Streptomyces* sp. and *Nocardiopsis* sp. lipases are used in the production of optically active compounds and intermediates for anti-inflammatory drugs (Rodríguez-Alonso et al., 2023). Moreover, non-steroidal anti-inflammatory drugs make use of esterases of the *Micromonospora* sp. to convert drug precursors (Hifnawy et al., 2020). β-glucosidases from *Streptomyces* sp. are utilized to activate glycosylated prodrugs at the target site in tissue, thereby reducing the toxicity of anticancer therapies (Martin et al., 2022).

The growing threat of antibiotic resistance has prompted the search for new medicines. Marine actinobacteria produce unique compounds and enzymes that are resistant to antibiotics. Marine *Streptomyces* sp. produces potent antibiotics and extracellular enzymes (amylases and proteases) that are involved in the production of novel antimicrobial compounds with antibiotic effects against multidrug-resistant pathogens, including methicillin-resistant *Staphylococcus aureus* and vancomycin-resistant *Enterococcus* (Norouzi et al., 2018; Pan et al., 2025). The high diversity of enzymes in marine actinobacteria enables the development of a new generation of medicines.

6.3 Textile Industry

The textile industry has evolved to produce a range of fibers, including viscose, cotton, and polyesters (Felgueiras et al., 2021). Due to the use of chemical processes in the textile industry, fabric quality is compromised, and these processes also generate toxic effluents (Madhav et al., 2018). To address these challenges, enzymes offer a sustainable alternative to conventional mechanical and chemical processes used in the textile industry. Currently, enzymes are used in bleaching, desizing, bio-polishing, dyeing, scouring, color stripping, and bio-washing (Rahman, 2021). *Dactylosporangium, Nonomuraea, Streptosporangium, Asanoa, Kitasatospora,*

Nocardia, and *Streptomyces* are the major actinobacterial producers of extremozymes (Putri & Setiawan, 2019; Meliani et al., 2022). Several actinobacterial enzymes, including chitinase, cellulase, and lipase, are used in the textile industry to degrade polymers (Ashok et al., 2022). These enzymes enhance product efficiency and quality by reducing chemical use, energy consumption, and water requirements (Behera et al., 2021; Boodhoo et al., 2022). Actinobacterial enzymes have potential for use in the treatment of textile effluents as decolorizers. For instance, laccase-producing *Streptomyces* decolorizes 84% of indigo carmine and 56% of diamond black (Cuebas-Irizarry & Grunden, 2023). Another actinobacterial enzyme, cellulase, is widely used in biofinishing of denim and cotton (Mukhtar et al., 2017). In biobleaching and scouring, marine actinobacterial laccase can be used. Moreover, laccase produced by *Streptomycese* aids in the removal of non-cellulosic impurities and lignin from fabrics of cotton (Jeyabalan et al., 2023).

6.4 Bioenergy

Actinobacteria have emerged as potential candidates for the production of sustainable bioenergy. In the conversion of biomass into biofuels, marine actinobacterial enzymes are used (Thanigaivel et al., 2022). By utilizing the lignocellulosic or lipid feedstocks via microbial fermentation, second-generation biofuels such as biodiesel and bioethanol are produced (Aslam & Rashid, 2021).

Marine actinobacteria are widely reported to produce enzymes such as cellulase, hemicellulase, and xylanase, which can be used for bioenergy production because they are biomass-degrading enzymes (Lewin et al., 2016). For example, *Nocardiopsis synnemataformans* produces cellulase enzyme that remains active over a wide range of pH, temperature, and salinity. This strain produces bean straw hydrolysate, which is subsequently used by *Saccharomyces cerevisiae* for fermentation of bioethanol (El-Sayed et al., 2024).

Recently published review emphasized that several marine actinobacterial genera, such as *Streptomyces*, *Micromonospora*, and *Rhodococcus*, produce enzyme cocktails that function under extreme conditions, making them suitable candidates for bioethanol production (John et al., 2022). Additionally, marine actinobacteria can produce cold-active cellulases that hydrolyze seagrass and yield bioethanol (Swain et al., 2017).

6.5 Detergent Industry

Extremozymes derived from marine actinobacteria have gained prominence in the detergent industry due to their resilience and efficiency under harsh washing conditions, including high temperatures, alkaline pH, and the presence of surfactants, oxidizing agents, and metal ions. Unlike conventional enzymes, these biocatalysts retain high catalytic activity and structural stability, making them ideal for modern, eco-conscious formulations for effective cleaning with reduced environmental impact (Adeoyo, 2020; El-Gendi et al., 2021; Al-Ghanayem et al., 2022; Singh et al., 2013; Gohel & Singh, 2018; Sharma et al., 2020).

As highlighted in the sections above, marine actinomycetes produce a diverse array of hydrolytic enzymes that function under extreme conditions, making them effective stain and dirt removers. For example, amylases hydrolyze starch residues in food, proteases degrade protein-based stains such as blood and dairy, and lipases effectively break down fats and grease. Cellulases and xylanases prevent fiber damage, enhance brightness, and reduce pilling (Salwan & Sharma, 2018). A thermostable amylase from *Actinomadura keratinilytica* sp. Cpt29 has potential in detergent formulations, as it functions at high temperatures and in the presence of commercial additives (Kherouf et al., 2021). Similarly, extremophilic lipases and proteases from *Streptomyces* species have proven their effectiveness in alkaline environments and in the presence of Ethylenediaminetetraacetic acid (EDTA), Sodium dodecyl sulfate (SDS), and hydrogen peroxide. These extremozymes improve the effectiveness of detergents in both cold and hot water, enabling shorter wash cycles with lower energy use. Thus, they are vital in developing sustainable, efficient, and environmentally friendly processes (El-Gendi et al., 2021).

6.6 Environmental Applications

Actinomycetes are a diverse and metabolically versatile group of bacteria with potential environmental and biotechnological applications. Biotransformation and biodegradation can convert underutilized urban and agricultural waste into valuable compounds (Devanshi et al., 2021). The actinomycetes produce many enzymes, including proteases, lipases, cellulases, ligninases, xylanases, and oxidoreductases that can degrade hydrocarbons, agrochemicals, and heavy metals. *Streptomyces*, *Rhodococcus*, *Arthrobacter*, and *Nocardia* can potentially treat complex pollutants (Chaturvedi & Khurana, 2019; Devanshiet al., 2021). Marine actinomycetes can remediate industrial and agricultural contaminants by degrading organic waste through their enzymes (Kim et al., 2012; Trincone, 2012). Oil spills pose serious threats to marine ecosystems. It can be mitigated using salt-tolerant, petroleum-degrading enzymes produced by marine actinomycetes, providing a sustainable, environmentally friendly alternative to conventional cleanup methods (Chaturvedi & Khurana, 2019). The need for efficient, environmentally friendly remediation strategies has gained attention amid ongoing industrialization and intensive agricultural practices. Actinomycetes have the potential to metabolize a wide range of pollutants into non-toxic by-products, supporting sustainable bioremediation and green technologies.

7 Conclusion

Marine actinobacteria represent a valuable and largely unexplored source of extremozymes adapted to function under high salinity, extreme temperatures, elevated pressure, and nutrient-limited conditions. These adaptations impart exceptional stability and catalytic efficiency, enabling robust performance under industrial processing environments where conventional enzymes often fail. Consequently, marine actinobacterial enzymes have attracted growing interest for applications in pharmaceuticals, bioremediation, bioenergy, food processing, and specialty chemical synthesis, particularly due to the functional diversity.

Despite their promise, systematic exploration of these extremozymes remains limited by challenges in strain isolation, cultivation, enzyme yield, and replication of marine growth conditions. Advances in genome mining, metagenomics, and computational enzyme discovery have accelerated identification of novel biocatalysts; however, their translation into industrially viable processes will require further progress in heterologous expression, protein engineering, and scalable downstream processing.

Acknowledgments RJ, SB, JR, and HB acknowledge the support provided by Atmiya University, Rajkot, India. SB and RJ acknowledge the ScHeme of Developing High Quality Research Fellowship by the Government of Gujarat. RS is grateful to Christ College, Rajkot, for its support. HB acknowledges Charotar University of Science and Technology, Anand for support. HB also acknowledges CSIR-Direct SRF and UGC Meritorious Fellowship at the Department of Biosciences, Saurashtra University. SPS is grateful for the Department of Biotechnology (DBT) Multi-Institutional Project, Ministry of Earth Science (MoES) Networking Project, University Grants Comission (UGC) - Basic Science Research (BSR) Faculty Fellowship, and International Travel Fellowships from Department of Science and Technology (DST) - Science and Engineering Research Board (SERB), Council of Scientific and Industrial Research (CSIR), Department of Biotechnology (DBT), and University Grants Comission (UGC). Furthermore, SPS acknowledges the facilities created under the University Grants Comission (UGC) - Center for Advanced Study (CAS) and Department of Science and Technology (DST) - Fund for improvement of Science and Technology (FIST), as well as those provided by Saurashtra University.

References

Adeoyo, O. R. (2020). Amylase and biosurfactant production from Bacillus species. *International Journal of Current Microbiology and Applied Sciences, 9*(5), 3183–3191.

Adiguzel, G., Faiz, O., Sisecioglu, M., Sari, B., Baltaci, O., Akbulut, S., et al. (2019). A novel endo-β-1, 4-xylanase from *Pediococcus acidilactici* GC25; purification, characterization and application in clarification of fruit juices. *International Journal of Biological Macromolecules, 129*, 571–578.

Aguiar, M. S., Maldonado, R. R., Carvalho, A. L., & Aguiar-Oliveira, E. (2022). Production of actinobacteria amylase by fermentation in solid state using residues of licuri palm (*Syagrus coronata*). In *Methods in actinobacteriology* (pp. 495–503). Springer.

Al-Ghanayem, A. A., Joseph, B., Alhussaini, M. S., & Ramteke, P. W. (2022). Current applications and future trends of extremozymes in detergent industries. In *Microbial extremozymes* (pp. 223–230). Academic.

An, T., Dong, Z., Lv, J., Liu, Y., Wang, M., Wei, S., Song, Y., Zhang, Y., & Deng, S. (2015). Purification and characterization of a salt-tolerant cellulase from the mangrove oyster, *Crassostrea rivularis*. *Acta Biochimica et Biophysica Sinica, 47*(4), 299–305.

Angelakis, G. N., Psarologaki, C., Pirintsos, S., & Kotzabasis, K. (2024). Extremophiles and extremophilic behaviour—New insights and perspectives. *Life, 14*(11), 1425.

Ashaolu, T. J., Malik, T., Soni, R., Prieto, M. A., & Jafari, S. M. (2025). Extremophilic microorganisms as a source of emerging enzymes for the food industry: A review. *Food Science & Nutrition, 13*(1), e4540.

Ashok, S., Siddanna, R., Sirasagar, R., Bhalerao, S., & Dayanand, A. (2022). Upsurge production of cellulase from maize stover under soildstate conditions mediated by *Streptomyces enissocaesilis* DQ026641. *Journal Of Applied Biology & Biotechnology, 10*(01), 136–144.

Aslam, M., & Rashid, N. (2021). Bioenergy production in extremophiles. In *Microbial extremozymes* (pp. 231–246). Elsevier eBooks.

Barzkar, N., Sohail, M., Tamadoni Jahromi, S., Gozari, M., Poormozaffar, S., Nahavandi, R., & Hafezieh, M. (2021). Marine bacterial esterases: Emerging biocatalysts for industrial applications. *Applied Biochemistry and Biotechnology, 193*, 1187–1214.

Beg, Q., Kapoor, M., Mahajan, L., & Hoondal, G. S. (2001). Microbial xylanases and their industrial applications: A review. *Applied Microbiology and Biotechnology, 56*(3), 326–338.

Behera, M., Nayak, J., Banerjee, S., Chakrabortty, S., & Tripathy, S. K. (2021). A review on the treatment of textile industry waste effluents towards the development of efficient mitigation strategy: An integrated system design approach. *Journal of Environmental Chemical Engineering, 9*(4), 105277. https://doi.org/10.1016/j.jece.2021.105277

Bhat, M. K. (2000). Cellulases and related enzymes in biotechnology. *Biotechnology Advances, 18*, 355–383.

Bhatt, H. B., & Singh, S. P. (2020). Cloning, expression, and structural elucidation of a biotechnologically potential alkaline serine protease from a newly isolated haloalkaliphilic *Bacillus lehensis* JO-26. *Frontiers in Microbiology, 11*, 941.

Bhatt, H. B., & Singh, S. P. (2022). Diversity of cultivable bacteria in a saline desert of Little Rann of Kutch, India: A phylogenetic perspective. *Frontiers in Marine Science, 9*, 769043.

Bhatt, H. B., Gohel, S. D., & Singh, S. P. (2018). Phylogeny, novel bacterial lineage and enzymatic potential of haloalkaliphilic bacteria from the saline coastal desert of Little Rann of Kutch, Gujarat, India. *3 Biotech, 8*(1), 53.

Bhatt, B. M., Trivedi, U. B., & Patel, K. C. (2020). Extremophilic amylases: Microbial production and applications. In *Microbial enzymes: Roles and applications in industries* (pp. 185–205). Springer Singapore.

Birolli, W. G., Lima, R. N., & Porto, A. L. M. (2019). Applications of marine-derived microorganisms and their enzymes in biocatalysis and biotransformation, the underexplored potentials. *Frontiers in Microbiology, 10*(AUG), 452098.

Bock, J. E. (2024). Enzymes in breadmaking. In *Improving and tailoring enzymes for food quality and functionality* (pp. 217–239). Woodhead Publishing.

Boodhoo, K., Flickinger, M., Woodley, J., & Emanuelsson, E. (2022). Bioprocess intensification: A route to efficient and sustainable biocatalytic transformations for the future. *Chemical Engineering and Processing Process Intensification, 172*, 108793.

Boran, R. (2018). Detergent compatible extracellular lipase from *Streptomyces cellulosae* Au-10: A green alternative for the detergent industry. *Journal of Surfactants and Detergents, 21*(4), 565–573.

Brininger, C., Spradlin, S., Cobani, L., & Evilia, C. (2018). The more adaptive to change, the more likely you are to survive: Protein adaptation in extremophiles. *Seminars in Cell and Developmental Biology, 84*, 158–169.

Bui, H. B. (2014). Isolation of cellulolytic bacteria, including actinomycetes, from coffee exocarps in coffee-producing areas in Vietnam. *International Journal of Recycling of Organic Waste in Agriculture, 3*, 48.

Cao, Y., Deng, D., Sun, A., Zhang, Y., & Hu, Y. (2016). Functional characterization of a novel marine microbial esterase and its utilization in the enantioselective preparation of (R)-methyl 2-chloropropionate. *Applied Biochemistry and Biotechnology, 180*(2), 210–227.

Casas-Godoy, L., Gasteazoro, F., Duquesne, S., Bordes, F., Marty, A., & Sandoval, G. (2018). Lipases: An overview. In *Lipases and phospholipases: Methods and protocols* (pp. 3–38). Humana Press.

Cera, G., Risdian, C., Pira, H., & Wink, J. (2022). Antimicrobial potential of culturable actinobacteria isolated from the Pacific oyster *Crassostrea gigas* (Bivalvia, Ostreidae). *Journal of Applied Microbiology, 133*(2), 1099–1114.

Chandra, P., Enespa, Singh, R., & Arora, P. K. (2020). Microbial lipases and their industrial applications: a comprehensive review. *Microbial Cell Factories, 19*(1), 169.

Chapla, D., Patel, H., Singh, A., Madamwar, D., & Shah, A. (2012). Production, purification and properties of a cellulase-free thermostable endoxylanase from newly isolated Paenibacillus sp. ASCD2. *Annals of Microbiology, 62*(2), 825–834.

Chatterjee, A., Puri, S., Sharma, P. K., Deepa, P. R., & Chowdhury, S. (2023). Nature-inspired Enzyme engineering and sustainable catalysis: Biochemical clues from the world of plants and extremophiles. *Frontiers in Bioengineering and Biotechnology, 11*, 1229300.

Chaturvedi, S., & Khurana, S. P. (2019). Importance of actinobacteria for bioremediation. In *Plant biotechnology: Progress in genomic era* (pp. 277–307). Springer.

Chauhan, J., Shukla, R. J., Bishoyi, A. K., Goyal, S., & Sanghvi, G. V. (2023). Investigation of physical, nutritional and sensory properties of wheat bread treated with purified thermostable cellulase and alpha amylase. *Cogent Food & Agriculture, 9*, 2261839.

Chen, P., Zhang, L., Guo, X., Dai, X., Liu, L., Xi, L., Wang, J., Song, L., Wang, Y., Zhu, Y., Huang, L., & Huang, Y. (2016). Diversity, biogeography, and biodegradation potential of actinobacteria in the deep-sea sediments along the southwest Indian ridge. *Frontiers in Microbiology, 7*(AUG), 1340.

Cherifa, L., Meissa, M., Uzel, A., & Allaoueddine, B. (2023). Thermophilic actinobacteria isolated from TLEGHMA hot spring: A potential source of thermostable α-amylase. *Carpathian Journal of Food Science and Technology, 15*(4), 168–182.

Collins, T., Gerday, C., & Feller, G. (2005). Xylanases, xylanase families and extremophilic xylanases. *FEMS Microbiology Reviews, 29*(1), 3–23.

Coral, G., Arikan, B. U. R. H. A. N., Unaldi, M. N., & Guvenmez, H. K. (2002). Some properties of thermostable xylanase from an *Aspergillus niger* strain. *Annals of Microbiology, 52*(3), 299–306.

Cuebas-Irizarry, M. F., & Grunden, A. M. (2023). Streptomyces spp. as biocatalyst sources in pulp and paper and textile industries: Biodegradation, bioconversion and valorization of waste. *Microbial Biotechnology, 17*(1), e14258.

Das, K. K., Pattnaik, S., & Behera, S. K. (2022). Application of actinomycetes in secondary metabolites production: An overview. *Journal of Emerging Technologies and Innovative Research, 9*(9), 150–169.

de Souza, H. F., Borges, L. A., Gonçalves, V. D. D. P., Dos Santos, J. V., Bessa, M. S., Carosia, M. F., et al. (2022). Recent advances in the application of xylanases in the food industry and production by actinobacteria: A review. *Food Research International, 162*, 112103.

Devanshi, S., Shah, K. R., Arora, S., & Saxena, S. (2021). Actinomycetes as an environmental scrubber. In *Crude oil—New technologies and recent approaches*. IntechOpen.

Doan, C. T., Tran, T. N., Pham, T. P., Tran, T. T. T., Truong, B. P., Nguyen, T. T., Nguyen, T. M., Bui, T. Q. H., Nguyen, A. D., & Wang, S. L. (2024). Production, purification, and characterization of a cellulase from *Paenibacillus elgii*. *Polymers, 16*(14), 2037.

Dobariya, A., Mankad, G. P., Ramavat, H., & Singh, S. P. (2023). Efficacy of the fruit and vegetable peels as substrates for the growth and production of α-amylases in marine actinobacteria. *Applied Biochemistry and Biotechnology, 195*(12), 7603–7623.

Dodge, T. (2010). Production of industrial enzymes. In R. J. Whitehurst & M. van Oort (Eds.), *Enzymes in food technology* (2nd ed., pp. 44–56). Wiley.

Dong, H., Secundo, F., Xue, C., & Mao, X. (2017). Whole-cell biocatalytic synthesis of cinnamyl acetate with a novel esterase from the DNA library of *Acinetobacter hemolyticus*. *Journal of Agricultural and Food Chemistry, 65*(10), 2120–2128.

Dutron, A., Georis, J., Genot, B., Dauvrin, T., Collins, T., Hoyoux, A., & Feller, G. (2004). *Use of family 8 enzymes with xylanolytic activity in baking*. World Intellectual Property Organization, PCT, WO 2004/023879 A1.

El-Gendi, H., Saleh, A. K., Badierah, R., Redwan, E. M., El-Maradny, Y. A., & El-Fakharany, E. M. (2021). A comprehensive insight into fungal enzymes: Structure, classification, and their role in mankind's challenges. *Journal of Fungi, 8*(1), 23.

Elleuche, S., Schröder, C., Sahm, K., & Antranikian, G. (2014). Extremozymes—Biocatalysts with unique properties from extremophilic microorganisms. *Current Opinion in Biotechnology, 29*(1), 116–123.

El-Naggar, N. E. A., & El-Shweihy, N. M. (2020). Identification of cholesterol-assimilating actinomycetes strain and application of statistical modeling approaches for improvement of cholesterol oxidase production by *Streptomyces anulatus* strain NEAE-94. *BMC Microbiology, 20*(1), 86.

El-Sayed, M. H., Elsayed, D. A., & Gomaa, A. E. R. F. (2024). *Nocardiopsis synnemataformans* NBRM9, an extremophilic actinomycete producing extremozyme cellulase, using lignocellulosic agro-wastes and its biotechnological applications. *AIMS Microbiology, 10*(1), 187.

Espina, G., Atalah, J., & Blamey, J. M. (2021). Extremophilic oxidoreductases for the industry: Five successful examples with promising projections. *Frontiers in Bioengineering and Biotechnology, 9*, 710035.

Fatokun, E. N., Nwodo, U. U., & Okoh, A. I. (2016). Classical optimization of cellulase and xylanase production by a marine Streptomyces species. *Applied Sciences, 6*(10), 286.

Felgueiras, C., Azoia, N. G., Gonçalves, C., Gama, M., & Dourado, F. (2021). Trends on the cellulose-based textiles: Raw materials and technologies. *Frontiers in Bioengineering and Biotechnology, 9*, 608826.

Fernandes, T. A. R., Da Silveira, W. B., Passos, F. M. L., & Zucchi, T. D. (2013). Characterization of a thermotolerant laccase produced by Streptomyces sp. SB086. *Annals of Microbiology, 64*(3), 1363–1369.

Fraatz, M. A., Rühl, M., & Zorn, H. (2014). Food and feed enzymes. *Advances in Biochemical Engineering/Biotechnology, 143*, 229–256.

Galante, Y. M., De Conti, A., & Monteverdi, R. (1998). Application of Trichoderma enzymes in food and feed industries. In G. E. Harman & C. P. Kubicek (Eds.), *Trichoderma and gliocladium—Enzymes, biological control and commercial applications* (pp. 327–342). Taylor and Francis.

Georis, J., Giannotta, F., De Buyl, E., Granier, B., & Frère, J. M. (2000). Purification and properties of three endo-β-1, 4-xylanases produced by Streptomyces sp. strain S38 which differ in their ability to enhance the bleaching of kraft pulps☆. *Enzyme and Microbial Technology, 26*(2–4), 178–186.

Gobalakrishnan, R., & Bhuvaneswari, R. (2019). Microbial fuel cells potential of marine actinobacteria Actinoalloteichus sp. MHA15 from the Havelock island of the Andamans, India. *Biotechnology Research and Innovation, 3*(1), 144–158.

Gohel, S. D., & Singh, S. P. (2018). Molecular phylogeny and diversity of the salt-tolerant alkaliphilic actinobacteria inhabiting coastal Gujarat, India. *Geomicrobiology Journal, 35*(9), 775–789.

Gohel, S. D., Sharma, A. K., Dangar, K. G., Thakrar, F. J., & Singh, S. P. (2015). Antimicrobial and biocatalytic potential of haloalkaliphilic actinobacteria. In *Halophiles biodiversity and sustainable exploitation* (pp. 59–71). Springer.

Gohel, S. D., Majithiya, V. R., & Singh, S. P. (2023). Genetic and physiological diversity of marine actinobacteria from the Okha Coast, Gujarat, India. *Geomicrobiology Journal, 40*(6), 590–604.

González, V., Vargas-Straube, M. J., Beys-Da-Silva, W. O., Santi, L., Valencia, P., Beltrametti, F., & Cámara, B. (2020). Enzyme bioprospection of marine-derived actinobacteria from the

Chilean coast and new insight in the mechanism of keratin degradation in Streptomyces sp. G11C. *Marine Drugs, 18*(11), 537.

Gupta, S. K., Kataki, S., Chatterjee, S., Prasad, R. K., Datta, S., Vairale, M. G., Sharma, S., Dwivedi, S. K., & Gupta, D. K. (2020). Cold adaptation in bacteria with special focus on cellulase production and its potential application. *Journal of Cleaner Production, 258*, 120351.

Gupta, N., Paul, J. S., & Jadhav, S. K. (2024). Biovalorizing agro-waste 'de-oiled rice bran' for thermostable, alkalophilic and detergent stable α-amylase production with its application as laundry detergent additive and textile desizer. *International Journal of Biological Macromolecules, 256*, 128470.

Hamedi, J., Kafshnouchi, M., & Ranjbaran, M. (2019). A study on actinobacterial diversity of Hampoeil cave and screening of their biological activities. *Saudi Journal of Biological Sciences, 26*(7), 1587–1595.

Hardiman, E., Gibbs, M., Reeves, R., & Bergquist, P. (2010). Directed evolution of a thermophilic betaglucosidase for cellulosic bioethanol production. *Applied Biochemistry and Biotechnology, 161*, 301–312.

Hassanein, W. A., Kotb, E., Awny, N. M., & El-Zawahry, Y. A. (2011). Fibrinolysis and anticoagulant potential of a metallo protease produced by *Bacillus subtilis* K42. *Journal of Biosciences, 36*, 773–779.

Hazari, S., Tabassum, U., Siddiqui, A. J., Hazari, S., Ravindar, A., Mandal, S. K., et al. (2022). Actinobacteria in natural product research: Avenues and challenges. In *Role of microbes in industrial products and processes* (pp. 171–193). Wiley.

Hifnawy, M. S., Fouda, M. M., Sayed, A. M., Mohammed, R., Hassan, H. M., AbouZid, S. F., Rateb, M. E., Keller, A., Adamek, M., Ziemert, N., & Abdelmohsen, U. R. (2020). The genus Micromonospora as a model microorganism for bioactive natural product discovery. *RSC Advances, 10*(35), 20939.

Huang, J., Zhang, Y., & Hu, Y. (2016). Functional characterization of a marine *Bacillus esterase* and its utilization in the stereo-selective production of D-methyl lactate. *Applied Biochemistry and Biotechnology, 180*(8), 1467–1481.

Jagannathan, S. V., Manemann, E. M., Rowe, S. E., Callender, M. C., & Soto, W. (2021). Marine actinomycetes, new sources of biotechnological products. *Marine Drugs, 19*(7), 365.

Jayasekara, S., Ratnayake, R., Jayasekara, S., & Ratnayake, R. (2019). Microbial cellulases: An overview and applications. In *Cellulose*. IntechOpen.

Jeyabalan, J., Veluchamy, A., Priyan, V. V., Kumar, A., Chandrasekar, R., & Narayanasamy, S. (2023). A review on the laccase assisted decolourization of dyes: Recent trends and research progress. *Journal of the Taiwan Institute of Chemical Engineers, 151*, 105081.

John, J. A., Samuel, M. S., Govarthanan, M., & Selvarajan, E. (2022). A comprehensive review on strategic study of cellulase producing marine actinobacteria for biofuel applications. *Environmental Research, 214*, 114018.

Jones, B. E., Wilhelmus, A. H. K., Solingen, P. V., & Weyler, W. (2004). *Cellulase producing actinomycetes, cellulase produced there from and method of producing same*. EP 1408108 B1.

Joo, H. S., & Chang, C. S. (2005). Oxidant and SDS stable alkaline protease from a halo-tolerant *Bacillus clausii* I-52: Enhanced production and simple purification. *Journal of Applied Microbiology, 98*(2), 491–497.

Kaneko, S., Kuno, A., Muramatsu, M., Iwamatsu, S., Kusakabe, I., & Hayashi, K. (2000). Purification and characterization of a family G/11 β-xylanase from *Streptomyces olivaceoviridis* E-86. *Bioscience, Biotechnology, and Biochemistry, 64*(2), 447–451.

Karan, R., Mathew, S., Muhammad, R., Bautista, D. B., Vogler, M., Eppinger, J., et al. (2020). Understanding high-salt and cold adaptation of a polyextremophilic enzyme. *Microorganisms, 8*(10), 1594.

Khalil, M. A., El-Shanshoury, A. E. R. R., Alghamdi, M. A., Sun, J., & Ali, S. S. (2022). *Streptomyces catenulae* as a novel marine actinobacterium mediated silver nanoparticles: Characterization, biological activities, and proposed mechanism of antibacterial action. *Frontiers in Microbiology, 13*, 833154.

Kherouf, M., Habbeche, A., Benamia, F., Saoudi, B., Kerouaz, B., & Ladjama, A. (2021). Statistical optimization of a novel extracellular alkaline and thermostable amylase production from thermophilic *Actinomadura keratinilytica* sp. Cpt29 and its potential application in detergent industry. *Biocatalysis and Agricultural Biotechnology, 35*, 102068.

Kikani, B., Patel, R., Thumar, J., Bhatt, H., Rathore, D. S., Koladiya, G. A., & Singh, S. P. (2023). Solvent tolerant enzymes in extremophiles: Adaptations and applications. *International Journal of Biological Macromolecules, 238*, 124051.

Kim, D. G., Moon, K., Kim, S. H., Park, S. H., Park, S., Lee, S. K., ... & Oh, D. C. (2012). Bahamaolides A and B, antifungal polyene polyol macrolides from the marine actinomycete Streptomyces sp. *Journal of Natural Products, 75*(5), 959–967.

Kulkarni, C. P., & Maurya, C. B. (2017). Characterization of the cellulase enzyme produced by actinomycetes isolated from the mangrove coastal areas. *Biosciences, Biotechnology Research Asia, 14*(2), 685–690.

Kumar, P., & Singh, J. (2024). Exploring enzyme biotechnology's role in green chemistry and advancements in pharmaceutical technologies. In *Enzyme biotechnology for environmental sustainability* (pp. 465–495). Academic.

Kumar, L., Singh, B., Adhikari, D. K., Mukherjee, J., & Ghosh, D. (2012). A thermoalkaliphilic halotolerant esterase from Rhodococcus sp. LKE-028 (MTCC 5562): Enzyme purification and characterization. *Process Biochemistry, 47*(6), 983–991.

Laye, V. J., Solieva, S., Voelz, V. A., & DasSarma, S. (2022). Effects of salinity and temperature on the flexibility and function of a polyextremophilic enzyme. *International Journal of Molecular Sciences, 23*(24), 15620.

Lee, G. L. Y., Zakaria, N. N., Futamata, H., Suzuki, K., Zulkharnain, A., Shaharuddin, N. A., Convey, P., Zahri, K. N. M., & Ahmad, S. A. (2022). Metabolic pathway of phenol degradation of a cold-adapted Antarctic bacteria, Arthrobacter sp. *Catalysts, 12*(11), 1422.

Lewin, G. R., Carlos, C., Chevrette, M. G., Horn, H. A., McDonald, B. R., Stankey, R. J., ... & Currie, C. R. (2016). Evolution and ecology of Actinobacteria and their bioenergy applications. *Annual Review of Microbiology, 70*(1), 235–254.

Li KaiChang, L. K., Azadi, P., Collins, R., Tolan, J., Kim, J. S., & Eriksson, K. E. L. (2000). Relationships between activities of xylanases and xylan structures. *Enzyme and Microbial Technology, 27*(1–2), 89–94.

Lobastova, T., Fokina, V., Tarlachkov, S., Shutov, A., Bragin, E., Kazantsev, A., & Donova, M. (2021). Steroid metabolism in thermophilic actinobacterium *Saccharopolyspora hirsuta* VKM Ac-666T. *Microorganisms, 9*(12), 2554.

Lv, Z., Yang, J., & Yuan, H. (2008). Production, purification and characterization of an alkaliphilic endo-β-1, 4-xylanase from a microbial community EMSD5. *Enzyme and Microbial Technology, 43*(4–5), 343–348.

Maat, J. (1992). Xylanases and their application in bakery. In *Xylans and xylanases* (pp. 349–360). Elsevier.

Madhav, S., Ahamad, A., Singh, P., & Mishra, P. K. (2018). A review of textile industry: Wet processing, environmental impacts, and effluent treatment methods. *Environmental Quality Management, 27*(3), 31–41.

Majithiya, V. R., Ghoghari, A. M., & Gohel, S. D. (2025). Purification, characterization, structural elucidation, and industrial applications of thermostable alkaline protease produced by seaweed-associated *Nocardiopsis dassonvillei* strain VCs-4. *International Journal of Biological Macromolecules, 305*, 141147.

Makarani, N., & Kaushal, R. S. (2025). Advances in actinobacteria-based bioremediation: Mechanistic insights, genetic regulation, and emerging technologies. *Biodegradation, 36*(2), 24.

Manivasagan, P., Venkatesan, J., Sivakumar, K., & Kim, S. K. (2013). Production, characterization and antioxidant potential of protease from Streptomyces sp. MAB18 using poultry wastes. *BioMed Research International, 2013*(1), 496586.

Mansour, S. R., Abdel-Azeem, A. M., & Abo-Deraz, S. S. S. (2015). A new record of Actinobacteria isolated from soil in Jerusalem and their enzymatic potential. *F1000Research, 4*, 11.

Martin, H., Lázaro, L. R., Gunnlaugsson, T., & Scanlan, E. M. (2022). Glycosidase activated prodrugs for targeted cancer therapy. *Chemical Society Reviews, 51*(23), 9694–9716.
Mathlouthi, N., Lalles, J. P., Lepercq, P., Juste, C., & Larbier, M. (2002). Xylanase and beta-glucanase supplementation improve conjugated bile acid fraction in intestinal contents and increase villus size of small intestine wall in broiler chickens fed a rye-based diet. *Journal of Animal Science, 80*, 2773–2779.
Mathlouthi, N., Juin, H., & Larbier, M. (2003). Effect of xylanase and beta-glucanase supplementation of wheat- or wheat- and barley-based diets on the performance of male turkeys. *British Poultry Science, 44*, 291–298.
Meena, B., Anburajan, L., Nitharsan, K., Vinithkumar, N. V., & Dharani, G. (2021). Existence in cellulose shelters: Industrial and pharmaceutical leads of symbiotic actinobacteria from ascidian Phallusia nigra, Andaman Islands. *World Journal of Microbiology and Biotechnology, 37*(7), 1–22.
Mehta, A., Guleria, S., Sharma, R., & Gupta, R. (2021). The lipases and their applications with emphasis on food industry. In *Microbial biotechnology in food and health* (pp. 143–164). Academic.
Meliani, M. F., Denis, F., Mohamed-Benkada, M., Gabed, N., Caruso, A., Callac, N., & Abi-Ayad, S. E. A. (2022). Characterization of actinomycetes strains isolated from Cheliff Estuary in the North-West of Algeria. *Jordan Journal of Biological Sciences, 15*(01), 7–14.
Mielenz, J. R. (2001). Ethanol production from biomass: Technology and commercialization status. *Current Opinion in Microbiology, 4*, 324–329.
Motta, F. L., Andrade, C. C. P., & Santana, M. H. A. (2013). A review of xylanase production by the fermentation of xylan: Classification, characterization and applications. In C. Anuj (Ed.), *Sustainable degradation of lignocellulosic biomass—Techniques, applications and commercialization*. InTechOpen.
Muffler, K., Sana, B., Mukherjee, J., & Ulber, R. (2015). Marine enzymes—Production & applications. In *Springer handbook of marine biotechnology* (pp. 413–429). Springer.
Mukhtar, S., Zaheer, A., Aiysha, D., Malik, K. A., & Mehnaz, S. (2017). Actinomycetes: A source of industrially important enzymes. *Journal of Proteomics & Bioinformatics, 10*(12), 316–319.
Musengi, A., Durrell, K., Prins, A., Khan, N., Agunbiade, M., Kudanga, T., Kirby-McCullough, B., Pletschke, B. I., Burton, S. G., & Roes-Hill, M. L. (2020). Production and characterisation of a novel actinobacterial DyP-type peroxidase and its application in coupling of phenolic monomers. *Enzyme and Microbial Technology, 141*, 109654.
Naveed, M., Nadeem, F., Mehmood, T., Bilal, M., Anwar, Z., & Amjad, F. (2021). Protease—A versatile and ecofriendly biocatalyst with multi-industrial applications: An updated review. *Catalysis Letters, 151*, 307–323.
Ninawe, S., Lal, R., & Kuhad, R. C. (2006). Isolation of three xylanase-producing strains of actinomycetes and their identification using molecular methods. *Current Microbiology, 53*(3), 178–182.
Nithya, K., Muthukumar, C., Biswas, B., Alharbi, N. S., Kadaikunnan, S., Khaled, J. M., & Dhanasekaran, D. (2018). Desert actinobacteria as a source of bioactive compounds production with a special emphases on Pyridine-2, 5-diacetamide a new pyridine alkaloid produced by Streptomyces sp. DA3-7. *Microbiological Research, 207*, 116–133.
Norouzi, H., Danesh, A., Mohseni, M., & Khorasgani, M. R. (2018). Marine actinomycetes with probiotic potential and bioactivity against multidrug-resistant bacteria. *International Journal of Molecular and Cellular Medicine, 7*(1), 44–52. https://doi.org/10.22088/IJMCM.BUMS.7.1.44
Oren, A. (2013). Life at high salt concentrations, intracellular KCl concentrations, and acidic proteomes. *Frontiers in Microbiology, 4*, 315.
Pal, A., & Khanum, F. (2010). Production and extraction optimization of xylanase from *Aspergillus niger* DFR-5 through solid-state-fermentation. *Bioresource Technology, 101*(19), 7563–7569.
Pan, C., ul Hassan, S. S., Ishaq, M., Yan, S., & Jin, H. (2025). Marine actinomycetes: A hidden treasure trove for antibacterial discovery. *Frontiers in Marine Science, 12*, 1558320.

Pandey, S., & Singh, S. P. (2012). Organic solvent tolerance of an amylase from haloalkaliphilic bacteria as a function of pH, temperature and salt concentrations. *Applied Biochemistry and Biotechnology, 166*, 1747–1757.

Pandey, A., Nigam, P., Soccol, C. R., Scoccol, V. T., Singh, D., & Mohan, R. (2000). Advances in microbial amylases. *Biotechnology and Applied Biochemistry, 31*, 135–152.

Panyachanakul, T., Lomthong, T., Lorliam, W., Prajanbarn, J., Tokuyama, S., Kitpreechavanich, V., & Krajangsang, S. (2020). New insight into thermo-solvent tolerant lipase produced by Streptomyces sp. A3301 for re-polymerization of poly (DL-lactic acid). *Polymer, 204*, 122812.

Parte, S., Sirisha, V. L., & D'Souza, J. S. (2017). Biotechnological applications of marine enzymes from algae, bacteria, fungi, and sponges. In *Advances in food and nutrition research* (Vol. 80, pp. 75–106). Academic.

Pasqualetti, M., Gorrasi, S., Giovannini, V., Braconcini, M., & Fenice, M. (2022). Polyextremophilic chitinolytic activity by a marine strain (IG119) of *Clonostachys rosea*. *Molecules, 27*(3), 688.

Pizarro, I. F., de Souza, H. F., dos Santos Ferreira, J., Maldonado, R. R., Kamimura, E. S., & Aguiar-Oliveira, E. (2022). Production of lipase by actinobacteria. In *Methods in actinobacteriology* (pp. 505–512). Springer.

Prade, R. A. (1995). Xylanases: From biology to biotechnology. *Biotechnology & Genetic Engineering Reviews, 13*, 100–131.

Prins, A., Dyani, S. S., Vreulink, J. M., Maldonado, L. A., & Le Roes-Hill, M. (2024). Actinobacteria diversity associated with marine sediments and a wetland system, Agulhas-South Africa. *Annals of Microbiology, 74*(1), 1–24.

Pupart, H., Jõul, P., Bramanis, M. I., & Lukk, T. (2023). Characterization of the ensemble of lignin-remodeling DyP-type peroxidases from Streptomyces coelicolor A3(2). *Energies, 16*(3), 1557. https://doi.org/10.3390/en16031557

Putri, A. L., & Setiawan, R. (2019). Isolation and screening of actinomycetes producing cellulase and xylanase from Mamasa soil, West Sulawesi. *IOP Conference Series: Earth and Environmental Science, 308*(1), 012035.

Ragauskas, A. J., Poll, K. M., & Cesternino, A. J. (1994). Effects of xylanase pretreatment procedures on nonchlorine bleaching. *Enzyme and Microbial Technology, 16*(6), 492–495.

Rahman, M. M. (2021). Application of enzymes in the textile industry: A review. *Southeast University Journal of Textile Engineering, 1*(1), 49–58.

Ramachandra, M., Crawford, D. L., & Pometto, A. L. (1987). Extracellular enzyme activities during lingocellulose degradation by *Streptomyces* sp. A comparative study of wild-type and genetically manipulated strains. *Applied and Environmental Microbiology, 32*, 2754–2760.

Ramírez, M. V., Calzadíaz, L., Ramírez, M. V., & Calzadíaz, L. (2016). Industrial enzymes and metabolites from actinobacteria in food and medicine industry. In *Actinobacteria—Basics and biotechnological applications*. https://doi.org/10.5772/61286

Rampelotto, P. H. (2013). Extremophiles and extreme environments. *Life, 3*(3), 482–485.

Rathore, D. S., & Singh, S. P. (2021). Kinetics of growth and co-production of amylase and protease in novel marine actinomycete, *Streptomyces lopnurensis* KaM5. *Folia Microbiologica, 66*(3), 303–316.

Rathore, D. S., Sheikh, M. A., Gohel, S. D., & Singh, S. P. (2021). Genetic and phenotypic heterogeneity of the *Nocardiopsis alba* strains of seawater. *Current Microbiology, 78*(4), 1377–1387.

Raval, V. H., Pillai, S., Rawal, C. M., & Singh, S. P. (2014). Biochemical and structural characterization of a detergent-stable serine alkaline protease from seawater haloalkaliphilic bacteria. *Process Biochemistry, 49*(6), 955–962.

Raval, V. H., Joshi, R. H., Bhatt, H. B., & Singh, S. P. (2022). Phenotypic characteristics, phylogenetic analysis and characterization of alkaline proteases of marine bacteria *Geomicrobium halophilum, Oceanobacillus oncorhynchi*, and *Oceanobacillus khimchii*. *Biologia, 77*(8), 2405–2422.

Rekadwad, B. N., Li, W. J., Gonzalez, J. M., Punchappady Devasya, R., Ananthapadmanabha Bhagwath, A., Urana, R., & Parwez, K. (2023). Extremophiles: The species that evolve and survive under hostile conditions. *3 Biotech, 13*(9), 316.

Ribeiro, I., Antunes, J. T., Alexandrino, D. A. M., Tomasino, M. P., Almeida, E., Hilário, A., Urbatzka, R., Leão, P. N., Mucha, A. P., & Carvalho, M. F. (2023). Actinobacteria from Arctic and Atlantic deep-sea sediments—Biodiversity and bioactive potential. *Frontiers in Microbiology, 14*, 1158441.

Rodríguez-Alonso, G., Toledo-Marcos, J., Serrano-Aguirre, L., Rumayor, C., Pasero, B., Flores, A., et al. (2023). A novel lipase from *Streptomyces exfoliatus* DSMZ 41693 for biotechnological applications. *International Journal of Molecular Sciences, 24*(23), 17071.

Román-Ponce, B., Millán-Aguiñaga, N., Guillen-Matus, D., Chase, A. B., Ginigini, J. G. M., Soapi, K., Feussner, K. D., Jensen, P. R., & Trujillo, M. E. (2020). Six novel species of the obligate marine actinobacterium salinispora, salinispora cortesiana sp. Nov., salinispora fenicalii sp. nov., salinispora goodfellowii sp. nov., salinispora mooreana sp. nov., salinispora oceanensis sp. nov. and salinispora vitiensis sp. nov., and emended description of the genus salinispora. *International Journal of Systematic and Evolutionary Microbiology, 70*(8), 4668–4682.

Rotter, A., Barbier, M., Bertoni, F., Bones, A. M., Cancela, M. L., Carlsson, J., et al. (2021). The essentials of marine biotechnology. *Frontiers in Marine Science, 8*, 629629.

Roy, S., Das, I., Munjal, M., Karthik, L., Kumar, G., Kumar, S., & Rao, K. V. B. (2014). Isolation and characterization of tyrosinase produced by marine actinobacteria and its application in the removal of phenol from aqueous environment. *Frontiers in Biology, 9*(4), 306–316.

Saha, B. C. (2003). Hemicellulose bioconversion. *Journal of Industrial Microbiology & Biotechnology, 30*, 279–291.

Salam, M. D., Tripathi, S., & Simran. (2025). Strategies for discovery and enhancement of enzyme function: Current developments and opportunities. *Microbial Enzymes: Production, Purification and Industrial Applications, 2*, 491–504.

Salama, S., Mostafa, H. S., Husseiny, S., & Sebak, M. (2024). Actinobacteria as microbial cell factories and biocatalysts in the synthesis of chiral intermediates and bioactive molecules; insights and applications. *Chemistry and Biodiversity, 21*(2), e202301205.

Salwan, R., & Sharma, V. (2018). The role of actinobacteria in the production of industrial enzymes. In *New and future developments in microbial biotechnology and bioengineering* (pp. 165–177). Elsevier.

Salwan, R., & Sharma, V. (2019). Trends in extracellular serine proteases of bacteria as detergent bioadditive: Alternate and environmental friendly tool for detergent industry. *Archives of Microbiology, 201*(7), 863–877.

Samanta, D., Govil, T., Saxena, P., Thakur, P., Sani, R. K., & Narayanan, A. (2022). Extremozymes and their applications. In *Extremozymes and their industrial applications* (pp. 1–39). Springer. https://doi.org/10.1016/B978-0-323-90274-8.00005-8

Sanjivkumar, M., Silambarasan, T., Palavesam, A., & Immanuel, G. (2017). Biosynthesis, purification and characterization of β-1, 4-xylanase from a novel mangrove associated actinobacterium *Streptomyces olivaceus* (MSU3) and its applications. *Protein Expression and Purification, 130*, 1–12.

Sanjivkumar, M., Brindhashini, A., Deivakumari, M., Palavesam, A., & Immanuel, G. (2018a). Investigation on saccharification and bioethanol production from pretreated agro-residues using a mangrove associated actinobacterium *Streptomyces variabilis* (MAB3). *Waste and Biomass Valorization, 9*(6), 969–984.

Sanjivkumar, M., Silambarasan, T., Balagurunathan, R., & Immanuel, G. (2018b). Biosynthesis, molecular modeling and statistical optimization of xylanase from a mangrove associated actinobacterium *Streptomyces variabilis* (MAB3) using Box-Behnken design with its bioconversion efficacy. *International Journal of Biological Macromolecules, 118*, 195–208.

Selvin, J., Kennedy, J., Lejon, D. P., Kiran, G. S., & Dobson, A. D. (2012). Isolation identification and biochemical characterization of a novel halo-tolerant lipase from the metagenome of the marine sponge *Haliclona simulans*. *Microbial Cell Factories, 11*(1), 72.

Sharma, A. K., Kikani, B. A., & Singh, S. P. (2020). Biochemical, thermodynamic and structural characteristics of a biotechnologically compatible alkaline protease from a haloalkaliphilic, *Nocardiopsis dassonvillei* OK-18. *International Journal of Biological Macromolecules, 153*, 680–696.

Sharma, A. K., Kikani, B. A., & Singh, S. P. (2021). Diversity and phylogeny of actinomycetes of Arabian Sea along the Gujarat Coast. *Geomicrobiology Journal, 38*(4), 347–364.

Shukla, R. J., & Singh, S. P. (2016). Structural and catalytic properties of immobilized α-amylase from *Laceyella sacchari* TSI-2. *International Journal of Biological Macromolecules, 85*, 208–216.

Shukla, R. J., & Singh, S. P. (2023). Purification and characterization of an amylase from a newly isolated *Geobacillus thermoleovorans* TTIO4 from the Hot Spring of Tuwa-Timba. *Starch-Stärke, 75*(9–10), 2200254.

Simpson, B. K., Rui, X., & Klomklao, S. (2012). Enzymes in food processing. In *Food biochemistry and food processing* (pp. 181–206). Wiley-Blackwell.

Singh, S. P., Thumar, J. T., Gohel, S. D., Kikani, B., Shukla, R., Sharma, A., & Dangar, K. (2013). Actinomycetes from marine habitats and their enzymatic potential. In *Marine enzymes for biocatalysis* (pp. 191–214). Woodhead Publishing.

Singh, S., Singh, V. K., Aamir, M., Dubey, M. K., Patel, J. S., Upadhyay, R. S., & Gupta, V. K. (2016). Cellulase in pulp and paper industry. In *New and future developments in microbial biotechnology and bioengineering: Microbial cellulase system properties and applications* (pp. 152–162). Elsevier.

Sivasankar, P., Poongodi, S., Sivakumar, K., Al-Qahtani, W. H., Arokiyaraj, S., & Jothiramalingam, R. (2022). Exogenous production of cold-active cellulase from polar Nocardiopsis sp. with increased cellulose hydrolysis efficiency. *Archives of Microbiology, 204*(4), 1–9. https://doi.org/10.1007/S00203-022-02830-Z/METRICS

Solanki, P., Putatunda, C., Kumar, A., Bhatia, R., & Walia, A. (2021). Microbial proteases: Ubiquitous enzymes with innumerable uses. *3 Biotech, 11*(10), 428.

Subramani, R., & Aalbersberg, W. (2013). Culturable rare Actinomycetes: diversity, isolation and marine natural product discovery. *Applied Microbiology and Biotechnology, 97*(21), 9291–9321.

Sujitha, P., & Shanthi, C. (2023). Importance of enzyme specificity and stability for the application of proteases in greener industrial processing—A review. *Journal of Cleaner Production, 425*, 138915.

Sundarram, A., & Murthy, T. P. K. (2014). α-amylase production and applications: A review. *Journal of Applied & Environmental Microbiology, 2*(4), 166–175.

Suriya, J., Bharathiraja, S., Krishnan, M., Manivasagan, P., & Kim, S. K. (2016). Extremozymes from marine actinobacteria. *Advances in Food and Nutrition Research, 79*, 43–66.

Suthindhiran, K., Jayasri, M. A., Dipali, D., & Prasar, A. (2014). Screening and characterization of protease producing actinomycetes from marine saltern. *Journal of Basic Microbiology, 54*(10), 1098–1109.

Swain, M. R., Natarajan, V., & Krishnan, C. (2017). Marine enzymes and microorganisms for bioethanol production. *Advances in Food and Nutrition Research, 80*, 181–197.

Swathy, R., Rambabu, K., Banat, F., Ho, S. H., Chu, D. T., & Show, P. L. (2020). Production and optimization of high grade cellulase from waste date seeds by *Cellulomonas uda* NCIM 2353 for biohydrogen production. *International Journal of Hydrogen Energy, 45*(42), 22260–22270.

Sysoev, M., Grötzinger, S. W., Renn, D., Eppinger, J., Rueping, M., & Karan, R. (2021). Bioprospecting of novel extremozymes from prokaryotes—The advent of culture-independent methods. *Frontiers in Microbiology, 12*, 630013.

Thakrar, F. J., Koladiya, G. A., & Singh, S. P. (2023). Heterologous expression and structural elucidation of a highly thermostable alkaline serine protease from haloalkaliphilic actinobacterium, Nocardiopsis sp. Mit-7. *Applied Biochemistry and Biotechnology, 195*(12), 7583–7602.

Thanigaivel, S., Priya, A. K., Dutta, K., Rajendran, S., & Vasseghian, Y. (2022). Engineering strategies and opportunities of next generation biofuel from microalgae: A perspective review on the potential bioenergy feedstock. *Fuel, 312*, 122827.

Thumar, J. T., & Singh, S. P. (2009). Organic solvent tolerance of an alkaline protease from salt-tolerant alkaliphilic *Streptomyces clavuligerus* strain Mit-1. *Journal of Industrial Microbiology and Biotechnology, 36*(2), 211.

Tiwari, S. P., Srivastava, R., Singh, C. S., Shukla, K., Singh, R. K., Singh, P., et al. (2015). Amylases: An overview with special reference to alpha amylase. *Journal of Global Biosciences, 4*(1), 1886–1901.
Trincone, A. (2012). Some enzymes in marine environment: Prospective applications found in patent literature. *Recent Patents on Biotechnology, 6*(2), 134–148.
Trivedi, N., Reddy, C. R. K., & Lali, A. M. (2016). Marine microbes as a potential source of cellulolytic enzymes. *Advances in Food and Nutrition Research, 79*, 27–41.
van Oort, M. (2010). Enzymes in food technology: Introduction. In R. J. Whitehurst & M. van Oort (Eds.), *Enzymes in food technology* (2nd ed., pp. 13–16). Wiley-Blackwell.
Viikari, L. (1994). Xylanases in bleaching: From an idea to the industry. *FEMS Microbiology Reviews, 13*, 335–350.
Walia, A., Mehta, P., Chauhan, A., Kulshrestha, S., & Shirkot, C. K. (2014). Purification and characterization of cellulase-free low molecular weight endo β-1, 4 xylanase from an alkalophilic *Cellulosimicrobium cellulans* CKMX1 isolated from mushroom compost. *World Journal of Microbiology and Biotechnology, 30*(10), 2597–2608.
Wang, X., Luo, H., Yu, W., Ma, R., You, S., Liu, W., Hou, L., Zheng, F., Xie, X., & Yao, B. (2016). A thermostable *Gloeophyllum trabeum* xylanase with potential for the brewing industry. *Food Chemistry, 199*, 516–523.
Wong, K. K. Y., & Saddler, J. N. (1993). Applications of hemicellulases in the food, feed and pulp and paper industries. In M. P. Coughlan & G. P. Hazlewood (Eds.), *Hemicelluloses and hemicellulases* (pp. 127–143). Portland Press.
Xu, Y., Lu, Y., Xing, Y., & Xi, T. (2015). Characterization of solvent, detergent and oxidizing agent stable protease from isolated Antarctic marine Streptomyces sp. XE-1. *Wei Sheng Wu Xue Bao, 55*, 311–320.
Yaradoddi, J. S., & Kontro, M. H. (2021). Actinobacteria: Basic adaptation to harsh environments. In *Rhizosphere biology* (pp. 69–88). Springer.
Yu, W., Liu, W., Huang, H., Zheng, F., Wang, X., Wu, Y., Li, K., Xie, X., & Jin, Y. (2014). Application of a novel alkali-tolerant thermostable DyP-type peroxidase from *Saccharomonospora viridis* DSM 43017 in biobleaching of eucalyptus kraft pulp. *PLoS One, 9*(10), e110319.
Yuan, D., Lan, D., Xin, R., Yang, B., & Wang, Y. (2016). Screening and characterization of a thermostable lipase from marine Streptomyces sp. strain W007. *Biotechnology and Applied Biochemistry, 63*(1), 41–50.
Zamora-Quintero, A. Y., Torres-Beltrán, M., Guillén Matus, D. G., Oroz-Parra, I., & Millán-Aguiñaga, N. (2022). Rare actinobacteria isolated from the hypersaline Ojo de Liebre Lagoon as a source of novel bioactive compounds with biotechnological potential. *Microbiology (United Kingdom), 168*(2), 001144.
Zhang, H., Li, F., Chen, H., Zhao, J., Yan, J., Jiang, P., et al. (2015). Cloning, expression and characterization of a novel esterase from a South China Sea sediment metagenome. *Chinese Journal of Oceanology and Limnology, 33*(4), 819–827.
Zhao, X. Q., Xu, X. N., & Chen, L. Y. (2016). Production of enzymes from marine actinobacteria. *Advances in Food and Nutrition Research, 78*, 137–151.

Enzyme Inhibitors From Marine Actinobacteria

P. V. Tharani and K. V. Bhaskara Rao

Abstract

Marine actinobacteria are increasingly recognized as prolific makers of bioactive secondary metabolites, including enzyme inhibitors (EIs) with promising therapeutic applications. These inhibitors, naturally generated as defense chemicals, work on many enzyme targets, including proteases, α-glucosidase (AGS), monoamine oxidases, chitinases, and angiotensin-converting enzyme. Their capacity to influence important metabolic pathways makes them intriguing candidates for therapeutic development against cancer, diabetes, hypertension, infectious illnesses, and neurodegenerative disorders. This chapter investigates the range of EIs obtained from marine actinobacteria, concentrating on their ecological origin, biochemical activities, inhibition mechanisms, and pharmaceutical importance. Rare and symbiotic actinobacterial taxa such as *Streptomyces, Salinispora, Nocardiopsis,* and *Actinomadura*, isolated from marine sediments, sponges, and mangrove environments, have been found to have potent inhibitors. Compounds from marine actinobacteria have been shown to block AGS and have antiplasmodial and antiviral activity, highlighting their potential as a source of new therapeutics.

Keywords

Marine actinobacteria · Enzyme inhibitors · Secondary metabolites · Streptomyces · α-Glucosidase inhibition · Drug discovery

P. V. Tharani · K. V. Bhaskara Rao (✉)
Marine Biotechnology Laboratory, Department of Bio-Medical Sciences, School of Bio Sciences and Technology, Vellore Institute of Technology, Vellore, Tamil Nadu, India
e-mail: kvbhaskararao@vit.ac.in

K. Loganathan, Z. Li (eds.), *Marine Actinomycetes*,
https://doi.org/10.1007/978-981-95-9849-6_14

1 Introduction

Enzymes are chemicals that catalyze metabolic and cellular processes, enabling all living species to maintain homeostasis and equilibrium. Enzymes are biocatalysts that can accelerate chemical reactions but are not impacted by the process. The term "enzyme" was coined by German scientist Wilhelm Kühne in 1878 to describe yeast's ability to make alcohol from carbohydrates; it is derived from the Greek words en (meaning "inside") and zyme (meaning "yeast"). Enzymes are classified based on their classes as oxidoreductases, transferases, hydrolases, lyases, ligases, and isomerases (Robinson, 2015).

Enzymes are a key therapeutic target because inhibiting their mode of action can slow or stop the progression of certain diseases. As a result, identifying or developing enzyme inhibitors (EIs) is crucial for treating significant medical conditions. Reversible inhibition and irreversible inhibition are the two main types of enzyme inhibition. All EIs function reversibly (weak, noncovalent interactions that are temporary) or irreversibly (covalently alter the structure of the enzyme and/or active site) (Patadiya et al., 2021; Ramsay & Tipton, 2017).

Reversible inhibitors, as the name implies, bind to enzymes through non-covalent interactions, allowing the enzyme's activity to be restored after the inhibitor is withdrawn. These are further classified into three types based on the binding mechanism: competitive, non-competitive, and uncompetitive inhibitors (Kuddus, 2019). Competitive inhibitors are structurally similar to the enzyme's natural substrate and compete directly for the same binding region, which is known as the active site. This competition inhibits the development of the enzyme–substrate (ES) complex, decreasing the overall reaction rate (Kuddus, 2019). Malonate is a classic example of a competitive inhibitor, as it competes with succinate for the enzyme succinate dehydrogenase's active site (Ramsay & Tipton, 2017). Increasing the substrate concentration can often counteract the action of a competitive inhibitor. Non-competitive inhibitors bind to an allosteric site, which is different from the active site. This binding event causes a conformational shift in the enzyme, modifying the shape of the active site and reducing its ability to bind to the substrate. As a result, the development of the ES complex is hampered, reducing the reaction rate (Kuddus, 2019). Unlike competitive inhibition, raising substrate concentration does not overcome non-competitive inhibition because the inhibitor's binding site differs from the substrates. Uncompetitive inhibitors are a separate family of reversible inhibitors. They bind only to the ES complex, not the free enzyme or substrate. This binding event disables the ES complex, blocking substrate conversion to product (Kuddus, 2019). Because this sort of inhibition requires the development of the ES complex, it is more effective at greater substrate concentrations.

The second primary type of enzyme inhibition is irreversible inhibition. This type of inhibition is distinguished by the persistent and irreversible inactivation of an enzyme. Irreversible inhibitors are distinguished by their capacity to create stable, strong connections, most commonly covalent bonds, with the enzyme's active site or another essential functional group. This strong bond physically or chemically alters the enzyme's structure, making it permanently inactive. Unlike reversible inhibitors, whose action can be reduced by increasing substrate concentration, Irreversible inhibitors cannot be reversed through simple reduction or the addition of more substrate. The term "suicide inhibitors" or "mechanism-based inactivators" refers to a well-known and specialized type of irreversible inhibitor. These inhibitors have a particularly ingenious mechanism. They are initially unreactive and mimic the enzyme's natural substrate, so they can bind to the active site. The enzyme then resumes its usual catalytic activity on the inhibitor molecule. However, during this process, the inhibitor is transformed into a highly reactive intermediate that quickly establishes a covalent connection with a critical amino acid residue in the active site, thus "committing suicide" by inactivating the enzyme it was intended to target. Serpins attach covalently to serine proteases and inhibit their action by altering or modifying their substrate binding sites (Ye et al., 2009). Monoamine oxidases (MAOs) are located on the mitochondrial outer membrane; consequently, metabolizing amines within the cell plays a crucial role in many neurodegenerative disorders. The primary MAO inhibitors were originally developed as antidepressants. Irreversible inhibitors such as clorgyline, L-deprenyl, and rasagiline are used as adjuvant therapy for Parkinson's disease (Ramsay & Tipton, 2017). Aspirin, a commonly used nonsteroidal anti-inflammatory medication, is another example of an irreversible inhibitor. Aspirin works by irreversibly blocking cyclooxygenase (COX) enzymes, specifically acetylating a serine residue in the enzyme's active site (Cipollone & Fazia, 2006). This covalent alteration inhibits the COX enzyme from making prostaglandins, which are essential mediators of pain and inflammation, resulting in a long-term therapeutic benefit that lasts the enzyme's lifetime. Because these inhibitors are irreversible, they are extremely effective in a wide range of therapeutic scenarios, from antibiotics like penicillin, which target enzymes essential for bacterial cell wall synthesis, to certain chemotherapy drugs that inhibit enzymes critical for cancer cell proliferation (Powers et al., 2002).

Natural inhibitors are frequently highly selective and potent, making them attractive pharmacological leads: diabetes treatment, antimicrobial resistance, and neuroprotection. EIs are naturally synthesized by a wide range of species, including bacteria, higher plants, and mammals, as part of their defense mechanisms. These bioactive compounds can be produced constantly as part of regular metabolism or stimulated in response to biotic stressors such as infection or insect infestations. Marine ecology, in particular, has emerged as a valuable and diverse source of EIs. EIs from marine organisms have been found to target a variety of enzymes, including proteases (Mourão & Schwartz, 2013), amylases (Raja et al., 2010), β-glucosidases (Pandey et al., 2013), angiotensin-converting enzyme (ACE) (Wijesekara & Kim, 2010), and chitinases (Imada, 2004). Since the biosynthesis of structurally novel secondary metabolites is driven by specific environmental pressures, including high salinity, limited nutrients, and intense interspecies competition, marine ecosystems in particular have emerged as rich and understudied sources of EIs. Actinobacteria are particularly abundant producers among marine microorganisms due to their genetic capacity to manufacture a wide variety of bioactive compounds via specialized clusters of biosynthetic genes. Several therapeutically significant enzyme groups are inhibited by enzymes derived from marine actinobacteria and other marine species. For instance, the discovery of anti-inflammatory, antiviral, and anticancer drugs is of interest in protease inhibitors originating from marine sources (Mourão & Schwartz, 2013).

2 Marine Actinobacteria

The ocean is home to a large microbial world, which accounts for nearly 98% of its productivity. These marine microorganisms play a crucial part in the bioactive systems of the ocean because they create a vast array of bioactive metabolites, whether they are free-living or in symbiosis (Mahapatra et al., 2020).

Growing recognition of the enormous potential of marine natural products (Haefner, 2003; König et al., 2006; Martins et al., 2014) has sparked a boom in studies aimed at utilizing marine microbial products (König et al., 2006). Among these, research on actinobacteria metabolites is especially well-known. Actinobacteria are a diverse collection of bacteria identified by their Gram-positive staining and ability to thrive in aerobic, facultative anaerobic, or anaerobic environments. They have a stiff cell wall that contains muramic acid. Phenotypically, they range from basic cocci to complicated, highly distinct mycelial forms (Goodfellow, 2015). Actinobacteria present in various marine habitats are the source of the majority of natural products identified to date (Tormo et al., 2015).

They have now been widely found in marine environments, but it was believed that these might just be pollutants from the land that have washed into the sea. On the other hand, actinomycetes have existed since the Eocene period (Waggoner, 1994). Nevertheless, according to compelling molecular and fossil data (Lam, 2006), presenting the unique genus *Salinospora* (in the family *Micromonosporaceae*) as a notable example of marine-exclusive species provided conclusive evidence of the widespread existence of obligate marine actinomycetes (Magarvey et al., 2004; Mincer et al., 2002). The notion of different marine actinobacterial lineages is further supported by studies that demonstrate that marine actinobacteria are more tolerant to salt concentrations than their terrestrial counterparts (Imada et al., 2010).

3 Natural Products of Marine Actinobacteria

Actinobacteria are well known for creating a diverse range of physiologically active natural chemicals, accounting for more than half of all bioactive molecules discovered. There is a growing global interest in natural chemicals obtained from rare marine actinobacteria. This is demonstrated by the geographical origins of these bacteria, the separation of novel compounds from them, and the precise bioactivities of these compounds.

Rare actinobacteria from the deep sea and their sediments, such as *Pseudonocardia* sp. and *Marinactinospora thermotolerans,* produce compounds such as Pseudonocardians and Marthiapeptide A, respectively, which have cytotoxic and antibacterial properties (Li et al., 2011; Zhou et al., 2012). Marine sediments from the coastal area species, such as *Verrucosispora* sp., *Saccharomonospora* sp., and *Marinispora* sp., produces antibacterial compounds such as Abyssomicins, Taromycin, and Lynamicins, respectively (Bister et al., 2004, Yamanaka et al., 2014; McArthur et al., 2008). Fijiolides act as an inhibitor of Tumor Necrosis Factor-alpha (TNF-α)-induced Nuclear Factor kappa-light-chain-enhancer of activated B cells (NFκB) activation obtained from *Nocardiopsis* isolated from a sediment sample (Nam et al., 2010). Dermacozines, Phenazine derivatives, which act as an antioxidant and cytotoxic compound, are also obtained from the *Dermacoccus abyssi* sp. isolated from the sediment sample from the Mariana Trench (Abdel-Mageed et al., 2010). *Actinomadura* sp., which resides as a symbiont in the ascidian *Ecteinascidia turbinata,* secretes Halomadurone, which is useful in the treatment of neurodegenerative diseases (Wyche et al., 2013).

4 EIs From Marine Actinobacteria

From land to sea, actinobacteria are known to produce a wide range of metabolites with potential medical applications (Watve et al., 2001). Among these, actinobacteria most frequently produce EIs as metabolites (Karthik et al., 2014).

4.1 Proteases and Protease Inhibitors

Proteases have a broad enzyme distribution. In humans, there are approximately 990 identified protease genes. Proteases are categorized into six types: cysteine, serine, threonine, glutamic acid, aspartate, and matrix metalloproteases. Proteases are a broad and diversified group of enzymes that catalyze the hydrolysis of peptide bonds, regulating protein turnover, activation, and destruction. They are required for several physiological activities, including cell signaling, immunological responses, and tissue remodeling. Because of their broad participation in normal and pathological processes, proteases have emerged as important molecular targets in drug discovery. Aberrant protease activity has been related to the progression of a variety of diseases, including cancer, neurological disorders, and infectious diseases, where it helps pathogens thrive and interact with hosts (López-Otín & Bond, 2008; Klein et al., 2017).

Proteases are also involved in tumor growth and progression, at both the primary and metastatic levels. Several tumors have been shown to have increased protease levels at an early stage, and these proteases are now known to be involved in many aspects of cancer, including proliferation, immune responses, inflammatory cell recruitment, tumor invasion, angiogenesis, metastasis, apoptosis, and epithelial-to-mesenchymal transition (Eatemadi et al., 2017). Infectious diseases frequently rely on proteases for pathogen proliferation and pathogenicity. For example, in viral infections like HIV, viral proteases are required to convert polyproteins into functional viral proteins, and their inhibition has resulted in the successful development of antiviral medicines. Similarly, in malaria, proteases such as falcipains and plasmepsins are essential for hemoglobin degradation and parasite growth inside red blood cells. Targeting these enzymes with particular inhibitors not only disturbs the parasite life cycle but also offers a promising antimalarial treatment option. These examples demonstrate how protease inhibition can be a useful therapeutic strategy for a variety of disorders (Sojka et al., 2021; Deu, 2017).

Inhibiting proteases is essential for treating inflammatory disorders such as pancreatitis, preventing tumor invasion in cancer (Eatemadi et al., 2017), and designing antiviral medicines that interrupt viral replication (Imada, 2004; de Leuw & Stephan, 2017). Similarly, ACE inhibition is critical in treating hypertension, heart failure, and chronic kidney disease because it regulates blood pressure and protects the cardiovascular and renal systems (Silva-Velasco et al., 2024). The isolation of *Streptomyces griseoincarnatus* from *Callyspongia* sp. serves as an example of the abundance of bioactive compounds found in marine sponges. This marine actinobacterium produced a new protein-based protease inhibitor. The inhibitor showed significant efficacy against papain and trypsin, making it the first natural protease inhibitor effective against the Chikungunya virus (Kamarudheen et al., 2021). Peptide protease inhibitors with significant effectiveness against trypsin and chymotrypsin were identified from marine streptomyces species on the Nicobar Islands. This irreversible protease inhibitor has shown substantial anti-plasmodial efficacy and could be used to produce malaria drugs (Karthik et al., 2014). This finding emphasizes the therapeutic potential of marine actinomycetes in antiviral medication development.

4.2 ACEs and ACE Inhibitors

ACE is a crucial enzyme in blood pressure regulation. It works in two ways to elevate blood pressure. First, ACE transforms the inactive decapeptide angiotensin I into the active octapeptide angiotensin II, a potent vasoconstrictor that narrows blood vessels and raises blood pressure. Second, ACE degrades bradykinin, a chemical that typically lowers blood pressure by inducing blood vessels to open. Because ACE boosts a vasoconstrictor and destroys a vasodilator, blocking its activity is an effective technique for decreasing angiotensin II production and increasing bradykinin levels, so controlling and reducing blood pressure (Heo et al., 2023; Ningrum et al., 2022).

Marine actinobacteria isolated from sponges and their habitats along India's western coast exhibit strong enzyme-inhibitory properties. *Streptomyces* isolates typically generate protease inhibitors, such as those targeting trypsin and chymotrypsin. *Nocardiopsis* isolates also exhibit strong protease inhibition, with some strains inhibiting ACE. Notably, an *Actinomycetospora* isolate has been found to be an ACE inhibitor. These findings underline the importance of marine actinobacteria in this region as a source of new EIs (Baig et al., 2018).

4.3 MAOs and MAO Inhibitors

MAO (monoamine oxygen oxidoreductase, EC 1.4.3.4) is an enzyme that catalyzes the oxidative deamination of monoamine neurotransmitters and xenobiotic amines, regulating their concentrations in the nervous system. It exists in two isoforms, type A monoamine oxidase (MAO-A) and type B monoamine oxidase (MAO-B), with different substrate specificity and inhibitor sensitivity. MAO-A is predominantly expressed in the presynaptic terminals of catecholaminergic neurons, whereas MAO-B is primarily localized in serotonergic and histaminergic neurons, astrocytes, and ventricular cells, reflecting their distinct functional roles in different neuronal and non-neuronal populations of the brain (Edmondson & Binda, 2018; Naoi et al., 2025).

MAO-A catabolizes monoamine transmitters, such as serotonin, norepinephrine, and dopamine, and plays an essential role in the development, progression, and treatment of neuropsychiatric diseases. Depressive disorders are thought to be caused by an increase in MAOA expression and a decrease in brain levels of serotonin and norepinephrine. Depression is linked to functional polymorphisms in the MAOA gene and genes involved in the serotonin signal pathway (Edmondson & Binda, 2018; Naoi et al., 2025). MAO-B is a mitochondrial enzyme that catalyzes the breakdown of monoamines, including dopamine and phenylethylamine, resulting in hydrogen peroxide and other reactive byproducts. This activity is necessary for neurotransmitter modulation, but it also causes oxidative stress, which can harm cells and tissues. Alzheimer's and Parkinson's diseases both have high MAO-B activity. Increased MAO-B levels in neurons and astrocytes in Alzheimer's disease are associated with Aβ synthesis and neurofibrillary tangle development, which contribute to disease progression (Schedin-Weiss et al., 2017; Behl et al., 2021).

MAO is involved in the processes of catecholamine and 5-hydroxytryptamine inactivation. Inhibition of MAO gives antidepressant effects and behavioral advantages in the brain, making MAO a key pharmaceutical target for drug design (Carradori et al., 2014; Ramsay & Albreht, 2018). Compound 1 (5,7-dihydroxy-2-isopropyl-4H-chromen-4-one) was a strong and specific inhibitor of MAO-A, with an inhibitory concentration of 50%. The compound identified from marine-derived *Streptomyces* sp. could be used as a possible treatment for neurological illnesses, such as depression and Alzheimer's disease (Oh et al., 2021).

4.4 α-Glucosidases and α-Glucosidase Inhibitors

Diabetes is a chronic metabolic condition characterized by consistently high blood glucose levels. It results from the body's inability to create enough insulin, poor insulin use, or a combination of the two. Diabetes is classified into three categories based on its underlying mechanism: type 1 diabetes, type 2 diabetes, and gestational diabetes. The most common kind of diabetes is type 2, which accounts for roughly 90% of all diabetes occurrences globally. It is mostly linked to insulin resistance and lifestyle factors such as obesity, poor food, and a lack of physical activity (Cho et al., 2018). α-Glucosidase (AGS) is a small intestine enzyme that catalyzes the final stage of carbohydrate digestion, converting complex carbs into glucose, which can be absorbed into the bloodstream. In diabetes, particularly type 2, this mechanism causes fast elevations in blood glucose after meals (postprandial hyperglycemia), which worsens glycemic control and raises the risk of complications (Khan et al., 2024).

AGS inhibitors slow down the synthesis and intestinal absorption of glucose by occupying the binding site of AGS, which prevents the breakdown of carbohydrates. Therefore, one of the most efficient approaches for controlling hyperglycemia (Santos et al., 2018). *Streptomyces* sp. isolated from mangrove sediment produces autolytimycin, which has the strongest AGS inhibitory activity; it inhibits the function of both Hsp90 and AGS and may serve as a multifunctional drug against cancer and diabetes (Lu et al., 2023). Similarly, chloroform extract of *Streptomyces coelicoflavus* from salterns in Kothapatnam, Ongole, Andhra Pradesh, India, significantly inhibited mammalian AGS by over 91%. This extract also showed significant anti-hyperglycemic effects in diabetic rats, lowering postprandial blood glucose by 82.25% (maltose) and 77.25% (sucrose), indicating its role in blood sugar regulation. The identified lead molecule, t-butyl-5-chloromethyl-3-methyl-4-oxoimidazolidine-1-carboxylic acid, t-butyl ester, shows the potential of actinomycete-derived compounds as innovative therapeutic agents in the treatment of diabetes (Kumar & Rao, 2018).

4.5 MurNAc-Pentapeptide Translocase and Inhibitors

MurNAc-pentapeptide translocase (MraY) is a critical enzyme for the survival of bacteria and acts as a great target for the development of novel antibacterial drugs (Bugg et al., 2006). MraY (translocase I) catalyzes the first membrane-associated process in peptidoglycan biosynthesis, converting phospho-MurNAc-pentapeptide from Uridine Diphosphate (UDP)-MurNAc-pentapeptide to undecaprenyl phosphate and producing Lipid I. This process is critical for bacterial growth because peptidoglycan maintains cell wall integrity in both Gram-positive and Gram-negative bacteria. The enzyme's activity is based on conserved aspartic acid and histidine residues, and it employs a nucleophilic attack mechanism necessitating dual substrate binding. Because of its critical role, MraY is a validated target for various natural product-derived antibiotics (Liu et al., 2016). MraY inhibitors inhibit a critical stage in peptidoglycan production, affecting bacterial cell wall construction and resulting in cell death. Many organic compounds are highly effective MraY inhibitors, with broad-spectrum antibacterial activity and unique modes of action. Structural studies show that these inhibitors bind to specific locations on MraY, and improving their architectures can improve potency and selectivity (Hering et al., 2018; Yamamoto et al., 2024).

A marine-derived actinomycete closely related to *Streptomyces bacillaris* was isolated from a marine sediment sample collected from a beach on Jeju Island,

Republic of Korea. Four tunicamycin-class compounds—tunicamycin VII, tunicamycin VIII, corynetoxin U17a, and tunicamycin IX—were recovered from the marine actinomycete Streptomyces sp. MBTG32, which was found to be closely related to *Streptomyces bacillaris* using 16S rDNA sequencing. Spectroscopic analyses were used to reveal structural details, which were then compared to previously reported NMR data. The compounds showed high antibacterial activity against Gram-positive bacteria, particularly *Staphylococcus aureus*, and inhibited MraY. These findings imply that tunicamycin derivatives from marine actinobacteria largely inhibit MraY and act as a promising drug for drug-resistant Gram-positive bacteria (Lee et al., 2024) (Table 1).

Table 1 EIs from marine actinobacteria and their key findings and therapeutic potential

Enzyme target	Inhibitor source (actinomycete genus/species)	Origin/habitat	Key findings and therapeutic potential	References
Proteases (papain, trypsin, Chikungunya virus protease)	*Streptomyces griseoincarnatus HK12*	Marine sponge (Callyspongia sp.)	Novel protein inhibitor active against papain, trypsin, and notably, the first natural inhibitor against Chikungunya virus, showing antiviral potential	Kamarudheen et al. (2021)
Proteases (trypsin, chymotrypsin)	Marine *Streptomyces* species	Nicobar Island (marine)	Irreversible peptide protease inhibitors with anti-plasmodial efficacy, suggesting malaria drug potential	Karthik et al. (2014)
Proteases (trypsin, chymotrypsin)	*Streptomyces* isolates	India's western coast (marine sponges and habitats)	Frequent production of protease inhibitors, highlighting broad enzyme-inhibitory properties	Baig et al. (2018)
Proteases	*Nocardiopsis* isolates	India's western coast (marine sponges and habitats)	Strong protease inhibition, contributing to new EI sources	Baig et al. (2018)
ACE	*Nocardiopsis* isolates	India's western coast (marine sponges and habitats)	ACE inhibition, indicating potential for cardiovascular and renal drug development	Baig et al. (2018) and Silva-Velasco et al. (2024)

(continued)

Table 1 (continued)

Enzyme target	Inhibitor source (actinomycete genus/species)	Origin/habitat	Key findings and therapeutic potential	References
ACE	*Actinomycetospora* isolate	India's western coast (marine sponges and habitats)	Identified as an ACE inhibitor, broadening the diversity of ACE-inhibiting marine actinobacteria	Baig et al. (2018) and Silva-Velasco et al. (2024)
MAO-A	*Streptomyces* sp.	Marine-derived	Produced compound 1 (MAO-A inhibitor), a strong and specific inhibitor with potential for neurological illnesses such as depression and Alzheimer's disease	Oh et al. (2021), Carradori et al. (2014) and Ramsay and Albreht (2018)
AGSs	*Streptomyces coelicoflavus*	Kothapatnam, Ongole, Andhra Pradesh, India (salterns)	Chloroform extract significantly inhibited mammalian AGS (91.27%) and showed anti-hyperglycemic effects in diabetic rats (82.25% reduction with maltose, 77.25% with sucrose). Lead compound identified.	Kumar and Rao (2018) and Santos et al. (2018)
AGSs	*Streptomyces* sp.	Mangrove sediment	Produces autolytimycin, a strong AGS inhibitor that also inhibits Hsp90, suggesting multi-functional drug potential against cancer and diabetes.	Lu et al. (2023) and Santos et al. (2018)
MraY	*Streptomyces bacillaris*	Jeju Island, Republic of Korea (marine sediment)	Four tunicamycin-class compounds identified with strong inhibitory effects on *Staphylococcus aureus* MraY, promising for drug-resistant Gram-positive bacteria.	Lee et al. (2024) and Bugg et al. (2006)

5 Future Perspectives

The future of EI discovery from marine actinobacteria is at the crossroads of genomics, metabolomics, and biotechnology, with rapid technological breakthroughs promising to transform how new bioactive chemicals are discovered and manufactured. Genome mining methods such as antibiotics & Secondary Metabolite Analysis SHell (anti-SMASH) and Prediction Informatics for Secondary Metabolome (PRISM) have revealed that marine actinobacteria include a large number of cryptic biosynthetic gene clusters (BGCs), many of which remain unexpressed under conventional laboratory conditions (Blin et al., 2021; Skinnider et al., 2017). These silent or weakly expressed BGCs are an untapped reservoir of EIs, and strategies such as heterologous expression, Clustered Regularly Interspaced Short Palindromic Repeats-C (CRISPR-C) as genome editing, and synthetic biology are increasingly being used to unlock their metabolic potential and increase metabolite yields (Zhang et al., 2017).

Furthermore, co-culture techniques, marine fermentation optimization, and deep-sea bioprospecting are anticipated to uncover new actinobacterial taxa and metabolite profiles, particularly in extreme or underexplored environments such as hydrothermal vents, hadal trenches, and polar regions. As antibiotic resistance and chronic diseases continue to develop globally, EIs from marine actinobacteria will remain crucial in the hunt for next-generation therapeutics. Combining such approaches with metabolomics-based platforms like Global Natural Products Social Molecular Networking allows researchers to swiftly dereplicate recognized molecules and detect novel scaffolds in complicated extracts, expediting discovery (Wang et al., 2016).

Another appealing strategy is to incorporate artificial intelligence and machine learning into drug discovery pipelines. These techniques enable predictive modeling of enzyme-inhibitor interactions, structural modification, and lead ranking based on drug similarity and pharmacokinetic features. *In silico* approaches can significantly reduce the time and cost of traditional screening while also discovering small structure–activity connections that would otherwise go undetected (Zhang et al., 2017). Simultaneously, co-culture approaches, fermentation optimization, and chemical elicitors are employed to imitate the natural interactions of marine actinobacteria with their environment, activating BGCs that would otherwise remain silent under monoculture circumstances. Such approaches not only widen the range of secondary metabolites acquired but also raise the chances of discovering novel EIs with distinct pharmacological characteristics.

Ecological exploration has huge promise for future discoveries. Deep-sea sediments, hydrothermal vents, hadal trenches, and polar marine environments are all understudied, even though these extreme habitats contain rare actinobacterial taxa with distinct evolutionary adaptations and biochemical machinery. Isolating actinobacteria from such environments has already resulted in the discovery of structurally new chemicals with antibacterial, anticancer, and enzyme-inhibitory properties (Nam et al., 2013; Abdel-Mageed et al., 2010). In the future, merging deep-sea bioprospecting with next-generation sequencing and metabologenomics techniques is predicted to increase the chemical diversity of marine actinobacteria.

EIs produced from marine actinobacteria are expected to have implications beyond established antibacterial and anticancer medicines. Inhibitors of MAOs and AGSs are being evaluated for their potential in treating neurodegenerative disorders and diabetes, respectively, while ACE inhibitors show promise for cardiovascular therapeutics (Baig et al., 2018). Future research is likely to concentrate on structure-guided optimization, nanoformulation, and targeted delivery systems to increase the inhibitors' stability, selectivity, and bioavailability in therapeutic settings. Furthermore, the development of sustainable marine bioprospecting strategies is critical to ensure that the exploration and exploitation of marine microbial diversity do not damage ecological equilibrium.

Finally, the combination of genomic mining, metabolomics, AI-driven screening, and ecological research establishes marine actinobacteria as a key source of new EIs for next-generation medicines. As the worldwide difficulties of antibiotic resistance, metabolic syndromes, and neurological illnesses worsen, the discovery of novel inhibitors from marine ecosystems is expected to play an important role in meeting urgent healthcare requirements.

6 Conclusion

EIs with a broad range of therapeutic applications are produced in large quantities by marine actinobacteria. These inhibitors are useful candidates for the treatment of cancer, diabetes, hypertension, infectious diseases, and neurological disorders because they target important enzymes, including ACE, AGSs, proteases, MAOs, and MraY. The biochemical specificity, ecological adaptability, and structural diversity of metabolites generated from actinobacteria highlight their importance in drug discovery.

EIs from genera such as *Streptomyces, Salinispora, Nocardiopsis*, and *Actinomadura* have not just notable biological activity but also promise for therapeutic translation, as this chapter highlights. The range of these bioactive substances will keep rising as a result of the expanding use of genome-guided discovery, improved metabolomics, synthetic biology, and computational drug design.

In conclusion, marine actinobacteria continue to be a valuable and mainly unexplored source of inhibitors of enzymes. Researchers may fully utilize contemporary biotechnological advancements and sustainable marine bioprospecting to tackle some of the most urgent medical issues of our present time, such as chronic metabolic and neurological diseases and antibiotic resistance. Thus, further research into their enzymatic inhibitory properties is both a scientific opportunity and a necessity for world healthcare.

References

Abdel-Mageed, W. M., Milne, B. F., Wagner, M., Schumacher, M., Sandor, P., Pathom-aree, W., et al. (2010). Dermacozines, a new phenazine family from deep-sea dermacocci isolated from a Mariana Trench sediment. *Organic & Biomolecular Chemistry, 8*(10), 2352–2362. https://doi.org/10.1039/c001445a

Baig, M. H., Ahmad, K., Rabbani, G., Danishuddin, M., & Choi, I. (2018). Computer aided drug design and its application to the development of potential drugs for neurodegenerative disorders. *Current Neuropharmacology, 16*(6), 740–748. https://doi.org/10.2174/1570159X15666171016163510

Behl, T., Kaur, D., Sehgal, A., et al. (2021). Role of monoamine oxidase activity in Alzheimer's disease: An insight into the therapeutic potential of inhibitors. *Molecules, 26*, 3724. https://doi.org/10.3390/molecules26123724

Bister, B., Bischoff, D., Ströbele, M., Riedlinger, J., Reicke, A., Wolter, F., et al. (2004). Abyssomicin C—A polycyclic antibiotic from a marine *Verrucosispora* strain as an inhibitor of the p-aminobenzoic acid/tetrahydrofolate biosynthesis pathway. *Angewandte Chemie International Edition, 43*(19), 2574–2576. https://doi.org/10.1002/anie.200353160

Blin, K., Shaw, S., Kloosterman, A. M., Charlop-Powers, Z., Van Wezel, G. P., Medema, M. H., & Weber, T. (2021). antiSMASH 6.0: Improving cluster detection and comparison capabilities. *Nucleic Acids Research, 49*(W1), W29–W35.

Bugg, T. D., Lloyd, A. J., & Roper, D. I. (2006). Phospho-MurNAc-pentapeptide translocase (MraY) as a target for antibacterial agents and antibacterial proteins. *Infectious Disorders Drug Targets, 6*(2), 85–106.

Carradori, S., D'Ascenzio, M., Chimenti, P., Secci, D., & Bolasco, A. (2014). Selective MAO-B inhibitors: A lesson from natural products. *Molecular Diversity, 18*(1), 219–243. https://doi.org/10.1007/s11030-013-9490-6

Cho, N. H., Shaw, J. E., Karuranga, S., et al. (2018). IDF Diabetes Atlas: Global estimates of diabetes prevalence for 2017 and projections for 2045. *Diabetes Research and Clinical Practice, 138*, 271–281. https://doi.org/10.1016/j.diabres.2018.02.023

Cipollone, F., & Fazia, M. L. (2006). Cyclooxygenase-2 inhibition: Vascular inflammation and cardiovascular risk. *Current Atherosclerosis Reports, 8*(3), 245–251. https://doi.org/10.1007/s11883-006-0080-2

de Leuw, P., & Stephan, C. (2017). Protease inhibitors for the treatment of hepatitis C virus infection. *GMS Infectious Diseases, 5*, Doc08. https://doi.org/10.3205/id000034

Deu, E. (2017). Proteases as antimalarial targets: Strategies for genetic, chemical, and therapeutic validation. *The FEBS Journal, 284*(16), 2604–2628. https://doi.org/10.1111/febs.14130

Eatemadi, A., Aiyelabegan, H. T., Negahdari, B., Mazlomi, M. A., Daraee, H., Daraee, N., et al. (2017). Role of protease and protease inhibitors in cancer pathogenesis and treatment. *Biomedicine & Pharmacotherapy, 86*, 221–231.

Edmondson, D. E., & Binda, C. (2018). Monoamine oxidases. *Subcellular Biochemistry, 87*, 117–139. https://doi.org/10.1007/978-981-10-7757-9_5

Goodfellow, M. (2015). Actinobacteria phyl. nov. In W. Whitman (Ed.), *Bergey's manual of systematics of archaea and bacteria*. Wiley.

Haefner, B. (2003). Drugs from the deep: Marine natural products as drug candidates. *Drug Discovery Today, 8*(12), 536–544.

Heo, S. Y., Kang, N., Kim, E. A., Kim, J., Lee, S. H., Ahn, G., Oh, J. H., Shin, A. Y., Kim, D., & Heo, S. J. (2023). Purification and molecular docking study on the angiotensin I-converting enzyme (ACE)-inhibitory peptide isolated from hydrolysates of the Deep-Sea Mussel *Gigantidas vrijenhoeki*. *Marine Drugs, 21*(8), 458. https://doi.org/10.3390/md21080458

Hering, J., Dunevall, E., Ek, M., & Brändén, G. (2018). Structural basis for selective inhibition of antibacterial target MraY, a membrane-bound enzyme involved in peptidoglycan synthesis. *Drug Discovery Today, 23*(7), 1426–1435. https://doi.org/10.1016/j.drudis.2018.05.020

Imada, C. (2004). Enzyme inhibitors of marine microbial origin with pharmaceutical importance. *Marine Biotechnology, 6*(3), 193–198.

Imada, C., Masuda, S., Kobayashi, T., Hamada-Sato, N., & Nakashima, T. (2010). Isolation and characterization of marine and terrestrial actinomycetes using a medium supplemented with NaCl. *Actinomycetologica, 24*(1), 12–17.

Kamarudheen, N., Khaparde, A., Gopal, S., & Rao, K. B. (2021). Unraveling a natural protease inhibitor from marine *Streptomyces griseoincarnatus* HK12 active against Chikungunya virus. *Microbiological Research, 252*, 126858. https://doi.org/10.1016/j.micres.2021.126858

Karthik, L., Kumar, G., Keswani, T., Bhattacharyya, A., Chandar, S. S., & Bhaskara Rao, K. V. (2014). Protease inhibitors from marine actinobacteria as a potential source for antimalarial compound. *PLoS One, 9*(3), e90972. https://doi.org/10.1371/journal.pone.0090972

Khan, F., Khan, M., Kumar, A., & Akhtar, S. (2024). Recent advances in the development of alpha-glucosidase and alpha-amylase inhibitors in type 2 diabetes management: Insights from in silico to in vitro studies. *Current Drug Targets*. https://doi.org/10.2174/0113894501313365240722100902

Klein, T., Eckhard, U., Dufour, A., Solis, N., & Overall, C. (2017). Proteolytic cleavage-mechanisms, function, and "omic" approaches for a near-ubiquitous posttranslational modification. *Chemical Reviews, 118*(3), 1137–1168. https://doi.org/10.1021/acs.chemrev.7b00120

König, G. M., Kehraus, S., Seibert, S. F., Abdel-Lateff, A., & Müller, D. (2006). Natural products from marine organisms and their associated microbes. *Chembiochem, 7*(2), 229–238.

Kuddus, M. (2019). Chapter 1—Introduction to food enzymes. In M. Kuddus (Ed.), *Enzymes in food biotechnology* (pp. 1–18). Academic. https://doi.org/10.1016/B978-0-12-813280-7.00001-3

Kumar, S. R. S., & Rao, K. V. B. (2018). Efficacy of alpha-glucosidase inhibitor from marine actinobacterium in the control of postprandial hyperglycaemia in streptozotocin (STZ) induced diabetic male albino Wister rats. *Iranian Journal of Pharmaceutical Research, 17*(1), 202–214.

Lam, K. S. (2006). Discovery of novel metabolites from marine actinomycetes. *Current Opinion in Microbiology, 9*(3), 245–251. https://doi.org/10.1016/j.mib.2006.03.004

Lee, J., Hwang, J. Y., Oh, D., et al. (2024). Tunicamycins from marine-derived *Streptomyces bacillaris* inhibit MurNAc-pentapeptide translocase in *Staphylococcus aureus*. *Marine Drugs, 22*(7), 293. https://doi.org/10.3390/md22070293

Li, S., Tian, X., Niu, S., Zhang, W., Chen, Y., Zhang, H., et al. (2011). Pseudonocardians A–C, new diazaanthraquinone derivatives from a deep-sea actinomycete *Pseudonocardia* sp. SCSIO 01299. *Marine Drugs, 9*(8), 1428–1439. https://doi.org/10.3390/md9081428

Liu, Y., Rodrigues, J. P., Bonvin, A. M., et al. (2016). New insight into the catalytic mechanism of bacterial MraY from enzyme kinetics and docking studies. *The Journal of Biological Chemistry, 291*(29), 15057–15068. https://doi.org/10.1074/jbc.M116.717884

López-Otín, C., & Bond, J. (2008). Proteases: Multifunctional enzymes in life and disease. *Journal of Biological Chemistry, 283*, 30433–30437. https://doi.org/10.1074/jbc.R800035200

Lu, X., Zhang, M., Qiu, Y., et al. (2023). α-Glucosidase inhibitors from two mangrove-derived actinomycetes. *Molecules, 28*(9), 3822. https://doi.org/10.3390/molecules28093822

Magarvey, N. A., Keller, J. M., Bernan, V., Dworkin, M., & Sherman, D. H. (2004). Isolation and characterization of novel marine-derived actinomycete taxa rich in bioactive metabolites. *Applied and Environmental Microbiology, 70*(12), 7520–7529.

Mahapatra, G. P., Raman, S., Nayak, S., Gouda, S., Das, G., & Patra, J. K. (2020). Metagenomics approaches in discovery and development of new bioactive compounds from marine actinomycetes. *Current Microbiology, 77*(4), 645–656.

Martins, A., Vieira, H., Gaspar, H., & Santos, S. (2014). Marketed marine natural products in the pharmaceutical and cosmeceutical industries: Tips for success. *Marine Drugs, 12*(2), 1066–1101.

McArthur, K. A., Mitchell, S. S., Tsueng, G., Rheingold, A., White, D. J., Grodberg, J., et al. (2008). Lynamicins A–E, chlorinated bisindole pyrrole antibiotics from a novel marine actinomycete. *Journal of Natural Products, 71*(10), 1732–1737. https://doi.org/10.1021/np800286d

Mincer, T. J., Jensen, P. R., Kauffman, C. A., & Fenical, W. (2002). Widespread and persistent populations of a major new marine actinomycete taxon in ocean sediments. *Applied and Environmental Microbiology, 68*(10), 5005–5011.

Mourão, C. B. F., & Schwartz, E. F. (2013). Protease inhibitors from marine venomous animals and their counterparts in terrestrial venomous animals. *Marine Drugs, 11*(6), 2069–2112.

Nam, S. J., Gaudêncio, S. P., Kauffman, C. A., Jensen, P. R., Kondratyuk, T. P., Marler, L. E., et al. (2010). Fijiolides A and B, inhibitors of TNF-α-induced NFκB activation, from a marine-derived sediment bacterium of the genus *Nocardiopsis*. *Journal of Natural Products, 73*(6), 1080–1086. https://doi.org/10.1021/np100087c

Nam, S. J., Kauffman, C. A., Paul, L. A., Jensen, P. R., & Fenical, W. (2013). Actinoranone, a cytotoxic meroterpenoid of unprecedented structure from a marine adapted Streptomyces sp. *Organic Letters, 15*(21), 5400–5403.

Naoi, M., Maruyama, W., Shamoto-Nagai, M., et al. (2025). Type A monoamine oxidase; its unique role in mood, behavior and neurodegeneration. *Journal of Neural Transmission, 132*, 387–406. https://doi.org/10.1007/s00702-024-02866-z

Ningrum, S., Sutrisno, A., & Hsu, J. L. (2022). An exploration of angiotensin-converting enzyme (ACE) inhibitory peptides derived from gastrointestinal protease hydrolysate of milk using a modified bioassay-guided fractionation approach coupled with in silico analysis. *Journal of Dairy Science, 105*(3), 1913–1928. https://doi.org/10.3168/jds.2021-21112

Oh, J. M., Lee, C., Nam, S. J., & Kim, H. (2021). Chromenone derivatives as monoamine oxidase inhibitors from marine-derived MAR4 clade *Streptomyces* sp. CNQ-031. *Journal of Microbiology and Biotechnology, 31*(7), 1022–1027.

Pandey, S., Sree, A., Dash, S. S., Sethi, D. P., & Chowdhury, L. (2013). Diversity of marine bacteria producing beta-glucosidase inhibitors. *Microbial Cell Factories, 12*, 35. https://doi.org/10.1186/1475-2859-12-35

Patadiya, N., Panchal, N., & Vaghela, V. (2021). A review on enzyme inhibitors. *International Research Journal of Pharmacy, 12*(6), 60–66. https://doi.org/10.7897/2230-8407.1206145

Powers, J. C., Asgian, J. L., Ekici, O. D., & James, K. E. (2002). Irreversible inhibitors of serine, cysteine, and threonine proteases. *Chemical Reviews, 102*(12), 4639–4750. https://doi.org/10.1021/cr010182v

Raja, S., Ganesan, S., Sivakumar, K., & Thangaradjou, T. (2010). Screening of marine actinobacteria for amylase enzymes inhibitors. *Indian Journal of Microbiology, 50*(2), 233–237.

Ramsay, R. R., & Albreht, A. (2018). Kinetics, mechanism, and inhibition of monoamine oxidase. *Journal of Neural Transmission, 125*(11), 1659–1683.

Ramsay, R. R., & Tipton, K. F. (2017). Assessment of enzyme inhibition: A review with examples from the development of monoamine oxidase and cholinesterase inhibitory drugs. *Molecules (Basel, Switzerland), 22*(7), 1192. https://doi.org/10.3390/molecules22071192

Robinson, P. K. (2015). Enzymes: Principles and biotechnological applications. *Essays in Biochemistry, 59*, 1–41. https://doi.org/10.1042/bse0590001

Santos, C. M. M., Freitas, M., & Fernandes, E. (2018). A comprehensive review on xanthone derivatives as α-glucosidase inhibitors. *European Journal of Medicinal Chemistry, 157*, 1460–1479. https://doi.org/10.1016/j.ejmech.2018.07.073

Schedin-Weiss, S., Inoue, M., Hromadkova, L., et al. (2017). Monoamine oxidase B is elevated in Alzheimer disease neurons, is associated with γ-secretase and regulates neuronal amyloid β-peptide levels. *Alzheimer's Research & Therapy, 9*. https://doi.org/10.1186/s13195-017-0279-1

Silva-Velasco, D. L., Cervantes-Pérez, L. G., & Sánchez-Mendoza, A. (2024). ACE inhibitors and their interaction with systems and molecules involved in metabolism. *Heliyon, 10*(2), e24655.

Skinnider, M. A., Merwin, N. J., Johnston, C. W., & Magarvey, N. A. (2017). PRISM 3: Expanded prediction of natural product chemical structures from microbial genomes. *Nucleic Acids Research, 45*(W1), W49–W54.

Sojka, D., Šnebergerová, P., & Robbertse, L. (2021). Protease inhibition—An established strategy to combat infectious diseases. *International Journal of Molecular Sciences, 22*(11), 5762. https://doi.org/10.3390/ijms22115762

Tormo, J. R., Oves, D., Lacret, R., Moreno, C., De La Cruz, M., Díaz, C., González, I., Martín, J., Vicente, F., Genilloud, O., & Reyes, F. (2015). Marine natural products drug discovery from actinomycetes at Fundación MEDINA. *Planta Medica, 81*(11), PK18. https://doi.org/10.1055/s-0035-1556296

Waggoner, B. M. (1994). Fossil actinomycetes in Eocene-Oligocene Dominican amber. *Journal of Paleontology, 68*(2), 398–401.

Wang, M., Carver, J. J., Phelan, V. V., Sanchez, L. M., Garg, N., Peng, Y., et al. (2016). Sharing and community curation of mass spectrometry data with Global Natural Products Social Molecular Networking. *Nature Biotechnology, 34*(8), 828–837.

Watve, M. G., Tickoo, R., Jog, M. M., & Bhole, B. D. (2001). How many antibiotics are produced by the genus Streptomyces? *Archives of Microbiology, 176*(5), 386–390. https://doi.org/10.1007/s002030100345

Wijesekara, I., & Kim, S.-K. (2010). Angiotensin-I-converting enzyme (ACE) inhibitors from marine resources: Prospects in the pharmaceutical industry. *Marine Drugs, 8*(4), 1080–1093.

Wyche, T. P., Standiford, M., Hou, Y., Braun, D., Johnson, D. A., Johnson, J. A., et al. (2013). Activation of the nuclear factor E2-related factor 2 pathway by novel natural products halomadurones A–D and a synthetic analogue. *Marine Drugs, 11*(12), 5089–5099. https://doi.org/10.3390/md11125089

Yamamoto, K., Sato, T., Hao, A., Asao, K., Kaguchi, R., Kusaka, S., Ruddarraju, R. R., Kazamori, D., Seo, K., Takahashi, S., & Ichikawa, S. (2024). Development of a natural product optimization strategy for inhibitors against MraY, a promising antibacterial target. *Nature Communications, 15*. https://doi.org/10.1038/s41467-024-49484-7

Yamanaka, K., Reynolds, K. A., Kersten, R. D., Ryan, K. S., Gonzalez, D. J., Nizet, V., et al. (2014). Direct cloning and refactoring of a silent lipopeptide biosynthetic gene cluster yields the antibiotic taromycin A. *Proceedings of the National Academy of Sciences, 111*(5), 1957–1962. https://doi.org/10.1073/pnas.1319584111

Ye, L., Zhou, Q., Liu, C., Luo, X., Na, G., & Xi, T. (2009). Identification and fermentation optimization of a marine-derived Streptomyces Griseorubens with anti-tumor activity. *IJMS, 38*(1). http://nopr.niscair.res.in/handle/123456789/4240

Zhang, M. M., Wong, F. T., Wang, Y., Luo, S., Lim, Y. H., Heng, E., et al. (2017). CRISPR–Cas9 strategy for activation of silent Streptomyces biosynthetic gene clusters. *Nature Chemical Biology, 13*(6), 607–609.

Zhou, X., Huang, H., Chen, Y., Tan, J., Song, Y., Zou, J., et al. (2012). Marthiapeptide A, an anti-infective and cytotoxic polythiazole cyclopeptide from a 60 L scale fermentation of the deep-sea-derived *Marinactinospora thermotolerans* SCSIO 00652. *Journal of Natural Products, 75*(12), 2251–2255. https://doi.org/10.1021/np300554f

Agricultural Application of Marine Actinobacteria: A 10-Year Overview

Pharada Rangseekaew and Wasu Pathom-aree

Abstract

Marine actinobacteria represent an underexplored yet highly attractive resource for sustainable agriculture. These actinobacteria, with unique adaptation to saline, oligotrophic, and environmentally fluctuating marine ecosystems, possess unique genomic and physiological properties that enable the production of diverse bioactive compounds, stress-protective metabolites, and hydrolytic enzymes. Members of genera such as *Dermacoccus*, *Micromonospora, Salinispora,* and *Streptomyces* have clearly demonstrated their potential in plant growth promotion, biological control of phytopathogens, and enhancement of plant tolerance to abiotic stresses, particularly salinity. Marine actinobacteria contribute to agricultural productivity through multiple mechanisms, ranging from phytohormone production (e.g., indole-3-acetic acid), siderophore-mediated iron acquisition, phosphate solubilization, nitrogen fixation, 1-aminocyclopropane-1-carboxylic acid deaminase activity, and the production of antimicrobial secondary metabolites effective against fungal pathogens. Their salt-tolerance strategies further strengthen the position of marine actinobacteria as promising bioinoculant candidates for saline degraded areas, which are expanding globally due to climate change and unsustainable agricultural practices. Despite these promising opportunities, challenges remain in strain formulation, large-scale cultivation, regulation of plant–microbe interactions, and validated field trial performance. This chapter reviews the current advancements in the agricultural applications of plant growth-promoting marine actinobacteria and their mechanisms from published literature reported within the past 10 years. Emphasis is also focused on their roles in enhancing plant growth under both biotic and abiotic stress conditions to improve crop productivity.

P. Rangseekaew · W. Pathom-aree (✉)
Department of Biology, Faculty of Science, Chiang Mai University, Chiang Mai, Thailand
e-mail: wasu.p@cmu.ac.th

K. Loganathan, Z. Li (eds.), *Marine Actinomycetes*,
https://doi.org/10.1007/978-981-95-9849-6_15

Keywords

Marine actinobacteria · Plant growth-promoting actinobacteria · Agricultural microbiology · Biotic stress · Abiotic stress · Salinity stress · Marine environments

1 Introduction

Plant growth promotion using microbes has attracted considerable interest due to its cost-effectiveness, accessibility, ease of use, crop productivity improvement, and potential for advancing biofertilizer technologies. Plant growth-promoting microorganisms (PGPMs) have long been recognized as playing a key role in enhancing plant development. Notably, several bacterial genera, including *Azospirillum, Arthrobacter*, *Bacillus*, *Klebsiella*, and *Pseudomonas,* are widely recognized for their plant growth-promoting abilities (Dhawi, 2023).

Although PGPMs have been widely studied, marine actinobacteria remain relatively underexplored in this context. Due to their ability to thrive in extreme environments, these actinobacteria hold promise as plant growth-promoting agents. Investigating both their direct and indirect mechanisms of promoting plant growth could pave the way for the development and commercialization of environmentally friendly and sustainable bioinoculants. This chapter focuses on the plant growth-promoting mechanisms of marine actinobacteria reported over the past 10 years and their roles in enhancing plant growth under both biotic and abiotic stress conditions.

2 Marine Actinobacteria With Plant Growth-Promoting Properties

Plant growth-promoting properties of actinobacteria can be broadly categorized into two major mechanisms: direct and indirect mechanisms. Direct plant growth promotion involves the ability of actinobacteria to enhance nutrient availability to plants through various processes. In contrast, indirect mechanisms refer to their role as biocontrol agents, whereby actinobacteria suppress plant pathogens through the production of antimicrobial compounds and extracellular enzymes.

2.1 Direct Plant Growth-Promoting Properties of Marine Actinobacteria

Direct plant growth promotion by actinobacteria primarily involves the acquisition of essential nutrients such as nitrogen, iron, and phosphorus, which are often present in forms unavailable to plants in natural environments. Plant growth-promoting actinobacteria employ different approaches to supply these nutrients to plants as follows.

2.1.1 Nitrogen Fixation

Nitrogen is a macronutrient essential for plant growth and development; however, most nitrogen in nature exists as atmospheric nitrogen gas (N_2), which cannot be directly utilized by plants or by most actinobacteria. Nitrogen fixation refers to the conversion of atmospheric nitrogen into ammonia, a process catalyzed by the enzyme nitrogenase (Sellstedt & Richau, 2013). Nitrogen fixation in actinobacteria occurs in two main forms: symbiotic nitrogen fixation, as seen in members of the genus *Frankia*, and free-living nitrogen fixation by non-symbiotic actinobacteria such as *Arthrobacter, Agromyces*, *Corynebacterium*, *Mycobacterium*, *Micromonospora*, *Propionibacteria,* and *Streptomyces,* which have nitrogen-fixing capacity (Sellstedt & Richau, 2013). There are some reports on nitrogen-fixing marine actinobacteria, such as *Streptomyces* sp. KLBMP S0051 and *Micromonospora* sp. KLBMP S0019 isolated from the rhizosphere soil of coastal salt marsh plants, China (Gong et al., 2018).

2.1.2 Siderophores

Siderophores are small, iron-chelating molecules with high affinity for binding iron (Vocciante et al., 2022). Although iron is an essential element for both plants and microorganisms, its availability in natural environments is limited because it predominantly occurs as insoluble ferric ions (Fe^{3+}), which form hydroxides and oxyhydroxides (Sathya et al., 2017). To overcome this limitation, microorganisms produce siderophores that bind Fe^{3+}, making it available for microbial metabolism. In addition, siderophores enhance iron uptake in plant cells, thereby contributing to an improvement in plant growth and health. Based on their chemical structures, siderophores are generally classified into three main types: catecholates, hydroxamates, and carboxylates. Moreover, certain siderophores are categorized as mixed types, containing features of more than one structural class (Aznar & Dellagi, 2015; Khan et al., 2018). Various marine actinobacteria have been reported for their ability to produce siderophores, as shown in Table 1. Approximately 51% of actinobacteria isolated from mangrove sediments and seawater in Ranong Province, Thailand, were found to produce siderophores on chrome azurol S agar. Notably, strains belonging to the genera *Nocardiopsis*, *Streptomyces*, and *Pseudonocardia* exhibited siderophore production under both non-saline conditions and salt stress of up to 300 mM NaCl (Suksaard et al., 2017). Additionally, a mangrove-derived *Streptomyces iranensis* isolated from Thailand was reported to produce hydroxamate-type siderophores (Pathom-aree et al., 2019). Deep-sea actinobacteria, including *Dermacoccus profundi* MT2.1^T and *Dermacoccus barathri* MT2.2^T (Rangseekaew et al., 2021), as well as *Dermacoccus abyssi* MT1.1^T (Rangseekaew et al., 2022), were found to produce both catecholate and hydroxamate-type siderophores.

2.1.3 Phosphate Solubilization

Phosphorus is an essential macronutrient required for metabolic processes in plants, such as cell division, energy generation, biosynthesis of macromolecules, membrane integrity, signal transduction, and photosynthesis (Rawat et al., 2021). It also plays an essential role in plant respiration. In nature, phosphorus is highly reactive

Table 1 Plant growth-promoting activity of marine actinomycetes

Location	Sample	Actinobacterial taxa	Plant growth-promoting activity	References
Coastal area, Egypt	Water, sediment, and shrimp	*Streptomyces* spp.	Nematicidal activity *in vitro* and *in vivo* against root-knot nematode, *Meloidogyne incognita* under greenhouse conditions Production of gibberellic acid, IAA, abscisic acid, kinetin, and benzyladenine Production of hydrolytic enzymes, for example, amylase, asparaginase, catalase, chitinase, lecithinase, lipase, pectinase, and protease	Rashad et al. (2015)
Sundarbans mangrove, India	Sediment	*Streptomyces* spp.	Antifungal activity against *Aspergillus niger*, *Rhizoctonia soloni*, *Macrophomina phaseolina*, and *Ustilago maydis*	Sengupta et al. (2015)
Southwest Indian Ridge, India site IR-CTD5 (16°59.9412′N 24°58.2958′E)	Sediment	*Janibacter* sp. L42	5-Enolpyruvylshikimate-3-phosphate synthase gene for the construction of transgenic glyphosate-tolerant rice	Yi et al. (2016)
Mangrove areas, Andaman Coastal Research Station for the Development of Kasetsart University Research and Development Institute, Ranong province, Thailand	Water and sediment	*Nocardiopsis yanglingensis* 1SM5-02	IAA and siderophore production under normal and salt stress conditions (0–300 mM NaCl) Promotion of rice seedling growth under 0, 100, and 200 mM NaCl	Suksaard et al. (2017)
		Streptomyces jiujiangensis 2SH3-07 *S. psammoticus* 3SH5-05	IAA and siderophore production and phosphate solubilization under normal and salt stress conditions (0–300 mM NaCl) Promotion of rice seedling growth under 0, 100, and 200 mM NaCl	
		Pseudonocardia oroxyli 3WH5-01	IAA and siderophore production and phosphate solubilization under normal and salt stress conditions (0–300 mM NaCl) (no phosphate solubilization at 100 mM NaCl) Promotion of rice seedling growth under 0, 100, and 200 mM NaCl	Suksaard et al. (2017)

Coastal region of Lianyungang city, Jiangsu province, China (34°34′15–34°45′61 N, 119°14′04–119°37′09S)	Rhizosphere soil	*Streptomyces* sp. KLBMP S0051 *Micromonospora* sp. KLBMP S0019	ACC deaminase, ammonia, hydrolytic enzymes, IAA production, and nitrogen fixation Promotion of wheat seed germination under 100 mM NaCl	Gong et al. (2018)
		Micromonospora sp. KLBMP S0019	Promotion of wheat seedling growth under 2% (w/v) NaCl in a pot experiment	
Solar saltern, India (8°05′04.35″N 77°31′17.07″E)	Soil	*Nocardiopsis lucentensis* AJ1	Antifungal activity against *Aspergillus niger*, *Fusarium* sp., and *Pythium* sp.	Adlin Jenifer et al. (2019)
Mangrove forest, Songkhla province, Thailand	Soil	*Streptomyces* sp. AMA49	Production of oligomycin A and diketopiperazines with antifungal activity against *Pyricularia oryzae*	Buatong et al. (2019)
Mangrove in Mai Po Reserve, Hong Kong	Rhizosphere of *Acanthus ilicifolius* Linn	*Streptomyces puniceus* L75	Antifungal activity against tomato early blight fungi (*Alternaria solani*)	Hao et al. (2019)
Seashore of Dachen Island, China	Mulberry shell (*Drupa granulate*)	*Streptomyces chumphonensis* AM4	Antifungal activity against citrus green mould (*Penicillium digitatum*) and citrus blue mould (*Penicillium italicum*)	Hu et al. (2019)
Mangrove, Chachoengsao province, Thailand	Sediment	*Streptomyces iranensis* S2-SC16	IAA and siderophore production and phosphate solubilization Inhibition of *Fusarium fujikuroi* conidia germination *in vitro* Promotion of rice seedling growth in a pot experiment Control of *Fusarium fujikuroi* infection in rice seeds	Pathom-aree et al. (2019)
Marchica Lagoon, Morocco (35.156468″ N-2.904342″ W)	Marine soil	*Saccharomonospora azurea* LNS-1	IAA and phosphate solubilization	Nafis et al. (2019)

(continued)

Table 1 (continued)

Location	Sample	Actinobacterial taxa	Plant growth-promoting activity	References
Challenger Deep, Mariana Trench, the western Pacific Ocean	Sediment	*Dermacoccus profundi* MT2.1^T *and D. barathri* MT2.2^T	IAA production under 0 and 150 mM NaCl Siderophore production and phosphate solubilization Promotion of tomato seedling growth under 150 mM NaCl	Rangseekaew et al. (2021)
Al Rams coast, Ras Al Khaimah, UAE (25°52′44″ N, 56°1′25″ E)	Soil	*S. tritolerans* RAK1	PAs Promotion of *Salicornia bigelovii* growth using seawater irrigation in a greenhouse experiment	Mathew et al. (2020)
		S. rochei RAMS1	ACC deaminase production *in vitro* Promotion of *Salicornia bigelovii* growth using seawater irrigation in a greenhouse experiment	
		S. chartreusis UAE1	IAA production Promotion of *Salicornia bigelovii* growth using seawater irrigation in a greenhouse experiment	
Tropical Central Pacific (19°5′55.21″ N, 104°23′24.47″ W and 19°3′28.87″ N, 104°15′40.25″ W)	Corals *Porites lobata* and *Porites panamensis*	*Salinispora arenicola*	Promotion of germination, growth, and photosynthetic response in wild tobacco (*Nicotiana attenuata*) under salt stress	Ocampo-Alvarez et al. (2020)
Khambhat city, Gujarat, India	Seawater	*Micrococcus luteus*	Ammonia, chitinase, glucanase, IAA production, potassium, and phosphate solubilization Antifungal activity against F.O.Ch, F.O.C, F.O.P, and F.O.V Promotion of chickpea growth	Patel et al. (2021)
East coast of Abu-Dhabi-UAE	Mangrove roots	*Streptomyces mutabilis* UAE1	Enhancing the growth of *Acaryochloris marina* (mangrove) under greenhouse conditions	El-Tarabily et al. (2021)

Challenger Deep, Mariana Trench, the western Pacific Ocean	Sediment	*Dermacoccus abyssi* MT1.1^T	IAA and siderophore production and phosphate solubilization Promotion of tomato seedling growth under 150 mM NaCl	Rangseekaew et al. (2022)
Pohang beach, Republic of Korea	Soil	*Arthrobacter* sp. S2T9	IAA and siderophore production and phosphate solubilization ACC deaminase production	Moon & Ali (2022)
Intertidal zone of Ilha do Mel, in the coastal region of the state of Parana, Brazil (25°20′ S, 48°20′ W and 25°35′ S, 48°35′ W)	Sediment	*Streptomyces* spp.	Production of chitinase Antagonistic activity against European pepper moth *Duponchelia fovealis* (Zeller) (Lepidoptera: Crambidae)	Porsani et al. (2022)
Coast of Oluhuta, Tomini Bay, Gorontalo, Indonesia (0°25′11.9″ N 123°08′31.8″ E)	Marine organism	*Kocuria palustris* 19C38A1	Fungicidal activity against *Fusarium oxysporum*	Setiawan et al. (2022)
Zacatecas estuary located in La Paz, Baja California Sur, Mexico (24°9′30″ N; 110°25′37″ W)	Mangrove sediment	*Nocardiopsis lucentensis* A15, *N. lucentensis* A20, and *Streptomyces* sp. A19 and *S. griseoflavus* A18	Antifungal activity against *Fusarium solani in vitro*	Torres-Rodriguez et al. (2022)
Intertidal zone in Dongfang, Hainan province, China	Sediment	*Streptomyces cinereoruber*	Production of aurone with fungicidal activity against powder mildew (*Botrytis cinerea*), cucurbits powder mildew (*Sphaerotheca fuliginea* (Schlecht ex Ff.) Poll), and rice blast (*Pyricularia oryzae*)	Liu et al. (2023)

(continued)

Table 1 (continued)

Location	Sample	Actinobacterial taxa	Plant growth-promoting activity	References
East coast of Abu Dhabi, UAE (24°26′48.5″ N; 54°26′40.6″ E)	*Avicennia marina* sediments	*Streptomyces griseorubens* UAE1	ACC deaminase, IAA, indole-3-pyruvic acid (IPA), PAs, putrescine, spermidine, spermine production, and phosphate solubilization	Alshamsi et al. (2024)
			Promotion of mangrove (*Avicennia marina*) growth in greenhouse and open-field nursery experiments	Alshamsi et al. (2024)
Zacatecas estuary located in La Paz, Baja California Sur, Mexico (24°9′30″ N; 110°25′37″ W)	Mangrove sediment	*Nocardiopsis lucentensis* A15, *Nocardiopsis lucentensis* A20, and *Streptomyces* sp. A19 and *Streptomyces griseoflavus* A18	Siderophore production VOC production Antifungal activity against *Fusarium solani* Biocontrol for tomato plants infected with *F. solani*	Torres-Rodriguez et al. (2024)

and typically occurs in combination with other elements, forming various phosphate compounds (Prabhu et al., 2019). However, most soil phosphates exist in the insoluble forms that cannot be directly utilized by plants or microorganisms. Actinobacteria are known as phosphate-solubilizing microorganisms via organic acid production to solubilize inorganic phosphate and release phosphorus in soil (Behera et al., 2014). Phosphate-solubilizing actinobacteria (PSA) have been found in various environments, including marine habitats. Nearly 39% of PSA have been recovered from mangrove sediments in Thailand. Among them, isolates identified as *Streptomyces jiujiangensis* 2SH3-07, *S. psammoticus* 3SH5-05, and *Pseudonocardia oroxyli* 3WH5-01 (Table 1) demonstrated the ability to solubilize insoluble tricalcium phosphate on Pikovskaya (PVK) agar (Suksaard et al., 2017). Similarly, another mangrove-derived strain, *Streptomyces iranensis* S2-SC16, with the ability to solubilize tricalcium phosphate, was also reported from Thailand (Pathom-aree et al., 2019). Additionally, *Saccharomonospora azurea* LNS-1, isolated from marine soil in Morocco, was able to solubilize tricalcium phosphate in broth culture. The phosphate-solubilizing activity of this strain was confirmed by a decrease in the culture broth pH from an initial 7.5 to below 6 after cultivation, indicating the production of organic acids (Nafis et al., 2019). Similarly, the pH of the culture broth of *Dermacoccus* spp. isolated from deep-sea sediment decreased from 7 to 4.62–5.1 (Rangseekaew et al., 2021, 2022). In particular, *D. abyssi* MT1.1^T demonstrated the ability to solubilize phosphate and release phosphorus under salt stress conditions of 150, 300, and 450 mM NaCl (Rangseekaew et al., 2022). *Micrococcus luteus* isolated from a marine water sample in Khambhat, Gujarat, India (Patel et al., 2021), and *Arthrobacter* sp. S2T9 from soil in Pohang beach, Republic of Korea, also showed phosphate-solubilizing activity (Moon & Ali, 2022).

2.1.4 Indole-3-Acetic Acid Production

Phytohormones play a crucial role in regulating growth, metabolism, and the capacity to withstand both biotic and abiotic stresses. Among them, indole-3-acetic acid (IAA) is a key auxin that is particularly important for cell division, elongation, fruit development, and senescence (Duca et al., 2014). Particularly in dicotyledon plants, IAA specifically induces lateral root formation, whereas in monocotyledon plants, it induces adventitious root formation (McSteen, 2010). IAA can be synthesized through several pathways, either tryptophan-dependent or tryptophan-independent. In the tryptophan-dependent pathway, L-tryptophan serves as the precursor (Duca et al., 2014; Spaepen et al., 2007). In addition to IAA synthesized by plants, many microorganisms, including actinobacteria, are also capable of producing this hormone. Numerous marine actinobacteria have been documented for their ability to produce IAA, highlighting their potential role in plant growth promotion. Diverse actinobacteria from different marine sources have been reported to possess IAA-producing capabilities. *Nocardiopsis yanglingensis* 1SM5-02 (94.3–100.2 μg/mL), *Streptomyces jiujiangensis* 2SH3-07 (14.9–29.6 μg/mL), and *Pseudonocardia oroxyli* 3WH5-01 (94.3–96.2 μg/mL), isolated from seawater and mangrove sediments in Thailand, produced IAA at comparable levels under both non-saline conditions and saline stress at 100, 200, and 300 mM NaCl. In contrast, *Streptomyces psammoticus* 3SH5-05 produced significantly higher levels of IAA under 200 mM NaCl (34.1 μg/mL) compared to non-saline conditions (9.2 μg/mL) (Suksaard et al., 2017). Moreover, *Streptomyces* sp. and *Micromonospora* sp., isolated from the rhizosphere soil of coastal salt marsh plants in China, were reported to produce low amounts of IAA, at 3.04 and 4.24 μg/mL, respectively (Gong et al., 2018). Similarly, *Streptomyces chartreusis* UAE1, isolated from rhizosphere soil on the Al Rams coast of Ras Al Khaimah, UAE, produced a low IAA level of 2.88 μg/mL (Mathew et al., 2020). *Saccharomonospora azurea* LNS-1 from marine soil in Morocco produces 49.46 μg/mL of IAA (Nafis et al., 2019). A high level of IAA production (81.3 μg/mL) was reported from *Micrococcus luteus* isolated from marine water in India (Patel et al., 2021).

An appropriate concentration of actinobacterial IAA is essential, as excessively high levels inhibit plant growth and development (Spaepen et al., 2007). However, the effects of IAA are influenced by several factors, including the physical association between plants and IAA-producing strains, the IAA biosynthetic pathway employed by the plant-associated bacteria, and the sensitivity of the host.

2.1.5 ACC Deaminase

Plants exposed to various abiotic stresses, such as high pH, temperature fluctuations, salinity, and limited nutrient availability, as well as biotic stresses like infections and weed competition, often increase their production of ethylene, "*stress ethylene*" to respond to the stresses (Boukhatem et al., 2022). Elevated ethylene levels negatively impact plant growth by causing stunted development, leaf drop, flower wilting, and premature fruit ripening.

Ethylene synthesis begins with the precursor, 1-aminocyclopropane-1-carboxylic acid (ACC). Certain plant growth-promoting bacteria that produce ACC deaminase can break down ACC into ammonia and α-ketobutyrate. This enzymatic activity helps reduce ethylene levels, thereby alleviating stress-induced damage and enhancing plant growth under such stress conditions. Some marine actinobacteria were reported to exhibit ACC deaminase activity, for example, *Streptomyces rochei* RAMS1 from Al Rams coastal soil, Ras Al Khaimah, UAE (Mathew et al., 2020), and *Arthrobacter* sp. S2T9 from rhizospheric soil in Pohang beach area, Republic of Korea (Moon & Ali, 2022).

2.2 Indirect Plant Growth-Promoting Properties of Marine Actinobacteria

Indirect action of plant growth-promoting properties refers to the biocontrol potential of actinobacteria. Biological control is generally defined as the use of living organisms to modify the agricultural ecosystem to control a crop disease or prevent the establishment of a pest (Dowling & O'Gara, 1994). Microorganisms with biocontrol activity are referred to as "*biocontrol agents.*" Actinobacteria are widely recognized for their effectiveness in biological control, particularly through antimicrobial activity and hydrolytic enzyme production, which are described in the following section.

2.2.1 Antimicrobial Compounds

Actinobacteria are well known for their capacity to produce secondary metabolites with antimicrobial properties, which account for approximately 60% of newly discovered antibiotics (Sathya et al., 2017). In agriculture, these compounds contribute to plant protection by suppressing pathogens and pests. Marine actinobacteria are one of the potential and promising biocontrol agents, as exemplified by the following examples. Marine-derived *Streptomyces* sp. AMA49 isolated from the Thai mangrove forest in Songkhla province is able to produce oligomycin A and diketopiperazines with antifungal activity against rice blast pathogen, *Pyricularia oryzae*. These compounds inhibit mycelial growth, conidial germination, and appressorium formation of *P. oryzae* (Buatong et al., 2019). *Streptomyces* spp. from Sundarbans mangrove sediment in India showed antimicrobial activity against both tested bacteria and three phytopathogens, *Ustilago maydis* SG200, *Rhizoctonia solani* AG1-1A, and *Macrophomina phaseolina* R9 (Sengupta et al., 2015). Particularly, *Streptomyces* sp. isolate MS_SU21 produced isoquinoline-1-carbonitrile, a compound previously known for antimicrobial activity (Hassaneen et al., 2012; Sengupta et al., 2015). Derivatives of 1,2,4-triazol [1,5-a] pyrimidine,5,7-dimethyl-2-phenyl and 4-dichloromethyl-5 6-epoxy-2-methoxy-4-methyl-2-cyclohexenone were also reported for antifungal activity against phytopathogens (Sengupta et al., 2015; Wang et al., 2014). Additionally, haloalkaliphilic *Nocardiopsis* sp. AJ1, isolated from solar salterns saline soil in India, produced metabolites that suppressed both bacterial and fungal pathogens (*Escherichia coli*, *Staphylococcus aureus*, *Pseudomonas aeruginosa*, *Vibrio parahaemolyticus*, *Aeromonas hydrophila*, *Aspergillus niger*, *Fusarium* sp., and *Pythium* sp.). The identified compounds included pyrrolo(1,2-a) pyrazine-1,4-dione, hexahydro-3-(2-methylpropyl), and actinomycin C2 (Adlin Jenifer et al., 2019). The extract fraction of *Kocuria palustris* 19C38A1 isolated from marine organisms at the Coast of Oluhuta, Tomini Bay, Gorontalo, Indonesia, displayed fungicidal activity against *Fusarium oxysporum*. The observed activity was due to the production of the benzimidazole compound (Setiawan et al., 2022). Several mangrove actinobacteria, *Nocardiopsis lucentensis* isolates A15 and A20, *Streptomyces* sp. A19 and *Streptomyces griseoflavus* A18 from Zacatecas estuary in La Paz, Baja California Sur, Mexico, exhibited antifungal activity against *Fusarium solani* (Torres-Rodriguez et al., 2022). Moreover, Marine *Streptomyces cinereoruber* from the intertidal zone of Dongfang city, Hainan province, China, produced a heterocyclic flavonoid, aurone with high fungicidal activity against powdery mildew (*Botrytis cinerea*), cucurbits powdery mildew (*Sphaerotheca fuliginea* (Schlecht ex Ff.) Poll), and rice blast (*Pyricularia oryzae*) under greenhouse conditions (Liu et al., 2023).

2.2.2 Production of Hydrolytic Enzymes

Actinobacteria possess strong potential for enzyme production, and several of these enzymes play key roles in protecting plants from the invading phytopathogens.

These extracellular or hydrolytic enzymes secreted by the actinobacteria can be detrimental to phytopathogens by degrading cell walls and other structural components of pathogenic fungi or pests. Various marine actinobacterial strains have been reported to possess antagonistic potential, being capable of degrading bacterial and fungal cell walls, cell membranes, cell membrane proteins, and extracellular virulence factors, all of which contribute to their role in the biocontrol of plant diseases (Boukhatem et al., 2022). A total of 112 marine *Streptomyces* spp. isolated from several marine samples in the Egyptian coastal area, secreted numerous hydrolytic enzymes, including those known cell wall degrading enzymes such as chitinase, lipase, and protease. Interestingly, cell-free culture filtrates containing metabolites from all *Streptomyces* strains demonstrated significant nematicidal activity against second-stage juveniles (J2) of the root-knot nematode and also exhibited nematicidal effects in vivo against *Meloidogyne incognita* in eggplants under greenhouse conditions. The surface coat of nematodes, which forms the outermost layer of the cuticle, is primarily composed of proteins, chitin, and lipids. Similarly, nematode eggshells consist of three layers: the outer vitelline layer (protein), the middle chitinous layer, and the inner lipid layer. The production of cell wall-degrading enzymes such as chitinase, lipase, and protease is regarded as an effective biocontrol strategy against plant-parasitic nematodes by targeting both the surface coat and the eggshells (Rashad et al., 2015). Additionally, *Streptomyces* spp. isolated from sediments in the intertidal zone of Ilha do Mel, located in the coastal region of Paraná State, Brazil, exhibited chitinase activity and showed antagonistic effects against the European pepper moth, *Duponchelia fovealis* (Zeller), a major pest of strawberry cultivation (Porsani et al., 2022). Marine *Micrococcus luteus* produced chitinase, glucanase, and exhibited antifungal activity against several *Fusarium* species, including *Fusarium oxysporum* f. sp. *chlamydosporum* (F.O.Ch), *F. oxysporum* f. sp. ciceris (F.O.C), *F. oxysporum* f. sp. *pallidoroseum* (F.O.P), and *F. oxysporum* f. sp. *vasinfectum* (F.O.V) (Patel et al., 2021).

3 Application of Marine Actinobacteria in Agriculture

Numerous studies have reported both direct and indirect plant growth-promoting effects of marine actinobacteria under *in vitro* conditions. However, their successful application in agriculture depends on several factors, including the interactions between actinobacteria and host plants. Therefore, *in planta* evaluation using actinobacterial inoculants to promote plant growth and development is essential to validate their potential. For example, the mangrove-derived *S. iranensis* S2-SC16, which is capable of producing IAA and siderophores as well as solubilizing phosphate, has been reported to promote rice seedling growth in pot experiments by increasing root length, shoot height, leaf width, and weight (Pathom-aree et al., 2019).

3.1 Role of Marine Actinobacteria in Biotic Stress

Biotic stress in plants is caused by living organisms, including viruses, bacteria, fungi, nematodes, insects, and weeds. These agents disrupt the normal metabolism of the plants. Various studies reported on the potential of marine actinobacteria in protecting plants from biotic stress. Several mangrove strains, *Nocardiopsis lucentensis* A15, *N. lucentensis* A20, and *Streptomyces* sp. A19 and *S. griseoflavus* A18 have antagonistic against *Fusarium solani,* which can cause disease in tomato (Torres-Rodriguez et al., 2024). *Streptomyces* sp. A19 inhibited the mycelial growth of *F. solani* via iron competition, as this strain can produce siderophores (Torres-Rodriguez et al., 2024). In addition to directly promoting plant growth by binding and supplying iron to plants, siderophores can also act indirectly by competing with phytopathogens for iron, thereby restricting iron availability to the pathogens (Boukhatem et al., 2022; Sathya et al., 2017). Additionally, *Streptomyces* sp. A19 and *S. griseoflavus* A18 exhibited antifungal activity through the production of volatile organic compounds (VOCs), which are known to inhibit phytopathogenic fungi. Moreover, *in vivo* assays demonstrated that *Streptomyces* sp. A19 reduced disease severity and incidence caused by *Fusarium solani* in tomato plants. In particular, the application of *Streptomyces* sp. A19 to tomato plants infested with *F. solani* enhanced the activities of defense-related enzymes (superoxide dismutase, peroxidase, catalase and phenylalanine ammonia-lyase) compared with plants treated only with *F. solani* or distilled water (control) (Torres-Rodriguez et al., 2024).

Marine *Micrococcus luteus* isolated from marine water in Khambhat city, Gujarat, India, showed biocontrol ability against *Fusarium oxysporum,* a causative agent of *Fusarium* wilt (Patel et al., 2021). Moreover, the chickpea plant treated with *M. luteus* exhibited a less severe disease index as compared to pathogen-treated plants without inoculated *M. luteus*, which could be due to its ability to produce cell wall-degrading enzymes, chitinase, and glucanase (Patel et al., 2021).

3.2 Role of Marine Actinobacteria in Abiotic Stress

Abiotic stress in plants is caused by environmental factors such as drought, salinity, heavy metals, extreme temperatures, and nutrient deficiencies, all of which pose major challenges to agricultural productivity. Marine actinobacteria are among the beneficial microbes that have gained increasing attention for agricultural applications, particularly for use in saline environments. For example, *S. mutabilis* UAE1 was recovered from inside mangrove root tissues of gray mangrove (*Avicennia marina*), which showed polyamines (PAs) producing activity (Table 1) (El-Tarabily et al., 2021). PAs, primarily putrescine (Put), spermidine (Spd), and spermine (Spm), function as signaling molecules that mediate plant responses to environmental stresses, particularly salt stress (Wu et al., 2018). PAs-producing *S. mutabilis* UAE1 significantly promote the growth of gray mangrove (*Avicennia marina*) under greenhouse conditions by improving dry weight, root and shoot length, numbers of branches, total leaf area, and photosynthetic pigments (chlorophyll and carotenoids) compared to non-PA-producing *Streptomyces* sp. Remarkably, one mechanism by which PA-producing *S. mutabilis* UAE1 promotes mangrove growth is through increasing the levels of Put, Spd, and Spm in shoots and roots while also balancing cytokinin and auxin levels in the plant tissues (El-Tarabily et al., 2021).

Moreover, plant growth-promoting deep-sea actinobacteria, *D. profundi* MT2.1^T and *D. barathri* MT2.2^T, have been reported to alleviate salt stress in tomato seedlings under 150 mM NaCl (Rangseekaew et al., 2021). Both strains exhibited IAA production under non-saline conditions as well as at NaCl concentrations of 150, 300, and 450 mM. In addition, they produced both hydroxamate- and catecholate-type siderophores and were able to solubilize tricalcium phosphate (Rangseekaew et al., 2021). Notably, *D. profundi* MT2.1^T alleviated salt stress in tomato seedlings by reducing H_2O_2 accumulation compared to uninoculated controls. Since H_2O_2 serves as a reactive oxygen species indicator of oxidative stress in plants under various conditions, including salinity, its lower levels in inoculated plants suggest an enhanced stress tolerance. This reduction was reflected in improved growth parameters, including shoot length, root length, fresh weight, dry weight, and relative water content (Rangseekaew et al., 2021).

Marine actinobacterium, *Salinispora arenicola,* was isolated from corals *Porites lobata* and *Porites panamensis* in coral reefs from the tropical central Pacific. Remarkably, two isolates, *S. arenicola* F51 and X29, significantly increased the germination of wild tobacco (*Nicotiana attenuata*) under salt stress at 100 mM NaCl compared to untreated seeds and seed -treated with plant growth-promoting *Azospirillum brasilense*. In addition, tobacco treated with both *S. arenicola* outperformed the control and *A. brasilense*-treated plants, as shown by higher photosystem II (PSII) quantum efficiency (F_v/F_m) and maximum electron transfer rate (ETR_{MAX}) values. These findings suggest that the photosynthetic rate was greater in the marine actinobacteria-treated plants than in the control. This positive effect is likely due to the ACC deaminase activity of *S. arenicola* (Ocampo-Alvarez et al., 2020).

4 Conclusion and Future Perspectives

Marine actinobacteria represent an emerging valuable resource for plant growth-promoting promotion that can be harnessed to advance sustainable agriculture practices. Over the past decade, numerous studies have demonstrated supporting evidence of their ability to produce antimicrobial metabolites, hydrolytic enzymes, siderophores, phytohormones, and other bioactive compounds that suppress phytopathogens and enhance plant resilience to both biotic and abiotic stresses. These findings highlight the remarkable versatility of marine actinobacteria as both biocontrol agents and bioinoculant candidates.

Despite these promising results, the translation of marine actinobacteria from laboratory research to field-scale agricultural applications remains limited. Future studies should prioritize *in vivo* and field trials to validate their efficacy under natural conditions. In addition, investigations into plant–microbe interactions are needed to improve understanding of compatibility and colonization dynamics. Omics-based approaches, including genomics, transcriptomics, and metabolomics, will be essential for uncovering novel genes and pathways involved in stress mitigation and metabolite biosynthesis. Equally important is the development of formulations that ensure stability and effectiveness in diverse agroecosystems. Integrating marine actinobacteria into sustainable crop management strategies has the potential to reduce chemical inputs, improve crop productivity, and support climate-resilient agriculture.

References

Adlin Jenifer, J. S. C., Michaelbabu, M., Eswaramoorthy Thirumalaikumar, C. L., Jeraldin Nisha, S. R., Uma, G., & Citarasu, T. (2019). Antimicrobial potential of haloalkaliphilic *Nocardiopsis* sp. AJ1 isolated from solar salterns in India. *Journal of Basic Microbiology, 59*(3), 288–301. https://doi.org/10.1002/jobm.201800252

Alshamsi, A. A. A., Sheteiwy, M. S., AbuQamar, S. F., & El-Tarabily, K. A. (2024). Enhancement of mangrove growth performance using fish emulsion and halotolerant plant growth-promoting actinobacteria for sustainable management in UAE. *Marine Pollution Bulletin, 199*, 115916.

Aznar, A., & Dellagi, A. (2015). New insights into the role of siderophores as triggers of plant immunity: What can we learn from animals? *Journal of Experimental Botany, 66*(11), 3001–3010. https://doi.org/10.1093/jxb/erv155

Behera, B. C., Singdevsachan, S. K., Mishra, R. R., Dutta, S. K., & Thatoi, H. N. (2014). Biocatalysis and agricultural biotechnology diversity, mechanism and biotechnology of phosphate solubilising microorganism in mangrove—A review. *Biocatalysis and Agricultural Biotechnology, 3*(2), 97–110. https://doi.org/10.1016/j.bcab.2013.09.008

Boukhatem, Z. F., Merabet, C., & Tsaki, H. (2022). Plant growth promoting actinobacteria, the most promising candidates as bioinoculants? *Frontiers in Agronomy, 4*, 849911. https://doi.org/10.3389/fagro.2022.849911

Buatong, J., Rukachaisirikul, V., Sangkanu, S., Surup, F., & Phongpaichit, S. (2019). Antifungal metabolites from marine-derived *Streptomyces* sp. AMA49 against *Pyricularia oryzae*. *Journal of Pure and Applied Microbiology, 13*(2), 653–665. https://doi.org/10.22207/JPAM.13.2.02

Dhawi, F. (2023). The role of plant growth-promoting microorganisms (PGPMs) and their feasibility in hydroponics and verticalfarming. *Metabolites, 13*(2), 247. https://doi.org/10.3390/metabo13020247

Dowling, D. N., & O'Gara, F. (1994). Metabolites of *Pseudomonas* involved in the biocontrol of plant disease. *Trends in Biotechnology, 12*(4), 133–141. https://doi.org/10.1016/0167-7799(94)90091-4

Duca, D., Lorv, J., Patten, C. L., Rose, D., & Glick, B. R. (2014). Indole-3-acetic acid in plant–microbe interactions. *Antonie Van Leeuwenhoek, 106*, 85–125. https://doi.org/10.1139/m96-032

El-Tarabily, K. A., Ramadan, G. A., Elbadawi, A. A., Hassan, A. H., Tariq, S., Ghazal, E. W., Abo Gamar, M. I., & AbuQamar, S. F. (2021). The marine endophytic polyamine-producing *Streptomyces mutabilis* UAE1 isolated from extreme niches in the Arabian gulf promotes the performance of mangrove (*Avicennia marina*) seedlings under greenhouse conditions. *Frontiers in Marine Science, 8*, 710200. https://doi.org/10.3389/fmars.2021.710200

Gong, Y., Bai, J., Yang, H., Zhang, W., Xiong, Y., & Ding, P. (2018). Phylogenetic diversity and investigation of plant growth-promoting traits of actinobacteria in coastal salt marsh plant rhizospheres from Jiansu, China. *Systematic and Applied Microbiology, 41*(5), 516–527. https://doi.org/10.1016/j.syapm.2018.06.003

Hao, L., Zheng, X., Wang, Y., Li, S., Shang, C., & Xu, Y. (2019). Inhibition of tomato early blight disease by culture extracts of a *Streptomyces puniceus* isolate from mangrove soil. *Phytopathology, 109*(7), 1149–1156. https://doi.org/10.1094/PHYTO-12-18-0444-R

Hassaneen, H. M., Abunada, N. M., Hassaneen, H. M. E., & Miqdad, O. A. (2012). Synthesis of new derivatives of heterocyclic compounds containing pyridine, pyrimidine and triazole ortho-fused to isoquinoline moiety. *International Journal of Chemistry, 4*(3), 86–97. https://doi.org/10.5539/ijc.v4n3p86

Hu, X., Cheng, B., Du, D., Huang, Z., Pu, Z., Chen, G., Peng, A., & Lu, L. (2019). Isolation and identification of a marine actinomycete strain and its control efficacy against citrus green and blue moulds. *Biotechnology & Biotechnological Equipment, 33*(1), 719–729. https://doi.org/10.1080/13102818.2019.1613175

Khan, A., Singh, P., & Srivastava, A. (2018). Synthesis, nature and utility of universal iron chelator—Siderophore: A review. *Microbiological Research, 212*, 103–111. https://doi.org/10.1016/j.micres.2017.10.012

Liu, B., Li, X., Wang, W., Wang, X., Aihaiti, P., Lin, T., Fu, Z., Xu, R., Wu, M., Li, Z., & Zhang, Y. (2023). A new method of preparing aurone by marine actinomycetes and its potential application in agricultural fungicides. *Molecules, 28*(1), 17. https://doi.org/10.3390/molecules28010017

Mathew, B. T., Torky, Y., Amin, A., Mourad, A. H. I., Ayyash, M. M., El-Keblawy, A., Hilal-Alnaqbi, A., AbuQamar, S. F., & El-Tarabily, K. A. (2020). Halotolerant marine rhizosphere-competent actinobacteria promote *Salicornia bigelovii* growth and seed production using seawater irrigation. *Frontiers in Microbiology, 11*, 552. https://doi.org/10.3389/fmicb.2020.00552

McSteen, P. (2010). Auxin and monocot development. *Cold Spring Harbor Perspectives in Biology, 2*(3), a001479–a001479. https://doi.org/10.1101/cshperspect.a001479

Moon, Y.-S., & Ali, S. (2022). Isolation and identification of multi-trait plant growth–promoting rhizobacteria from coastal sand dune plant species of Pohang beach. *Folia Microbiologica, 67*(3), 523–533. https://doi.org/10.1007/s12223-022-00959-4

Nafis, A., Raklami, A., Bechtaoui, N., El Khalloufi, F., El Alaoui, A., Glick, B. R., Hafidi, M., Kouisni, L., Ouhdouch, Y., & Hassani, L. (2019). Actinobacteria from extreme niches in Morocco and their plant growth-promoting potentials. *Diversity, 11*(8), 139. https://doi.org/10.3390/d11080139

Ocampo-Alvarez, H., Meza-Canales, I. D., Mateos-Salmón, C., Rios-Jara, E., Rodríguez-Zaragoza, F. A., Robles-Murguía, C., Muñoz-Urias, A., Hernández-Herrera, R. M., Choix-Ley, F. J., & Becerril-Espinosa, A. (2020). Diving into reef ecosystems for land-agriculture solutions: Coral microbiota can alleviate salt stress during germination and photosynthesis in terrestrial plants. *Frontiers in Plant Science, 11*, 648. https://doi.org/10.3389/fpls.2020.00648

Patel, P., Patel, K., Dhandhukia, P., & Thakker, J. N. (2021). Plant growth promoting traits of marine *Micrococcus* sp. with bio-control ability against *Fusarium* in chickpea plant. *Vegetos, 34*(1), 94–101. https://doi.org/10.1007/s42535-021-00191-4

Pathom-aree, W., Kreawsa, S., Kamjam, M., Tokuyama, S., Yoosathaporn, S., & Lumyong, S. (2019). Potential of selected mangrove streptomyces as plant growth promoter and rice bakanae disease control agent. *Chiang Mai Journal of Science, 46*(2), 261–276.

Porsani, M. V., Poitevin, C. G., Tralamazza, S. M., de Souza, M. T., de Souza, M. T., da Silva, É. D. B., Zawadneak, M. A. C., Pimentel, I. C., & de Melo, I. S. (2022). *Streptomyces* spp. isolated from marine and Caatinga biomes in Brazil for the biological control of *Duponchelia fovealis*. *Neotropical Entomology, 51*(2), 299–310. https://doi.org/10.1007/s13744-022-00942-1

Prabhu, N., Borkar, S., & Garg, S. (2019). Phosphate solubilization by microorganisms. In *Advances in biological science research* (pp. 161–176). Elsevier. https://doi.org/10.1016/B978-0-12-817497-5.00011-2

Rangseekaew, P., Barros-Rodríguez, A., Pathom-aree, W., & Manzanera, M. (2021). Deep-sea actinobacteria mitigate salinity stress in tomato seedlings and their biosafety testing. *Plants, 10*, 1687. https://doi.org/10.3390/plants10081687

Rangseekaew, P., Barros-Rodríguez, A., Pathom-aree, W., & Manzanera, M. (2022). Plant beneficial deep-sea actinobacterium, *Dermacoccus abyssi* MT1.1^T. *Biology, 11*, 191. https://doi.org/10.3390/biology11020191

Rashad, F. M., Fathy, H. M., El-Zayat, A. S., & Elghonaimy, A. M. (2015). Isolation and characterization of multifunctional *Streptomyces* species with antimicrobial, nematicidal and phytohormone activities from marine environments in Egypt. *Microbiological Research, 175*, 34–47. https://doi.org/10.1016/j.micres.2015.03.002

Rawat, P., Das, S., Shankhdhar, D., & Shankhdhar, S. C. (2021). Phosphate-solubilizing microorganisms: Mechanism and their role in phosphate solubilization and uptake. *Journal of Soil Science and Plant Nutrition, 21*, 49–68. https://doi.org/10.1007/s42729-020-00342-7

Sathya, A., Vijayabharathi, R., & Gopalakrishnan, S. (2017). Plant growth-promoting actinobacteria: A new strategy for enhancing sustainable production and protection of grain legumes. *3. Biotech, 7*(102), 1–10. https://doi.org/10.1007/s13205-017-0736-3

Sellstedt, A., & Richau, K. H. (2013). Aspects of nitrogen-fixing actinobacteria, in particular free-living and symbiotic Frankia. *FEMS Microbiology Letters, 342*(2), 179–186. https://doi.org/10.1111/1574-6968.12116

Sengupta, S., Pramanik, A., Ghosh, A., & Bhattacharyya, M. (2015). Antimicrobial activities of actinomycetes isolated from unexplored regions of Sundarbans mangrove ecosystem. *BMC Microbiology, 15*(1), 170. https://doi.org/10.1186/s12866-015-0495-4

Setiawan, A., Setiawan, F., Juliasih, N. L. G. R., Widyastuti, W., Laila, A., Setiawan, W. A., Djailani, F. M., Mulyono, M., Hendri, J., & Arai, M. (2022). Fungicide activity of culture extract from *Kocuria palustris* 19C38A1 against *Fusarium oxysporum*. *Journal of Fungi, 8*(3), 280. https://doi.org/10.3390/jof8030280

Spaepen, S., Vanderleyden, J., & Remans, R. (2007). Indole-3-acetic acid in microbial and microorganism-plant signaling. *FEMS Microbiology Reviews, 31*(4), 425–448.

Suksaard, P., Pathom-aree, W., & Duangmal, K. (2017). Diversity and plant growth promoting activities of actinomycetes from mangroves. *Chiang Mai Journal of Science, 44*(4), 1210–1223.

Torres-Rodriguez, J., Reyes-Pérez, J., Castellanos, T., Quiñones-Aguilar, E., & Hernandez-Montiel, L. (2022). Identification and morphological characterization of marine actinomycetes as biocontrol agents of *Fusarium solani* in tomato. *Revista de La Facultad de Agronomía, Universidad Del Zulia, 39*(1), e223915. https://doi.org/10.47280/RevFacAgron(LUZ).v39.n1.15

Torres-Rodriguez, J. A., Reyes-Pérez, J. J., Hernandez-Adame, L., Llerena-Fuentes, B. L., & Hernandez-Montiel, L. G. (2024). Marine actinomycetes for biocontrol of *Fusarium solani* in tomato plants: *In vitro* and *in vivo* studies. *Notulae Botanicae Horti Agrobotanici Cluj-Napoca, 52*(2), 13562. https://doi.org/10.15835/nbha52213562

Vocciante, M., Grifoni, M., Fusini, D., Petruzzelli, G., & Franchi, E. (2022). The role of plant growth-promoting rhizobacteria (PGPR) in mitigating plant's environmental stresses. *Applied Sciences, 12*(3), 1231. https://doi.org/10.3390/app12031231

Wang, Y., Xu, K., Bai, G., Huang, L., Wu, Q., Pan, W., & Yu, S. (2014). Synthesis and antifungal activity of novel triazole compounds containing piperazine moiety. *Molecules, 19*(8), 11333–11340. https://doi.org/10.3390/molecules190811333

Wu, J., Shu, S., Li, C., Sun, J., & Guo, S. (2018). Spermidine-mediated hydrogen peroxide signaling enhances the antioxidant capacity of salt-stressed cucumber roots. *Plant Physiology and Biochemistry, 128*, 152–162. https://doi.org/10.1016/j.plaphy.2018.05.002

Yi, S., Cui, Y., Zhao, Y., Liu, Z., Lin, Y., & Zhou, F. (2016). A novel naturally occurring class I 5-enolpyruvylshikimate-3-phosphate synthase from *Janibacter* sp. confers high glyphosate tolerance to rice. *Scientific Reports, 6*(1), 19104. https://doi.org/10.1038/srep19104

Synthetic Biology Tools for Marine Actinobacteria

Muneer Oladipupo Yaqub, Ravali Arugonda, Aashika Jain, and Lekshmi K. Edison

Abstract

Marine actinobacteria represent a vast, yet underexplored, reservoir of bioactive natural products with immense potential for pharmaceutical and industrial applications. Despite their genetic richness, the majority of biosynthetic gene clusters (BGCs) in these organisms remain silent under standard laboratory conditions, limiting the discovery of novel metabolites. Recent advances in synthetic biology, encompassing genome mining, clustered regularly interspaced short palindromic repeats-Cas-based editing, modular cloning, and heterologous expression, have revolutionized our ability to decode and engineer these cryptic pathways. This chapter provides a comprehensive overview of synthetic biology tools tailored for marine actinobacteria, including bioinformatics platforms such as antibiotics and Secondary Metabolite Analysis Shell, Prediction Informatics for Secondary Metabolomes, Bacteriocin Genome Mining Tool, and DeepBGC for BGC prediction, as well as genetic toolkits for promoter engineering, DNA assembly, and the design of regulatory circuits. It highlights the integration of multi-omics approaches and machine learning in pathway optimization and discusses emerging marine-adapted chassis systems, such as *Salinispora tropica*, for scalable biotechnological applications. Finally, this chapter addresses the

M. O. Yaqub · R. Arugonda
Department of Biological Sciences, School of Natural Science and Mathematics, University of Texas at Dallas, Richardson, TX, USA

A. Jain
Manipal School of Life Sciences, Manipal Academy of Higher Education, Manipal, Karnataka, India

L. K. Edison (✉)
Department of Comparative, Diagnostic, and Population Medicine, College of Veterinary Medicine, University of Florida, Gainesville, FL, USA
e-mail: edison.le@ufl.edu

K. Loganathan, Z. Li (eds.), *Marine Actinomycetes*,
https://doi.org/10.1007/978-981-95-9849-6_16

persistent challenges in transformation efficiency, vector stability, and cultivation of marine isolates, outlining future directions for establishing standardized synthetic biology platforms that harness the full biosynthetic potential of marine actinobacteria.

Keywords

Marine actinobacteria · Synthetic biology · Biosynthetic gene clusters (BGCs) · Genome mining · CRISPR-Cas · Heterologous expression · Modular cloning · Promoter engineering · Metabolomics · Machine learning · *Salinispora tropica* · Natural product discovery

1 Introduction

1.1 Marine Actinobacteria as a Biotechnological Treasure Trove

Marine actinobacteria represent one of the most prolific yet underexplored reservoirs of natural products with pharmaceutical and industrial significance. Traditionally associated with terrestrial soils, members of the phylum *Actinobacteria* have long been recognized as prolific producers of antibiotics, antitumor agents, immunosuppressants, and enzyme inhibitors (El-Shatoury et al., 2009; Bérdy, 2012). The discovery of marine-derived *Streptomyces*, *Micromonospora*, *Salinispora*, and *Nocardiopsis* species has expanded this frontier, unveiling structurally novel compounds such as salinosporamide A, marinomycins, and abyssomicins, which exhibit unique scaffolds and bioactivities distinct from their terrestrial analogs (Fenical & Jensen, 2006; Fenical et al., 2009; Manivasagan et al., 2014; Subramani & Aalbersberg, 2012). Table 1 shows the list of bioactive metabolites produced by the marine actinobacteria. The extreme marine environment, characterized by high salinity, pressure, and low nutrient availability, has driven the evolution of metabolic pathways that are rare or absent in terrestrial microbes (Bull & Stach, 2007). However, most biosynthetic gene clusters (BGCs) within marine actinobacteria remain silent under laboratory conditions (Baltz, 2017). Traditional cultivation and screening approaches have proven inadequate in unlocking this hidden chemical diversity, prompting the integration of systems biology, genome mining, and synthetic biology as new strategies for activating pathways and discovering novel metabolites.

1.2 The Emergence of Synthetic Biology in Marine Microbiology

Synthetic biology applies engineering principles, such as design, build, test, and learn, to biological systems, allowing precise reprogramming of cellular functions

Table 1 Representative marine actinobacteria and their bioactive metabolites

Genus/species	Isolation source	Representative metabolite(s)	Bioactivity/ application	References
Salinispora tropica (e.g., strain CNB-440)	Tropical marine sediment	Salinosporamide A (also called NPI-0052, Marizomib)	Potent 20S proteasome inhibitor; anticancer lead candidate	Udwary et al. (2007)
Salinispora arenicola	Marine sediment	Salinosporamide analogs (e.g., D–J)	Proteasome inhibition	Jensen et al. (2015)
Micromonospora sp.	Marine sponge/ sediment	Micromonohalimanes A and B	Antibacterial activities	Zhang et al. (2016)
Verrucosispora sp.	Marine sediment	Abyssomicin C	Anti-MRSA (Methicillin-resistant Staphylococcus aureus) and anti-tuberculosis activities	Keller et al. (2007)
Nocardiopsis sp. (rare actinomycete)	Marine sediment	α-pyrones and other novel metabolites	Antibacterial, cytotoxic, and anti-inflammatory activities	Kim et al. (2014)
Actinomadura sp.	Marine sediment	Madurastatin C1	Antifungal, iron-chelating	Mazzei et al. (2012)
Verrucosispora maris AB-18-032	Marine sediment	abyssomicin C, atrop-abyssomicin, abyssomicin H, and proximicin A	Antibacterial (especially Gram-positive bacteria via pABA/ folate pathway inhibition	Keller et al. (2007)
Nocardiopsis lucentensis	Marine sediment	Lucentamycins A–D	*In vitro* cytotoxicity (HCT-116 human colon carcinoma)	Cho et al. (2007)
Streptomyces carnosus AZS17	Marine sponge	Lobophorin C and D	*In vitro* cytotoxicity (human liver cancer cell line 7402, breast cancer MDA-MB-435)	Wei et al. (2011)
Rhodococcus fascians	Deep-sea sediment	Rhodostreptomycin A and B	Antibacterial (broad Gram-positive/negative)	Kurosawa et al. (2008)

(Cameron et al., 2014). The overall synthetic biology framework used for engineering marine actinobacteria is illustrated in Fig. 1, showing the progression from genome mining and pathway design to artificial intelligence (AI)-assisted optimization and large-scale application. In marine microbiology, this approach has transitioned the field from descriptive genome annotation to active genome engineering,

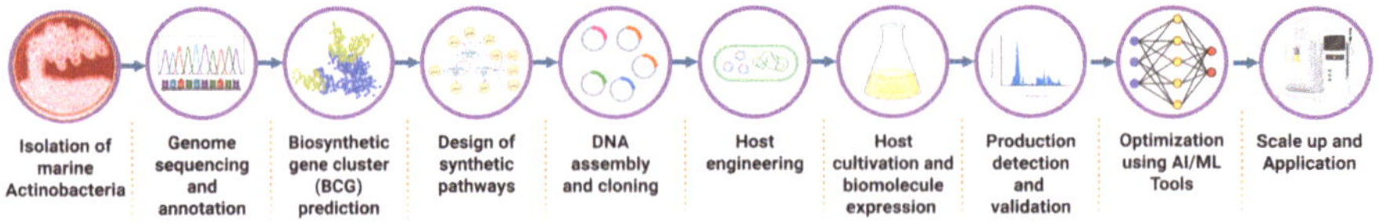

Fig. 1 Schematic overview of the synthetic biology workflow for marine actinobacteria. The stepwise pipeline begins with the isolation of marine actinobacteria and proceeds through genome sequencing, BGC prediction, and design of synthetic pathways. Engineered constructs are assembled and cloned, introduced into suitable host strains via CRISPR modification or transformation, and cultivated for biomolecule expression. The resulting products are detected and validated, followed by data-driven optimization using AI and ML tools. Finally, optimized strains are scaled up for industrial or pharmaceutical applications. (Image created in BioRender.com)

enabling the targeted activation of cryptic BGCs, construction of hybrid biosynthetic pathways, and expression of marine genes in heterologous hosts (Medema & Fischbach, 2015; Nielsen & Keasling, 2016). Recent advances, including modular cloning (MoClo) platforms, clustered regularly interspaced short palindromic repeats (CRISPR)-Cas genome editing, and high-throughput DNA assembly, have accelerated efforts to harness marine actinobacteria for biotechnological applications (Zhang et al., 2020). For example, refactoring silent gene clusters from *Salinispora tropica* using synthetic regulatory parts has yielded novel secondary metabolites (Amos et al., 2017). Moreover, DNA synthesis and biofoundry automation are reducing the barriers to constructing large-scale genetic circuits, allowing the rational design of marine microbial strains optimized for metabolite production and stress adaptation (Smanski et al., 2016).

1.3 Synthetic Biology Versus Classical Genetic Engineering

While classical genetic engineering pioneered the manipulation of individual genes or operons using restriction enzymes and cloning vectors, synthetic biology introduced a paradigm of standardization, modularity, and computational predictability (Purnick & Weiss, 2009). Tools such as Golden Gate and Gibson assembly, CRISPR-Cas systems, and modular DNA libraries have enabled the simultaneous assembly and editing of multiple genetic elements (Engler et al., 2009; Doudna & Charpentier, 2014). These methods now permit multiplexed editing, promoter engineering, and pathway transplantation in marine actinobacteria, which have historically been recalcitrant to transformation (Li et al., 2021). Thus, synthetic biology expands the scope of genetic manipulation from "gene modification" to "genome design," combining computational modeling and machine learning (ML) to optimize gene expression and metabolic flux (Carbonell et al., 2016).

1.4 Applications of Synthetic Biology in Marine Systems

Synthetic biology has enabled diverse applications in marine actinobacterial studies. One major frontier is bioactive metabolite production, where refactored or heterologously expressed BGCs yield antibiotics, anticancer agents, and enzyme inhibitors that are otherwise silent in native hosts (Chevrette et al., 2019; Zhang et al., 2023). Another emerging application involves bioremediation, in which engineered marine actinobacteria are designed to degrade hydrocarbons, microplastics, and heavy metals (Das et al., 2021). Synthetic constructs conferring tolerance to oxidative stress and salinity further enhance the performance of these organisms in marine environments. In addition to bioremediation, biosensor development leverages the sensory and regulatory proteins of marine microbes to detect pollutants, toxins, or environmental changes with high sensitivity (Gupta et al., 2020). Collectively, these innovations mark a transition from natural product "discovery" to programmable "design," positioning synthetic biology as a key driver for translating marine microbial potential into tangible biotechnological outcomes.

2 Genome Mining and BGC Prediction Tools

The advent of high-throughput genome sequencing has transformed natural product discovery from serendipitous screening to genome-guided predictions. Marine actinobacteria, with their large and GC-rich genomes, often harbor dozens of BGCs that encode polyketides, nonribosomal peptides, terpenes, and ribosomally synthesized and post-translationally modified peptides (RiPPs) (Ziemert et al., 2016). However, many of these clusters remain silent under standard laboratory conditions, emphasizing the need for *in silico* genome-mining tools that can identify, annotate, and prioritize BGCs for their activation or refactoring. Several bioinformatics platforms, including antibiotics and Secondary Metabolite Analysis Shell (anti-SMASH), Prediction Informatics for Secondary Metabolomes (PRISM), Bacteriocin Genome Mining Tool (BAGEL), ClusterFinder, and DeepBGC, have become central to this effort, each offering unique capabilities for decoding the biosynthetic potential of marine actinobacteria. Figure 2 shows the overall workflow of BGC prediction and novel metabolite identification.

2.1 Anti-SMASH—Identification and Annotation of BGCs

The anti-SMASH is the most widely used platform for BGC identification and annotation (Blin et al., 2021). It integrates multiple algorithms to detect hallmark biosynthetic domains, such as ketosynthases, adenylation domains, and condensation motifs characteristic of polyketide synthases (PKSs) and nonribosomal peptide synthetases (NRPSs). Anti-SMASH also predicts hybrid clusters and regulatory elements, providing detailed annotations of domain organization, tailoring enzymes, and potential chemical products. In marine actinobacteria, such as *Salinispora*

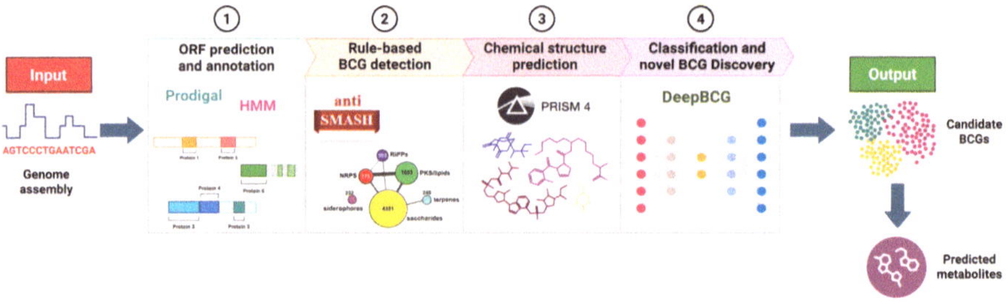

Fig. 2 Schematic workflow for BGC identification and metabolite prediction. Starting with a genome assembly (Input), (1) open reading frames are predicted and annotated (e.g., Prodigal and Hidden Markov Model (HMM)-based) and (2) BGCs are detected via rule-based methods (such as anti-SMASH). Next, (3) chemical structure prediction is carried out using tools like PRISM to infer likely metabolite scaffolds, and (4) classification and novel cluster discovery are achieved via deep learning frameworks (e.g., DeepBGC). The Output comprises prioritized candidate BGCs and their predicted small-molecule products, ready for downstream synthetic biology exploitation. (Image created in BioRender.com)

tropica and *Micromonospora marina*, anti-SMASH has uncovered dozens of putative BGCs beyond those responsible for known metabolites, such as salinosporamide A and marinomycin A (Letzel et al., 2017). Its visualization tools allow comparative BGC analyses across strains, revealing strain-specific pathways that may account for the chemical diversity within marine environments. The platform's continuous updates, integration with databases such as MIBiG (Minimum Information about a Biosynthetic Gene cluster), and the addition of ClusterBlast and SubClusterBlast modules make it indispensable for both genome annotation and metabolic engineering.

2.2 PRISM—Structure Prediction of Natural Products From Genomic Sequences

The PRISM suite extends genome mining by linking BGC architecture to *in silico* structure prediction (Skinnider et al., 2015, 2020). PRISM uses rule-based logic and ML to infer the chemical scaffolds of secondary metabolites encoded within BGCs, predicting monomer specificity, cyclization patterns, and tailoring modification. For marine actinobacteria, PRISM enables researchers to anticipate the chemical novelty of uncharacterized BGCs prior to experimental validation. For example, the analysis of *Streptomyces* and *Salinispora* genomes using PRISM has revealed cryptic NRPS–PKS hybrids with unusual amino acid or extender unit compositions, guiding targeted activation and heterologous expression efforts (Tracanna et al., 2017). Combined with anti-SMASH, PRISM provides an integrated pipeline that bridges genomic potential with structural hypotheses, accelerating the prioritization of high-value BGCs in synthetic biology workflows.

2.3 BAGEL—Specialized Tool for Bacteriocins and RiPPs

The BAGEL focuses on RiPPs, including bacteriocins, lanthipeptides, and thiopeptides (van Heel et al., 2013, 2018). BAGEL identifies small, conserved motifs associated with leader peptides and modification enzymes, distinguishing RiPP clusters that might be overlooked by general BGC predictors. Bacteriocin-like compounds have been increasingly reported as potent antimicrobials in marine actinobacteria, effective against multidrug-resistant pathogens (Sivakumar et al., 2020). BAGEL's curated database of RiPP families allows the rapid classification and functional annotation of such clusters, supporting the downstream design of synthetic analogs or expression cassettes. Moreover, the integration of BAGEL 5 with HMM-based detection and visualization tools enables high-confidence prediction of precursor peptides and modification enzymes, thus guiding cloning and engineering strategies for marine-derived RiPPs.

2.4 ClusterFinder and DeepBGC-ML-Based Tools for Identifying Cryptic or Novel BGCs

While rule-based tools, such as anti-SMASH and PRISM, rely on known biosynthetic motifs, ClusterFinder and DeepBGC utilize ML to identify novel or cryptic gene clusters that lack canonical domain signatures. ClusterFinder utilizes a hidden Markov model trained on Pfam domain distributions to identify regions with BGC-like characteristics, even when their functions are uncharacterized (Cimermancic et al., 2014). This approach has been particularly valuable for marine actinobacteria, whose genomes often encode atypical hybrid or minimal BGCs that are not captured by traditional motif-based algorithms. Building on this framework, DeepBGC employs deep neural networks to predict BGC boundaries, classify cluster types, and infer associated chemical classes with improved precision (Hannigan et al., 2019a, b). By learning from both known and predicted clusters, DeepBGC can recognize entirely new biosynthetic architectures, such as non-canonical NRPS-like or hybrid terpene–RiPP clusters, which are abundant in marine genomes. The integration of these machine-learning predictors with synthetic biology enables researchers to systematically mine, design, and reconstruct uncharted biosynthetic pathways from marine actinobacteria, thereby expanding the chemical space available for drug discovery.

3 CRISPR–Cas Systems in Marine Actinobacteria

3.1 Genome Editing With CRISPR–Cas9 and Cas12a: Precise Gene Modifications in Streptomyces and Challenges in Marine Isolates

The discovery of the CRISPR-Cas9 genome editing system has revolutionized modern biology, enabling unprecedented precision in gene manipulation across diverse organisms (Barrangou & Horvath, 2017; Lander, 2016). While this technology has been extensively applied in model bacteria and eukaryotes for gene therapy, crop improvement, and vaccine development, its use in *Actinobacteria*, renowned producers of antibiotics, immunosuppressants, and other secondary metabolites, remains relatively underexplored (Tong et al., 2015). In terrestrial *Streptomyces*, CRISPR-Cas9-mediated genome manipulation has been successfully employed for targeted gene insertions, deletions, and promoter replacements, leading to the discovery or overproduction of rare antibiotics, such as amicetin, 5-chloro-3-formylindole, thiolactomycin, and several anticancer agents (Cobb et al., 2015; Zeng et al., 2015). The Type II-A Cas9 system has been particularly useful for large-scale deletion of BGCs and activation of silent pathways via promoter knock-ins (Culp et al., 2019; Zhang et al., 2017). However, the constitutive expression of Cas9 can be cytotoxic in *Streptomyces*, often resulting in low transformation efficiencies and growth inhibition due to nonspecific double-strand breaks (Ye et al., 2020).

To overcome these issues, Cas9 variants have been engineered to reduce their toxicity. One such example, Cas9-BD, a Cas9 fused to a polyaspartate domain, demonstrates reduced nonspecific DNA binding and lower cellular stress while maintaining high editing efficiency (Kim et al., 2025). Optimized nucleases are particularly advantageous for genetically fragile marine isolates that exhibit slow growth and heightened sensitivity to osmotic stress. Beyond Cas9, Cas12a (Cpf1) from Type V-A CRISPR systems has emerged as an attractive alternative in actinomycetes (Li et al., 2018). Cas12a recognizes a distinct protospacer adjacent motif (Thymine-Thymine-Valine-TTV rather than Guanine rich ie., NGG) and possesses intrinsic RNase activity, allowing it to autonomously process CRISPR RNA arrays (Abudayyeh et al., 2016). These features facilitate multiplex genome editing under a single promoter, which is an important advantage for high-GC-content genomes, such as those of marine *Actinobacteria*.

Recent advances have demonstrated the feasibility of CRISPR editing in marine organisms. For example, a CRISPR-Cas9 system was successfully applied to *Streptomyces griseoincarnatus* HNS054, a marine isolate, to delete a 20-kb BGC, resulting in the emergence of cryptic metabolites (Wang et al., 2024). Similarly, *Salinispora tropica* CNB-440 was genetically engineered through double-crossover recombination and ΦC31 phage integration to introduce an *attB* site and delete the salinosporamide BGC, which was the first demonstration of large-scale genome engineering in a marine actinomycete (Zhang et al., 2018). These studies established that salt-adapted strains are genetically tractable and can serve as chassis organisms for future CRISPR–Cas-based editing (Xu et al., 2022). Remarkably, *S. tropica* expressing a heterologous thiolactomycin BGC produced a threefold higher yield than *Streptomyces coelicolor* M1152, underscoring the biosynthetic advantages conferred by the native marine physiology (Wietz et al., 2014). Nevertheless, CRISPR-Cas applications in marine *Streptomyces* and related genera remain limited because of persistent transformation barriers. These challenges include low transformation efficiency, osmotic fragility, and poorly characterized and replicative plasmids. Moreover, native CRISPR-Cas immune systems in taxa, such as *Salinispora*, can degrade exogenous DNA, further complicating editing efforts (Qi et al., 2013).

3.2 CRISPR Interference: Targeted Gene Repression for Functional Genomics

CRISPR interference (CRISPRi) employs catalytically inactive Cas proteins (dCas9 or dCas12a) guided by RNA molecules to bind to specific DNA targets and block transcription without inducing double-strand breaks (Yao et al., 2016; Li et al., 2018). In *Streptomyces*, CRISPRi has become an essential tool for functional genomics and metabolic engineering, offering reversible repression of essential genes, redirection of metabolic fluxes, and activation of cryptic BGCs (Zhao et al., 2018). Three major CRISPRi architectures have been established for *S. coelicolor*. The first method used a pGM1190 plasmid containing a temperature-sensitive pSG5 replicon for facile curing, expressing *dCas9* under a thiostrepton-inducible promoter (Tong et al., 2015). Subsequent systems achieved more stable repression through the chromosomal integration of *dCas9*–sgRNA modules via ΦC31-derived vectors. The later development of dCas12a-based interference (ddCpf1) introduced greater flexibility in target design and multiplex regulation (Li et al., 2018).

In marine actinomycetes, which often harbor large, silent BGCs tightly repressed under standard cultivation, CRISPRi provides a powerful means of reactivating cryptic metabolic pathways. By silencing transcriptional repressors or modulating flux-controlling enzymes, CRISPRi can stimulate metabolite production with minimal stress on the host cells (Bull et al., 2005; Fenical & Jensen, 2006). Recent developments in CRISPR-based activator systems have further extended this potential by enabling the programmable upregulation of target genes, a strategy successfully applied in terrestrial *Streptomyces* and now being adapted for marine isolates such as *Salinispora* and *Marinactinospora* (Ameruoso et al., 2022). Despite its promise, CRISPRi applications in obligate marine actinomycetes are still scarce. Most current efforts focus on deploying CRISPRi in heterologous hosts carrying cloned marine BGCs or using transcriptomic analyses to map regulatory networks before reintroducing genetic modifications into native marine backgrounds (Kim & Cha, 2020).

3.3 Technical Barriers: Limitations in DNA Delivery Methods

A persistent bottleneck in applying CRISPR technologies to marine *Actinobacteria* lies not in the activity of Cas nucleases but in the efficiency of DNA delivery. Transformation protocols optimized for terrestrial *Streptomyces* often fail in marine isolates because of osmotic fragility, unique cell wall compositions, and the presence of active restriction or CRISPR systems that degrade foreign DNA (Kieser et al., 2000; Wietz et al., 2014). In *Salinispora* species, osmotic down-shock sensitivity during transformation has been linked to the activation of mechanosensitive channels, leading to cell lysis when standard low-salt electroporation buffers are used (Udwary et al., 2007). Recent methodological advances have addressed this issue by simulating seawater ionic strength during transformation, which substantially improves electrocompetence and plasmid uptake (Arnli et al., 2025). When direct transformation is unfeasible, an alternative strategy is heterologous expression, which involves cloning marine BGCs into genetically amenable terrestrial *Streptomyces* chassis for refactoring, expression, and product analysis (Bonet et al., 2015; Zhang, 2019). This two-tier approach allows the functional validation of marine biosynthetic pathways, while the ongoing optimization of transformation protocols continues to expand the genetic accessibility of marine actinobacteria.

4 Heterologous Expression Systems

Heterologous expression is crucial for the functional characterization and production of secondary metabolites from marine Actinobacteria, whose genetic intractability often hinders the direct manipulation of their pathways. By transferring BGCs into well-characterized and genetically amenable hosts, researchers can validate gene function, optimize yields, and access silent metabolites. Recent advances in genome-reduced chassis design, synthetic BGC refactoring, and alternative expression systems have collectively expanded the toolkit for investigating marine microbial chemistry.

4.1 Chassis Hosts for BGC Expression

4.1.1 *Streptomyces coelicolor* (M1146/M1152 and Derivatives)

The model organism *S. coelicolor* remains the foundation for heterologous expression in actinomycetes because of its extensively characterized genetics, robust transcriptional machinery, and compatibility with a broad range of vectors. Strains M1146 and M1152 were engineered by deleting four endogenous antibiotic BGCs—*act*, *red*, *cpk*, and *cda*—thereby creating a "clean" background that enhances the detection of heterologous products (Gomez-Escribano & Bibb, 2011). Further optimization, such as introducing an *rpoB* mutation (C1298T), has been shown to increase global transcriptional activity and secondary metabolite yield, making M1152 one of the most versatile expression chassis available.

4.1.2 *Streptomyces avermitilis*

The industrially important *S. avermitilis*, known for avermectin production, has been re-engineered into genome-reduced SUKA chassis strains through the systematic deletion of nonessential genomic islands and endogenous BGCs (Komatsu et al., 2013). These SUKA strains display improved precursor flux, reduced metabolic burden, and enhanced stability for large BGC integrations. Their metabolic efficiency and industrial scalability make them suitable hosts for the expression of high-GC-content marine gene clusters.

4.1.3 *Streptomyces albus* (Del14 and J1074)

S. albus is another popular host recognized for its rapid growth, efficient conjugation, and high transformation rates. The Del14 variant, generated by deleting 15 native BGCs, provides a streamlined metabolic background conducive to the expression of diverse pathways (Myronovskyi et al., 2018). *S. albus* J1074 and its derivatives are routinely used to test heterologous expression of marine BGCs due to their robustness and well-validated plasmid systems.

4.1.4 *Salinispora tropica* (CNB-4401)

The first marine-adapted chassis, *S. tropica* CNB-4401, represents a milestone in the expansion of synthetic biology to marine actinomycetes. This strain was engineered by deleting the *salA–C* genes responsible for salinosporamide biosynthesis and inserting an *attB* site to enable ΦC31-mediated integration (Zhang et al., 2018). Expression of a thiolactomycin BGC in CNB-4401 achieved titers threefold higher than in *S. coelicolor* M1152, demonstrating that native marine physiology—characterized by optimized ion homeostasis and membrane composition—can outperform terrestrial hosts (Xu et al., 2022). These advances position *S. tropica* as a next-generation marine chassis for pathway refactoring and metabolic optimization.

4.2 Refactored and Synthetic BGCs

Native BGCs often remain silent when introduced into heterologous hosts because of incompatible promoters, codon usage biases, or regulatory constraints. Refactoring, the process of rebuilding genetic clusters with standardized promoters, ribosome-binding sites (RBSs), and transcriptional regulators, enables the predictable and tunable expression of such pathways. Transformation-Associated Recombination (TAR) Cloning in *Saccharomyces cerevisiae* allows the direct capture and refactoring of large, high-GC BGCs without extensive subcloning, facilitating the reconstruction of complex marine clusters (Yamanaka et al., 2014). The CAPTURE (Cas12a-Assisted Precise Targeted Cloning) platform further enhances this by achieving near-100% cloning efficiency across 10–113 kb regions, regardless of GC content (Enghiad et al., 2021). Similarly, CAT-FISHING (CRISPR/Cas12a-mediated fast direct biosynthetic gene cluster cloning) integrates Cas12a-guided cleavage with bacterial artificial chromosome cloning for the rapid retrieval of BGCs from marine genomes (Liang et al., 2022).

For direct sequence optimization, CRISPR-BEST (CRISPR-Base Editing SysTem) utilizes deaminase-fused Cas9 nickases for targeted base editing, thereby avoiding the induction of double-stranded breaks, a feature particularly valuable for genetically fragile actinomycetes (Tong et al., 2019). In parallel, modular synthetic refactoring, which involves the combination of standardized promoters, terminators, and operon structures, has significantly boosted metabolite yields in terrestrial *Streptomyces* hosts and is now being adapted for marine-derived pathways (Culp et al., 2019). Collectively, these synthetic biology-driven strategies have transformed BGC discovery from a descriptive to a design-based process, enabling the systematic optimization of marine metabolic potential.

4.3 Non-Traditional Hosts (Yeast and *Escherichia coli*)

4.3.1 Yeast (*Saccharomyces cerevisiae*)

Yeast serves as a powerful eukaryotic "DNA foundry" for assembling large BGCs through homologous recombination. TAR-based cloning in *S. cerevisiae* enables the efficient capture and *in vivo* assembly of high-GC marine clusters, whereas CRISPR-based activation systems in yeast facilitate promoter characterization and regulatory network analysis (Shao et al., 2013; Klein et al., 2014). Yeast platforms have also proven to be effective for modular pathway engineering, precursor supply balancing, and combinatorial biosynthesis.

4.3.2 *E. coli*

Although *E. coli* is limited in its capacity to express complex secondary metabolites, it remains indispensable for reconstituting pathways and characterizing enzymes. Its genetic tractability and rapid growth make it ideal for validating enzymatic steps within modular PKSs and NRPSs, forming the foundation for downstream engineering in *Streptomyces* hosts (Pfeifer et al., 2001). The convergence of optimized terrestrial chassis, synthetic BGC cloning, and marine-adapted hosts, such as *S. tropica* CNB-4401, heralds a new era in rational strain design. This integrative approach accelerates the pace of natural product discovery and enables the sustainable exploitation of marine actinobacterial diversity.

5 Modular DNA Assembly and Cloning Platforms

Modern synthetic biology in marine *Actinobacteria* relies on reliable modular DNA assembly systems that can handle large GC-rich BGCs. These cloning platforms enable the precise construction, refactoring, and heterologous expression of pathways, thereby accelerating the discovery of novel natural products. Among the most widely used approaches are Golden Gate-based systems (including MoClo and EcoFlex), Gibson Assembly, and BioBrick standardization, each of which offers distinct advantages for constructing complex genetic circuits and metabolic pathways (Weber et al., 2011; Gibson et al., 2009).

5.1 Golden Gate, MoClo, and EcoFlex Systems—Modular and Standardized Assembly

Golden Gate assembly utilizes Type IIS restriction enzymes (such as *Bsa*I or *Bsm*BI), which cleave outside their recognition sites to create user-defined four-base overhangs. These overhangs guide the fragment orientation, while the recognition sites are removed during ligation, producing scarless constructs in a single cyclic reaction (Engler et al., 2009). This one-pot approach allows the seamless joining of multiple DNA fragments, including promoters, coding sequences, and terminators, under standardized rules, making it highly suitable for assembling large modular BGCs typical of marine *Actinobacteria* (Casini et al., 2015). The MoClo framework builds upon the Golden Gate principle to enable hierarchical DNA assemblies. Genes or regulatory parts are first cloned into Level 0 vectors and then combined into Level 1 and Level 2 constructs, which represent full pathways or operons (Weber et al., 2011). This hierarchical grammar enables the rapid construction and swapping of parts, such as promoters, RBSs, and terminators, facilitating the systematic optimization of gene expression and metabolic flux.

The EcoFlex system extends the MoClo architecture for bacterial hosts, providing standardized vectors and part libraries compatible with *E. coli* and actinomycete chassis (Moore et al., 2016). For marine isolates with high GC content, these modular systems allow flexible promoter replacement and pathway balancing without extensive subcloning procedures. Collectively, Golden Gate, MoClo, and EcoFlex have become core technologies for refactoring marine BGCs, enabling combinatorial biosynthesis and accelerating hybrid pathway engineering.

5.2 Gibson Assembly—Seamless, Scar-Free DNA Fusion

Gibson Assembly provides a restriction-independent, scar-free alternative for constructing large genetic assemblies. It relies on sequence overlaps (20–40 bp) between adjacent fragments and a single isothermal reaction containing an exonuclease, polymerase, and ligase (Gibson et al., 2009). The exonuclease creates complementary single-stranded overhangs that anneal, whereas the polymerase and ligase fill and seal the joints, respectively. This method is particularly valuable for marine actinomycete systems, where internal restriction sites or repetitive motifs hinder Golden Gate assembly. Gibson Assembly enables the direct fusion of high-GC DNA fragments, refactoring of large (>100 kb) BGCs, and construction of CRISPR-compatible editing cassettes for *in vivo* manipulation (Yamanaka et al., 2014). It also integrates seamlessly with next-generation cloning strategies, such as CAPTURE and CAT-FISHING, for the precise retrieval of marine biosynthetic pathways (Enghiad et al., 2021; Liang et al., 2022). The simplicity and flexibility of Gibson Assembly make it a mainstay in synthetic biology pipelines for marine *Actinobacteria*, complementing modular approaches such as MoClo for applications where restriction site independence and high-fidelity joins are essential.

5.3 BioBrick Standardization—Use of Interchangeable DNA Parts in Synthetic Constructs

The BioBrick framework introduced the concept of standardized, interchangeable DNA parts with compatible restriction sites, enabling modular assembly and reproducibility across laboratories (Knight, 2003; Shetty et al., 2008). Each BioBrick part, such as a promoter, coding sequence, or terminator, conforms to defined prefix and suffix sequences, allowing the predictable and reusable construction of complex circuits. Although initially developed for *E. coli*, BioBrick standardization provides conceptual and practical benefits for engineering marine *actinobacteria*. It facilitates the gradual buildup of large biosynthetic pathways from smaller, validated components, promoting interoperability between research groups and integration with other modular systems (e.g., Golden Gate or Gibson systems). When adapted for high-GC templates, BioBrick parts support pathway optimization and regulatory fine-tuning while maintaining a community-based approach to part sharing and documentation (Casini et al., 2015). The combination of BioBrick principles with modern high-efficiency assembly methods ensures that synthetic biology in marine *Actinobacteria* remains reproducible, scalable, and collaborative, which are critical features for advancing biotechnological innovation in marine natural product discovery.

6 Promoter Libraries and Regulatory Elements

Fine-tuning of gene expression is crucial for the success of synthetic biology in marine *Actinobacteria*. Precise transcriptional and translational control enables the predictable activation of BGCs, balanced metabolic flux, and reduced metabolic burden on the host. Synthetic promoters, RBS libraries, and transcription terminators provide the regulatory architecture necessary to design robust and modular genetic circuits. The adaptation of these elements—originally developed for terrestrial *Streptomyces*—to marine systems has markedly expanded the capacity to engineer complex metabolic pathways and activate silent gene clusters (Luo et al., 2015; Xu et al., 2022).

6.1 Synthetic and Inducible Promoters—Control Systems Such as *tet*, *tipA*, and *ermE*

Promoters determine transcription initiation and represent the first level of gene expression control. Several constitutive and inducible promoters have been characterized for use in synthetic biology in Streptomyces and related marine actinomycetes. The *ermE* promoter, derived from the erythromycin resistance gene, is among the most widely used constitutive promoters, driving strong expression in multiple Actinomycete hosts (Bibb et al., 1985; Hong et al., 2005). Its consistent activity and broad compatibility have made it a standard for the heterologous expression of marine BGCs.

Inducible systems, such as *tipA* (thiostrepton-inducible) and *tet* (tetracycline-inducible), allow temporal and tunable expression control, minimizing toxicity associated with the overexpression of biosynthetic or regulatory genes (Horinouchi & Beppu, 1992; Blin et al., 2019). *tipA*-driven expression is triggered by thiostrepton binding to the TipA transcriptional activator, whereas *tet* systems use tetracycline or anhydrotetracycline to regulate TetR-controlled promoters. Such inducible promoters are particularly valuable for manipulating complex marine gene clusters, whose products may affect cell viability or require controlled timing for expression.

Recent developments in synthetic promoter libraries have further expanded this toolkit. Libraries built from *ermE∗*, *kasO*, and synthetic consensus sequences offer graded expression spanning several orders of magnitude, enabling the fine-tuning of gene dosage in refactored BGCs (Li et al., 2017; Wang et al., 2020). The integration of these promoters into the Golden Gate or MoClo frameworks allows for the rapid testing of promoter–gene combinations for optimized metabolite yields in both terrestrial and marine chassis.

6.2 RBS Engineering—Translation Tuning for Gene Expression Balance

At the translational level, RBS sequences govern initiation efficiency and are key to achieving balanced multi-gene expression in synthetic circuits. RBS engineering strategies leverage predictive algorithms, such as the RBS Calculator, to design sequences that modulate translation initiation rates across several orders of magnitude (Salis et al., 2009).

In high-GC *Actinobacteria*, differences in codon bias and mRNA secondary structures can significantly affect translation. Synthetic RBS libraries, including those derived from the *hrdB* sigma factor consensus, have been applied to fine-tune the expression in *Streptomyces coelicolor* and extended to marine isolates (Xu et al., 2022). By adjusting the RBS strength in PKS or NRPS operons, researchers can optimize precursor flux and cofactor balance, thereby enhancing the yield of marine natural products.

Moreover, combinatorial RBS assembly using MoClo systems enables the rapid screening of translation variants, helping to identify optimal configurations for both pathway performance and host viability (Horbal & Luzhetskyy, 2016). When combined with synthetic or inducible promoters, RBS tuning provides an additional layer of quantitative control over the engineered gene networks.

6.3 Transcription Terminators—Elements to Ensure Proper Gene Circuit Insulation

Transcription terminators ensure proper termination of mRNA synthesis and prevent transcriptional read-through, which can interfere with neighboring genes or regulatory modules. In complex biosynthetic systems, especially those assembled from multiple transcriptional units, the choice of terminator is critical for maintaining circuit insulation and avoiding cross-talk between pathways.

Intrinsic (rho-independent) terminators, characterized by GC-rich hairpin structures followed by poly U tracts, are frequently used in actinomycetes for efficient transcript termination. Synthetic terminator libraries have been developed to evaluate termination strength across diverse *Streptomyces* species, identifying robust candidates, such as *trrnB* and *hupA* terminators, that function reliably across GC-rich genomes (Luo et al., 2015; Xu et al., 2022).

The combination of well-characterized terminators with synthetic promoters and RBS libraries allows the modular tuning of entire pathways with minimal regulatory interference. In marine actinomycetes, where genetic instability and metabolic cross-regulation can hinder pathway productivity, precise terminator selection enhances construct stability and expression. Together, these elements form the foundation of predictable gene expression systems for advanced metabolic engineering applications.

7 Transcriptomic, Proteomic, and Metabolomic Validation

The validation of synthetic constructs in marine *Actinobacteria* requires a system-level understanding of how engineered components translate into functional cellular outcomes. While the design and assembly of genetic parts form the foundation of synthetic biology, multi-omics validation ensures that the resulting pathways behave predictably under the dynamic physicochemical conditions of marine environments. The integration of transcriptomics, proteomics, and metabolomics enables researchers to identify regulatory bottlenecks, off-target effects, and adaptive stress responses that may influence pathway performance and metabolite yields (Vogel & Marcotte, 2012; Subramani & Sipkema, 2019). Figure 3 illustrates the integration of multi-omics for pathway validation.

7.1 Transcriptomics

High-throughput RNA sequencing (RNA-Seq) offers a quantitative view into genome-wide transcriptional activity following the introduction of a pathway and environmental perturbation. Modern workflows combine spliced alignment tools, such as Spliced Transcripts Alignment to a Reference (STAR) (Dobin et al., 2013), with efficient read-counting algorithms, such as featureCounts (Liao et al., 2014), and statistical packages, such as DESeq2 (Love et al., 2014), to measure differential expression. In marine *Actinobacteria*, RNA-Seq can reveal how synthetic pathways reshape the native transcriptional networks involved in salinity tolerance, redox regulation, and the biosynthesis of secondary metabolites. Mapping reads to both chromosome and plasmid constructs enables the evaluation of promoter activity, transcriptional leakage, and host stress responses. Complementary RT-qPCR assays—performed under MIQE (Minimum Information for Publication of Quantitative Real-Time PCR Experiments)-compliant conditions (Bustin et al., 2009)—allow precise quantification of specific genes within synthetic circuits. Together, these transcriptomic tools verify that engineered promoters and riboregulators maintain the expected activity across fluctuating marine conditions.

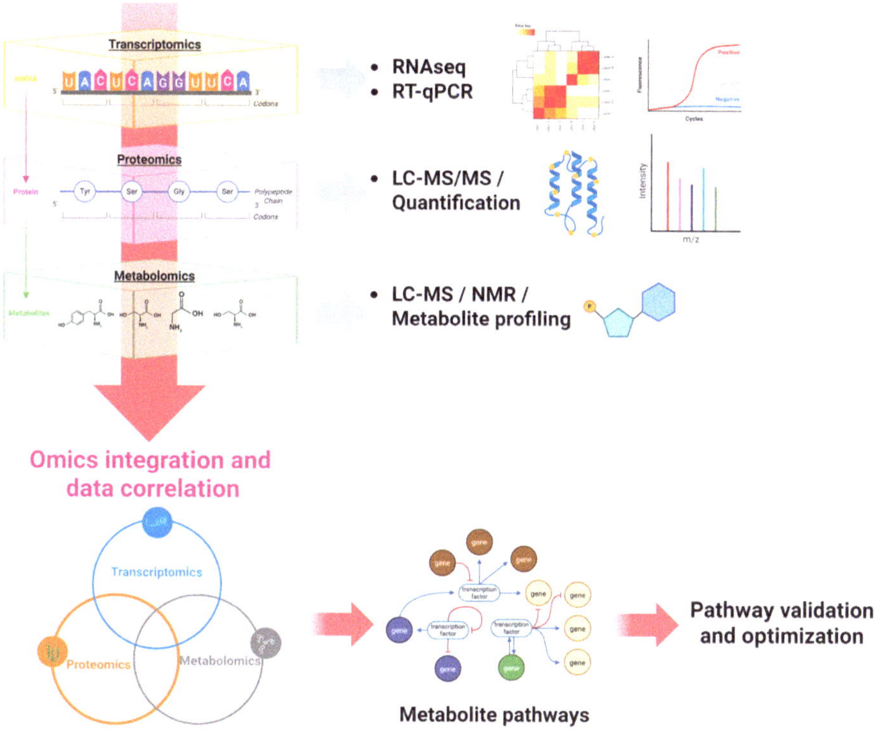

Fig. 3 Multi-omics integration for pathway validation. The schematic illustrates the integration of transcriptomic, proteomic, and metabolomic datasets to validate and optimize biosynthetic pathways. Transcriptomic analysis such as RNA-seq and Real-Time quantitative PCR or (RT-qPCR) identifies differential gene expression. Proteomic profiling (liquid chromatography-tandem mass spectrometry (LC-MS/MS)-based quantification) determines protein abundance and enzyme activity levels. Metabolomic analysis such as LC-MS or Nuclear Magnetic Resonance (NMR) detects and quantifies pathway intermediates and metabolites. Data from these omics layers are integrated and correlated to reveal relationships among genes, proteins, and metabolites. This integrative approach enables mapping of metabolite pathways and supports pathway validation, flux analysis, and iterative optimization of metabolic networks for improved biosynthetic performance

7.2 Proteomics

Because mRNA levels do not always correlate with protein abundance, proteomic validation is essential for confirming translation efficiency and post-translational stability (Aebersold & Mann, 2016). Quantitative LC-MS/MS identifies and measures synthetic enzymes, transporters, and regulatory proteins within complex cellular extracts. Label-free quantification and stable isotope labeling workflows enable the capture of changes in protein abundance between engineered and wild-type strains (Aebersold & Mann, 2003). In marine *Actinobacteria*, proteomic analyses confirm whether heterologous enzymes involved in halometabolite or polyketide synthesis are correctly expressed and folded under osmotic or oxidative stress conditions. The integration of proteomic and transcriptomic data highlights rate-limiting steps or host stress signatures, such as elevated chaperone or efflux-pump expression, that may constrain synthetic pathway performance.

7.3 Metabolomics

Metabolomics closes the validation loop by linking molecular changes to functional metabolite outputs. Untargeted LC-MS or NMR-based profiling detects thousands of small molecules, enabling the discovery of novel products generated by synthetic pathways or perturbations to native metabolic fluxes (Johnson et al., 2016). Targeted metabolomics further quantifies pathway intermediates and end products, directly linking enzymatic activity to metabolite yields. In marine *Actinobacteria*, metabolomic analyses are invaluable for assessing the biosynthesis of halogenated or nitrogen-containing secondary metabolites, which often display potent antimicrobial or anticancer properties (Subramani & Sipkema, 2019). When integrated with transcriptomic and proteomic datasets, metabolomics provides a comprehensive systems view that supports iterative redesign to enhance production while minimizing the host burden.

8 Synthetic Regulatory Circuits and Biosensors

Synthetic regulatory circuits and biosensors enable the adaptive and fine-tuned control of gene expression in engineered marine microbes. Given the environmental variability of marine habitats, such as fluctuations in salinity, pH, and nutrient levels, these dynamic systems are crucial for maintaining productivity and metabolic balance (Brophy & Voigt, 2014).

8.1 Design of Environmental-Response Circuits

The design of circuits that respond to environmental cues involves the coupling of sensory modules with logic-based genetic controllers. Synthetic constructs can incorporate osmoresponsive or pH-sensitive promoters from *Streptomyces* relatives to regulate the expression of halotolerance genes, while quorum-sensing modules derived from *LuxR/LuxI* or *LasR/LasI* systems coordinate population-level responses (Papenfort & Bassler, 2016). Circuit-design automation tools, such as Cello, translate high-level logic into DNA sequences with predictable dynamics (Nielsen et al., 2016). In marine *Actinobacteria*, modular environmental circuits allow the expression of biosynthetic pathways only under favorable salinity or nutrient conditions, conserving cellular resources and improving yield consistency.

8.2 Biosensor Integration

Biosensors convert intracellular signals into measurable outputs (fluorescent, colorimetric, or electrical), providing real-time feedback on circuit function. RNA-based toehold switches are programmable biosensors capable of detecting specific metabolites or nucleic acids (Green et al., 2014). Modern biosensor design increasingly leverages deep learning to enhance both the dynamic range and specificity (Hartline et al., 2021; Liu et al., 2015). In marine *Actinobacteria*, biosensors have the potential to monitor metabolite formation, redox state, and quorum-signal accumulation under fluctuating salinity. Coupling biosensors with microfluidic platforms enables the high-throughput screening of variants (Dehghan et al., 2025), accelerating the optimization of production strains.

8.3 Feedback-Regulated Synthetic Pathways

Self-regulating circuits employ feedback loops to maintain homeostasis and prevent the buildup of toxic metabolites. Negative feedback through metabolite-responsive repressors or riboswitches can stabilize production fluxes, whereas positive feedback can sustain the bistable activation of secondary metabolite pathways. This dynamic regulation mitigates resource competition among synthetic modules (Qian et al., 2017). For marine *Actinobacteria*, where environmental stress frequently perturbs cellular physiology, feedback-controlled designs enhance robustness and maintain a steady output of target compounds even under osmotic or nutrient fluctuations.

9 ML and AI in Synthetic Design

AI and ML are reshaping the design–build–test–learn cycle of synthetic biology. For marine *Actinobacteria*, these tools accelerate the discovery and optimization of biosynthetic pathways by integrating sequence data, metabolic models, and structure predictions into a unified framework (Woo et al., 2022).

9.1 ML-Guided Promoter and RBS Optimization

Promoter and RBS strength dictate transcriptional and translational efficiency. Predictive models, such as the RBS Calculator (Salis et al., 2009) and standardized part libraries (Mutalik et al., 2013), allow for the rational tuning of gene expression. ML algorithms trained on sequence–function datasets can infer complex motif–activity relationships, enabling the de novo design of regulatory elements suited to the high-GC genomes of marine actinobacteria. These models reduce experimental trial-and-error and help balance the expression across multigene clusters to optimize flux through biosynthetic pathways.

9.2 AI-Based Metabolic Modeling

Genome-scale metabolic reconstructions, supported by AI optimization, can predict how genetic modifications influence flux distributions and metabolite yields (Monk, 2022). Databases such as BiGG Models standardize these reconstructions and facilitate data sharing (King et al., 2016). Deep-learning-enhanced flux balance analysis incorporates nonlinear kinetics and environmental parameters, improving the predictions for non-model marine strains. Such hybrid approaches identify optimal knock-ins or deletions to redirect carbon flow toward the desired metabolites while maintaining growth under marine stress conditions.

9.3 Tools for Functional and Structural Prediction: DeepBGC and AlphaFold

ML frameworks have also transformed gene cluster discovery and protein structure prediction. DeepBGC, a deep neural network trained on known BGCs, accurately identifies and classifies cryptic clusters in actinobacterial genomes (Hannigan et al., 2019a, b). When applied to marine isolates, it uncovers the hidden potential of halogenated polyketides and nonribosomal peptides. Structural AI models, such as AlphaFold and its proteome-scale extension (Jumper et al., 2021; Tunyasuvunakool et al., 2021), predict three-dimensional enzyme structures with near-experimental accuracy, revealing catalytic motifs and substrate-binding features unique to marine enzymes. Integrating DeepBGC-based annotation with AlphaFold-derived structural insights enables the rational engineering of biosynthetic enzymes and assembly-line tailoring for novel compound production.

10 Technical Challenges and Future Directions

Despite remarkable progress in developing genetic and synthetic biology tools for *Actinobacteria*, translating these advances into marine systems remains challenging. Marine *Actinobacteria*, including *Salinispora*, *Marinactinospora*, and *Streptomyces* species isolated from marine sediments, sponges, and deep-sea environments, possess unique physiological traits that complicate laboratory cultivation and genetic manipulation. Overcoming these barriers is critical to unlocking their full potential as sources of novel natural products and next-generation platforms for biotechnological innovation.

10.1 Cultivation Barriers—Unique Physiological Requirements (Salinity, Pressure, and Temperature)

Marine *Actinobacteria* are adapted to saline, high-pressure, and often nutrient-limited environments. Standard terrestrial media, such as ISP (International

Streptomyces Project) or TSB (Tryptic Soy Broth), often fail to support optimal growth, resulting in poor biomass yield and low metabolite production (Fenical & Jensen, 2006; Bull & Stach, 2007). Many species require seawater-based media containing specific ionic compositions (e.g., Mg^{2+}, Ca^{2+}, and Na^{+}) and trace elements that influence membrane stability and enzymatic activity (Maldonado et al., 2005).

High hydrostatic pressure and variable temperature tolerance further complicate the cultivation of deep-sea isolates. These strains frequently exhibit slow growth rates, forming compact colonies and resistant spores, which reduce transformation efficiency and DNA uptake (Subramani & Aalbersberg, 2012). Recent advances in microfluidic cultivation systems and adaptive evolution approaches have enabled the partial mitigation of these issues by mimicking marine microenvironments and promoting faster laboratory adaptation (Zengler et al., 2019). However, the large-scale cultivation of obligate marine actinomycetes remains a limiting factor in their genetic and metabolic engineering.

10.2 Genetic Tool Adaptation—Porting Systems From Terrestrial to Marine Actinobacteria

Most molecular tools, such as vectors, promoters, CRISPR systems, and selectable markers, were originally developed for terrestrial *Streptomyces* and often perform suboptimally in their marine counterparts (Zhang et al., 2018). Differences in codon usage, GC content, restriction–modification systems, and native CRISPR–Cas immunity can hinder the functionality of imported systems. For example, although *ermE** and *tipA* promoters are widely used in *Streptomyces coelicolor*, their activity in *Salinispora tropica* or *Marinactinospora thermotolerans* is inconsistent and requires re-optimization (Xu et al., 2022).

The development of marine-adapted vectors with stable replication origins and selectable markers compatible with seawater ionic conditions remains an ongoing challenge. Moreover, although CRISPR-Cas tools are powerful, they require optimization of guide RNA processing and expression systems to match the transcriptional machinery of the marine hosts (Wang et al., 2024). Efforts to port orthogonal systems, such as Cas12a or CRISPRi activators, have shown promise but remain strain-specific. Expanding these molecular platforms is essential for achieving routine, high-efficiency editing across diverse marine *Actinobacteria.*

10.3 Transformation Limitations—Low Efficiency and Lack of Universal Vectors

Low transformation efficiency remains a significant barrier to the genetic engineering of marine *Actinobacteria.* Conventional methods, such as conjugation from *E. coli*, protoplast transformation, and electroporation, often fail due to osmotic sensitivity, thickened cell walls, and stress responses triggered by ionic imbalances during transformation (Udwary et al., 2007). In *Salinispora* species, for instance, rapid osmotic downshock following cell wall digestion leads to lysis, limiting the recovery of viable transformants (Wietz et al., 2014).

Another constraint is the lack of universal shuttle vectors capable of stable replication across marine and terrestrial hosts. While phage-based systems such as ΦC31 and pSET152 vectors have been successfully applied in *S. tropica* (Zhang et al., 2018), replication remains unstable in other marine isolates. Recent studies have sought to optimize transformation buffers that mimic seawater ionic strength, dramatically improving the transformation efficiency (Arnli et al., 2025). Complementary strategies, such as conjugation via methylation-matched donor strains or nanoparticle-assisted DNA delivery, are emerging as innovative solutions (Das et al., 2021).

10.4 Toward Scalable Engineering Platforms—Needs for Standardized Tools and Marine-Specific Synthetic Biology Chassis

To achieve scalable and predictable engineering of marine *Actinobacteria*, future research must focus on standardizing synthetic biology tools, including modular vectors, promoters, and transformation protocols, adapted to the unique physiological context of marine microbes (Baltz, 2017; Xu et al., 2022). Establishing marine-specific chassis strains, analogous to *Streptomyces coelicolor* M1152 or *S. albus* Del14, will provide stable and high-yield hosts for the heterologous expression of marine BGCs.

The integration of genome-scale modeling, multi-omics analysis, and ML will further enable the rational design of metabolic networks, predictive regulation of pathway flux, and automated optimization of gene expression. Coupled with advances in automation and biofoundry technologies, these tools will allow high-throughput screening of BGC variants, accelerating the discovery of novel antibiotics and bioactive metabolites.

Ultimately, the future of marine actinobacterial engineering lies in bridging the gap between environmental genomics and synthetic biology, transforming these unique marine microbes into versatile, programmable platforms for sustainable biomanufacturing and drug discovery.

References

Abudayyeh, O. O., Gootenberg, J. S., Konermann, S., Joung, J., Slaymaker, I. M., Cox, D. B. T., & Zhang, F. (2016). C2c2 is a single-component programmable RNA-guided RNA-targeting CRISPR effector. *Science, 353*(6299), aaf5573.

Aebersold, R., & Mann, M. (2003). Mass spectrometry-based proteomics. *Nature, 422*(6928), 198–207. https://doi.org/10.1038/nature01511

Aebersold, R., & Mann, M. (2016). Mass-spectrometric exploration of proteome structure and function. *Nature, 537*(7620), 347–355.

Ameruoso, A., Hoskisson, P. A., & Kaczmarczyk, A. (2022). CRISPRa for Streptomyces: Tunable activation of secondary metabolite pathways. *ACS Synthetic Biology, 11*(6), 2088–2100.

Amos, G. C. A., Zhang, L., Hawkey, P. M., Gaze, W. H., & Wellington, E. M. H. (2017). Functional metagenomic analysis reveals rivers are a reservoir for diverse antibiotic resistance genes. *Frontiers in Microbiology, 8*, 974.

Arnli, R. T., Wu, L., & Zhang, W. (2025). Optimized electroporation conditions for marine actinobacteria. *Marine Biotechnology, 27*, 88–101.

Baltz, R. H. (2017). Gifted microbes for genome mining and natural product discovery. *Journal of Industrial Microbiology & Biotechnology, 44*, 573–588.

Barrangou, R., & Horvath, P. (2017). A decade of discovery: CRISPR–Cas systems in adaptive immunity and genome engineering. *Science, 357*(6351), eaan5174.

Bérdy, J. (2012). Thoughts and facts about antibiotics: Where we are now and where we are heading. *Journal of Antibiotics, 65*, 385–395.

Bibb, M. J., Ward, J. M., & Hopwood, D. A. (1985). Transformation of plasmid DNA into Streptomyces at high frequency. *Nature, 274*, 398–400.

Blin, K., Shaw, S., Kloosterman, A. M., et al. (2019). The antiSMASH database version 2. *Nucleic Acids Research, 47*(D1), D625–D630.

Blin, K., Shaw, S., Kloosterman, A. M., Charlop-Powers, Z., van Wezel, G. P., Medema, M. H., & Weber, T. (2021). antiSMASH 6.0. *Nucleic Acids Research, 49*(W1), W29–W35.

Bonet, B., Teufel, R., Crüsemann, M., Ziemert, N., & Moore, B. S. (2015). Direct cloning and refactoring of natural product pathways from marine bacteria. *Nature Chemical Biology, 11*, 435–437.

Brophy, J. A. N., & Voigt, C. A. (2014). Principles of genetic circuit design. *Nature Methods, 11*(5), 508–520.

Bull, A. T., & Stach, J. E. M. (2007). Marine actinobacteria: New opportunities for natural product search and discovery. *Trends in Microbiology, 15*, 491–499.

Bull, A. T., Ward, A. C., & Goodfellow, M. (2005). Search and discovery strategies for biotechnology: The paradigm shift. *Microbiology and Molecular Biology Reviews, 69*, 86–108.

Bustin, S. A., Benes, V., Garson, J. A., Hellemans, J., Huggett, J., Kubista, M., Mueller, R., Nolan, T., Pfaffl, M. W., Shipley, G. L., Vandesompele, J., & Wittwer, C. T. (2009). The MIQE guidelines: Minimum information for publication of quantitative real-time PCR experiments. *Clinical Chemistry, 55*(4), 611–622.

Cameron, D. E., Bashor, C. J., & Collins, J. J. (2014). A brief history of synthetic biology. *Nature Reviews Microbiology, 12*, 381–390.

Carbonell, P., Radivojevic, T., & Garcia Martin, H. (2016). Opportunities at the interface of machine learning and metabolic engineering. *Biotechnology Journal, 11*, 731–742.

Casini, A., MacDonald, J. T., De Jonghe, J., Christodoulou, G., Freemont, P. S., Baldwin, G. S., & Ellis, T. (2015). MODAL strategy. *Nucleic Acids Research, 43*(4), e45.

Chevrette, M. G., Carlson, C. M., Ortega, H. E., Thomas, C., Ananiev, G. E., Barns, K. J., & Currie, C. R. (2019). The antimicrobial potential of marine actinomycetes. *Nature Microbiology, 4*, 1454–1463.

Cho, J. Y., Williams, P. G., Kwon, H. C., Jensen, P. R., & Fenical, W. (2007). Lucentamycins A-D, cytotoxic peptides from the marine-derived actinomycete Nocardiopsis lucentensis. *Journal of Natural Products, 70*(8), 1321–1328. https://doi.org/10.1021/np070101b

Cimermancic, P., Medema, M. H., Claesen, J., Kurita, K., Brown, L. C. W., Mavrommatis, K., & Fischbach, M. A. (2014). Global analysis of biosynthetic gene clusters. *Cell, 158*, 412–421.

Cobb, R. E., Wang, Y., & Zhao, H. (2015). High-efficiency multiplex genome editing of Streptomyces species using an engineered CRISPR/Cas system. *ACS Synthetic Biology, 4*(6), 723–728.

Culp, E. J., Yim, G., Waglechner, N., Wang, W., & Wright, G. D. (2019). Evolution-guided discovery of antibiotics that inhibit peptidoglycan remodeling. *Nature Microbiology, 4*, 1120–1130.

Das, S., Lee, J.-H., & Kim, Y. (2021). Microbial degradation of marine pollutants. *Marine Drugs, 19*(7), 375.

Dehghan, A., Kiani, M. J., Gholizadeh, A., Aminizadeh, J., Rahi, A., Zare, I., Pishbin, E., & Heli, H. (2025). Electrochemical genosensors on-a-chip: Applications in early diagnosis of pathogens. *Sensors and Actuators Reports, 9*, 100335. https://doi.org/10.1016/j.snr.2025.100335

Dobin, A., Davis, C. A., Schlesinger, F., Drenkow, J., Zaleski, C., Jha, S., Batut, P., Chaisson, M., & Gingeras, T. R. (2013). STAR: Ultrafast universal RNA-seq aligner. *Bioinformatics, 29*(1), 15–21.

Doudna, J. A., & Charpentier, E. (2014). Genome engineering with CRISPR-Cas9. *Science, 346*, 1258096.

El-Shatoury, S. A., El-Shenawy, N. S., & Abd El-Salam, I. M. (2009). Antimicrobial, antitumour and in vivo cytotoxicity of actinomycetes inhabiting marine shellfish. *World Journal of Microbiology and Biotechnology, 25*, 1547–1555.

Enghiad, B., et al. (2021). Cas12a-assisted precise targeted cloning using CAPTURE. *Nature Communications, 12*, 1171.

Engler, C., Kandzia, R., & Marillonnet, S. (2009). A one pot, one step, precision cloning method. *PLoS One, 4*(11), e5553.

Fenical, W., & Jensen, P. R. (2006). Developing a new resource for drug discovery: Marine actinomycete bacteria. *Nature Chemical Biology, 2*, 666–673.

Fenical, W., Jensen, P. R., Palladino, M. A., Lam, K. S., Lloyd, G. K., & Potts, B. C. (2009). Discovery and development of the anticancer agent salinosporamide A (NPI-0052). *Bioorganic & Medicinal Chemistry, 17*(6), 2175–2180. https://doi.org/10.1016/j.bmc.2008.10.075

Gibson, D. G., et al. (2009). Enzymatic assembly of DNA molecules. *Nature Methods, 6*(5), 343–345.

Gomez-Escribano, J. P., & Bibb, M. J. (2011). Engineering Streptomyces coelicolor for heterologous expression of secondary metabolite gene clusters. *Microbial Biotechnology, 4*(2), 207–215.

Green, A. A., Silver, P. A., Collins, J. J., & Yin, P. (2014). Toehold switches: De novo-designed regulators of gene expression. *Cell, 159*(4), 925–939.

Gupta, A., Gupta, R., & Singh, R. L. (2020). Biosensors for environmental monitoring: A review. *Environmental Science and Pollution Research, 27*, 24887–24902.

Hannigan, G. D., Prihoda, D., Palicka, A., Soukup, J., Klempir, O., Rampula, L., Durcak, J., Wurst, M., Kotowski, J., Chang, D., Wang, R., Piizzi, G., Temesi, G., Hazuda, D. J., Woelk, C. H., & Bitton, D. A. (2019a). A deep-learning genome-mining strategy for biosynthetic gene-cluster prediction. *Nucleic Acids Research, 47*(18), e110.

Hannigan, G. D., Prihoda, D., Palicka, A., Soukup, J., Klempir, O., Rampula, L., & Kostic, A. D. (2019b). Deep learning genome mining. *Nature Communications, 10*, 3611.

Hartline, C. J., Schmitz, A. C., Han, Y., & Zhang, F. (2021). Dynamic control in metabolic engineering. *Metabolic Engineering, 63*, 126–140.

Hong, S. K., Kim, Y. G., & Choi, C. Y. (2005). Synthetic promoter library for Streptomyces. *Applied Microbiology and Biotechnology, 67*, 641–646.

Horbal, L., & Luzhetskyy, A. (2016). Synthetic regulatory elements for antibiotic production. *Current Opinion in Microbiology, 33*, 113–120.

Horinouchi, S., & Beppu, T. (1992). Regulation of secondary metabolism and cell differentiation in Streptomyces. *Microbiological Reviews, 56*(3), 462–476.

Jensen, P. R., Moore, B. S., & Fenical, W. (2015). The marine actinomycete genus Salinispora: A model organism for secondary metabolite discovery. *Natural Product Reports, 32*(5), 738–751. https://doi.org/10.1039/c4np00167b

Johnson, C. H., Ivanisevic, J., & Siuzdak, G. (2016). Metabolomics: Beyond biomarkers and toward mechanisms. *Nature Reviews Molecular Cell Biology, 17*(7), 451–459.

Jumper, J., Evans, R., Pritzel, A., Green, T., Figurnov, M., Ronneberger, O., Tunyasuvunakool, K., Bates, R., Zidek, A., Potapenko, A., Bridgland, A., Meyer, C., Kohl, S. A. A., Ballard, A. J., Cowie, A., Romera-Paredes, B., Nikolov, S., Jain, R., Adler, J., & Hassabis, D. (2021). Highly accurate protein structure prediction with AlphaFold. *Nature, 596*(7873), 583–589.

Keller, S., Nicholson, G., Drahl, C., Sorensen, E., Fiedler, H.-P., & Süssmuth, R. D. (2007). Abyssomicins G and H and atrop-abyssomicin C from the marine Verrucosispora strain AB-18-032. *The Journal of Antibiotics, 60*(6), 391–394. https://doi.org/10.1038/ja.2007.54
Kieser, T., Bibb, M. J., Buttner, M. J., Chater, K. F., & Hopwood, D. A. (2000). *Practical streptomyces genetics*. John Innes Foundation.
Kim, D., & Cha, C. J. (2020). Functional genomics in marine actinobacteria: Approaches and challenges. *Frontiers in Microbiology, 11*, 2740.
Kim, Y., Kajihara, K., Imada, C., Yasuda, I., & Oikawa, H. (2014). α- and γ-Pyrones from a marine-derived Nocardiopsis sp. *Marine Drugs, 12*(7), 4113–4118. https://doi.org/10.3390/md12074110
Kim, J., Lee, S. H., & Park, J. H. (2025). Development of low-toxicity Cas9 variants for GC-rich actinomycetes. *Biotechnology Journal, 20*(1), 2300452.
King, Z. A., Lu, J., Dräger, A., Miller, P., Federowicz, S., Lerman, J. A., Ebrahim, A., Palsson, B. O., & Lewis, N. E. (2016). BiGG models: A platform for integrating and sharing genome-scale models. *Nucleic Acids Research, 44*(D1), D515–D522.
Klein, J., Healey, R. D., & Wolfe, K. H. (2014). Promoter engineering in yeast for heterologous biosynthesis. *ACS Synthetic Biology, 3*, 132–140.
Knight, T. (2003). *Idempotent vector design for BioBricks*. MIT Technical Reports.
Komatsu, M., Komatsu, K., Koiwai, H., Yamada, Y., Kozone, I., Izumikawa, M., & Takahashi, Y. (2013). Engineered *Streptomyces avermitilis* host for heterologous expression of biosynthetic gene clusters. *Proceedings of the National Academy of Sciences USA, 110*(19), 7850–7855.
Kurosawa, K., Ghiviriga, I., Sambandan, T. G., Lessard, P. A., Barbara, J. E., Rha, C., & Sinskey, A. J. (2008). Rhodostreptomycins, antibiotics biosynthesized following horizontal gene transfer from *Streptomyces padanus* to *Rhodococcus fascians*. *Journal of the American Chemical Society, 130*(4), 1126–1127. https://doi.org/10.1021/ja077821p
Lander, E. S. (2016). The heroes of CRISPR. *Cell, 164*(1–2), 18–28.
Letzel, A. C., Pidot, S. J., & Hertweck, C. (2017). Genome mining from marine actinomycetes. *Marine Drugs, 15*(2), 42.
Li, S., Zhang, H., & Zhao, H. (2017). Promoter engineering. *Biotechnology Advances, 35*, 106–120.
Li, L., Wei, K., Zheng, G., Liu, X., Chen, S., & Jiang, W. (2018). CRISPR–Cpf1–assisted multiplex genome editing of Streptomyces species. *Applied and Environmental Microbiology, 84*(16), e00827–e00818.
Li, X., Zhang, C., & Zhang, W. (2021). CRISPR/Cas-based genome editing in Streptomyces. *Frontiers in Microbiology, 12*, 705265.
Liang, J., Liu, X., & Zhang, C. (2022). CAT-FISHING: Cas12a-assisted targeting for rapid retrieval of biosynthetic gene clusters. *Nucleic Acids Research, 50*(11), e63.
Liao, Y., Smyth, G. K., & Shi, W. (2014). featureCounts: An efficient general-purpose read-summarization program. *Bioinformatics, 30*(7), 923–930.
Liu, D., Evans, T., & Zhang, F. (2015). Applications and advances of metabolite biosensors for metabolic engineering. *Metabolic Engineering, 31*, 35–43.
Love, M. I., Huber, W., & Anders, S. (2014). Moderated estimation of fold change and dispersion for RNA-seq data with DESeq2. *Genome Biology, 15*(12), 550.
Luo, Y., Zhang, L., Barton, K. W., & Zhao, H. (2015). Identification of strong promoters from Streptomyces. *ACS Synthetic Biology, 4*, 1001–1010.
Maldonado, L. A., Stach, J. E., Pathom-aree, W., Ward, A. C., Bull, A. T., & Goodfellow, M. (2005). Diversity of cultivable actinobacteria. *Antonie Van Leeuwenhoek, 87*, 11–18.
Manivasagan, P., Venkatesan, J., Sivakumar, K., & Kim, S.-K. (2014). Marine actinobacterial metabolites. *Microbiological Research, 169*, 262–278.
Mazzei, E., Iorio, M., Maffioli, S. I., Sosio, M., & Donadio, S. (2012). Characterization of madurastatin C1, a novel siderophore from Actinomadura sp. *The Journal of Antibiotics, 65*(5), 267–269. https://doi.org/10.1038/ja.2012.10
Medema, M. H., & Fischbach, M. A. (2015). Computational approaches to natural product discovery. *Nature Chemical Biology, 11*, 639–648.

Monk, J. M. (2022). Genome-scale metabolic network reconstructions of diverse Escherichia strains reveal strain-specific adaptations. *Philosophical Transactions of the Royal Society B, 377*(1861), 20210236.

Moore, S. J., et al. (2016). EcoFlex kit. *ACS Synthetic Biology, 5*(10), 1059–1069.

Mutalik, V. K., Guimaraes, J. C., Cambray, G., Lam, C., Christoffersen, M. J., Mai, Q. A., Tran, A. B., Paull, M., Keasling, J. D., Arkin, A. P., & Endy, D. (2013). Standard transcription and translation initiation elements. *Nature Methods, 10*(4), 354–360.

Myronovskyi, M., Rosenkränzer, B., Nadmid, S., Pujic, P., Normand, P., & Luzhetskyy, A. (2018). Streptomyces albus Del14: A genome-reduced host for heterologous expression of secondary metabolite clusters. *Microbial Cell Factories, 17*, 37.

Nielsen, J., & Keasling, J. D. (2016). Engineering cellular metabolism. *Cell, 164*, 1185–1197.

Nielsen, A. A. K., Der, B. S., Shin, J., Vaidyanathan, P., Paralanov, V., Strychalski, E. A., Ross, D., Densmore, D., & Voigt, C. A. (2016). Genetic circuit design automation. *Science, 352*(6281), aac7341.

Papenfort, K., & Bassler, B. L. (2016). Quorum-sensing signal–response systems in Gram-negative bacteria. *Nature Reviews Microbiology, 14*(9), 576–588.

Pfeifer, B. A., Admiraal, S. J., Gramajo, H., Cane, D. E., & Khosla, C. (2001). Biosynthesis of complex polyketides in a metabolically engineered strain of E. Coli. *Science, 291*(5509), 1790–1792.

Purnick, P. E. M., & Weiss, R. (2009). The second wave of synthetic biology. *Nature Reviews Molecular Cell Biology, 10*, 410–422.

Qi, L. S., Larson, M. H., Gilbert, L. A., Doudna, J. A., Weissman, J. S., Arkin, A. P., & Lim, W. A. (2013). Repurposing CRISPR as an RNA-guided platform for sequence-specific control of gene expression. *Cell, 152*(5), 1173–1183.

Qian, Y., Huang, H. H., Jiménez, J. I., & Del Vecchio, D. (2017). Resource competition shapes the response of genetic circuits. *ACS Synthetic Biology, 6*(7), 1263–1272.

Salis, H. M., Mirsky, E. A., & Voigt, C. A. (2009). Design of synthetic RBSs. *Nature Biotechnology, 27*(10), 946–950.

Shao, Z., Luo, Y., & Zhao, H. (2013). Rapid construction and screening of combinatorial biosynthetic libraries in yeast using TAR cloning. *Nature Protocols, 8*, 2319–2331.

Shetty, R. P., Endy, D., & Knight, T. F. (2008). BioBrick vectors from BioBrick parts. *Journal of Biological Engineering, 2*, 5.

Sivakumar, K., Sahu, M. K., Manivel, P. R., & Thangaradjou, T. (2020). Bacteriocins from marine actinobacteria. *Frontiers in Microbiology, 11*, 557.

Skinnider, M. A., Dejong, C. A., Rees, P. N., Johnston, C. W., Li, H., Webster, A. L. H., & Magarvey, N. A. (2015). PRISM 3. *Nucleic Acids Research, 43*(W1), W441–W448.

Skinnider, M. A., Merwin, N. J., Johnston, C. W., & Magarvey, N. A. (2020). PRISM 4. *Nucleic Acids Research, 48*(W1), W541–W547.

Smanski, M. J., Zhou, H., Claesen, J., Shen, B., Fischbach, M. A., & Voigt, C. A. (2016). Synthetic biology to access and expand nature's chemical diversity. *Nature Reviews Microbiology, 14*, 135–149.

Subramani, R., & Aalbersberg, W. (2012). Marine actinomycetes: Novel bioactive metabolites. *Microbiological Research, 167*, 571–580.

Subramani, R., & Sipkema, D. (2019). Marine rare actinomycetes: A promising source of novel natural products. *Marine Drugs, 17*(5), 269.

Tong, Y., Charusanti, P., Zhang, L., Weber, T., & Lee, S. Y. (2015). CRISPR–Cas9 based engineering of actinomycetal genomes. *ACS Synthetic Biology, 4*(9), 1020–1029.

Tong, Y., Whitford, C. M., Robertsen, H. L., Blin, K., & Weber, T. (2019). CRISPR-BEST: A highly efficient base editing system for actinomycetes. *ACS Synthetic Biology, 8*(10), 2336–2343.

Tracanna, V., de Jong, A., Medema, M. H., & Kuipers, O. P. (2017). Mining prokaryotes for antimicrobial compounds. *FEMS Microbiology Reviews, 41*, 417–429.

Tunyasuvunakool, K., Adler, J., Wu, Z., Green, T., Zielinski, M., Žídek, A., Bridgland, A., Cowie, A., Meyer, C., Laydon, A., Velankar, S., Kleywegt, G. J., Bateman, A., Evans, R., Pritzel, A.,

Figurnov, M., Ronneberger, O., Bates, R., Kohl, S. A. A., & Hassabis, D. (2021). Highly accurate protein structure prediction for the human proteome. *Nature, 596*(7873), 590–596.

Udwary, D. W., Zeigler, L., Asolkar, R. N., Singan, V., Lapidus, A., Fenical, W., Jensen, P. R., & Moore, B. S. (2007). Genome sequencing reveals complex secondary metabolome in *Salinispora tropica*. *Proceedings of the National Academy of Sciences of the United States of America, 104*(25), 10376–10381. https://doi.org/10.1073/pnas.0700962104

van Heel, A. J., de Jong, A., Montalbán-López, M., Kok, J., & Kuipers, O. P. (2013). BAGEL3. *Nucleic Acids Research, 41*(W1), W448–W453.

van Heel, A. J., Hendriks, D., Kuipers, O. P., & de Jong, A. (2018). BAGEL4. *Nucleic Acids Research, 46*(W1), W278–W281.

Vogel, C., & Marcotte, E. M. (2012). Insights into the regulation of protein abundance. *Nature Reviews Genetics, 13*(4), 227–232.

Wang, W., Li, S., Li, Z., Zhang, J., Fan, K., & Tan, G. (2020). A promoter library in Streptomyces. *Nature Chemical Biology, 16*(7), 779–787.

Wang, H., Zhang, C., & Wu, J. (2024). CRISPR–Cas9 deletion of a 20-kb biosynthetic cluster in marine Streptomyces griseoincarnatus activates cryptic metabolite production. *Marine Drugs, 22*(3), 156.

Weber, E., Engler, C., Gruetzner, R., Werner, S., & Marillonnet, S. (2011). Modular cloning system. *PLoS One, 6*(2), e16765.

Wei, R. B., Xi, T., Li, J., Wang, P., Li, F. C., Lin, Y. C., & Qin, S. (2011). Lobophorin C and D, new kijanimicin derivatives from a marine sponge-associated actinomycetal strain AZS17. *Marine Drugs, 9*(3), 359–368. https://doi.org/10.3390/md9030359

Wietz, M., Millán-Aguiñaga, N., Jensen, P. R., & Fenical, W. (2014). Genetic manipulation of marine Salinispora: Heterologous expression of thiolactomycin pathway. *Applied and Environmental Microbiology, 80*, 1430–1437.

Woo, D. J., Kim, G. B., Kim, Y., & Lee, S. Y. (2022). Applications of artificial intelligence to enzyme and pathway design for metabolic engineering. *Current Opinion in Biotechnology, 73*, 101–107.

Xu, Y., Fu, P., & Ding, L. (2022). Engineering marine Salinispora as a chassis. *Synthetic and Systems Biotechnology, 7*(2), 480–490.

Yamanaka, K., Reynolds, K. A., Kersten, R. D., Ryan, K. S., Gonzalez, D. J., Nizet, V., & Moore, B. S. (2014). Direct cloning and refactoring of a silent lipopeptide biosynthetic gene cluster from a marine bacterium by TAR cloning. *Nature Chemical Biology, 10*, 431–433.

Yao, L., Cengic, I., Anfelt, J., & Hudson, E. P. (2016). Multiple gene repression in cyanobacteria using CRISPRi. *ACS Synthetic Biology, 5*(3), 207–212.

Ye, X., Luo, Y., & Zhang, X. (2020). Addressing Cas9 toxicity in Streptomyces: Strategies for improved editing efficiency. *Frontiers in Microbiology, 11*, 614.

Zeng, H., Wen, S., Xu, W., He, Z., Zhai, G., Liu, Y., & Deng, Z. (2015). Highly efficient editing of the actinorhodin polyketide chain length factor gene using CRISPR/Cas9. *Applied Microbiology and Biotechnology, 99*(24), 10575–10585.

Zengler, K., Palsson, B. Ø., & Zaramela, L. S. (2019). Microbial systems biology. *Nature Reviews Microbiology, 17*(9), 587–600.

Zhang, W. (2019). Synthetic biology approaches for marine natural product discovery. *Marine Biotechnology, 21*, 157–167.

Zhang, Y., Adnani, N., Braun, D. R., Ellis, G. A., Barns, K. J., Parker-Nance, S., Guzei, I. A., & Bugni, T. S. (2016). Micromonohalimanes A and B: Antibacterial halimane-type diterpenoids from a marine Micromonospora species. *Journal of Natural Products, 79*(11), 2968–2972. https://doi.org/10.1021/acs.jnatprod.6b00555

Zhang, M. M., Wong, F. T., Wang, Y., Luo, S., & Zhao, H. (2017). CRISPR–Cas9 editing of Streptomyces to activate cryptic biosynthetic gene clusters. *Nature Chemical Biology, 13*(10), 1065–1072.

Zhang, W., Hertweck, C., & Moore, B. S. (2018). Large-scale genome engineering of *Salinispora tropica* via site-specific recombination. *Nature Communications, 9*, 1101.

Zhang, M. M., Wang, Y., Ang, E. L., & Zhao, H. (2020). Engineering microbial hosts. *Natural Product Reports, 37*, 1971–2000.
Zhang, Q., Blin, K., & Medema, M. H. (2023). Unlocking silent biosynthetic gene clusters. *Trends in Biotechnology, 41*(2), 150–163.
Zhao, Y., Wang, L., & Liu, Y. (2018). CRISPR–Cas-based transcriptional repression in Streptomyces coelicolor enables pathway refactoring. *Metabolic Engineering, 47*, 87–96.
Ziemert, N., Alanjary, M., & Weber, T. (2016). The evolution of genome mining in microbes. *Natural Product Reports, 33*, 988–1005.

Future Perspectives of Marine Actinobacteria

A. Sai Ramesh, A. Kowsalya, B. G. Pavana, and G. Sowmiya

Abstract

Marine actinobacteria are among the most diverse and versatile microbial groups inhabiting the oceans, renowned for their ability to produce structurally unique and bioactive secondary metabolites. Their remarkable adaptability to extreme ecological niches has made them a reservoir of antibiotics, anticancer agents, immunosuppressants, extremozymes, and functional nanomaterials, with applications spanning medicine, agriculture, aquaculture, industry, and environmental management. Despite their immense promise, much of their potential remains unexplored due to challenges in cultivation, limited discovery, and transcriptionally silent biosynthetic gene clusters. Recent advances in metagenomics, genome mining, synthetic biology, and high-throughput culturing, coupled with artificial intelligence and bioinformatics, are revolutionizing approaches to unlock these hidden pathways. By integrating computational tools with experimental microbiology, marine actinobacteria are transitioning from understudied curiosities to strategic biotechnological resources, offering sustainable solutions to global challenges such as antimicrobial resistance, food security, industrial innovation, and environmental resilience.

A. Sai Ramesh (✉)
Department of BioEngineering, Vels Institute of Science, Technology and Advanced Studies, Chennai, Tamil Nadu, India

A. Kowsalya
Anna University, Chennai, Tamil Nadu, India

B. G. Pavana · G. Sowmiya
Department of Biotechnology, Prathyush Engineering College, Tiruvallur, Chennai, Tamil Nadu, India

K. Loganathan, Z. Li (eds.), *Marine Actinomycetes*,
https://doi.org/10.1007/978-981-95-9849-6_17

Graphical Abstract

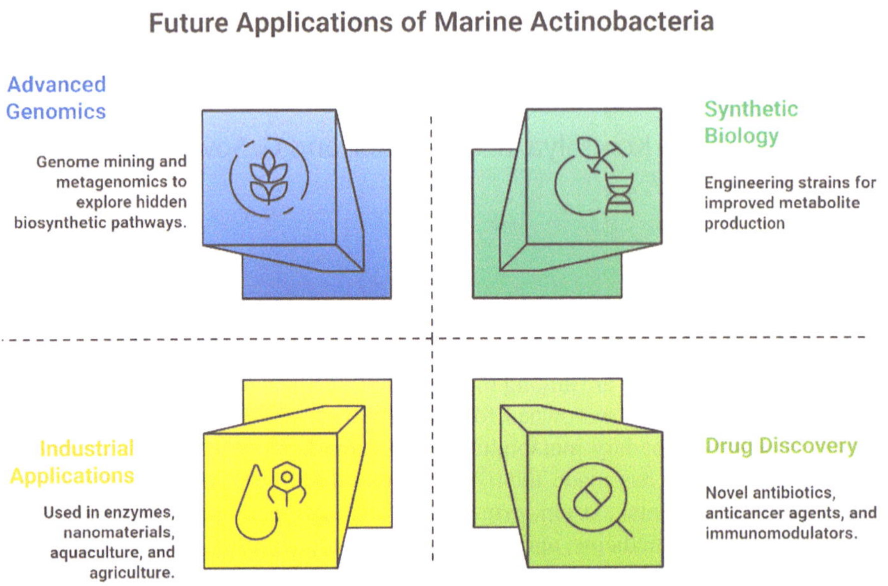

Keywords

Bioactive metabolites · Drug discovery · Environmental bioremediation · Metagenomics and synthetic biology

17.1 Introduction

Marine actinobacteria are a diverse phylogenetically and metabolically separated class of Gram-positive filamentous bacteria that have high GC content. Classes such as *Streptomyces*, *Micro-monospora*, *Nocardiopsis*, and *Salinispora* have now become some of the largest components of marine microbiomes and are arguably considered a new biotechnology and drug discovery frontier (Manivasagan et al., 2014; Subramani & Aalbersberg, 2012). Their ecological roles include the degradation of organic matter, nitrogen cycling, symbiosis with marine invertebrates, and competitive inhibition of pathogens by antibiotic production, which makes a significant proportion of bioactive metabolites attributed to actinobacteria. Some marine actinobacteria have been reported to produce a range of bioactive compounds that progressed through preclinical or clinical evaluation (Newman & Cragg, 2020; Abdelmohsen et al., 2014). An example is *Salinispora tropica*, the source of salinosporamide A, which has been isolated and clinically evaluated, increasing the pharmaceutical significance, and has also been FDA-approved as a proteasome inhibitor for cancer treatment (Feling et al., 2003). In addition to pharmaceuticals, marine actinobacteria are also producers of remarkable industrial enzymes that benefit from their extremophilic properties, which can be advantageous in situations where conventional microbial enzyme activity is compromised that have potential application in biofuels, detergents, food processing, textiles, and bioremediation (Trincone, 2011). However, there are multiple effects inhibiting the sustainable exploitation of marine actinobacteria. To start with cultivability, more than 90% of marine microbial taxa, including these rare actinobacteria, remain uncultured by conventional laboratory methods (Vartoukian et al., 2010). Second, rediscovery of well-known compounds puts limits on novelty rates in many screening programs, with high costs and low yields for unique metabolites often being the end result (Subramani & Aalbersberg, 2013). Finally, many biosynthetic gene clusters (BGCs) are silent under standard culture conditions, presenting unique methods required for their activation (Rutledge & Challis, 2015). There have arisen tools for genome mining, such as anti-Secondary Metabolite Analysis SHell (anti-SMASH) and PRediction Informatics for Secondary Metabolomes (PRISM), that have revolutionized the identification of cryptic BGCs in marine actinobacteria (Blin et al., 2021). With synthetic and metabolic engineering strategies, the heterologous expression of these gene clusters into different organisms induced an increase in yield and alteration in form (Smanski et al., 2016). Experimental evolution and co-culturing with artificial intelligence (AI) and machine learning (ML) have given a push to a much-increased metabolite production (Abdelmohsen et al., 2014). Using AI for predicting the metabolite structures, prioritization of candidate strains, modeling protein/ligand interactions, and optimizing these pipelines for smart screening is expected to tap into immense reservoirs of unexplored chemical diversity (Stokes et al., 2020).

This review aims to give a systematic overview of the diverse applications of marine actinobacteria on their current advancements, problem statements, and future perspectives. Furthermore, marine actinobacteria's contributions to industrial biotechnology, aquaculture and agriculture, environmental applications, genomics, metagenomics, and synthetic biology, and the integration of AI and smart screening into current discovery workflows were studied.

17.2 Emerging and Future Applications of Marine Actinobacteria

Marine actinobacteria are increasingly viewed as vital resources for future scientific and industrial advancements. With the aid of genomics and synthetic biology, they are being explored for the development of novel antibiotics, anticancer compounds, antivirals, and patient-specific therapeutics. Their role in probiotics is also gaining attention for maintaining human health. In industry, engineered strains may enable large-scale production of valuable metabolites, extremozymes, and eco-friendly chemicals while supporting innovations in nanotechnology. Agriculture can benefit through sustainable biofertilizers and crop protection strategies. Beyond this, they offer solutions for bioremediation, renewable energy, ecosystem restoration, and the creation of biodegradable and smart materials. These insights collectively emphasize the potential of marine actinobacteria to contribute significantly to forthcoming scientific discoveries and industrial solutions (Fig. 1).

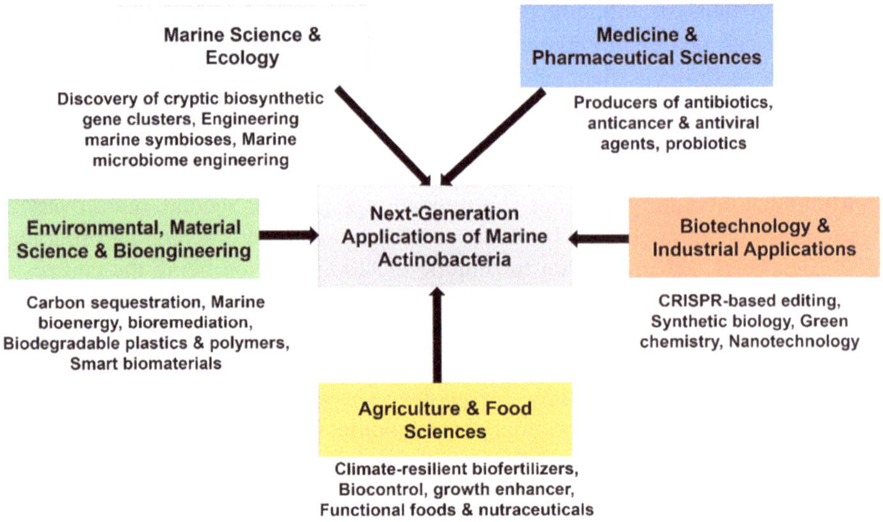

Fig. 1 Emerging and future applications of marine actinobacteria

17.3 Marine Actinobacteria in Drug Discovery

Based on the taxonomic and chemical diversity, among marine actinobacteria, *Streptomyces* dominates (~70% of producers), followed by the genera *Micromonospora, Nocardiopsis*, and allied genera. These genera and allied ones have been implicated in the discovery of hundreds of novel bioactive compounds, mostly polyketides (≈33%), alkaloids (≈37%), and peptides (~15%) (Chen et al., 2021). The notable bioactive compounds involve *Salinispora tropica* produced salinosporamide A, which is a potent proteasome inhibitor, currently in clinical anticancer trials, together with halogenated macrolides, including sporolide A and B (Gulder & Moore, 2010). Marine *Streptomyces* strains have produced anthracimycin, a macrolide antibiotic extremely effective in fighting *Bacillus anthracis* and Methicillin-Resistant Staphylococcus aureus (MRSA), discovered from Californian and Singaporean coasts (Jang et al., 2013). This also unlocked novel BGCs in marine *Streptomyces,* particularly MAR4 clades, for the discovery of molecules like indanopyrrole A, a new halogenated antibiotic with potent activity against Gram-positive and Gram-negative bacteria (Sweeney et al., 2024). At first, growing more kinds of bacteria involves the problem statement, "Most marine actinobacteria cannot grow in the lab using normal methods." The solution is to grow the bacteria with the help of a tool like the iChip, which allows bacteria to grow within their own environment before coming into the lab. Study bacteria that cannot be grown yet via DNA sequencing, straight from the seawaters. It is like making a mini ocean right inside the lab for those rare bacteria to feel at home (Hug et al., 2016). Thus, to avoid repeated discoveries, here, the problem is that scientists find the same compounds time and time again. For this, the better solution is to employ smart-computer tools (AI-based) for studying new compounds and comparing them to the ones already known. This way, these tools can quickly say whether this is a new molecule or has not been produced before.

Third is to turn on "silent" genes, where the problem is that some bacteria have genes that can make medicine, but those genes are "turned off" in the lab. An effective solution is to use Clustered Regularly Interspaced Short Palindromic Repeats; a DNA sequence family used in genome editing (CRISPR) or chemicals to "wake up" those sleeping genes. Co-culturing bacteria with other microbes can also sometimes trigger production of new chemicals (Liang et al., 2025).

Further maximizing the utility of the medicine, here, the problem is that even when we discover a new compound, it will be in a small amount, so could not always have enough to use it. The solution is to move the useful genes into a more easily cultivated bacterium (like *Escherichia coli*) to produce more product. Use smart machines to figure out the optimal manner and conditions to grow your bacteria to make the most amount of compound possible (Katz et al., 2018), finally, ensuring that new drugs work in humans. The challenge lies in the poor absorption and potential toxicity of some marine-derived compounds, which can be addressed by using AI-based molecular docking to predict their behavior in the human body. Encapsulate the drug in nanoparticles (tiny little packaging) to allow for safer delivery and packaging (Jumper et al., 2021). Hidden metabolomes and BGCs are being investigated to discover new bioactive compounds, enabling innovative applications of marine actinobacteria in medicine, industry, and environmental sustainability (Table 1).

17.4 Marine Actinobacteria in Industrial Biotechnology (Enzymes)

Marine actinobacterial organisms produced various hydrolytic and synthetic enzymes predominantly active at alkaline pH, tolerant of high salinity, detergent-resistant, enhancing biowaste valorization, and with a wide range of applications in various industries (Fig. 2) (Xin et al., 2015). L-asparaginase is a clinically significant enzyme used in the treatment of acute lymphoblastic leukemia (Cachumba et al., 2016). Alkaline protease from *Streptomyces* sp. (marine sediments) is stable at pH 10 and 60 °C, with potential detergent application (Al-Askar et al., 2015). Salt-tolerant lipase from *Nocardiopsis* sp., active in high $Na+$ concentration, can effectively degrade marine oil waste (Joseph et al., 2017). It has revealed many carbohydrate-active enzymes (CAZymes) and protease gene clusters that will not be expressed under normal laboratory conditions (Chen et al., 2021). Metagenomic testing of marine sediments and the study of symbiotic hosts (e.g., those associated with corals or sponges) have uncovered new and novel extremozymes that exhibited unique catalytic properties (Kennedy et al., 2011).

Table 1 BGCs and hidden metabolomes used to bring advancements in marine actinobacterial applications

Tools	Functions	References
Cryptic BGC activation	Activation of silent clusters utilizing stressors, small-molecule elicitors, and pathway regulators	Pepler et al. (2022) and Yan et al. (2025)
Plasmid-encoded BGCs	HGT, rapid and adaptive evolution, effects of plasmids being found in different locations than a genome (chromosomal loci)	Saati-Santamaría (2023) and Yan et al. (2025)
Genome mining (anti-SMASH, Biosynthetic Gene Similarity Clustering And Prospecting Engine (BiG-SCAPE))	Toolkits for BGC prediction and classification, similarity networking across genomes of marine Actinobacteria	Blin et al. (2021)
CRISPR Cas12a cloning (CRISPR/Cas12a-mediated Advanced Targeted Fast In-gel Selection Heterologous expression and Isolation of New Gene clusters (CAT-FISHING))	*In vitro* CRISPR aided direct cloning of large high GC BGCs, and heterologous expression	Liang et al. (2022)
BAC library screening (Library EXpression Analysis System (LEXAS))	Heterologous expression of large insert clones (BGCs) into Streptomyces hosts to find new metabolites	Xu et al. (2016)
Metabologenomics	Metabolomics (Liquid Chromatography–Mass Spectrometry (LC-MS/MS)), along with mining BGCs, to help link orphan clusters to metabolite features	Goering et al. (2016), Yan et al. (2025) and Zheng et al. (2024)
Halogenated compound mining	Targeted mining for halogenase-rich clusters to find marine halo metabolite structures (e.g., indanopyrrole A)	Sweeney et al. (2024)
Shotgun metagenomics	Culture-independent profiling of environmental BGCs in marine microbiomes discovers rare or novel taxa diversity of BGCs	Ghurye et al. (2016) and Yan et al. (2025)
Transcriptomics (RNA-seq)	Conditional expression of or profiled BGC; uses comparative gene expression to analyze mechanisms of expression.	Papini et al. (2010)

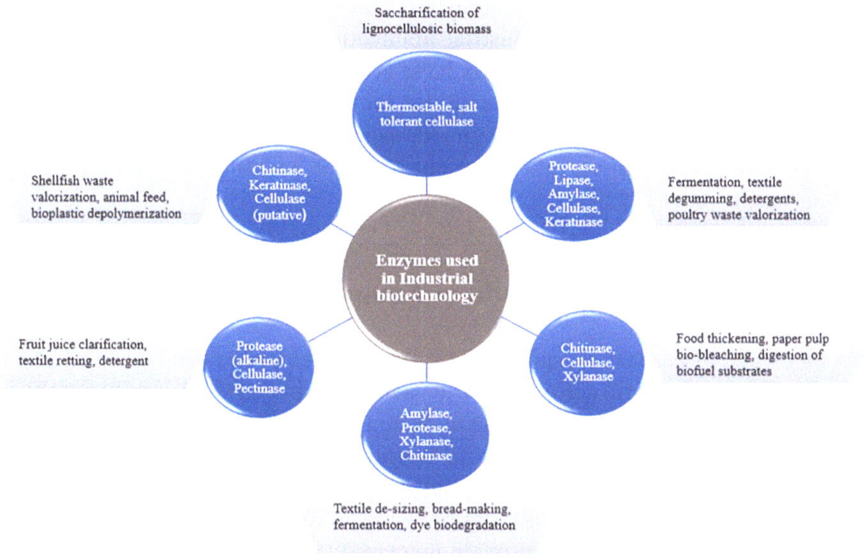

Fig. 2 Applications of enzymes produced by marine actinobacteria in industrial biotechnology

17.4.1 Future of Marine Actinobacteria in Industrial Enzyme Applications

First, unlocking silent enzyme genes has a problem statement where many marine actinobacteria possess gene clusters encoding novel enzymes, but they remain silent due to limitations of laboratory cultivation methods. For solution, we can use CRISPR activation systems or chemical elicitors to induce activity in these genes. Co-culturing with other microbes may activate the expression of dormant enzymes (Liang et al., 2025). Second, the problem is enhancing enzyme yield, where the production processes of short seasons of industrial production are often limited by the low yields in native marine strains, not capable of increasing productivity and production capacity. For a better solution, transfer the relevant genes to high-yield industrial hosts such as *E. coli, Streptomyces lividans,* or *Bacillus subtilis.* Optimizing fermentation conditions, which may include pH, salinity, and temperature, will increase enzyme yield (Katz et al., 2018).

The main challenge is to improve enzyme properties for difficult conditions; some enzymes will lose activity under difficult industrial conditions, such as temperature, solvents, or alkaline pH. The best solution is to use protein engineering and directed evolution to improve the stability of the enzymes (thermal, solvent, and even pH range) (Arnold, 1998). For processes that require the enzyme(s) continuously, immobilize them to re-use (Sheldon & van Pelt, 2013). Thus, in order to avoid redundant discoveries, where different research teams are frequently rediscovering the same types of enzymes, which slows down innovation. Using AI-based comparative enzyme databases that allow researchers to quickly recognize if there is a novel enzyme or if it has previously been described will be a better solution. Furthermore, expanding the industrial application portfolios problem statement, where many marine-derived enzymes have not been investigated to their potential beyond the natural substrates. The solutions for this problem are to formulate multi-enzyme cocktails to break down lignocellulose, use marine-derived enzymes for green-chemistry syntheses, and adapt enzymes for environmental bioremediation (Manivasagan et al., 2014).

17.5 Marine Actinobacteria in Aquaculture and Agriculture

Probiotic and disease control potential of Marine *Streptomyces* and other actinobacteria produce anti-Vibrio compounds (i.e., actinomycin D and other antimicrobials), and some possible strains are capable of decreasing biofilm formation and virulence in Vibrio. For example, actinonin from *Streptomyces* sp. NHF165 inhibited *Vibrio anguillarum*, a primary fish pathogen that potentially exists for controlling vibriosis using either probiotics or natural extracts (Yang & Sun, 2016). Experimental studies show that these actinobacteria provide increased survival as a feed additive or co-treatment (Liu et al., 2023). Microbiome and host benefits show that a number of indigenous marine isolates contribute to the digestion of crustacean insect meal and valorize by-products with active research areas for feeding formulation, which will suggest actionability as host-adapted probiotics (Ghosh et al., 2023; Foysal & Gupta, 2022). In agriculture, biocontrol and plant growth-promoting agents such as rhizosphere and halotolerant actinobacteria present in the soil will suppress and mitigate soil-borne fungal pathogens (e.g., *Fusarium*) and increase plant growth through hormone production, siderophores, and nutrient solubilization, also showing effectiveness in greenhouse trials (Ribeiro & Van Der Sand, 2024; Torres-Rodríguez et al., 2024).

17.5.1 Future of Marine Actinobacteria in Aqua and Agricultural Applications

Marine actinobacteria also enhance feed efficiency and plant nutrient uptake in agriculture by producing phytohormones. For instance, auxins (e.g., Indole-3-Acetic Acid (IAA)) produced by a tryptophan-independent pathway with *Micrococcus aloeverae* and cytokinins were produced, but not as extensively as auxins. *Streptomyces* species such as *Streptomyces olivaceoviridis*, *Streptomyces rochei*, and *Streptomyces rimosus* were good gibberellin producers in culture conditions that stimulated root growth, seed germination, and shoot proliferation. Marine actinobacteria are shown to solubilize phosphorus and secrete siderophores that would benefit nutrient availability (Boukhatem et al., 2022).

Marine actinobacteria can contribute to reducing the use of antibiotics in aquaculture by acting as natural therapeutics instead of antibiotics. For example, actinonin displayed potent anti-Vibrio activity and reduced reliance on traditional antibiotics. Additionally, quorum-quenching compounds from marine *Streptomyces* and *Nocardiopsis* interrupt communication between pathogens and virulence (Yang & Sun, 2016; Zhang et al., 2023).

Even though some barriers to the effective use of marine actinobacteria in farms and fields are antibiotic resistance, these bacteria must carry antibiotic resistance genes (ARGs) for self-defense; horizontal gene transfer (HGT) could be an ecological concern; fortunately, linear plasmids have reduced the probability of HGT for these bacteria. Whole-genome sequencing (WGS) is equivalent to reading a microbial genome, like getting one "manual" to read in its entirety. When using WGS on a marine actinobacteria strain, we can determine if a strain contains any "bad" genes from a human/animal/environmental health perspective that could compromise environmental health. ARGs could transfer to pathogenic bacteria. Virulence genes could make the strain unsafe to fish, plants, or humans (Panel, 2017). Marine actinobacteria can also promote plant growth under stress by environmental robustness, while many actinobacteria are not marine-specific, a substantial proportion are stress-tolerant (e.g., salinity or drought) and can be able to produce ACC deaminase (e.g., *Streptomyces, Amycolatopsis, Nocardia, Mycobacterium, Rhodococcus*, and others) to diminish stress ethylene in plants and enhance resilience. Inoculation of legumes with actinobacteria such as taxa in the Streptomycete lineage and rhizobium improved nodulation and growth in salinity (Sathya et al., 2017).

17.6 Marine Actinobacteria in Environmental Bioremediation

Marine actinobacteria can degrade petroleum hydrocarbons using enzymes like alkane hydroxylases and monooxygenases. They also produce biosurfactants from natural or synthetic sources, which enhance the bioavailability of hydrocarbons, heavy metals, and plastics for more effective degradation.

17.6.1 Future of Marine Actinobacteria in Oil and Hydrocarbon Degradation

Certain marine actinobacteria (*Streptomyces*, *Nocardiopsis*, and *Rhodococcus* sp.) can be used to bio-degrade oil, either as a single strain or in consortia in seawater or sediments, or salt-affected soils (Fernandes et al., 2025). In this, natural degradation rates are inherently very low or limited by oil bioavailability and environmental stress (salinity and temperature). The oil biodegradation can be enhanced by using biosurfactant-producing actinobacteria or adding biosurfactants, along with nutrients, oxygen, and locally adapted microbial consortia. Pilot optimization tests have shown high degradation when a consortium with biosurfactants and changing optimized conditions is used (Farooq et al., 2024; Aparicio et al., 2018).

17.6.2 Future of Marine Actinobacteria in Biosurfactants Production for Remediation

Actinobacteria have trehalolipids, lipoproteins, and other biosurfactants that decrease surface/interfacial tension, can take up oil, and stimulate microbial hydrocarbon-degrading processes (Farooq et al., 2024; Patiño et al., 2021). Biosurfactant-based dispersants are potential ecologically substitutes for synthetic dispersants that are more expensive to produce at an industrial scale and are difficult to purify downstream. So, cultivate in cheap feed stocks for waste substrates prior to fermentation, work with partially purified crude biosurfactants, and engineer high-yielding producing strains or use cell-free enzyme systems to reduce cost and test low-toxicity blends as drop-in dispersants (Miao et al., 2024; Datta & Chattopadhyay, 2024).

17.6.3 Future of Marine Actinobacteria in Heavy Metal Removal

Many marine actinobacterial taxa display an ability to tolerate and immobilize heavy metals through biosorption, precipitation, bioaccumulation, excretion of chelators/exopolysaccharides (EPSs), or enzymatic transformation (e.g., Cr^{6+} to Cr^{3+} enhances precipitation). These attributes allow them to be well remediate contaminated seawater, sediments, and industrial effluents (Alabssawy & Hashem, 2024). Metal remediation in complexity matrices is inefficient and may be remobilized. So, employ immobilized cell systems, biofilms, or EPS-producing strains to stabilize sequestration of metals and then add downstream phyto or electrochemical recovery to biomass recovery for metal reclamation, and mixed microbial consortia of species with complementary mechanisms would also work to reduce the chances of re-mobilization (Alvarado-Campo et al., 2023; Qattan, 2025).

17.6.4 Future of Marine Actinobacteria in the Degradation of Plastics and Microplastics

Recent research indicates that several marine actinobacteria produce enzymes that chemically target individual plastics or bioplastics components, resulting in surface erosion, weight loss, or fragmentation applicable for waste treatment or composting applications (Oliveira et al., 2022; Lv et al., 2024). Widespread degradation rates of familiar plastics such as Polyethylene (PE), Polypropylene (PP), and Polyvinyl Chloride (PVC) remain slow and inconsistent in nature. For that, develop and engineer highly active polymerases/esterases from marine actinobacteria that employ pre-treatment (UV and chemical oxidation) to render plastics more vulnerable to enzymatic degradation and design bioaugmentation reactors with controlled temperature, nutrient stratification for specific concentrated waste streams, not for open sea deployment (Yang et al., 2023; Wichatham et al., 2024).

17.6.5 Future of Marine Actinobacteria in Integrated Bioremediation

Mixed microbial consortia utilize complementary degradation pathways; for example, one strain may solubilize a contaminant, while another mineralizes an intermediate product, increasing the robustness of a bioremediation application across mixed pollutants (Qattan, 2025; Chen et al., 2024). In this, creating stable, effective consortia is difficult, and maintaining them during field use is a challenge. So, design and create synthetic consortia that match functional capabilities using genomic/metabolomic tools to identify and combine complementary functions and then stabilize by immobilization through biofilm or beads and protection through nutrient release formulations with pilot projects in mesocosms before open-field experimentation (Qattan, 2025).

17.7 Marine Actinobacteria in Smart-Screening Solutions

Modern deep-learning genome mining tools and community MS/MS resources have guided the discovery of complex extracts that have rapidly advanced the ability to predict BGCs and tie them to molecules (Fig. 3) (Gangwal & Lavecchia, 2025). GNPS can now support large collaborative workflows for molecular networking and the searching of spectral libraries (Wang et al., 2016). Also, experimental enabling technologies have made improvements in *in situ* cultivation systems such as iChip variants and microfluidic cultivation, and metagenome-assembled genomes have increased recoveries of previously unculturable marine microbes; heterologous expression toolkits make it feasible to express large marine BGCs in tractable hosts (Nazipi Bushi et al., 2025; Xu et al., 2022). ML and AI are playing an increasingly important role in marine actinobacterial research. These approaches help predict and analyze the production of bioactive compounds and enable the exploration of genomic and metagenomic data to discover previously unrecognized BGCs with industrial and pharmaceutical relevance. Furthermore, computational modeling and systems biology facilitate the design of optimized strains, supporting sustainable applications in medicine, agriculture, and environmental management (Table 2).

Many marine actinobacteria have low culturability and lost biodiversity, showing little to no growth on lab media, which limits classical isolation and risks missing cryptic taxa and their BGCs. Combine a field-deployable cultivation platform such as iChip, microdroplets, and AI-sourced sampling like predictive noticing of niches likely to take novel taxa using environmental metadata with previously assembled metagenomic assemblies. By constructing an active-learning model that would suggest the best sampling locations and devices, we could potentially draw out positive notions of novelty. As far as we know, this hybrid solution would most likely yield a greater chance of novel producers (Nazipi Bushi et al., 2025).

Traditional screens are limited by an increase in rediscovery rates (dereplication burden). Couple MS/MS molecular networking (GNPS) with existing ML dereplication and spectral to structure predictors, whereby remote pipelines automatically

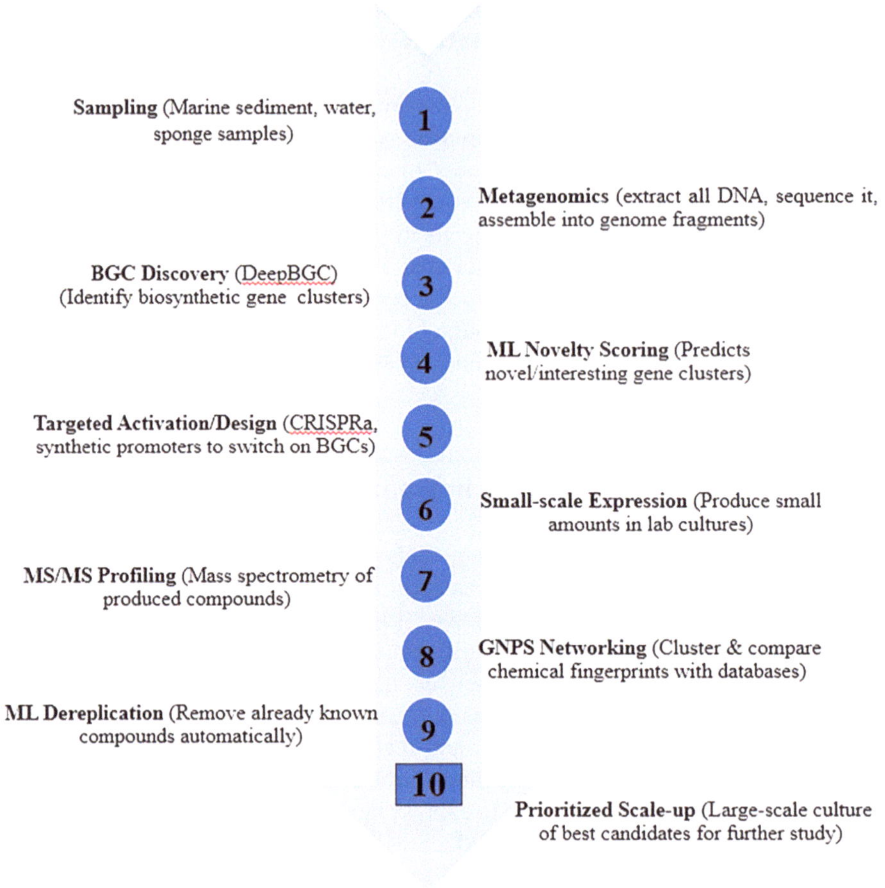

Fig. 3 Advanced knowledge in smart-screening solutions

examine every spectrum linked to a known chemical scaffold, like a steroid, and suggest sampling and prioritization. In this way, we could reduce the hit list, prior to scale-up processes (Wang et al., 2016). Many BGCs remain transcriptionally silent in culture conditions. Leverage AI-driven promoter and regulator prediction for designing CRISPRa guides or synthetic promoters to activate specific BGCs, and rank targets by predicted product novelty from sequence features with DeepBGC-like embeddings. Include co-culture predictions that partner with the co-culture to make induction of expression learned from experiments (Zhang et al., 2017).

Predicting chemical structure from BGC sequence is not perfect, while having experimental valuation is slow. Combine BGC to structure models with deep learning that predict tailoring enzymes and likely substructures with MS/MS predicted fragmentation, and the molecular networking loop becomes closed. Give priority to BGCs whose predicted product has no spectral library matches. This step enhances de-risking before heterologous expression.

Table 2 Tools in AI and ML used for the advancement of marine actinobacterial applications

Tools	Functions	Applications	Future perspective	Referencccs
ML for metagenomic BGC detection (e.g., e DeepBGC, Deep BGCpred, BGC Prophet, and AutoBGC/ GEne Cluster prediction with COnditional random fields (GECCO))	Use of deep learning and transformer models to detect BGCs from genomic and metagenomic data	Applied to marine actinobacterial metagenomes to identify rare or cryptic secondary metabolite clusters	Adapt models to sequence traits specifically for the ocean, expand marine-specific training datasets, and integrate with culture-guided validation.	Yan et al. (2025) and Goering et al. (2016)
Automated BGC assembly pipeline: BiosyntheticSPAdes	Reconstruction of complete BGCs from metagenomic reads using assembly graph-based methods	Has the ability to recover full BGCs from marine environmental samples, assisting with the discovery of novel actinomycete pathways	Integrate ML-based error correction and prediction of BGC function for marine datasets	Mtemeli et al. (2022), Xu et al. (2016), Pepler et al. (2022) and Saati-Santamaría (2023)
ML in enzyme engineering (e.g., ProtDCal, generative protein models, AlphaFold-Multimer, and variational autoencoders)	ML models that are sequence- and structure-informed in order to predict function or design new enzyme variants	Used to engineer marine-derived enzymes, such as halotolerant cellulases or proteases to improve activity or stability	ML-guided mutagenesis of marine CAZymes scaffolds, optimizing for thermostability or halotolerance	Hawkins-Hooker et al. (2021)
MetaEnzyme/ EnzyGen, EnzymeFlow generative models	New ML frameworks (ArXiv preprints) that design reaction-specific catalytic pockets and redesign enzymes for tasks such as mutation or sequence generation	Ideal for tailoring active sites of marine enzymes to specific industrial reactions (e.g., biofuel precursors)	Use Meta-Enzyme or Enzyme-Flow to redesign marine enzyme scaffolds like salt-tolerant cellulases for high-efficiency biomass processing	Zheng et al. (2024)

(continued)

Table 2 (continued)

Tools	Functions	Applications	Future perspective	References
Computational metabolic modelling (PathPred, Global Natural Products Social Molecular Networking + Mass Spectrometry Second Stage - Latent Dirichlet Allocation (GNPS + MS2LDA)) and strain engineering tools (COBRApy/ OptKnock)	Integration of ML with constraint-based modeling to design microbial hosts for optimized production of enzymes or metabolites	Genetic modifications to marine actinobacteria to overproduce desired enzymes (e.g., proteases and CAZymes) or natural products from BGCs	Build predictive strain engineering pipelines for marine actinobacteria, combining omics, BGC data, and ML-guided design	Markus et al. (2023) and Goering et al. (2016)

17.8 Conclusion

Marine actinobacteria are positioned at the intersection of the three areas of biodiversity, biotechnology, and innovation. Their remarkable metabolic promiscuity and ecological adaptability are what make them such abundant sources of bioactive compounds, extremozymes, and nanomaterials that can be used in the fields of pharmaceuticals, industrial biotechnology, aquaculture, agriculture, and environmental cleanup. Formally, however, the considerable translational potential of marine actinobacteria has been hampered by issues associated with uncultivability, redundantly rediscovering the same targets, and large swathes of transcriptionally silent gene clusters. Recent developments in metagenomics, genome mining, synthetic biology, and smart screening enabled by AI have paved the way for a reconceptualization of the notion of cryptic biosynthetic pathways, which involves accessing these hidden reservoirs of stable, unexplored chemical space. Moreover, certainly, the combination of computational intelligence-based solutions and experimental microbiology-based approaches represents arguably the biggest paradigm shift to the single biology paradigm, in both accelerated discovery pipelines and cooperative supply chains along with scale in terms of culture, and accounts for reducing redundancies. In conclusion, marine actinobacteria illustrate a frontier where nature's creativity meets our creativity as humans. In doing so, they provide us with the ability to not only study microbial diversity as a curiosity, but more meaningfully as our partners in co-creating sustainable solutions to the world's various problems in health, food security, industry, and the environment.

References

Abdelmohsen, U. R., Yang, C., Horn, H., Hajjar, D., Ravasi, T., & Hentschel, U. (2014). Actinomycetes from Red Sea sponges: Sources for chemical and phylogenetic diversity. *Marine Drugs, 12*(5), 2771–2789.

Alabssawy, A. N., & Hashem, A. H. (2024). Bioremediation of hazardous heavy metals by marine microorganisms: A recent review. *Archives of Microbiology, 206*(3), 103.

Al-Askar, A. A., Rashad, Y. M., Hafez, E. E., Abdulkhair, W. M., Baka, Z. A., & Ghoneem, K. M. (2015). Characterization of alkaline protease produced by *Streptomyces griseorubens* E44G and its possibility for controlling Rhizoctonia root rot disease of corn. *Biotechnology & Biotechnological Equipment, 29*(3), 457–462.

Alvarado-Campo, K. L., Quintero, M., Cuadrado-Cano, B., Montoya-Giraldo, M., Otero-Tejada, E. L., Blandón, L., et al. (2023). Heavy metal tolerance of microorganisms isolated from coastal marine sediments and their lead removal potential. *Microorganisms, 11*(11), 2708.

Aparicio, J. D., Saez, J. M., Raimondo, E. E., Benimeli, C. S., & Polti, M. A. (2018). Comparative study of single and mixed cultures of actinobacteria for the bioremediation of co-contaminated matrices. *Journal of Environmental Chemical Engineering, 6*(2), 2310–2318.

Arnold, F. H. (1998). Design by directed evolution. *Accounts of Chemical Research, 31*(3), 125–131.

Blin, K., Shaw, S., Kloosterman, A. M., Charlop-Powers, Z., Van Wezel, G. P., Medema, M. H., & Weber, T. (2021). antiSMASH 6.0: Improving cluster detection and comparison capabilities. *Nucleic Acids Research, 49*(W1), W29–W35.

Boukhatem, Z. F., Merabet, C., & Tsaki, H. (2022). Plant growth promoting actinobacteria, the most promising candidates as bioinoculants? *Frontiers in Agronomy, 4*, 849911.

Cachumba, J. J. M., Antunes, F. A. F., Peres, G. F. D., Brumano, L. P., Santos, J. C. D., & Silva, S. S. D. (2016). Current applications and different approaches for microbial L-asparaginase production. *Brazilian Journal of Microbiology, 47*, 77–85.

Chen, J., Xu, L., Zhou, Y., & Han, B. (2021). Natural products from actinomycetes associated with marine organisms. *Marine Drugs, 19*(11), 629.

Chen, W., Sun, J., Ji, R., Min, J., Wang, L., Zhang, J., et al. (2024). Crude oil biodegradation by a biosurfactant-producing bacterial consortium in high-salinity soil. *Journal of Marine Science and Engineering, 12*(11), 2033.

Datta, M., & Chattopadhyay, I. (2024). Applications of microbial biosurfactants in human health and environmental sustainability: A narrative review. *Discovery Medicine, 1*(1), 160.

Farooq, U., Szczybelski, A., Ferreira, F. C., Faria, N. T., & Netzer, R. (2024). A novel biosurfactant-based oil spill response dispersant for efficient application under temperate and arctic conditions. *ACS Omega, 9*(8), 9503–9515.

Feling, R. H., Buchanan, G. O., Mincer, T. J., Kauffman, C. A., Jensen, P. R., & Fenical, W. (2003). Salinosporamide A: A highly cytotoxic proteasome inhibitor from a novel microbial source, a marine bacterium of the new genus Salinospora. *Angewandte Chemie International Edition, 42*(3), 355–357.

Fernandes, C. F., da Silva Iúdice, T. N., Bezerra, N. V., & Pontes, A. N. (2025). Biodegradation of oil-derived hydrocarbons by marine actinobacteria: A systematic review. *Environmental Pollution, 367*, 125509.

Foysal, M. J., & Gupta, S. K. (2022). A systematic meta-analysis reveals enrichment of Actinobacteria and Firmicutes in the fish gut in response to black soldier fly (*Hermetica illucens*) meal-based diets. *Aquaculture, 549*, 737760.

Gangwal, A., & Lavecchia, A. (2025). Artificial intelligence in natural product drug discovery: Current applications and future perspectives. *Journal of Medicinal Chemistry, 68*(4), 3948–3969.

Ghosh, K., Harikrishnan, R., Mukhopadhyay, A., & Ringø, E. (2023). Fungi and actinobacteria: Alternative probiotics for sustainable aquaculture. *Fishes, 8*(12), 575.

Ghurye, J. S., Cepeda-Espinoza, V., & Pop, M. (2016). Metagenomic assembly: Overview, challenges and applications. *The Yale Journal of Biology and Medicine, 89*(3), 353.

Goering, A. W., McClure, R. A., Doroghazi, J. R., Albright, J. C., Haverland, N. A., Zhang, Y., & Kelleher, N. L. (2016). Metabologenomics: Correlation of microbial gene clusters with metabolites drives discovery of a nonribosomal peptide with an unusual amino acid monomer. *ACS Central Science, 2*(2), 99–108.

Gulder, T. A., & Moore, B. S. (2010). Salinosporamide natural products: Potent 20 S proteasome inhibitors as promising cancer chemotherapeutics. *Angewandte Chemie International Edition, 49*(49), 9346–9367.

Hawkins-Hooker, A., Depardieu, F., Baur, S., Couairon, G., Chen, A., & Bikard, D. (2021). Generating functional protein variants with variational autoencoders. *PLoS Computational Biology, 17*(2), e1008736.

Hug, L. A., Baker, B. J., Anantharaman, K., Brown, C. T., Probst, A. J., Castelle, C. J., et al. (2016). A new view of the tree of life. *Nature Microbiology, 1*(5), 1–6.

Jang, K. H., Nam, S. J., Locke, J. B., Kauffman, M. C. A., Beatty, M. D. S., Paul, M. L. A., & Fenical, W. (2013). Anthracimycin, a potent anthrax antibiotic from a marine-derived actinomycete. *Angewandte Chemie (International Ed. in English), 52*(30), 10–1002.

Joseph, B. A. B. U., Kuddus, M., & Ramteke, P. W. (2017). Role of cold active lipases in bioremediation. *Biotechnology Trends and Application, 2017*, 246–257.

Jumper, J., Evans, R., Pritzel, A., Green, T., Figurnov, M., Ronneberger, O., et al. (2021). Highly accurate protein structure prediction with AlphaFold. *Nature, 596*(7873), 583–589.

Katz, L., Chen, Y. Y., Gonzalez, R., Peterson, T. C., Zhao, H., & Baltz, R. H. (2018). Synthetic biology advances and applications in the biotechnology industry: A perspective. *Journal of Industrial Microbiology and Biotechnology, 45*(7), 449–461.

Kennedy, J., O'leary, N. D., Kiran, G. S., Morrissey, J. P., O'Gara, F., Selvin, J., & Dobson, A. D. W. (2011). Functional metagenomic strategies for the discovery of novel enzymes and biosurfactants with biotechnological applications from marine ecosystems. *Journal of Applied Microbiology, 111*(4), 787–799.

Liang, M., Liu, L., Xu, F., Zeng, X., Wang, R., Yang, J., et al. (2022). Activating cryptic biosynthetic gene cluster through a CRISPR–Cas12a-mediated direct cloning approach. *Nucleic Acids Research, 50*(6), 3581–3592.

Liang, Y., Tong, S., Zhang, J., Tan, G. Y., Zhang, L., Lee, S. Y., & Tong, Y. (2025). Expanding horizons of CRISPR applications beyond genome editing. *Trends in Genetics, 41*(10), 934–953.

Liu, W., Ou, P., Tian, F., Liao, J., Ma, Y., Wang, J., & Jin, X. (2023). Anti-Vibrio parahaemolyticus compounds from Streptomyces parvus based on Pan-genome and subtractive proteomics. *Frontiers in Microbiology, 14*, 1218176.

Lv, S., Li, Y., Zhao, S., & Shao, Z. (2024). Biodegradation of typical plastics: From microbial diversity to metabolic mechanisms. *International Journal of Molecular Sciences, 25*(1), 593.

Manivasagan, P., Kang, K. H., Sivakumar, K., Li-Chan, E. C., Oh, H. M., & Kim, S. K. (2014). Marine actinobacteria: An important source of bioactive natural products. *Environmental Toxicology and Pharmacology, 38*(1), 172–188.

Markus, B., Gruber Christian, C., Andreas, K., Arkadij, K., Stefan, L., Gustav, O., & Radka, S. (2023). Accelerating biocatalysis discovery with machine learning: A paradigm shift in enzyme engineering, discovery, and design. *ACS Catalysis, 13*(21), 14454–14469.

Miao, Y., To, M. H., Siddiqui, M. A., Wang, H., Lodens, S., Chopra, S. S., et al. (2024). Sustainable biosurfactant production from secondary feedstock—Recent advances, process optimization and perspectives. *Frontiers in Chemistry, 12*, 1327113.

Mtemeli, F. L., Ndlovu, J., Mugumbate, G., Makwikwi, T., & Shoko, R. (2022). Advances in schistosomiasis drug discovery based on natural products. *All Life, 15*(1), 608–623.

Nazipi Bushi, S., Lund, M. B., Sandfeld, T., Nørskov, S. S., Fruergaard, S., Glasius, M., et al. (2025). A modified iChip for in situ cultivation of bacteria in arid environments. *Applied and Environmental Microbiology, 91*(2), e01325–e01324.

Newman, D. J., & Cragg, G. M. (2020). Natural products as sources of new drugs over the nearly four decades from 01/1981 to 09/2019. *Journal of Natural Products, 83*(3), 770–803.

Oliveira, J., Almeida, P. L., Sobral, R. G., Lourenço, N. D., & Gaudêncio, S. P. (2022). Marine-derived actinomycetes: Biodegradation of plastics and formation of PHA bioplastics—A circular bioeconomy approach. *Marine Drugs, 20*(12), 760.

Panel, E. B. (2017). Scientific opinion on the update of the list of QPS-recommended biological agents intentionally added to food or feed as notified to EFSA. *EFSA Journal, 15*(3), 4664.

Papini, M., Salazar, M., & Nielsen, J. (2010). Systems biology of industrial microorganisms. In *Biosystems engineering I: Creating superior biocatalysts* (pp. 51–99). Springer Berlin Heidelberg.

Patiño, A. D., Montoya-Giraldo, M., Quintero, M., López-Parra, L. L., Blandón, L. M., & Gómez-León, J. (2021). Dereplication of antimicrobial biosurfactants from marine bacteria using molecular networking. *Scientific Reports, 11*(1), 16286.

Pepler, M. A., Zhang, X., Hindra, & Elliot, M. A. (2022). Inducing global expression of actinobacterial biosynthetic gene clusters. In *Engineering natural product biosynthesis: Methods and protocols* (pp. 157–171). Springer US.

Qattan, S. Y. (2025). Harnessing bacterial consortia for effective bioremediation: Targeted removal of heavy metals, hydrocarbons, and persistent pollutants. *Environmental Sciences Europe, 37*(1), 85.

Ribeiro, H. G., & Van Der Sand, S. T. (2024). Exploring the trends in actinobacteria as biological control agents of phytopathogenic fungi: A (mini)-review. *Indian Journal of Microbiology, 64*(1), 70.

Rutledge, P. J., & Challis, G. L. (2015). Discovery of microbial natural products by activation of silent biosynthetic gene clusters. *Nature Reviews Microbiology, 13*(8), 509–523.

Saati-Santamaría, Z. (2023). Global map of specialized metabolites encoded in prokaryotic plasmids. *Microbiology Spectrum, 11*(4), e01523–e01523.

Sathya, A., Vijayabharathi, R., & Gopalakrishnan, S. (2017). Plant growth-promoting actinobacteria: A new strategy for enhancing sustainable production and protection of grain legumes. *3 Biotech, 7*(2), 102.

Sheldon, R. A., & van Pelt, S. (2013). Enzyme immobilisation in biocatalysis: Why, what and how. *Chemical Society Reviews, 42*(15), 6223–6235.

Smanski, M. J., Zhou, H., Claesen, J., Shen, B., Fischbach, M. A., & Voigt, C. A. (2016). Synthetic biology to access and expand nature's chemical diversity. *Nature Reviews Microbiology, 14*(3), 135–149.

Stokes, J. M., Yang, K., Swanson, K., Jin, W., Cubillos-Ruiz, A., Donghia, N. M., et al. (2020). A deep learning approach to antibiotic discovery. *Cell, 180*(4), 688–702.

Subramani, R., & Aalbersberg, W. (2012). Marine actinomycetes: An ongoing source of novel bioactive metabolites. *Microbiological Research, 167*(10), 571–580.

Subramani, R., & Aalbersberg, W. (2013). Culturable rare actinomycetes: Diversity, isolation and marine natural product discovery. *Applied Microbiology and Biotechnology, 97*(21), 9291–9321.

Sweeney, D., Bogdanov, A., Chase, A. B., Castro-Falcón, G., Trinidad-Javier, A., Dahesh, S., & Jensen, P. R. (2024). Pattern-based genome mining guides discovery of the antibiotic indanopyrrole A from a marine Streptomycete. *Journal of Natural Products, 87*(12), 2768–2778.

Torres-Rodríguez, J. A., Reyes-Pérez, J. J., Adame, L. H., Llerena-Fuentes, B. L., & Hernandez-Montiel, L. G. (2024). Marine actinomycetes for biocontrol of *Fusarium solani* in tomato plants: In vitro and in vivo studies. *Notulae Botanicae Horti Agrobotanici Cluj-Napoca, 52*(2), 13562–13562.

Trincone, A. (2011). Marine biocatalysts: Enzymatic features and applications. *Marine Drugs, 9*(4), 478–499.

Vartoukian, S. R., Palmer, R. M., & Wade, W. G. (2010). Strategies for culture of 'unculturable' bacteria. *FEMS Microbiology Letters, 309*(1), 1–7.

Wang, M., Carver, J. J., Phelan, V. V., Sanchez, L. M., Garg, N., Peng, Y., et al. (2016). Sharing and community curation of mass spectrometry data with Global Natural Products Social Molecular Networking. *Nature Biotechnology, 34*(8), 828–837.

Wichatham, K., Piyaviriyakul, P., Boontanon, N., Surinkul, N., Visvanathan, C., Fujii, S., & Boontanon, S. K. (2024). Biodegradation of polypropylene plastics in vitro and natural condition by Streptomyces sp. isolated from plastic-contaminated sites. *Environmental Technology & Innovation, 35*, 103681.

Xin, Y., Sun, Z., Chen, Q., Wang, J., Wang, Y., Luogong, L., et al. (2015). Purification and characterization of a novel extracellular thermostable alkaline protease from Streptomyces sp. M30. *Journal of Microbiology and Biotechnology, 25*(11), 1944–1953.

Xu, M., Wang, Y., Zhao, Z., Gao, G., Huang, S. X., Kang, Q., et al. (2016). Functional genome mining for metabolites encoded by large gene clusters through heterologous expression of a whole-genome bacterial artificial chromosome library in Streptomyces spp. *Applied and Environmental Microbiology, 82*(19), 5795–5805.

Xu, Y., Du, X., Yu, X., Jiang, Q., Zheng, K., Xu, J., & Wang, P. (2022). Recent advances in the heterologous expression of biosynthetic gene clusters for marine natural products. *Marine Drugs, 20*(6), 341.

Yan, B., Nam, Y., Li, L., Deek, R. A., Li, H., & Ma, S. (2025). Recent advances in deep learning and language models for studying the microbiome. *Frontiers in Genetics, 15*, 1494474.

Yang, N., & Sun, C. (2016). The inhibition and resistance mechanisms of actinonin, isolated from marine streptomyces sp. NHF165, against *Vibrio anguillarum*. *Frontiers in Microbiology, 7*, 1467.

Yang, X. G., Wen, P. P., Yang, Y. F., Jia, P. P., Li, W. G., & Pei, D. S. (2023). Plastic biodegradation by in vitro environmental microorganisms and in vivo gut microorganisms of insects. *Frontiers in Microbiology, 13*, 1001750.

Zhang, M. M., Wong, F. T., Wang, Y., Luo, S., Lim, Y. H., Heng, E., et al. (2017). CRISPR–Cas9 strategy for activation of silent Streptomyces biosynthetic gene clusters. *Nature Chemical Biology, 13*(6), 607–609.

Zhang, Z., Sun, Y., Yi, Y., Bai, X., Zhu, L., Zhu, J., et al. (2023). Screening and identification of a streptomyces strain with quorum-sensing inhibitory activity and effect of the crude extracts on virulence factors of *Pseudomonas aeruginosa*. *Microorganisms, 11*(8), 2079.

Zheng, J., Zhang, H., Xu, Q., Zeng, A. P., & Li, S. Z. (2024, October). Metaenzyme: Meta pan-enzyme learning for task-adaptive redesign. In *Proceedings of the 32nd ACM international conference on multimedia* (pp. 1721–1730).

MIX
Papier aus verantwortungsvollen Quellen
Paper from responsible sources
FSC® C105338

If you have any concerns about our products,
you can contact us on
ProductSafety@springernature.com

In case Publisher is established outside the EU,
the EU authorized representative is:
Springer Nature Customer Service Center GmbH
Europaplatz 3, 69115 Heidelberg, Germany

Printed by Libri Plureos GmbH
in Hamburg, Germany